NEUROMETHODS □ 5

Neurotransmitter Enzymes

NEUROMETHODS

Program Editors: Alan A. Boulton and Glen B. Baker

Series I: Neurochemistry

1. **General Neurochemical Techniques**
   Edited by *Alan A. Boulton and Glen B. Baker,* 1985
2. **Amines and Their Metabolites**
   Edited by *Alan A. Boulton, Glen B. Baker, and Judith M. Baker,* 1985
3. **Amino Acids**
   Edited by *Alan A. Boulton, Glen B. Baker, and James D. Wood,* 1985
4. **Receptor Binding Techniques**
   Edited by *Alan A. Boulton, Glen B. Baker, and Pavel D. Hrdina,* 1986
5. **Neurotransmitter Enzymes**
   Edited by *Alan A. Boulton, Glen B. Baker, and Peter H. Yu,* 1986
6. **Peptides**
   Edited by *Alan A. Boulton, Glen B. Baker, and Quentin Pittman,* 1986
7. **Lipids and Related Compounds**
   Edited by *Alan A. Boulton, Glen B. Baker, and Lloyd A. Horrocks,* 1986
8. **Physicochemical Techniques**
   Edited by *Alan A. Boulton, Glen B. Baker, and Donald P. Boisvert,* 1986
9. **Brain Electrolytes and Water Spaces**
   Edited by *Alan A. Boulton, Glen B. Baker, and Wolfgang Walz,* 1986
10. **Analysis of Psychiatric Drugs**
    Edited by *Alan A. Boulton, Glen B. Baker, and Ronald T. Coutts,* 1986
11. **Carbohydrates and Energy Metabolism**
    Edited by *Alan A. Boulton, Glen B. Baker, and Roger F. Butterworth*
12. **Drugs as Tools in Neurotransmitter Research**
    Edited by *Alan A. Boulton, Glen B. Baker, and Augusto V. Juorio*
13. **Psychopharmacology**
    Edited by *Alan A. Boulton, Glen B. Baker, and Andrew J. Greenshaw*
14. **Neurophysiology**
    Edited by *Alan A. Boulton, Glen B. Baker, and Case H. Vanderwolf*
15. **Neuropsychology**
    Edited by *Alan A. Boulton, Glen B. Baker, and Merril Hiscock*

NEUROMETHODS

Series 1: Neurochemistry

Program Editors: Alan A. Boulton and Glen B. Baker

Neurotransmitter Enzymes

Edited by

Alan A. Boulton, Glen B. Baker, and Peter H. Yu

Humana Press • Clifton, New Jersey

Library of Congress Cataloging in Publication Data
Main entry under title:

Neurotransmitter enzymes.
(Neuromethods ; 5. Series I, Neurochemistry)
Includes bibliographies and index.
1. Neurotransmitters—Metabolism. 2. Enzymes.
I. Boulton, A. A. (Alan A.) II. Baker, Glen B.,
1947– . III. Yu, Peter H. IV. Series: Neuromethods;
5. V. Series: Neuromethods. Series I, Neurochemistry.
QP364.7.N466 1986 599'.0188 86–10309
ISBN 0–89603–079–2

Crescent Manor
PO Box 2148
Clifton, NJ 07015

Printed in the United States of America

Foreword

Techniques in the neurosciences are evolving rapidly. There are currently very few volumes dedicated to the methodology employed by neuroscientists, and those that are available often seem either out of date or limited in scope. This series is about the methods most widely used by modern-day neuroscientists and is written by their colleagues who are practicing experts.

Volume 1 will be useful to all neuroscientists since it concerns those procedures used routinely across the widest range of subdisciplines. Collecting these general techniques together in a single volume strikes us not only as a service, but will no doubt prove of exceptional utilitarian value as well. Volumes 2 and 3 describe all current procedures for the analyses of amines and their metabolites and of amino acids, respectively. These collections will clearly be of value to all neuroscientists working in or contemplating research in these fields. Similar reasons exist for Volume 4 on receptor binding techniques since experimental details are provided for all types of ligand-receptor binding, including chapters on general principles, drug discovery and development, and a most useful appendix on computer programs for Scatchard, nonlinear, and competitive displacement analyses. Volume 5 provides procedures for the assessment of enzymes involved in biogenic amine synthesis and catabolism.

Volumes in the NEUROMETHODS series will be useful to neurochemists, -pharmacologists, -physiologists, -anatomists, psychopharmacologists, psychiatrists, neurologists, and chemists (organic, analytical, pharmaceutical, medicinal); in fact, everyone involved in the neurosciences, both basic and clinical.

Preface to the Series

When the President of Humana Press first suggested that a series on methods in the neurosciences might be useful, one of us (AAB) was quite skeptical; only after discussions with GBB and some searching both of memory and library shelve did it seem that perhaps the publisher was right. Although some excellent methods books have recently appeared, notably in neuroanatomy, it is a fact that there is a dearth in this particular field, a fact attested to by the alacrity and enthusiasm with which most of the contributors to this series accepted our invitations and suggested additional topics and areas. After a somewhat hesitant start, essentially in the neurochemistry section, the series has grown and will encompass neurochemistry, neuropsychiatry, neurology, neuropathology, neurogenetics, neuroethology, molecular neurobiology, animal models of nervous disease, and no doubt many more "neuros." Although we have tried to include adequate methodological detail and in many cases detailed protocols, we have also tried to include wherever possible a short introductory review of the methods and/or related substances, comparisons with other methods, and the relationship of the substances being analyzed to neurological and psychiatric disorders. Recognizing our own limitations we have invited a guest editor to join with us on most volumes in order to ensure complete coverage of the field and to add their specialized knowledge and competencies. We anticipate that this series will fill a gap; we can only hope that it will be filled appropriately and with the right amount of expertise with respect to each method, substance or group of substances, and area treated.

Alan A. Boulton
Glen B. Baker

Preface

Life depends on a complex network of chemical reactions mediated by specific enzymes. Different enzymes actively catalyze the synthesis and degradation of a variety of neurotransmitters and neuromodulators. A dynamic balance and intricate regulation of the rates of metabolism of these compounds are required to maintain neuronal homeostasis. Any alteration in this homeostasis may affect neurotransmission and thus give rise to neurological and psychiatric disorders. The neurotransmitter enzymes have been and will continue to be the subject of intense research in order to understand the nature of neurotransmission, to discover their links to neurological and psychiatric illnesses, and to develop some therapeutic strategies. Manipulation of some enzyme activities, such as inhibition of GABA-transaminase or monoamine oxidase, have been used in the treatment of epilepsy and depression, respectively. Such research relies upon the adoption of reliable and reproducible methodologies. A wide variety of enzyme techniques has been developed in the last two decades. In this volume, we have assembled, and condensed, information on seventeen important brain enzymes that play significant roles in neurotransmitter metabolism.

Each chapter includes a concise description of the mechanism of the enzyme reaction; its substrates, function, occurrence, and distribution, as well as its relationship to neurological and/or mental disorders. There are, in addition, comprehensive discussions on each enzyme's properties, specific inhibitors, purification, and determination of activity. The commonly used assay methods are reviewed in detail. The best method should be simple, selective, sensitive, and economic, and not require extensive or expensive laboratory facilities.

The first seven chapters deal with enzymes involved in the metabolism of the biogenic amines, i.e., phenylalanine hydroxylase, tyrosine hydroxylase, tryptophan hydroxylase, dopamine-β-hydroxylase, aromatic amino acid decarboxylase, phenylethanolamine *N*-methyltransferase, catechol-*O*-methyltransferase, monoamine oxidase, histamine *N*-methyltransferase, and histidine decarboxylase. The key enzymes of the cholinergic system, choline acetyltransferase and acetylcholine esterase, are described in Chapters 8 and 9. The principal synthetic and catabolic enzymes, glutamate decarboxylase and GABA transaminase, respectively, of the GABA system are discussed in Chapters 10 and 11, and Chapter 12 describes a group of proteolytic enzymes related to neuropeptide metabolism. Three important enzymes involved in ion transport, receptor binding, and subsequent protein phosphorylation are reviewed in Chapters 13 (NA^+/K^+ ATPase), 14 (adenylate cyclase), and 15 (protein kinase). The last two chapters describe two metabolic enzymes that have recently received much attention—*S*-adenosyl-homocysteine hydroxylase and phenolsulfotransferase.

All the contributors to this volume possess expertise, and all have recently been, or are, active in the field. This volume will be useful to those who are searching for important and up-to-date information about neurotransmitter enzymes, as well as to those who wish to use the reported methods directly in their laboratory research or in teaching.

Peter H. Yu

Contributors

CORINNE ABATE • *Department of Neurology, Cornell University Medical College, New York, New York*

RONALD R. BOWSHER • *Lilly Laboratory for Clinical Research, Eli Lilly and Company, Indianapolis, Indiana*

RAMESH C. GUPTA • *Department of Pharmacology and Cell Biophysics, University of Cincinnati Medical Center, Cincinnati, Ohio*

DAVID P. HENRY • *Lilly Laboratory for Clinical Research, Eli Lilly and Company, Indianapolis, Indiana*

KOON-SEA HUI • *Neurochemistry Division, The Nathan S. Kline Institute for Psychiatric Research, Orangeburg, New York*

ONYOU HWANG • *Department of Neurology, Cornell University Medical College, New York, New York*

GOWRAGANAHALLI JAGADEESH • *Department of Pharmaceutics, Banaras Hindu University, Varanasi, India*

TONG H. JOH • *Department of Neurology, Cornell University Medical College, New York, New York*

ABEL LAJTHA • *Department of Psychiatry, New York University Medical Center, New York, New York*

DAVID L. MARTIN • *Wadsworth Center for Laboratories and Research, New York State Department of Health, Albany, New York*

HENRY G. MAUTNER • *Department of Biochemistry and Pharmacology, Tufts University School of Medicine, Boston, Massachusetts*

EISHICHI MIYAMOTO • *Department of Pharmacology, Kumamoto University Medical School, Kumamoto, Japan*

TOSHIHARU NAGATSU • *Department of Biochemistry, Nagoya University School of Medicine, Nagoya, Japan*

DONG H. PARK • *Laboratory of Neurobiology, Cornell University Medical College, New York, New York*

HÉLÈNE POLLARD • *Centre Paul Broca de l'INSERM, Paris, France*

ZOLTÁN RAKONCZAY • *Central Research Laboratory, Medical University, Szeged, Hungary*

JEROME A. ROTH • *Department of Pharmacology and Therapeutics, State University of New York at Buffalo, Buffalo, New York*

JEAN-CHARLES SCHWARTZ • *Centre Paul Broca de l'INSERM, Paris, France*

O. Z. SELLINGER • *Laboratory of Neurochemistry, Mental Health Research Institute, University of Michigan Medical Center, Ann Arbor, Michigan*

PRAKASH V. SULAKHE • *Department of Physiology, University of Saskatchewan, Saskatoon, Saskatchewan, Canada*

GODFREY TUNNICLIFF • *Laboratory of Neurochemistry, Indiana University School of Medicine, Evansville, Indiana*

KENNETH M. VERBURG • *Departments of Medicine and Pharmacology and Toxicology, Indiana University School of Medicine, Indianapolis, Indiana*

P. H. WU • *Clinical Pharmacology Program, Addiction Research Foundation, Toronto, Ontario, Canada*

PETER H. YU • *Psychiatric Research Division, University of Saskatchewan, Saskatoon, Saskatchewan, Canada*

Contents

CHAPTER 2
AROMATIC L-AMINO ACID DECARBOXYLASE: BIOCHEMISTRY AND FUNCTIONAL SIGNIFICANCE
Ronald R. Bowsher and David P. Henry

CHAPTER 3
DOPAMINE β-HYDROXYLASE
Toshiharu Nagatsu

CHAPTER 4
MONOAMINE METHYLTRANSFERASES: PHENYLETHANOLAMINE *N*-METHYLTRANSFERASE AND CATECHOL *O*-METHYLTRANSFERASE
Dong H. Park

CHAPTER 8
CHOLINE ACETYLTRANSFERASE
Henry G. Mautner

CHAPTER 10
BRAIN GLUTAMATE DECARBOXYLASE
David L. Martin

CHAPTER 11
4-AMINOBUTYRATE TRANSAMINASE
Godfrey Tunnicliff

CHAPTER 12
PROTEINASES AND PEPTIDASES
Koon-Sea Hui and Abel Lajtha

CHAPTER 13
NA^+/K^+-ATPASE IN NERVOUS TISSUE
P. H. Wu

Chapter 1
Phenylalanine Hydroxylase, Tyrosine Hydroxylase, and Tryptophan Hydroxylase

Tong H. Joh, Onyou Hwang, and Corinne Abate

1. Introduction

There are four enzymes that utilize reduced pteridine as an electron donor and incorporate one atom of oxygen into their substrates. They are phenylalanine 4-monooxygenase (phenylalanine hydroxylase, PAH, EC 1.14.16.1), tyrosine 3-monooxygenase (tyrosine hydroxylase, TH, EC 1.14.16.2), anthranilate 3-monooxygenase (EC 1.14.16.3), and tryptophan 5-monooxygenase (tryptophan hydroxylase, TPH, EC 1.14.16.4). All four enzymes are also known as iron-proteins since they require iron for enzymatic activity. With the exception of anthranilate hydroxylase, these enzymes have been extensively studied. The present chapter concentrates on PAH, TH, and TPH, stressing recent work on biochemical and molecular biological properties, regulation, purification procedures, and methods for determination of enzyme activity. Since a number of excellent reviews for these enzymes have been published recently (Hamon et al., 1981; Kuhn and Lovenberg, 1982; Masserano and Weiner, 1983; Shiman, 1985), the present chapter avoids unnecessary repetition of data already presented. Classical reaction mechanisms, kinetic properties, anatomical and histochemical analyses, and pharmacological studies on the reaction products, e.g., DOPA, dopamine, 5-hydroxytryptophan, and serotonin, are not included in this chapter.

2. Structure of Enzymes

2.1. Molecular Weight

The molecular weights of the monomeric forms of PAH, TH, and TPH are reported to be 50,000–51,000 (Kaufman and Fisher, 1970; Shiman et al., 1979; Nakata and Fujisawa, 1980), 60,000–62,000 (Joh and Reis, 1975; Joh et al., 1978; Vulliet et al., 1980; Markey et al., 1980; Edelman a. et al., 1981), and 55,000 (Cash et al., 1985), respectively. It has been reported (Smith et al., 1984) that the phosphorylated and nonphosphorylated forms of PAH have molecular weights of 50,000 and 49,000. Native forms of the enzymes have higher molecular weights, probably as a result of the formation of polymeric or aggregated structures.

2.2. Multigene Hypothesis

Recent rapid advancement in recombinant DNA technology allows for the determination of the primary structure of proteins by elucidating the sequence of DNA complementary to the mRNA. The complete amino acid sequences of rat PAH and TH have been thus determined (Robson et al., 1984; Grima et al., 1985). It was anticipated that these two enzymes are structurally related, since PAH-cDNA probes hybridized to a mRNA in adrenal medulla where PAH is not expressed (Robson et al., 1979), and antibodies to liver PAH cross-reacted with a 60-kdalton protein, probably TH, in PC12 cells (Chikaraishi et al., 1983). By comparing their amino acid sequences (Robson et al., 1984; Grima et al., 1985), we have observed that the primary structures of PAH and TH are actually more than 75% homologous. This surprisingly high degree of homology suggests that these enzymes are members of a multigene family. Numerous examples of the existence of multigene families are well documented (Beltz et al., 1984). Recently, Joh and colleagues (Joh and Ross, 1983; Joh et al., 1985; Baetge et al., 1985) have shown that three of the four enzymes involved in catecholamine biosynthesis (Fig. 1) may be structurally related, and that these three enzymes, TH, dopamine β-hydroxylase (DBH), and phenylethanolamine *N*-methyltransferase (PNMT), share common protein domains. Although the sequence homologies among the three enzymes appear to be less than those observed for PAH and TH (Baetge et al., 1985), the enzymes mentioned above may belong to a multigene family. Although it is not clear whether we can assume that proteins that share similarities in biological function, substrate, or ligand binding sites or active sites always comprise multigene families, if this is so, it is probable that there are other proteins or enzymes that belong to the same gene family. One example is tryptophan hydroxylase (TPH), which catalyses the conversion of tryptophan to 5-hydroxytryptophan in the biosynthetic pathway of 5-hydroxytryptamine (5-HT, serotonin). As illustrated in Fig. 2, PAH, TH, and TPH require the same cofactors, and their respective substrates, i.e., phenylalanine, tyrosine, and tryptophan, are similar in structure. Since TPH has not been well characterized in terms of its biochemical and molecular biological properties, it is premature to conclude that TPH belongs to the same multigene family. It should be noted that there are obvious differences among PAH, TH, and TPH. Expression of TH is usually limited to catecholamine-containing neurons and chromaffin cells, TPH to serotonergic neurons in the central nervous system, and PAH to nonneuronal cells of the liver and kidney. There is no evidence that PAH is expressed in the central nervous system. Although phenotypic expression of these three enzymes is different, biochemical and structural evidence suggests that PAH, TH, and possibly TPH have evolved from a common ancestral precursor.

3. Phenylalanine Hydroxylase

3.1. Phenylalanine Hydroxylation Reactions

Phenylalanine hydroxylase catalyzes the conversion of phenylalanine to tyrosine. The reaction requires tetrahydrobiopterin (H_4-biopterin) and molecular oxygen, and has been well characterized by the excellent work of Kaufman (1957, 1959, 1963, 1971). Stoichiometry of the reaction is as follows:

$$\text{phenylalanine} + O_2 + H_4\text{-biopterin} \xrightarrow{\text{PAH}} H_2\text{-biopterin} + \text{tyrosine} + H_2O$$

Dihydrobiopterin (H_2-biopterin) is reduced to H_4-biopterin by the enzyme dihydrobiopterin reductase (DHPR) (Kaufman, 1963):

$$H_2\text{-biopterin} + \text{NADH (or NADPH)} \xrightarrow{\text{DHPR}} \text{NAD} + (\text{or } NADP^+) + H_4\text{-biopterin}$$

Phenylalanine hydroxylase is an iron-containing enzyme (Fisher et al., 1972). The stoichiometry of the reaction and detailed reaction mechanisms have been thoroughly studied by Shiman and col-

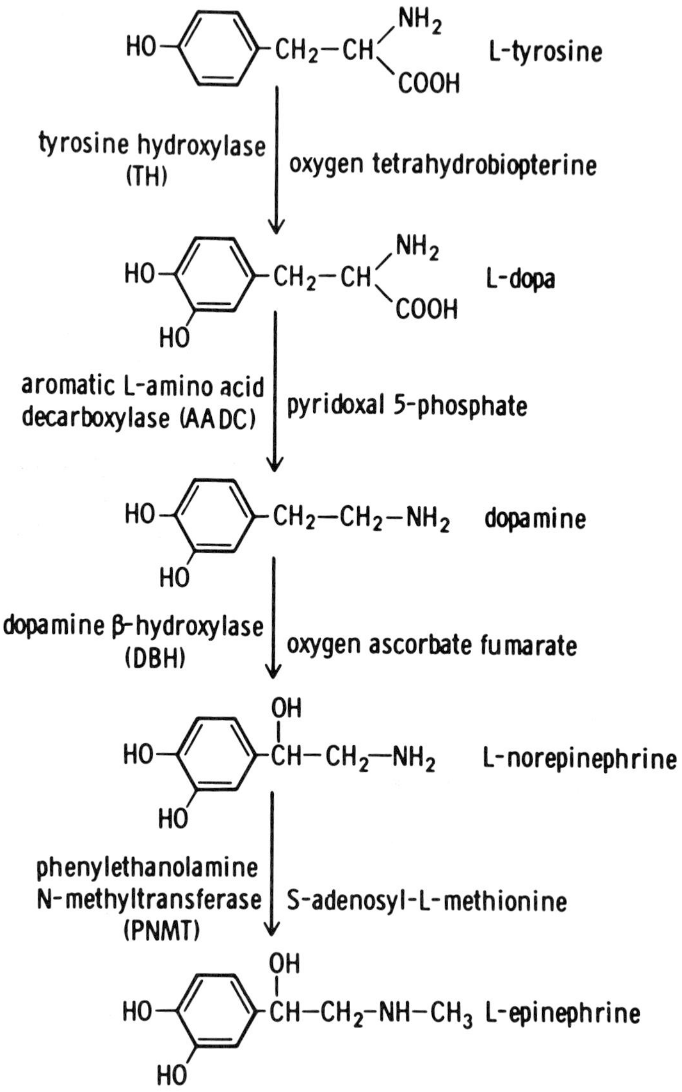

Fig. 1. Catecholamine biosynthetic pathway. Tyrosine hydroxylase is the first and presumably the rate-limiting enzyme in the pathway.

leagues (Gottshell et al., 1982; Lazarus et al., 1982). Gottshell et al. (1982) demonstrated that there is one iron atom per subunit, that the enzymatic activity is proportional to iron content, and that the presence of iron is required for enzymatic activity of the rat liver PAH. Iron-chelating reagents inhibit the enzyme activity (Bublitz, 1971; Fisher et al., 1972), but the actual role of the enzyme-bound

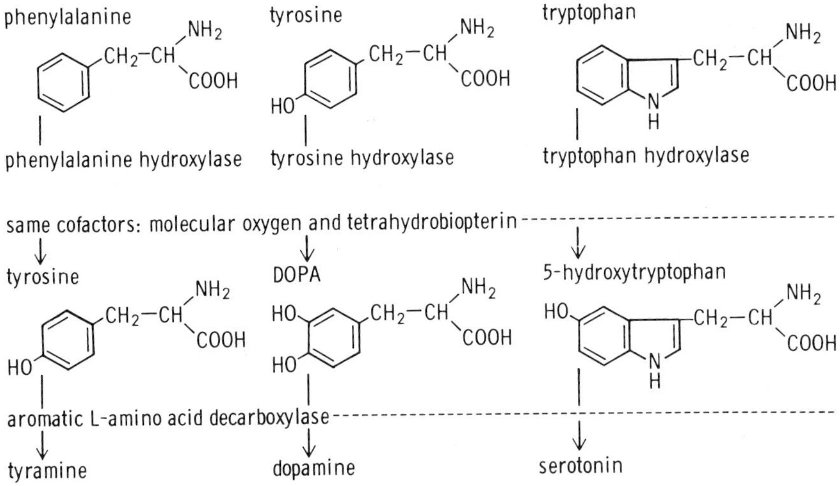

Fig. 2. Similarity of three monooxygenase reactions. The three oxygenases (phenylalanine, tyrosine, and tryptophan hydroxylases) require molecular oxygen and tetrahydrobiopterin as cofactors in their catalytic reaction. It should also be mentioned that the three substrates phenylalanine, tyrosine, and tryptophan are similar in their structures.

iron in the PAH is unknown (Shiman, 1985). In order to be catalytically active, PAH must be not only activated with phenylalanine (Tourian, 1971; Shiman and Gray, 1980) or a detergent such as lysolecitin (Fisher and Kaufman, 1972a; Miller et al., 1975), but also reduced (Kaufman, 1971; Fisher et al., 1972; Marota and Shiman, 1984). Marota and Shiman (1984) have recently demonstrated that the enzyme can be reduced in the absence of phenylalanine. One molecule of H_4-pterin is oxidized per PAH subunit in the presence of molecular oxygen to form one molecule of quinonoid H_2-pterin and reduced enzyme.

3.2. Tetrahydropterin and Dihydropterin Reductase

Since PAH, TH, and TPH all require a common cofactor (Fig. 2), tetrahydropterin, the reaction of dihydropterin reductase is important. The enzyme is relatively nonspecific with respect to its substrate, utilizing either NADH or NADPH.

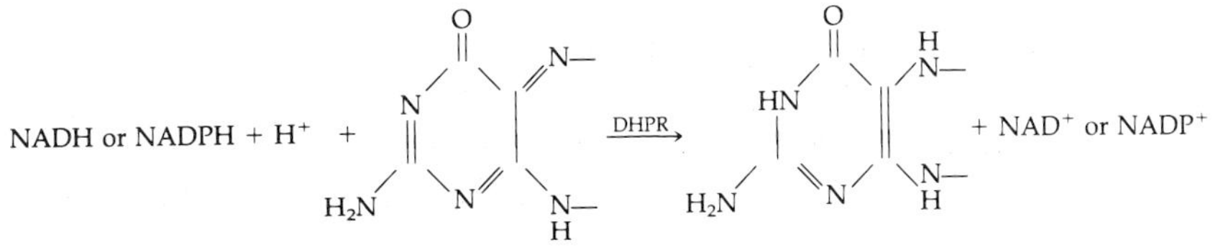

Dihydropterin reductase has a molecular weight of 25,000–26,000 (Figaira et al., 1981; Webber and Whiteley, 1981), and accepts a variety of quininoid dihydropterins as substrates, including 6-methyl- (Kaufman, 1963), 7-methyl- (Lind, 1972), and 6,7-dimethyl- (Kaufman, 1963) quinonoid dihydropterin. Quinonoid dihydrobiopterin (William et al., 1976), dihydroneopterin (William et al., 1976), and dihydrofolate (Lind, 1972) may also serve as substrates. The enzyme is rich in tissues containing PH, TH, and TPH.

3.3. Regulation of Phenylalanine Hydroxylase

3.3.1. Phenylalanine Activation

The best understood regulation of the enzyme in vivo is its activation of PAH by phenylalanine. At normal plasma phenylalanine concentrations, PAH in rat liver is slightly activated (1–4%) (Shiman et al., 1982). When a phenylalanine concentration of 50 mg/100 g body weight is administered, the degree of activation of the enzyme increases to 40% of maximum (Shiman et al., 1982).

3.3.2. Activation by Phosphorylation

Direct phosphorylation of PAH results in an increase in catalytic activity of the enzyme. Rat liver PAH has 0.2–0.3 molecules of covalently bound phosphate per subunit of enzyme (Abita et al., 1976). Glucagon administration increases the phosphate content to 1–1.2 molecules per subunit (Donlon and Kaufman, 1978). The increase in phosphate is also observed in vitro by incubation of enzyme with beef heart muscle protein kinase and ATP (Abita et al., 1976). Kinetic analyses and chemical mechanisms for the phosphorylation activation of the enzyme are detailed in the recent review by Shiman (1985) and related articles (Abita et al., 1976; Shiman et al., 1982; Phillips and Kaufman, 1984; Doskeland et al., 1984). Glucagon-induced phosphorylation of rat liver PAH is maximal and complete at about 30 min after the hormone is injected and falls rapidly thereafter (Donlon and Kaufman, 1978), and the phosphorylation is believed to activate PAH in vivo (Abita et al., 1976; Milstein et al., 1976; Donlon and Kaufman, 1980). However, Shiman (1985) has questioned whether the phosphorylated form of PAH is inherently more active than nonphosphorylated PAH (Shiman et al., 1982). He states that the apparent activation is almost certainly caused by the assay system employed in these studies, and that no data have yet been reported to support the notion that in a mechanistic sense phosphorylation activates PAH in vivo (Shiman, 1985). However, Fisher et al. (1984) suggest that phosphorylation is associated with changes in catalytic activity under physiologically relevant conditions.

3.3.3. Inhibition

Tetrahydrobiopterin is an inhibitor of PAH activity in vivo (Shiman and Gray, 1980; Shiman et al., 1982), and this inhibition may play a significant role in regulation of the enzyme. The cofactor appears to act by competing with phenylalanine activation (Shiman and Gray, 1980). The catecholamines 3,4-dihydroxyphenylalanine (DOPA), norepinephrine (noradrenaline, NA), and epinephrine (adrenaline, A) inhibit PAH activity *in situ* in H_4 cell cultures (Miller and Shiman, 1976). However, the concentration of NA required for 50% inactivation in H_4 cells is about 5 μM (Miller and Shiman, 1976), a concentration of the hormone several orders of magnitude greater than normally found in blood. Thus it is not clear whether catecholamines are regulators of the enzyme in liver in vivo (Shiman, 1985).

3.4. Molecular Biology

Robson et al. (1979) have purified mRNA for PAH from total rat liver mRNAs, where PAH-mRNA constitutes less than 0.25%. A greater than 10% purity was obtained in a single step by polysome immunoprecipitation. Complementary DNA to PAH-mRNA was synthesized and cloned. The rat cDNA clone was found to be capable of hybridizing with human PAH-mRNA, an important first step in the isolation of the human PAH gene. A thorough molecular analysis of phenylketonuria, a hereditary disorder in phenylalanine metabolism that causes permanent mental retardation in humans (Robson et al., 1979), is now underway. Ledley et al. (1985) have demonstrated by a gene transfer technique that a single gene contains all the necessary genetic information to code for functional PAH. The PAH gene has already been assigned to human chromosome 12 (Ladsky et al., 1984). Thus it is possible now to use the human PAH probe in prenatal diagnosis and the detection of carrier states. Furthermore, it provides new opportunities for the biochemical characterization of normal and mutant enzymes, and the investigation of alternative genetic therapies for phenylketonuria (Ledley et al., 1985).

3.5. Enzyme Assays

A recent review of this subject by Shiman (1985) detailed methodological problems and the issue of the stability of PAH in tissue preparations.

3.5.1. Continuous Assays

Continuous assays have been widely used to monitor the PAH reaction (Nielson et al., 1969; Kaufman, 1959, 1971; Shiman et al., 1971; Ayling et al., 1973; Shiman and Gray, 1980). These assay procedures have a clear advantage in enzyme characterization including determination of kinetic parameters, reaction mechanisms, and substrate specificity.

1. Direct measurement of the oxidation of H_4-pterin to H_2-pterin (Kaufman, 1971; Shiman and Kaufman, 1970; Shiman and Gray, 1980).
2. Indirect measurement of the oxidation of H_4-pterin by using the quinonoid H_2-pterin formed to oxidize NADPH (Kaufman, 1959, 1971) or NADH (Nielson et al., 1969) in the presence of dihydropterin reductase.
3. Direct measurement of tyrosine formed by monitoring at 275 nm (Shiman and Gray, 1980).

3.5.2. Commonly Used Assays

3.5.2.1. Fluorometric Determination of Tyrosine Conjugated With Nitrosonaphthol (Waalkes and Udenfriend, 1957). Fluorometric determination is one of the original assay procedures. To a tissue preparation in 0.1*M* Tris-Cl, pH 7.4, containing 1 μmol phenylalanine and 3000 units catalase/mL, 0.2 m*M* 6-methyl-H_4-pterin and 50 m*M* mercaptoethanol, final concentration, are added. The mixture is incubated at 37°C for 20 min and the reaction is stopped by the addition of 6*M* trichloroacetic acid for a final concentration of 0.55*M*. Protein is removed by centrifugation at 10,000*g* for 5 min. To 2 mL of the deproteinized reaction mixture in a glass-stoppered centrifuge tube are added 1 mL of nitrosonaphthol reagent (0.1% 1-nitroso-2-naphthol in 95% ethanol) and 1 mL in nitric acid reagent (24.5 mL of 1 : 5 nitric acid is mixed with 0.5 mL of 2.5% $NaNO_2$). The tube is shaken and placed in a waterbath at 55°C for 30 min. After cooling, 10 mL of ethylene dichloride is added, and the tube is shaken to extract the unreacted nitrosonaphthol reagent. After the tube is centrifuged, the aqueous layer is transferred to a cuvet and read in a spectrophotofluorometer (activation at 460 nm and measurement at 570 nm).

3.5.2.2. Fluorometric Determination of Intrinsic Fluorescence of Tyrosine (Bailey and Ayling, 1980). The PAH reaction is similar to that described in section 3.5.2.1. Aliquots of deproteinized reaction mixture are directly injected into a reverse-phase chromatographic column (50 and 100 μL samples into 10 and

25 cm columns, respectively). Elution of tyrosine is optimal with 0.1*M* ammonium acetate, pH 4.1–4.7. Tyrosine is detected using a high-pressure mercury lamp, at 282 and 300 nm of the monochromators.

3.5.2.3. Radiochemical Assays. The following methods are the most sensitive assay procedures.

1. *Thin layer chromatography (Kaufman, 1979).* The reaction mixture contains the following components (in μmol): 50 K-phosphate, pH 7.2; 0.15 DPNH; 2.0 L-phenylalanine; 0.5 μCi L-,U-^{14}C-phenylalanine; 0.4 H_4-biopterin; the enzyme preparation; DHPR and excess catalase, in a volume of 1 mL. The mixtures are incubated for 30 min at 25°C, and the reaction is stopped by the addition of 50 μL of 2.4*M* trichloroacetic acid. Tyrosine formation is determined by measurement of the conversion of ^{14}C-phenylalanine to ^{14}C-tyrosine. The tyrosine is separated from the phenylalanine by chromatography on sheets of Eastman precoated cellulose, No. 6064, with the use of a solvent system 2-propanol: H_2O:NaOH (8:1:1 by volume). After development in the solvent, strips are placed in 10 mL of H_2O and 10 mL of Aquasol (New England Nuclear). The amount of radioactivity is then determined.
2. *Ion exchange (sulfonic acid ion-exchanger, Bio-Rad Aminex-Q153) method (Shiman et al., 1982).* The deproteinized reaction mixture, 1.0 mL, is applied to a column (0.5 × 4.5 cm) of Q153 resin equilibrated with 0.7 HCl containing 4% ethanol by volume. The column is developed in the same buffer, and 0.3 mL fractions are collected every 10 min. Tyrosine elutes before phenylalanine, and ^{14}C-tyrosine is counted.
3. Other radioisotope assay methods, including an isotope dilution method (Miller et al., 1975) and iodination of tyrosine (Guroff et al., 1967), are also used.

3.6. Purification

Several purification procedures have been reported (Kaufman and Fisher, 1970; Nakata and Fujisawa, 1980; Shiman et al., 1979, 1982; Donlon and Kaufman, 1980; Al-Janabi, 1980; Webber et al., 1980; Webber and Whiteley, 1980; Cotton, 1974; Cotton and Grattan, 1975; Cotton and Jenning, 1978; Shiman, 1980; Doskeland et al., 1982).

3.6.1. Classical Fractionation Procedures

Classical fractionation procedures for the purification of PAH from rat liver are still used in several laboratories. The most commonly used procedures (Shiman et al., 1979; Shiman, 1980) are detailed in Table 1.

Table 1
Purification of Phenylalanine Hydroxylase[a]

Fresh minced rat livers, 50g	
Homogenization	200 mL, 0.2*M* KCl, 0.02*M* phenylalanine, 0.03*M* Tris, pH 7.25
Centrifugation	40,000*g* for 30 min
Incubation at 25°C for 20 min	
Centrifugation	40,000*g* for 20 min. Add Tween 80, 0.1% final, just before next column chromatography step
Phenyl Sepharose[b]	30% Volume of enzyme preparation, prewashing with homogenizing buffer containing 0.1% Tween 80
Wash and elution	1. 1 Bed vol of homogenization buffer 2. 10 Bed vol of homogenization buffer containing 4.8% dimethylformamide 3. 10 Bed vol of a buffer containing 15% glycerine, 10 m*M* phenylalanine, 50 μ*M* EDTA, 0.037% Tween 80, 0.03*M* Tris, pH 7.25. PAH is eluted with 1 bed vol of this buffer
Phenyl Sepharose[b]	10% Bed vol of the initial column
DEAE-cellulose[b]	Equilibrated with 15% glycerine, 50 μ*M* EDTA, 0.3*M* Tris, pH 7.25
Linear gradient elution	0–0.4*M* KCl in the equilibrium buffer

[a]From Kaufman and Fisher, 1970; Shiman et al., 1979; Shiman and Jefferson, 1982.

[b]If necessary, repeat phenyl Sepharose and DEAE-cellulose columns.

3.6.2. Affinity Chromatography

Affinity chromatography has been used for the purification of PAH. Dimethyl-H_4-pterin (6,7,-dimethyltetrahydropteridin)-linked Sephrose is commonly used, and homogenous enzyme with maximal specific activity has been obtained. Al-Janabi (1980) used an affinity matrix consisting of a succinylated diaminodipropylamine arm linked to Sepharose 4B, to which the dimethyl-H_4-pterin is covalently linked. The enzyme is eluted with buffered 50% glycerol, 1*M* KCl in one step. Webber et al. (1980) used a matrix containing dimethyl-H_4-pterin bound to Sepharose via a butanediol diglycidyl ether linkage. This procedure uses catalase to protect the enzyme during purification. In order to eliminate catalase contamination in the final product, rabbit IgG anti-catalase Sepharose matrix is used (Webber and Whiteley, 1980).

4. Tyrosine Hydroxylase

Early in the 1950s, Udenfriend and colleagues (Udenfriend et al. 1953, Udenfriend and Wyngaarden, 1956) established that tyrosine is converted to DOPA by an enzymatic reaction. Direct demonstration of enzymatic hydroxylation of tyrosine to DOPA came 10 yr later by his group (Nagatsu et al., 1964). Since then, a massive volume of research on this enzyme has been produced. This article concentrates on recent advances in the study of the reaction, regulation, and recent molecular biology of this enzyme. Detailed recent reviews include those by Kuhn and Lovenberg (1982) and Messarano and Weiner (1983). The molecular weight of the TH subunit is known to be 60,000. However, there is controversy regarding the molecular weight of the native form of the enzyme. Reports range from 60,000 to 220,000 or higher (Shiman and Kaufman, 1970; Shiman et al., 1971; Fisher and Kaufman, 1972; Mussachio et al., 1971; Joh and Reis, 1975). It is generally believed that the native form of TH is a tetramer of identical subunits of 60,000. It is not clearly established that the active species of TH is tetrameric, or that monomeric and dimeric forms do not possess catalytic activity. It is also not clear whether the active form in the assay reaction mixture is tetrameric. However, a high molecular weight enzyme of bovine adrenal medulla eluted near the void volume of a Sepharose 4B column possesses TH activity (Joh, unpublished observation). These high molecular weight forms of TH are probably the aggregates, and their molecular weights range to several million. It is not known whether these aggregated forms consist of monomeric or tetrameric structures. When a high-speed centrifugation sediment of TH containing tissue homogenates is digested with trypsin (Petrack et al., 1972; Mussachio et al., 1971) or chymotrypsin (Shiman et al., 1971), a small fragment of TH is obtained, the molecular weight of which is 34,000. This protease-digested form of TH is enzymatically active, and the entire active site of TH, therefore, probably resides in this fragment.

4.1. Tyrosine Hydroxylase Reactions

Tyrosine hydroxylase catalyzes the conversion of tyrosine to DOPA. The reaction is the first and probably the rate-limiting step (Levitt et al., 1965) in the biosynthesis of catecholamines, DA, NA, and A (Fig. 1).

$$\text{L-tyrosine} + O_2 + H_4\text{-biopterin} \xrightarrow{\text{TH}} H_2\text{-biopterin} + \text{L-DOPA} + H_2O$$

Participation of oxido-reduction reaction of biopterin or pterin derivatives in the hydroxylation reaction is similar to that discussed for phenylalanine hydroxylation (*see* section 3.2). The kinetic mechanism of the tyrosine hydroxylation reaction is not yet well established, although numerous publications on this subject are available (Joh et al., 1969; Mandell, 1978; Bullard and Capson, 1983). There have been many obstacles in determining the exact reaction mechanisms and kinetics of TH. First, it is difficult to obtain enough purified TH for this purpose. Partially purified enzyme has been used, but the reaction is significantly affected by the presence of inhibitors, activators, and other effectors. Recently, however, several excellent purification procedures have been reported (*see* section 4.4), and it may now be possible to investigate the exact reaction mechanisms governing the tyrosine hydroxylation reaction. Second, kinetic patterns of the TH reaction are complex. A Lineweaver-Burke plot of velocity vs H_4-pterin or H_4-biopterin concentration is not linear and shows a cooperative kinetic pattern. This is probably a result of the different states of phosphorylation-mediated activation of the enzyme (*see* section 4.2). The structural correlation between phosphorylated and activated states of the enzyme must be fully understood in order to establish its reaction mechanisms based on enzyme kinetics and reaction chemistry.

4.2. Regulation

Since TH is the presumed rate-limiting enzyme in catecholamine biosynthesis and catecholamines are major neurotransmitters in the central and peripheral nervous systems and chromaffin cells, studies on regulation of TH are highly popular in the fields of pharmacology and neuroscience. Accordingly, a large volume of research activity on this subject is reported each year. The regulation of the activity of TH in vivo can be attributed to two basic regulatory mechanisms. First, an increase or decrease in TH activity is caused by an increase or decrease in the amount of TH with similar specific activity (activity/protein weight). Any mechanisms that alter specific turnover of TH result in an increase or decrease of available TH in individual cells. These include a faster or slower rate of TH synthesis regulated at either the transcriptional and/or translational level, and an enhanced or slower rate of TH degradation. This type of regulation is a long-term process. Under the assumption that all of the TH in these cells has a similar specific activity and immunoreactivity, these changes in levels of enzyme can be measured by immunotitration, immunoprecipitation, and other means of quantitative immunoreaction. Determination of changes in V_m under similar reaction conditions is sometimes performed, but this method can yield false conclusions, since TH kinetics are complex. The second regulatory process is that of an increase or decrease in the specific activity of TH. The best known example of this type of short-term regulation is the activation of TH by phosphorylation. In this case, determination of enzyme kinetic parameters has been employed (*see* section 4.2.2)

4.2.1. Long-Term Regulation

Following prolonged stress of the animal or treatment with various drugs, TH activity in the central nervous system, adrenal medulla, and peripheral ganglia slowly increases severalfold and returns to control level in 2–3 wk. For instance, TH activity in the adrenal gland, superior cervial ganglia, and locus ceruleus is increased severalfold by a number of stresses (Thoenen, 1970; Chuang et al., 1975; Palkovits, 1975; Zigmond et al., 1974; Zigmond and Chalazonitis, 1974; Gilad and McCarty, 1981; Kvetansky et al., 1970; Masserano et al., 1981; Stone et al., 1978; Weiss et al., 1975). Treatment with drugs alters TH activity. For instance, reserpine, which depletes stores of catecholamines in the central and peripheral nervous systems, increases TH activity in rat adrenal medulla, superior ganglia, and locus ceruleus (Mueller et al., 1969; Zigmond et al., 1980; Zigmond and Ben-Ari, 1977; Joh et al., 1973, Reis et al., 1975; Ross et al., 1978). Massarano and Weiner (1983) have reviewed this subject in detail.

4.2.2. Short-Term Regulation

There is general agreement that TH can be phosphorylated, and that phosphorylation is coupled with the activation of TH.

Since this subject alone deserves a full review, it is not possible to introduce all published results in this chapter. This section concentrates on reports of recent developments in the area of different modes of TH phosphorylation and, thereby, activation mechanisms.

Early observations that cAMP and its analogs are able to increase TH activity in crude tissue preparations, in the presence of added ATP and Mg^{2+}, triggered a new field of research for the regulation of TH. It is believed that cAMP-dependent protein kinase phosphorylates the TH subunit of 60,000 (Yamauchi and Fujisawa, 1978, 1979; Joh et al., 1978; Vulliet et al., 1980; Raese et al., 1979; Waymire et al., 1979) (*see* Andrews et al., 1983, refs. 14–18). Meligeni et al. (1982) presented data that support a direct relationship between cAMP-dependent phosphorylation and activation of the enzyme in vitro and demonstrated that similar processes can regulate catecholamine biosynthesis *in situ*. Mechanisms of the phosphorylation-mediated TH activation in vivo are explained by Okuno and Fujisawa (1985) who suggested that TH activity may be suppressed by its products in a normal state and that it may be stimulated by cAMP-dependent protein kinase as the occasion demands.

Phosphorylation-mediated activation of TH is not limited to cAMP-dependent protein kinase activity. Haycock et al. (1982a,b) showed that acetylcholine also activates TH via phosphorylation in adrenal chromaffin cells. Many other findings indicate that TH is phosphorylated by other kinds of protein kinases. These include calcium-dependent protein kinase (George et al., 1985), calcium-calmodulin-dependent protein kinase (Yamauchi and Fujishima, 1981), and protein kinase C (calcium/phospholipid protein kinase) (Albert et al., 1984). Yamauchi and Fujisawa (1981) suggested that the activation of TH occurs in a two-step reaction, phosphorylation by calcium-calmodulin-dependent protein kinase and activation by activator protein.

It should be mentioned here that phosphorylation-mediated TH activation, in general, is accompanied by the kinetic observation of a lowered K_m for H_4-pterin cofactor. As mentioned in a previous section, the Lineweaver-Burke plot of the TH reaction is not linear, showing cooperative kinetic patterns. This suggests that there is more than one species of enzyme catalyzing the same reaction. Since multiple site phosphorylation of TH has been demonstrated, we may assume that there are multiple species of active TH with different degrees of phosphorylation and that these multiple active species of TH produce a cooperative pattern of enzyme kinetics. This in turn indicates that there may be multiple

K_m values for H_4-pterin, rather than a fixed low and a fixed high value.

It is of interest to note here that the enzymatically active, protease-digested form of TH can not be activated by phosphorylation (Vigny and Henry, 1982; Joh, unpublished observation). This suggests that phosphorylation-mediated activation of TH involves conformational changes of the enzyme and that phosphorylation may not be directed to the active sites of TH.

4.3. Molecular Biology

The first studies showing the RNA level of TH regulation were reported by Baetge et al. (1981). In this report, polysomal polyAmRNA was purified from a cloned cell line of rat pheochromocytoma, PC 12, and translated in vitro in a reticulocyte cell-free protein-synthesizing system. TH was identified by immunoprecipitation. Using this technique, they demonstrated that dexamethasone enhances translation of TH in PC 12 cells. Lamouroux et al. (1982) identified complementary DNA clones coding for TH in PC 12 cell lines. They also showed that blot hybridization analysis of phosphorylated RNA from PC 12 cells indicated a major mRNA species of 1.9 kilobases (kb), and a species of the same size was identified from human pheochromocytoma tumor. The same group subsequently identified a complete amino acid sequence of TH by elucidating the cDNA sequence. The presence of many serine and threonine residues in the sequence will facilitate the investigation of phosphorylation sites of the enzyme. Meanwhile, Lewis et al. (1983) prepared a partial cDNA clone to TH from rat PC 12 cell lines. An important aspect of these reports is that it is now possible to investigate molecular mechanisms governing induction or action of TH by use of TH-cDNA probes (Mallet, 1983; Chikaraishi, 1983).

4.4. Enzyme Assays

There are three commonly used TH activity assays: determination of tritium-labeled H_2O released from the tritium-labeled 3' position of tyrosine (Nagatsu et al., 1964; Leighton and Waggoner, 1981), measurement of $^{14}CO_2$ release from the ^{14}C-labeled carboxyl group of DOPA by aromatic L-amino acid decarboxylase (Waymire et al., 1971), and direct measurement of ^{14}C-DOPA separated from tyrosine by an alumina column (Coyle, 1972; Reis et al., 1975). Several other procedures are also available (Haavick and Flatmark, 1980; Nagatsu et al., 1979; Kato et al., 1981).

4.4.1. Determination of Tritium-Labeled H_2O (Nagatsu et al., 1964; Leighton and Waggoner, 1981; Levin et al., 1984)

The assay mixture contains (in a total volume of 0.5 mL) 100 μmol Na-acetate buffer, pH 6.1; 0.5 μmol ferrous sulfate; 1.0 μmol 6-methyl-H_4-pterin; 20 μmol 2-mercaptoethanol; 10 μmol Na-phosphate; 0.05 μmol 3,5-^{3}H-L-tyrosine (Nagatsu et al., 1964) or 4-^{3}H-L-tyrosine (Leighton and Waggoner, 1981), and enzyme preparation. The mixture is incubated for 30 min at 37°C and the reaction is stopped by addition of 0.5 mL 10% trichloroacetic acid or glacial acetic acid. The protein is sedimented by centrifugation. The supernatant is passed through a Dowex 50 × 4 H^+ column (0.5 × 3–5 cm). Each column is washed with 2 mL water. The combined effluents are taken up to 5 mL and washed three times with anhydrous ether to eliminate deaminated tyrosine metabolites. An aliquot of 0.5 mL is counted. For the small scale assay in a total volume of 50–100 μL, the reaction is stopped with 50 μL glacial acetic acid. Each column is washed with 1 mL water and the effluent is counted. Recently, Levin et al. (1984) used an ion exchange column containing, bottom to top, Dowex 50-H^+ (3.4 cm), a small layer of activated charcoal, and Dowex 1-acetate (0.5 cm). The eluate and each of two column washes (1.5 mL total) are collected in glass scintillation vials and counted. Levin et al. (1984) also used an organic solvent extraction procedure in order to simplify quantitation of ^3H_2O. In this procedure, the enzyme reaction is arrested by the addition of 50 μL of 3.0*M* $NaCO_3$ (pH 11.6). Subsequently, 5 mL of toluene-based scintillant [25% isoamyl alcohol, toluene, 2,5-diphenyloxazole, and *p*-bis-(2-[5-phenyloxazolyl]) benzene] is added, and the mixture is blended vigorously for 10 s. The organic and aqueous phases are allowed to separate and the organic phase is counted.

4.4.2. Measurement of $^{14}CO_2$ Released by Decarboxylase (Waymire et al., 1971)

The reaction mixture is similar, except for the use of 1-^{14}C-L-tyrosine instead of ^{3}H-tyrosine. To the reaction mixture, 5 mmol pyridoxal 5-phosphate and hog kidney aromatic L-amino acid decarboxylase are added. The mixture is incubated for 30 min at 37°C in a 10-mL Erlenmeyer flask or test tube, covered by a rubber injection vial cap. A plastic well (Kontes Glass Co.) is suspended from the rubber stopper and contains a folded paper wick and 0.2 mL NCS Solubilizer. The reaction is stopped by injection of 0.5 mL 10% trichloroacetic acid. The mixture is allowed to stand for 2 h and trapped $^{14}CO_2$ in the paper wick is counted.

4.4.3. Measurement of ^{14}C-DOPA Separated by an Alumina Column (Coyle, 1972; Reis et al., 1975)

The assay mixture, which contains 50 μL of enzyme, 10 μL of 30 m*M* ferrous ammonium sulfate (or 1000 units catalase), 20 μL of 6 m*M* 6-methyl-H_4pterin in 140 m*M* mercaptoethanol, 10 μL of 1*M* Na-acetate buffer, pH 5.8, 5 μL of 4 m*M* tyrosine, 2.5 μL of 1*N* NaOH, and 2.5 μL of L-(U-^{14}C)-tyrosine (400,000 cpm), is incubated for 20–30 min at 37°C. The reaction is stopped by addition of 1 mL of 0.4*N* perchloric acid containing 10 μg of carrier L-DOPA and centrifuged. To the supernatant, 6.5 mL of the following mixture is added: 5 vol 2% Na_2-EDTA and 1.5 vol 0.35*M* KH_2PO_4 (monobasic). After adjusting the pH to 8.6 with 1*N* NaOH, the mixture is passed through 400 mg of aluminum oxide in a column (0.5 × 3 cm). Each column is washed with 20 mL H_2O, and L-DOPA is eluted with 4 mL of 0.5*N* acetic acid. The effluent is counted.

4.4.4. Direct Measurement of L-DOPA by HPLC (Haavik and Flatmark, 1980; Kato et al., 1981; Messripour and Clark, 1982; Hirata et al., 1983)

This direct measurement of L-DOPA formed in incubation mixtures of brain slices or synaptosomal preparations is also used for tissue homogenates. Brain synaptosomes (2 mg protein in 200 μL) are incubated in the presence of 40 μ*M* tyrosine and 100 μ*M* pargyline at 37°C for 15 min (Messripour and Clark, 1982). The reaction is stopped by freezing the samples over dry ice. Two milliliters of ice-cold 0.05*M* perchloric acid and 100 ng (50 μL) of internal standard solution (dihydroxybenzylamine) are added to the frozen samples and the mixture is homogenized and centrifuged. The clear supernatant is transferred to a vial containing 0.1 g alumina and 1 mL of 3*M* Tris buffer, pH 8.6. The mixture is shaken for 15 min and the alumina is washed with ice-cold Tris buffer, pH 8.6, followed by ice-cold H_2O. After the addition of 0.5 mL of 0.05*M* perchloric acid, the mixture is filtered. A 200-μL sample of the extract is injected into an HPLC column (Altex 15 cm × 4.6 mm, intraspher-IP with the mobile phase containing 90% 0.1*M* KH_2PO_4, 0.1 m*M* K-EDTA, 0.3 m*M* sodium octylsulfate, and 10% methanol, pH 3.0).

Brain slices or minces (Hirata et al., 1983) are incubated in 1.0 mL of medium containing NSD 1055 (DOPA decarboxylase inhibitor) under saturated pressure of 95% O_2/5% CO_2 for 60 min at 37°C, with occasional gassing and shaking. The samples are then homogenized first in perchloric acid (0.2*M* final concentration) containing alpha-methylDOPA, 0.25% EDTA, and 0.5% sodium metabisulfite. After centrifugation, Tris-HCl, pH 9.0 (0.7*M* final concentration) is added to the clear supernatant (final pH, 8.0–8.5). The mixture is passed through double columns, 0.5 mL of Amberlite CG-50 attached to a column of 100 mg alumina. Both columns are washed twice with 4 mL of H_2O. The bottom column (alumina) is then washed twice with 3 mL H_2O, followed by 100 μL of 0.5*M* HCl. L-DOPA and alph-methylDOPA are eluted with 250 μL of 0.5*M* HCl. A 50 μL aliquot is injected into an HPLC column (Nucleosil $7C_{18}$). The mobile phase is 0.1*M* phosphoric acid containing 8.5% methanol.

4.5. Purification

There are numerous reports on the purification of TH. It is difficult to recommend one specific method for this purpose, since different tissues require different purification procedures. For instance, TH in bovine sources, adrenal medulla, and brain easily aggregates in tissue homogenates and it is practically impossible to purify the native form of the enzyme. Yet, Nagatsu's group (Oka et al., 1981, 1982) has been successful in purifying the native form of TH from these sources. Purification of the protease-digested form of TH from bovine adrenal medulla was initially reported by Petrack et al. (1972) and Shiman et al. (1971). There are several procedures for the purification of TH from rat tissues and PC 12 cell lines (Joh et al., 1983; Markey et al., 1980; Vulliet et al., 1980; Richtand et al., 1985). Joh and Ross (1983) have detailed procedures for the production of antisera to TH. The purification procedures for TH from rat brain are detailed in Table 2.

5. Tryptophan Hydroxylase

Tryptophan hydroxylase (TPH) is the least characterized enzyme among the three monooxygenases mentioned in this chapter, as a result of the difficulty in obtaining purified enzyme. The other two enzymes, PAH and TH, are abundant in the rat or bovine liver and adrenal gland, respectively, and also in cancer cell lines that express them. TPH, on the other hand, is found only in the central nervous system, mainly in the midbrain and pineal gland. Studies on reaction mechanisms, biochemical and physical characteristics, and activation of the enzyme have been carried out mainly with crude enzyme preparations.

Table 2
Purification of Tyrosine Hydroxylase[a]

Rat striata (500*g*) or rat adrenal glands (1000*g*)	
Homogenization	500 mL 10 m*M* K-phosphate buffer, pH 7.4
Centrifugation	100*g* for 10 min
Centrifugation	100,000*g* for 60 min
Ammonium sulfate precipitation (50% saturation)	
Dialysis and equilibration with 20% ammonium sulfate in 5 m*M* K-phosphate buffer, pH 7.0	
Centrifugation	1000*g* for 10 min
Phenyl Sepharose[b]	Equilibrated with 20% ammonium sulfate in 5 m*M* K-phosphate buffer, pH 7.0
Wash and elution	1. Wash with equilibrium buffer 2. Wash with 5% ammonium sulfate in 5 m*M* K-phosphate buffer 3. 5 m*M* K-phosphate buffer, pH 7.0. 95% of TH activity is eluted in this buffer 4. H_2O
Heparin[b]	*See* details in Richtandt et al., 1985

[a]From Joh et al., 1978; Joh and Ross, 1983; Richtandt et al., 1985.
[b]Phenyl-Sepharose and heparin column may be repeated if necessary.

5.1. Tryptophan Hydroxylase Reactions

Tryptophan hydroxylase catalyzes the initial and probably rate-limiting step in the biosynthesis of serotonin (5-HT).

$$\text{L-tryptophan} + H_4\text{-biopterin} + O_2 \xrightarrow{\text{TPH}} \text{5-hydroxytryptophan} + H_2\text{-biopterin} + H_2O$$

Like PAH and TPH, the oxido-reduction reaction involves dihydropterin reductase, which reduces the quinonoid form of H_2-pterin (*see* section 3.2.). Unlike the TH reaction, however, the TPH reaction does not indicate a cooperative kinetic pattern, and is not inhibited by its product, 5-HT (Freedman et al., 1972; Macon et al., 1972; Kuhn et al., 1980a). However, catecholamines are potent inhibitors in vitro (Kuhn et al., 1980a). The reaction mechanism probably resembles that observed with TH. Some variables, including pH optimum, requirement of Fe^{2+} for activity, and activation of the enzyme in vitro by several reagents are detailed in the recent review by Hamon et al. (1981).

A limited exposure of crude TPH preparation to trypsin results in a significant increase in TPH activity (Hamon et al., 1978), and this may be caused by the partial proteolysis of the enzyme (Hamon et al., 1977). The activation kinetics are a slight increase in V_m, a decrease of K_m for H_4-biotperin to one-half of the control value, and a shift of optimal pH from 7.6 to 7.9.

5.2. Regulation

5.2.1. Effects of Tryptophan and 5-HT In Vivo

Although 5-HT does not inhibit TPH activity in vitro, it appears that an increase in 5-HT concentration in vivo reduces *de novo* synthesis of 5-HT in brain slices and intact synaptosomes (Hamon et al., 1973). There are several reports that indicate fluctuation of 5-HT production in vivo, caused by the availability of tryptophan and 5-HT. Depletion of 5-HT from the tissues appears to increase 5-HT synthesis (Hamon et al., 1977), and tissue concentrations of tryptophan may be involved in the regulation of 5-HT production in vivo (Gessa and Tagliamonte, 1974). However, no evidence has been presented for direct activation of TPH by these factors (Hamon et al., 1981).

5.2.2. Ion Influx, Availability of Tryptophan and 5-HT, and Enzyme Phosphorylation

Boadle-Biber (1978, 1979) and Elks et al. (1979) demonstrated that increases in Ca^{2+} influx across membranes or release of Ca^{2+} from intracellular storage sites increases 5-HT synthesis. This may have a significant effect on biosynthesis mechanisms for 5-HT. Boadle-Biber (1978) and Hamon et al. (1981) suggested that the increase in the rate of 5-HT synthesis in stimulated neurons is associated with an increased intrinsic activity of TPH. The increase in TPH activity may be a result of Ca^{2+}-dependent phosphorylation-mediated activation of TPH (Hamon et al., 1979, 1981). Phosphorylation-mediated activation of TPH has been postulated (Hamon et al., 1978; Kuhn et al., 1978; Lysz and Sze, 1978). These workers demonstrated that cAMP-dependent protein kinase-mediated phosphorylation activates TPH. The effect of phosphorylation on TPH activation is controversial, and changes in K_m and V_m values are not constant in each experiment. Recently, Ca^{2+}- or Ca^{2+}/calmodulin-dependent protein kinase was proposed to participate in TPH phosphorylation (Yamauchi and Fujisawa, 1978, 1979; Kuhn et al., 1980b). A recent report from Fujisawa's laboratory (Yamauchi and Fujisawa, 1981) identified the activator protein that activates TPH as well as TH in the presence of ATP, Mg^{2+}, Ca^{2+}, calmodulin, and Ca^{2+}/calmodulin-dependent protein kinase.

5.3. Assays

There are two generally used assay methods: determination of 5-hydroxytryptophan (5-HTP) by fluorometry (Freedman et al., 1972) and measurement of $^{14}CO_2$ released from the ^{14}C-labeled carboxyl group of 5-HTP by aromatic L-amino acid decarboxylase (Ichiyama et al., 1970).

5.3.1. Fluorometric Determination (Freedman et al., 1972)

The complete reaction mixture contains the following components (in μmol) in a final volume of 0.36 mL: Tris-acetate, pH 7.5; ^{14}C-L-tryptophan, 0.3; 6-methyl-H_4-pterin, 0.1; NADPH, 0.05; glucose 6-phosphate, 0.5; dithiothreitol, 0.5; and TPH. The mixture is incubated for 25 min at 37°C, and the reaction is stopped by the addition of 30 μL 70% perchloric acid. After centrifugation, a 0.2-mL aliquot of the supernatant fraction is added to 60 μL of concentrated HCl. The fluorescence of the solution at 538 nm is determined at an excitation wave length of 295 nm. The amount of 5-HTP formed is calculated from a standard curve.

5.3.2. Measurement of $^{14}CO_2$ (Ichiyama et al., 1970)

The assay mixture contains 200 μL of enzyme preparation, 50 μL of 1*M* Tris-acetate buffer, pH 7.6, 10 μL of 30 m*M* 6-methyl-H_4-pterin in 120 m*M* dithiothretol, 66 units of catalase in 10 μL, 10 μL of 1 m*M* ferrous ammonium sulfate, and 20 μL of 150 μ*M* L-(1-^{14}C)-tryptophan. The mixture is incubated for 30 min at 37°C, and 20 μL of aromatic L-amino acid decarboxylase and 40 μL of 25 m*M* pyridoxal 5-phosphate are added. The assay tube is then covered by a rubber vial cap in which is suspended a plastic well containing a folded filter paper wick and 250 μL of NCS Tissue Solubilizer. After an additional 30 min incubation at 37°C, 100 μL of 30% perchloric acid is injected through the rubber cap and $^{14}CO_2$ is trapped at 37°C for 90 min. The plastic well is removed and the radioactivity is counted.

5.4. Purification

In 1975, Joh et al. purified TPH from rat midbrain and produced antibodies against rat TPH. Final purification was achieved by polyacrylamide gel electrophoresis of the partially purified enzyme. Recent publications claim homogeneity of the purified enzyme from rat brain (Nakata and Fujisawa, 1982; Cash et al., 1985). Instead of classical fractionation procedures, these authors applied purification procedures similar to those used for PAH. Chromatography with a dimethyl-H_4-pterin affinity column, a phenylSepharose column, and a hydroxyapatite column in the presence of excess catalase in buffer was used (Cash et al., 1985).

Acknowledgments

The author thanks Dr. Michelle Ehrlich for her excellent advice in the preparation of this manuscript. The work is supported by NIH grants MH24285 and NS19002.

References

Abita J., Milstein S., Chang N., and Kaufman S. (1976) In vitro activation of rat liver phenylalanine hydroxylation. *J. Biol. Chem.* **251,** 5310–5314.

Albert K. A., Helmer-Matyjek E., Nairn A. C., Muller T. H., Haycock J. W., Green L. A., Goldstein M., and Greengard P. (1984) Calcium/phospholipid-dependent protein kinase (protein kinase C) phosphorylates and activates tyrosine hydroxylase. *Proc. Natl. Acad. Sci. USA* **81,** 7713–7717.

Al-Janabi J. M. (1980) Purification of rat liver phenylalanine hydroxylase by affinity chromatography. *Arch. Biochem. Biophys.* **200,** 603–608.

Andrews D. W., Langan T. A., and Weiner N. (1983) Evidence for the involvement of a cAMP-independent protein kinase in the activation of soluble TH from rat striatum. *Proc. Natl. Acad. Sci. USA* **80,** 2097–2101.

Ayling J., Person R., Pirson W., and Boehm G. (1973) A specific kinetic assay for phenylalanine hydroxylase. *Anal. Biochem.* **51,** 80–90.

Baetge E. E., Kaplan B. B., Reis D. J., and Joh T. H. (1981) Translation of tyrosine hydroxylase from Poly(A)mRNA in pheochromocytoma cells (PC12) is enhanced by dexamethasone. *Proc. Natl. Acad. Sci. USA* **78,** 1269–1273.

Baetge E. E., Suh Y. H., and Joh T. H. (1985) Complete nucleotide and deduced amino acid sequence of bovine phenylethanolamine *N*-methyltransferase: Sequence homology with rat tyrosine hydroxylase. *Science,* submitted for publication.

Bailey S. and Ayling J. (1980) An assay for picomole levels of tyrosine and related phenols and its application to the measurement of phenylalanine hydroxylase activity. *Anal. Biochem.* **107,** 156–164.

Beltz G. A., Jacobs K. A., Eickbush T. H., Cherbas P. T., and Kafatos F. C. (1984) Isolation of multigene families and determination of homologies by filter hybridization methods. *Meth. Enzymol.* **10B,** 266–284.

Boadle-Biber M. C. (1978) Activation of tryptophan hydroxylase from central serotonergic neurons by calcium and depolarization. *Biochem. Pharmacol.* **27,** 1069–1079.

Boadle-Biber M. C. (1979) Activation of tryptophan hydroxylase from slices of rat brain stem incubated with agents which promote calcium uptake or intraneuronal release. *Biochem. Pharmacol.* **28,** 2129–2138.

Bublitz C. (1971) Two mechanisms for the inhibition of phenylalanine hydroxylase by catecholamines. *Biochem. Pharmacol.* **20,** 2543–2553.

Bullard W. P. and Capson T. L. (1983) Steady-state kinetics of bovine striatal tyrosine hydroxylase. *Mol. Pharmacol.* **23,** 104–111.

Cash C. D., Vayer P., Mandel P., and Maitre M. (1985) Tryptophan hydroxylase. *Eur. J. Biochem.* **149,** 239–245.

Chikaraishi D. M., Brilliant M. H., and Lewis E. J. (1983) Cloning and Characterization of Rat Brain-Specific Transcripts: Rare, Brain-Specific Transcripts and Tyrosine Hydroxylase, in *Molecular Neurobiology* Cold Spring Harbor Symposia on Quantitative Biology, Vol. 48, (Watson J. D. and McKay R., eds.), Cold Spring Harbor Laboratory, New York.

Chuang D., Zsilla G., and Costa E. (1975) Turnover rate of tyrosine hydroxylase during transsynaptic induction. *Mol. Pharmacol.* **11,** 784–794.

Cotton, R. G. H. (1974) A pteridine absorbent for affinity chromatography. *FEBS Lett.* **44,** 290–292.

Cotton R. G. H. and Grattan P. J. (1975) Phenylalanine hydroxylase of *Macaca irus:* A simple purification by affinity chromatography. *Eur. J. Biochem.* **60,** 427–430.

Cotton R. G. H. and Jenning I. (1978) Affinity chromatography of phenylalanine hydroxylase: The structure of a pteridine adsorbent. *Eur. J. Biochem.* **85,** 357–363.

Coyle J. T. (1972) Tyrosine hydroxylase in rat brain—cofactor requirements, regional and subcellular distribution. *Biochem. Pharmacol.* **21,** 1935–1940.

Donlon J. and Kaufman S. (1978) Glucagon stimulation of rat hepatic phenylalanine hydroxylase through phosphorylation in vivo. *J. Biol. Chem.* **253,** 6657–6659.

Donlon J. and Kaufman S. (1980) Relationship between the multiple forms of rat hepatic phenylalanine hydroxylase and degree of phosphorylation. *J. Biol. Chem.* **255,** 2146–2152.

Doskland A. P., Doskland S. O., Ogreid D., and Flatmark T. (1984) The effects of ligands of phenylalanine 4-monooxygenase on the cAMP-dependent phosphorylation of the enzyme. *J. Biol. Chem.* **259,** 11242–11248.

Doskeland A., Ljones T., Skotland T., and Flatmark T. (1982) Phenylalanine 4-monooxygenase from bovine and rat liver. *Neurochem. Res.* **7,** 407–421.

Edelman A., Raese J., Lazal M., and Barchas S. (1981) Tyrosine hydroxylase: Studies on the phosphorylation of a purified preparation of the brain enzyme by cyclic AMP dependent protein kinase. *J. Pharmacol Exp. Ther.* **216,** 647–655.

Elks M. L., Youngblood W. W., and Kizer J. S. (1979) Synthesis and release of serotonin by brain slices: Effect of ionic manipulations and cationic ionophores. *Brain Res.* **172,** 461–469.

Firgaira F., Cotton G. H., and Danks D. M. (1981) Isolation and characterization of dihydropteridin reductase from human lever. *Biochem. J.* **197,** 31–43.

Fisher D. and Kaufman S. (1972) The stimulation of rat liver phenylalanine hydroxylase by phospholipids. *J. Biol. Chem.* **247,** 2250–2252.

Fisher D. B., Kirkwood R., and Kaufman S. (1972) Rat liver phenylalanine hydroxylase, an iron enzyme. *J. Biol Chem.* **247,** 5161–5167.

Fisher M. J., Santina M. A., and Pogson C. I. (1984) Effects of adrenergic agents, vasopressin and ionophore A23187 on the phosphorylation of, and flux through, phenylalanine hydroxylase in rat liver cells. *Biochem. J.* **219,** 87–90.

Freedman P. A., Kappelman A. H., and Kaufman S. (1972) Partial purification and characterization of tryptophan hydroxylase from rabbit hindbrain. *J. Biol. Chem.* **247,** 4165–4173.

George R. J., Haycock J. W., and Waymire J. C. (1985) Personal communication.

Gessa G. L. and Tagliamonte A. (1974) Serum Free Tryptophan: Control of Brain Concentrations of Tryptophan and of Synthesis of 5-Hydroxytryptamine, in *Aromatic Amino Acids in the Brain* (Wolstenholm G. E. W. and Fitzsimons D. W., eds.), CIBA Found. Symp. 22, Elsevier, Ecerpta Medica, North-Holland, Amsterdam.

Gilard G. M. and McCarty R. (1981) Differences in choline acetyltransferase but similarities in catecholamine biosynthetic enzymes in brains of two rat strains differing in their response to stress. *Brain Res.* **206,** 239–243.

Gottshell D. W., Dietrich R. F., Benkovic S. J., and Shiman R. (1982) Phenylalanine hydroxylase. *J. Biol. Chem.* **257,** 845–849.

Grima B., Lamouraoux A., Blanet F., Biguet N. F., and Mallet J. (1985) Complete coding sequence of rat tyrosine hydroxylase mRNA. *Proc. Natl. Acad. Sci. USA* **82,** 617–621.

Guroff G., Rhoads C. A., amd Abramowitz A. (1967) A simple radioisotope assay for phenylalanine hydroxylase cofactor. *Anal. Biochem.* **21,** 273–278.

Haavik J. and Flatmark T. (1980) Rapid and sensitive assay of tyrosine 3-monooxygenase activity by high performance liquid chromatography using the native fluorescence of dopa. *J. Chromatogr.* **198,** 511–515.

Hamon M., Bourgoin S., and Glowinski J. (1973) Feedback regulation of 5HT synthesis in rat striatal slices. *J. Neurochem.* **20,** 1727–1745.

Hamon M., Bourgoin S., Hery F. and Glowinski J. (1977) Regulation of Tryptophan Hydroxylase, in *Structure and Function of Monoamine Enzymes,* Vol. 10, (Usdin E., Weiner N., and Youdim M. B. H., eds.), Marcel Dekker, New York.

Hamon M., Bourgoin S., Hery F., and Simonnet G. (1978) Activation of tryptophan hydroxylase by adenosine triphosphate, magnesium and calcium. *Mol. Pharmacol.* **14,** 99–110.

Hamon M., Bourgoin S., Artaud F., and Glowinski J. (1979) The role of intraneuronal 5HT and of tryptophan hydroxylase activation in the control of 5HT synthesis in rat brain slices incubated in K^+-enriched medium. *J. Neurochem.* **33,** 1031–1042.

Hamon M., Bourgoin S., Artaud F., and Nelson D. (1981) Regulatory properties of neuronal tryptophan hydroxylase. *Adv. Exp. Med. Biol.* **133,** 231–251.

Haycock J. W., Meligeni J. A., Bennett W. F., and Waymire J. C. (1982a) Phosphorylation and activation of TH mediate the acetylcholine-induced increase in catecholamine biosynthesis in adrenal chromaffin cells. *J. Biol. Chem.* **257,** 12641–12648.

Haycock J. W., Bennett W. F., George R. J., and Waymire J. C. (1982b) Multiple site phosphorylation of tyrosine hydroxylase. *J. Biol. Chem.* **257,** 13699–13703.

Hirata Y., Togari A., and Nagatsu T. (1983) Studies on tyrosine hydroxylase system in rat brain slices using high-performance liquid chromatography with electrochemical detection. *J. Neurochem.* **40,** 1585–1589.

Ichiyama A., Nakamura S., Nishizuka Y., and Hayaishi O. (1970) Enzymatic studies on the biosynthesis of serotonin in mammalian brain. *J. Biol. Chem.* **245,** 1699–1709.

Joh T. H. and Reis D. J. (1975) Different forms of tyrosine hydroxylase in central dopaminergic and noradrenergic neurons and sympathetic ganglia. *Brain Res.* **85,** 146–151.

Joh T. H. and Ross M. E. (1983) Preparation of Catecholamine-Synthesizing Enzymes as Immunogens for Immunohistochemistry, in *Immunohistochemistry* (Cuello A. C., ed.), IBRO Handbook.

Joh T. H., Baetge E. E., Ross M. E., and Reis D. J. (1983) Evidence for the Existence of Homologous Gene Coding Regions for the Catecholamine Biosynthetic Enzymes, in *Molecular Neurobiology*, Cold Spring Harbor Symposia on Quantitative Biology, Vol. 48 (Watson J. D. and McKay R., eds.), Cold Spring Harbor Laboratory, New York.

Joh T. H., Baetge E. E., Ross M. E., Lai C-Y, Docherty M., Bradford H., and Reis D. J. (1985) Genes for neurotransmitter synthesis, storage and uptake. *Fed. Proc.* (in press).

Joh T. H., Geghman C., and Reis D. J. (1973) Immunochemical demonstration of increased accumulation of tyrosine hydroxylase protein in sympathetic ganglia and adrenal medulla elicited by reserpine. *Proc. Natl. Acad. Sci. USA* **70,** 2767–2771.

Joh T. H., Kapit R., and Goldstein M. (1969) A kinetic study of particulate bovine adrenal tyrosine hydroxylase. *Biochim. Biophys. Acta* **171,** 378–380.

Joh T. H., Park D. H., and Reis D. J. (1978) Direct phosphorylation of brain tyrosine hydroxylase by cyclic AMP-dependent protein kinase: Mechanism of enzyme activation. *Proc. Natl. Acad. Sci. USA* **75,** 4744–4748.

Joh T. H., Shikimi T., Pickel V., and Reis D. J. (1975) Brain tryptophan hydroxylase: Purification of, production of antibodies to, and cellular and ultrastructural localization of serotonergic neurons of rat midbrain. *Proc. Natl. Acad. Sci. USA* **72,** 3575–3579.

Kato T., Horiuchi S., Togari A., and Nagatsu T. (1981) A sensitive and inexpensive high performance liquid chromatographic assay for tyrosine hydroxylase. *Experientia* **37,** 809–811.

Kaufman S. (1957) The enzymatic conversion of phenylalanine to tyrosine. *J. Biol. Chem.* **226,** 511–526.

Kaufman S. (1959) Studies on the mechanism of the enzymatic conversion of phenylalanine to tyrosine. *J. Biol. Chem.* **234,** 2677–2682.

Kaufman S. (1963) The structure of the phenylalanine hydroxylase cofactor. *Proc. Natl. Acad. Sci. USA* **50,** 1085–1093.

Kaufman S. (1971) The phenylalanine hydroxylating system from mammalian liver. *Adv. Enzymol.* **35,** 245–319.

Kaufman S. (1979) The activity of 2,4,5-triamino-6-hydroxypyrimidine in the phenylalanine hydroxylase system. *J. Biol. Chem.* **254,** 5150–5154.

Kaufman S. and Fisher D. B. (1970) Purification and some physical properties of phenylalanine hydroxylase from rat liver. *J. Biol. Chem.* **245,** 4745–4750.

Kaufman S. and Lovenberg W. (1959) Further studies on the phenylalanine hydroxylation cofactor. *J. Biol. Chem.* **234,** 2683–2688.

Kuhn D. and Lovenberg W. (1982) Hydroxylases, in *Handbook of Neurochemistry,* 2nd Ed., Vol. 4 (Lajtha A., ed.), Plenum, New York.

Kuhn D. M., Meyer M. A., and Lovenberg W. (1980a) Tryptophan hydroxylase from a malignant murine mast cell tumor and rat mesencephlic tegmentum. *Arch. Biochem. Biophys.* **199,** 355–361.

Kuhn D. M., O'Callaghan J. P., Juskevich J., and Lovenberg W. (1980b) Activation of brain tryptophan hydroxylase by ATP-Mg^{++}: Dependence on calmodulin. *Proc. Natl. Acad. Sci. USA* **77,** 4688–4691.

Kuhn D. M., Vogel R. L., and Lovenberg W. (1978) Calcium-dependent activation of tryptophan hydroxylase by ATP and magnesium. *Biochem. Biophys. Res. Commun.* **82,** 759–763.

Kvetansky R., Weise V. K., and Kopin I. J. (1970) Tyrosine hydroxylase and phenylethanolamine *N*-methyltransferase by repeated immobilization in rats. *Endocrinology* **87,** 744–749.

Ladsky A. S., Robson K. J. H., Thirumalachary C., Marker P. E., Ruddle F. H., and Woo S. L. C. (1984) The penylketonuria locus in man is on chromosome 12. *J. Human Genet.* **36,** 527–573.

Lamouroux A., Faucon Bignet N., Samolyk D., Privat A., Salomon J. C., Pujol J. C., and Mallet J. (1982) Identification of cDNA clones for rat tyrosine hydroxylase antigen. *Proc. Natl. Acad. Sci. USA* **79,** 3881–3885.

Lazarus R. A., Wallick D. E., Dietrich R. F., Gotschall D. W., Gaffney B. J., and Shiman R. (1982) The mechanism of phenylalanine hydroxylase. *Fed. Proc.* **41,** 2605–2607.

Ledley F. D., Grenett H. E., DiLella A. G., Kwok S. C. M., and Woo S. L. C. (1985) Gene transfer and expression of human phenylalanine hydroxylase. *Nature* **228,** 77–79.

Leighton H. J. and Waggoner W. G. (1981) Measurement of tyrosine hydroxylase activity in nerve terminals of rat hippocampus, hypothalamus, and striatum using an improved assay for tyrosine hydroxylase. *Life Sci.* **28,** 385–391.

Levin R. A., Pollard H. B., and Kuhn D. M. (1984) A rapid and simplified assay method for tyrosine hydroxylase. *Anal. Biochem.* **143,** 205–208.

Levitt M., Spector S., Sjoerdsma A., and Udenfriend S. (1965) Elucidation of rate-limiting step of norepinephrine biosynthesis in the perfused guinea pig heart. *J. Pharmacol. Exp. Ther.* **148,** 1–8.

Lewis E. J., Tank A. W., Weiner N., and Chikaraishi D. M. (1983) Regulation of tyrosine hydroxylase mRNA by glucocorticoid and cyclic AMP in a rat pheochromocytoma cell line. *J. Biol. Chem.* **258,** 14632–14637.

Lind K. (1972) Investigation of the specificity for quinonoid dihydropteridine and the inhibition by 2,4-diaminopteridines. *Eur. J. Biochem.* **25,** 560–562.

Lysz T. W. and Sze P. Y. (1978) Activation of brain tryptophan hydroxylase by phosphorylating system. *J. Neurosci. Res.* **3,** 411–416.

Macon J. R., Sokoloff L., and Glowinski J. (1972) Feedback control of rat brain 5-hydroxtryptamine synthesis. *J. Neurochem.* **18,** 323–331.

Mallet J., Biquet N. F., Buda M., Lamouroux A., and Samolyk D. (1983) Detection and regulation of the tyrosine hydroxylase mRNA level in rat adrenal medulla and brain tissues. *Cold Spring Harbor* **48,** 305–308.

Mandell A. J. (1978) Redundant mechanisms regulating brain tyrosine and tryptophan hydroxylases. *Ann. Rev. Pharmacol. Toxicol.* **18,** 461–493.

Markey K. A., Kondo S., Shenkman L., and Goldstein M. (1980) Purification and characterization of tyrosine hydroxylase from cloned pheochromocytoma cell line. *Mol. Pharmacol.* **17,** 79–85.

Marota S. and Shiman R. (1984) Stoichiometric reduction of phenylalanine hydroxylase by its cofactor: A requirement of enzymatic activity. *Biochemistry* **23,** 1303–1313.

Masserano J. M. and Weiner N. (1983) Tyrosine hydroxylase regulation in the central nervous system. *Mol. Cell. Biochem.* **53/54,** 129–152.

Masserano J. M., Takimoto G. S., and Weiner N. (1981) Electroconvulsive shock increases tyrosine hydroxylase activity in the brain and adrenal gland of the rat. *Science* **214,** 662–665.

Meligeni J. A., Haycock J. W., Bennett W. F., and Waymire J. C. (1982) Phosphorylation and activation of tyrosine hydroxylase mediate the cAMP-induced increase in catecholamine biosynthesis in adrenal chromaffin cells. *J. Biol. Chem.* **257,** 12632–12640.

Messripour M. and Clark J. B. (1982) Tyrosine hydroxylase activity in rat brain synaptosomes: Direct measurement using high performance liquid chromatography. *J. Neurochem.* **38,** 1139–1143.

Mestikawy S. E., Glowinski J., and Hamon M. (1983) Tyrosine hydroxylase activation in depolarized dopaminergic terminals: Involvement of Ca^{++}-dependent phosphorylation. *Nature* **302,** 830–832.

Miller M. R. and Shiman R. (1976) Reversible inactivation of phenylalanine hydroxylase by catecholamines in cultured hepatoma cells. *J. Biol Chem.* **251,** 3671–3676.

Miller M. R., McClure D., and Shiman R. (1975) *p*-Chlorophenylalanine effect on phenylalanine hydroxylase in hepatoma cells in culture. *J. Biol. Chem.* **250,** 1132–1140.

Milstein S., Abita J., Chang N., and Kaufman S. (1976) Hepatic phenylalanine 4-monooxygenase is a phosphoprotein. *Proc. Natl. Acad. Sci. USA* **73,** 1591–1595.

Mueller R. A., Thoenen H., and Axelrod J. (1969) Increase in tyrosine hydroxylase activity after reserpine administration. *J. Pharmacol. Exp. Ther.* **169,** 74–79.

Mussachio J., Wulzberger R. J., and D'Angelo G. L. (1971) Different molecular forms of bovine adrenal tyrosine hydroxylase. *Mol. Pharmacol.* **7,** 136–146.

Nagatsu T., Levitt M., and Udenfriend S. (1964) Tyrosine hydroxylase: The initial step in norepinephrine biosynthesis. *J. Biol. Chem.* **239,** 2910–2917.

Nagatsu T., Oka K., Numata Y., and Kato T. (1979) A simple and sensitive fluorescent assay for tyrosine hydroxylase activity. *Anal. Biochem.* **93,** 82–87.

Nakata H. and Fujisawa H. (1980) Purification and characterization of phenylalanine 4-monooxygenase from rat liver. *Biochim. Biophys. Acta* **614,** 315–327.

Nakata H. and Fujisawa H. (1982) Purification and properties of tryptophan 5-monooxygenase from rat brainstem. *Eur. J. Biochem.* **122,** 41–47.

Nielson K. H., Simonsen V., and Lind K. E. (1969) Dihydropteridine reductase: A method for the measurement of activity and investigations of specificity for NADH and NADPH. *Eur. J. Biochem.* **9,** 497–502.

Oka K., Ashiba G., Sugimoto T., Matsuura S., and Nagatsu T. (1982) Kinetic properties of tyrosine hydroxylase purified from bovine adrenal medulla and bovine caudate nucleus. *Biochim. Biophys. Acta* **706,** 188–196.

Oka K., Kato T., Sugimoto T., Matsuura S., and Nagatsu T. (1981) Kinetic properties of tyrosine hydrozylase with natural tetrahydrobiopterin as cofactor. *Biochim. Biophys. Acta* **661,** 45–53.

Okuno S. and Fujisawa H. (1985) A new mechanisms for regulation of tyrosine 3-monooxygenase by end product and cAMP-dependent protein kinase. *J. Biol. Chem.* **260,** 2633–2635.

Palkovits M., Kobayashi R. M., Kizer J. S., Jacobowitz D. M., and Kopin I. J. (1975) Effects of stress on catecholamines and tyrosine hydroxylase activity of individual hypothalamic nucleus. *Neuroendocrinology* **18,** 144–153.

Petrack P., Sheppy E., Fetzer V., Manning T., Chertock H., and Ma D. (1972) Effect of ferrous iron on tyrosine hydrosine hydroxylase of bovine adrenal medulla. *J. Biol. Chem.* **247,** 4872–4878.

Phillips R. S. and Kaufman S. (1984) Ligand effects on the phosphorylation site of hepatic phenylalanine hydroxylase. *J. Biol. Chem.* **259,** 2474–2479.

Reis D. J., Joh T. H., and Ross R. A. (1975) Effects of reserpine on activities and amounts of tyrosine hydroxylase and dopamine β-hydroxylase in catecholamine neuronal system in rat brain. *J. Pharmacol. Exp. Ther.* **193,** 775–784.

Richtandt N. M., Inagami T., Misono K., and Kuczenski R. (1985) Purification and characterization of rat striatal tyrosine hydroxylase. *J. Biol. Chem.* **260,** 8465–8473.

Robson K. J. H., Beattie W., James R. J., Cotton R. C. H., Morgan F. J., and Woo S. L. C. (1984) Sequence comparison of rat liver phenylalanine hydroxylase and its cDNA clones. *Biochemistry* **23,** 3671–3675.

Robson K. L. H., Chandra T., MacGillary R., and Woo S. L. C. (1979) Polysome immunoprecipitation of phenylalanine hydroxylase mRNA from rat liver and cloning its cDNA. *Proc. Natl. Acad. Sci. USA* **79,** 4701–4705.

Ross R. A., Joh T. H., and Reis D. J. (1978) Increase in the relative rate of synthesis of dopamine-hydroxylase in the nucleus locus ceruleus elicited by reserpine. *J. Neurochem.* **31,** 1491–1500.

Shiman R. (1980) Relationship between the substrate activation site and catalytic site of phenylalanine hydroxylase. *J. Biol. Chem.* **255,** 10029–10032.

Shiman R. (1985) Phenylalanine Hydroxylase and Dihydropterin Reductase, in *Chemistry and Biochemistry of Pterins,* Vol. 2 (Benkovic S. J. and Blakely R. L., eds.), John Wiley, New York.

Shiman R. and Gray D. W. (1980) Substrate activation of phenylalanine hydroxylase. *J. Biol. Chem.* **255,** 4793–4800.

Shiman R. and Jefferson L. S. (1982) Iron-dependent regulation of rat liver phenylalanine hydroxylase activity in vivo, in vitro, and in perfused liver. *J. Biol. Chem.* **257,** 839–844.

Shiman R. and Kaufman S. (1970) Tyrosine hydroxylase (bovine adrenal gland). *Meth. Enzymol.* **17A,** 609–615.

Shiman R., Akino M., and Kaufman S. (1971) Solubilization and partial purification of tyrosine hydroxylase from bovine adrenal medulla. *J. Biol. Chem.* **246,** 1330–1349.

Shiman R., Gray D., and Peter A. (1979) A simple purification of phenylalanine hydroxylase by substrate-induced hydrophobic chromatography. *J. Biol. Chem.* **254,** 11300–11306.

Shiman R., Mortimors G., Schworer C. M., and Gray D. W. (1982) Regulation of phenylalanine hydroxylase activity by phenylalanine in vivo, in vitro, and in perfused rat liver. *J. Biol. Chem.* **257,** 11213–11216.

Smith S. C., Kemp B. E., McAdams W. J., Mercer J. F. B., and Cotton R. G. H. (1984) Two apparent molecular weight forms of human and monkey phenylalanine hydroxylase are due to phosphorylation. *J. Biol. Chem.* **259,** 11284–11289.

Stone E. A., Freedman L. S., and Margario L. E. (1978) Brain and adrenal tyrosine hydroxylase activity after chronic footshock stress. *Pharmacol. Biochem. Behav.* **9,** 551–553.

Thoenen H. (1970) Induction of tyrosine hydroxylase in periphral and central adrenergic neurons by cold exposure of rats. *Nature* **228,** 861–862.

Tissari A. H., Argiolas A., Fadda F., Serra G., and Gessa G. L. (1979) Foot-shock stress accelerates non-striatal dopamine synthesis without activating tyrosine hydroxylase. *Naunyn Schmiedebergs Arch. Pharmacol.* **308,** 155–157.

Tourian A. (1971) Activation of phenylalanine hydroxylase by phenylalanine. *Biochim. Biophys. Acta* **242,** 345–354.

Udenfriend S. and Wyngaarden J. B. (1956) Precursors of adrenal epinephrine and norepinephrine. *Biochim. Biophys. Acta* **20,** 48–52.

Udenfriend S., Cooper J. R., Clark C. T., and Baer J. E. (1953) Rate of turnover of epinephrine in the adrenal medulla. *Science* **117,** 663–665.

Vigney A. and Henry J. P. (1982) Mechanism of tyrosine hydroxylase activation by phosphorylation. *Biochem. Biophys. Res. Commun.* **106,** 1–7.

Vulliet P. R., Langan T. A., and Weiner N. (1980) Tyrosinc hydroxylase: A substrate of cyclic AMP-dependent protein kinase. *Proc. Natl. Acad. Sci. USA* **77,** 92–96.

Waalkes T. P. and Udenfriend S. (1957) A fluorometric method for the estimation of tyrosine in plasma and tissues. *J. Lab. Clin. Med.* **50,** 733–736.

Waymire J. C., Bjur R., and Weiner N. (1971) Assay of tyrosine hydroxylase by coupled decarboxylation of Dopa formed from 1-^{14}C-L-tyrosine. *Anal. Biochem.* **43,** 588–600.

Waymire J. C., Haycock J. W., and George R. J. (1984) Stimulation-Dependent Regulation of Catecholamine Biosynthesis in Purified Adrenal Chromaffin Cells: Phosphorylation and Activation of Tyrosine Hydroxylase, in *Catechoamines* Part A, *Basic and Peripheral Mechanisms* (Usdin E., Carlsson A., Dahlstrom A., and Engel J., eds.), Alan Liss, New York.

Waymire J. C., Haycock J., Meligeni J., and Browning M. (1979) Activation and Phosphorylation of Tyrosine Hydroxylase by cAMP, in *Catecholamines: Basic and Clinical Frontiers* (Usdin E., Kopin I., and Barchas J., eds.), Pergamon, New York.

Webber S. and Whiteley J. M. (1980) Separation of catalase from phenylalanine hydroxylase: Use of a Sepharose anti-catalase IgG affinity matrix. *Biochem. Biophys. Res. Commun.* **97,** 17–21.

Webber S. and Whiteley J. (1981) The effect of specific amino acid modifications on the catalytic properties of rat liver dihydropteridine reductase. *Arch. Biochem. Biophys.* **206,** 145–152.

Webber S., Harzer G., and Whiteley J. M. (1980) Isolation of rat liver phenylalanine hydroxylase using a novel pteridine matrix. *Anal. Biochem.* **106,** 63–72.

Weiss J. M., Glazier H. I., Pohorecky L. A., Brick J., and Miller N. E. (1975) Effects of chronic exposure to stressors on avoidance-escape behavior and on brain norepinephrine. *Psychomatic Med.* **37,** 522–534.

William C. D., Dickensm G., Letendre C. H., Guroff G., Haines C., and Shiota T. (1976) Isolation and characterization of dihydropteridine reductase from pseudomonas species. *J. Bact.* **127,** 1197–1207.

Yamauchi T. and Fujisawa H. (1978) Evidence for phosphorylation of bovine adrenal tyrosine hydroxylase by cAMP-dependent protein kinase. *Biochem. Biophys. Res. Commun.* **82,** 514–517.

Yamauchi T. and Fujisawa H. (1979) In vitro phosphorylation of bovine adrenal tyrosine hydroxylase by adenosine-3',5'-monophosphate-dependent protein kinase. *J. Biol. Chem.* **254,** 503–507.

Yamauchi T. and Fujisawa H. (1981) Tyrosine 3-monooxygenase is phosphorylated by Ca^{++}-calmodulin dependent protein kinase, followed by activation by activator protein. *Biochem. Biophys. Res. Commun.* **100,** 807–813.

Zigmond R. E. and Ben-Ari Y. (1977) Electrical stimulation of preganglionic nerve increases tyrosine hydroxylase activity in sympathetic ganglia. *Proc. Natl. Acad. Sci. USA* **74,** 3078–3080.

Zigmond R. E. and Chalazonitis A. (1974) Long-term effects of preganglionic nerve stimulation on tyrosine hydroxylase activity in the rat superior cervical ganglion. *Brain Res.* **164,** 137–152.

Zigmond R. E., Charazonitis A., and Joh T. H. (1980) Preganglionic nerve stimulation increases the amount of tyrosine hydroxylase in the rat superior cervical ganglion. *Neurosci. Lett.* **20,** 61–65.

Zigmond R. E., Schon F., and Iversen L. L. (1974) Increased tyrosine hydroxylase activity in the locus coeruleus of rat brainstem after reserpine treatment and cold stress. *Brain Res.* **70,** 547–552.

Chapter 2
Aromatic L-Amino Acid Decarboxylase
Biochemistry and Functional Significance

Ronald R. Bowsher and David P. Henry

1. Introduction

Monamines derived from the decarboxylation of aromatic L-amino acids are formed by every type of living organism from bacteria to plants and animals. These compounds and their derivatives are important regulatory substances. Although consensus has been difficult to reach, it appears that only two enzymes catalyze the decarboxylation of aromatic amino acids in mammals: first, histidine decarboxylase (EC 4.1.1.22), a very specific enzyme; and second, aromatic L-amino acid decarboxylase (EC 4.1.1.28) (AADC), a very nonspecific enzyme. The notable features of AADC are as follows. The enzyme is found in tissues where a functional role is obvious, such as catecholaminergic neurons, serotonergic neurons, the adrenal medulla, pineal gland, and enterochromaffin tissue. The highest levels of AADC, however, are found in tissues such as the kidney, liver, and gastrointestinal tract where its functional status is unclear. In those tissues in which AADC is clearly functional, it is assumed the enzyme is present in massive excess relative to the activity of other monoamine enzymes. Thus, the prevailing concept has been that AADC is not rate-limiting in monoamine synthesis. The enzyme does not appear to be under metabolic control. In fact, the Michaelis constant of the enzyme for even its best substrates is several orders of magnitude greater than the endogenous concentration of those amino acids. The functional significance of the broad substrate specificity of the enzyme is not known. The role of AADC in the biosynthesis of catecholamines, serotonin (5-hydroxytryptamine, 5-HT), and melatonin, is as widely accepted as is its lack of any role in the synthesis of histamine. The role of AADC in the synthesis of other biogenic amines is just beginning to be described. Furthermore, no known disease states or pathophysiologic conditions have been associated with selective abnormalities of AADC function. However, pharmacologic manipulation of this enzyme has resulted in a number of therapeutic successes, including the development of α-methylDOPA, an antihypertensive agent, and the use of L-DOPA and carbiDOPA in the control of the symptoms of Parkinson's disease. The literature on AADC is so extensive that numerous excellent reviews have been published (Sourkes, 1966, 1977; Aures et al., 1970; Nagatsu, 1973). This review will, therefore, be selective and emphasize more recent observations regarding AADC enzymology and its functional significance.

2. Historical Perspective

The presence of L-3,4-dihydroxphenylalanine (DOPA) decarboxylase activity in extracts of mammalian tissues was first reported by Holtz et al. in 1938. The vigorous rate of carbon dioxide evolution from DOPA permitted direct quantification of this activity by manometric analysis using a Warburg apparatus (Blaschko, 1942). This led Blaschko (1939) to postulate a role for DOPA decarboxylase in the biosynthesis of catecholamines. However, the insensitive manometric technique, despite demonstrating the decarboxylation of *o*-tyrosine (Blaschko and Holton, 1950), *m*-tyrosine (Blaschko et al., 1949), 3,4-dihydroxyphenylserine (Blaschko et al., 1950; Sourkes et al., 1952), and the various positional isomers of DOPA (Blaschko et al., 1949; Blaschko, 1950; Ferrini and Glasser, 1964), failed to detect the decarboxylation of the naturally occurring aromatic amino acids, histidine, tryptophan, phenylalanine, and *p*-tyrosine (Blaschko, 1942; Blaschko et al., 1949; Clark et al., 1954; Holtz, 1959; Hagen, 1962). This observation directly contradicted earlier studies that suggested the presence of aromatic L-amino acid decarboxylase activities in mammalian tissues assayed using bioassay methodology (Blaschko, 1945). Blaschko probably best explained this paradox in 1945: "The fact that neither tyrosine or histidine led to the formation of carbon dioxide in detectable amounts is particularly interesting because it stresses the great difference in the rate of decarboxylation, which is slow with these amino acids as compared with DOPA." The large disparity in the rate of DOPA decarboxylation with respect to other aromatic amino acids led to the general belief that DOPA was the only aromatic amino acid whose decarboxylation was of physiologic significance. However, Udenfriend et al. (1953) subsequently reported that 5-hydroxytryptophan (5-HTP) was readily decarboxylated by extracts of mammalian kidney and postulated a role for 5-HTP decarboxylase in the synthesis of serotonin. Langemann (1958) later hypothesized DOPA decarboxylase also catalyzed the decarboxylation of 5-HTP and this hypothesis was widely supported by a number of subsequent studies. Accordingly, DOPA decarboxylase and 5-HTP decarboxylase had similar tissue distributions (Westermann et al., 1958; Bertler and Rosengren, 1959a; Davis and Awapara, 1960; Kuntzman et al., 1961), DOPA and 5-HTP were mutually inhibitory (Bertler and Rosengren, 1959a,b; Fellman, 1959; Yuwiler et al., 1959; Rosengren, 1960), both decarboxylase activities were inhibited by the same compounds (Yuwiler et al., 1959; 1960; Rosengren, 1960), and the ratio of decarboxylase activities remained constant through various protein purification schemes (Clark et al., 1954; Fellman, 1959; Yuwiler et al., 1959; Awapara et al., 1962). In 1962, Lovenberg et al., using improved analytical methodology, expanded the substrate specificity of DOPA decarboxylase to include a number of aromatic amino acids and their α-methylated derivatives, and proposed the new name aromatic L-amino acid decarboxylase (AADC) to reflect this newer broad substrate specificity. Nevertheless, the Commission on Biochemical Nomenclature in 1964 designated DOPA decarboxylase (4.1.1.26) and 5-HTP decarboxylase (4.1.1.28) as separate enzymes, and recognized the presence of histidine decarboxylase (4.1.1.22), tyrosine decarboxylase (4.1.1.25), and tryptophan decarboxylase (4.1.1.27) in mammalian tissues (Florkin and Stotz, 1964). Christenson et al. (1970) corroborated the existance of aromatic L-amino acid decarboxylase using an apparently homogeneous enzyme preparation from porcine kidney. This conclusion prompted the renaming of DOPA decarboxylase as aromatic L-amino acid decarboxylase in 1972 (Florkin and Stotz, 1973). However, the exact substrate specificity of AADC with respect to certain amino acids, such as *p*-tyrosine, remained controversial (Awapara et al., 1964; Gonnard and Camier, 1966; Coulson et al., 1968; Lancaster and Sourkes, 1972; Corgier and Pacheco, 1975; Srinivasan and Awapara, 1978), and in 1978 the Commission on Biochemical Nomenclature still recognized the presence of a separate tyrosine decarboxylase enzyme in mammalian tissues (Nomenclature Commission, 1979). Recently, Bowsher and Henry (1983) reexamined the substrate specificity of a homogeneous preparation of AADC from rat kidney. They concluded that all naturally occurring aromatic L-amino acids are decarboxylated by AADC and presented reasons why some previous studies may have failed to detect the in vitro decarboxylation of aromatic amino acids. First, the exceedingly large disparity in decarboxylation rates necessitated the use of highly sensitive and specific analytical methods. Second, the lack of optimization of in vitro reaction conditions, such as pH, may have hindered decarboxylase quantification. Finally, the presence of a high level of aromatic amino acid transaminase activity in mammalian tissues may also interfere with the quantification of aromatic amino acid decarboxylation. In conclusion, mammalian tissues possess a decarboxylase enzyme with broad substrate specificity that includes all aromatic L-amino acids and is appropriately identified as AADC.

Mammalian tissues have been postulated to contain other decarboxylases for aromatic amino acids. The presence of a distinct and specific histidine decarboxylase (4.1.1.22) in mammals is well documented (Schayer, 1957; Weissbach et al., 1961; Aures and Hakanson, 1971; Tran and Snyder, 1981). A number of studies have suggested that multiple enzymes may catalyze the decarboxylation of DOPA and 5-HTP. Accordingly, individual in vitro decarboxylation rates are highly influenced by reaction parameters, such as pH and the presence of exogenous pyridoxal-5'-phosphate (PLP) and organic solvents. Also, several studies have shown that the distribution of DOPA and 5-HTP decarboxylase in tissues, although very similar, is not always a constant ratio (Davis and Awapara, 1960; Kuntzman et al., 1961; Sims et al., 1973; Rahman et al., 1981; Rahman and Nagatsu, 1982). Additionally, in brain, a greater portion of 5-HTP decarboxylase is reported to be synaptosomal in contrast to DOPA decarboxylase, which is mainly soluble (Sims et al., 1973). Rahman and Nagatsu (1982) have recently postulated the existence of multiple isoenzymes of AADC in human brain.

Two experimental approaches have been used to determine whether DOPA decarboxylase and 5-HTP decarboxylase are the same protein. The first approach has been to prepare antisera to purified decarboxylase preparations. Immunotitration experiments involving specific decarboxylase antisera have demonstrated proportional precipitation of DOPA and 5-HTP decarboxylase activities (Christenson et al., 1972), as well as proportional precipitation of 5-HTP and *p*-tyrosine decarboxylase activities (Bowsher and Henry, 1983). Additionally, antisera to bovine adrenal AADC crossreacted with cell bodies in the substantia nigia and raphe regions, thus suggesting the presence of an antigenically indistinguishable enzyme in known dopamine and serotonin systems, respectively (Goldstein et al., 1972; Hokfelt et al., 1973a,b). Most electrophoretic experiments involving mammalian tissues have also supported the unitary nature of 5-HTP decarboxylase (Cavalli-Sforza et al., 1974; Landon, 1977). The other experimental approach used to characterize the identity of DOPA decarboxylase and 5-HTP decarboxylase has been intracisternal injections of neurotoxic agents. An early study by Sims and Bloom (1973) showed preferential loss of brain DOPA decarboxylase activity after 6-hydroxydopamine treatment. However, subsequent investigation showed treatment with either 6-hydroxydopamine or 5,6-dihydroxytryptamine yielded parallel decreases in both DOPA and 5-HTP decarboxylase activities, therefore supporting the unitary nature of AADC (Dairman et al., 1975; Melamed et al., 1980). In summary, some previous studies using unpurified enzyme preparations have suggested that multiple decarboxylases may exist for DOPA and 5-HTP. However, the AADC protein may contain a complex catalytic site with adjacent binding sites for fixing substrates (Bender and Coulson, 1972, 1977; Borri Voltattorni et al., 1983), and in vitro decarboxylase activities are highly influenced by the reaction milieu. Thus, the data concerning the multiplicity of DOPA and 5-HTP decarboxylases should be regarded as unproven until another decarboxylase is successfully purified from a mammalian tissue.

3. Characteristics of the AADC Reaction

3.1. Substrate Specificity

The exact substrate specificity of AADC has remained an ongoing controversy, however, current evidence indicates purified AADC can catalyze the in vitro decarboxylation of the following L-amino acids: DOPA, 5-HTP, 3,4-dihydroxyphenylserine, *o*-tyrosine, *m*-tyrosine, *p*-tyrosine, phenylalanine, histidine, and tryptophan (Fig. 1) (Lovenberg et al., 1962; Christenson et al., 1970; Bowsher and Henry, 1983). The corresponding D-stereoisomers are not decarboxylated by AADC (Borri Voltattorni et al., 1983). The various α-methyl analogs of aromatic amino acids are also substrates for AADC (Weissbach et al., 1960; Lovenberg et al., 1962).

3.2. Kinetic Characterization

The Michaelis constants of AADC for DOPA and 5-HTP have been estimated frequently. As shown in Table 1, studies have

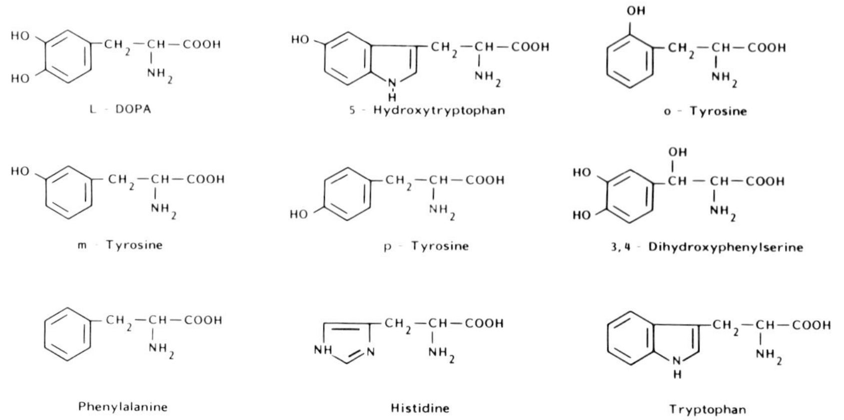

Fig. 1. Amino acid substrates for purified aromatic L-amino acid decarboxylase.

routinely yielded very similar K_m estimates despite using different enzyme sources, incubation parameters, and analytical methodologies. Estimates of V_{max} for DOPA and 5-HTP have consistently resulted in a substantially larger value for L-DOPA. Borri Voltattorni et al. (1983) have recently reported that D-stereoisomers of aromatic amino acids can bind to the active site of AADC and inhibit decarboxylation of their corresponding L-amino acids. Thus, the estimation of Michaelis constants using racemic mixtures of aromatic amino acids should be avoided. These authors have reported K_ms of 180 and 230 μM for L-DOPA and L-5-HTP, respectively, which suggests the Michaelis constant for 5-HTP may be somewhat higher than previously reported.

Kinetic estimates for other aromatic amino acid substrates have been reported less frequently. These studies have also been hampered by the limited solubility of substrates near their respective K_ms. Nevertheless, Lovenberg et al. (1962) and Christenson et al. (1970) have reported Michaelis constants for the following naturally occurring amino acids: phenylalanine, 42 mM; tryptophan, 10 mM; tyrosine, 8.4 mM; and histidine, 22 mM. Thus, these substrates have Michaelis constants that are substantially higher than DOPA and 5-HTP. Also, the V_{max} values for these substrates are considerably lower than the estimates for DOPA and 5-HTP. *o*-Tyrosine and *m*-tyrosine, the two positional isomers of *p*-tyrosine, are reported to be excellent substrates for AADC (Awapara et al., 1962; Srinivasan and Awapara, 1978; Borri Voltattorni et al., 1983). These compounds are decarboxylated at rates greater than L-DOPA, and a recent investigation reported the K_m values for *o*- and *m*-tyrosine to be 120 μM and 520 μM, respectively (Borri

Table 1
Michaelis Constants for DOPA and 5-HTP

Enzyme source	Analytical method	K_m(DOPA), mM	K_m(5-HTP), mM	Reference
Guinea pig kidney	Fluorometric	0.4	0.02	Lovenberg et al., 1962
Hog kidney	Radiometric	0.19	0.1	Christenson et al., 1970
Rat liver	Radiometric	0.1	0.043	Bender and Coulson, 1972
Rat brain	Radiometric	0.6	0.016	Sims et al., 1973
Rat brain	HPLC	0.045		Nagatsu et al., 1979
Cat brain	Fluorometric		0.0076	Bouchard and Roberge, 1979
Rat brain	HPLC		0.0095	Rahman et al., 1980
Rat adrenal	HPLC	0.102	0.037	Rahman et al., 1981
Hog kidney	Radiometric	0.04		Okuno and Fujisawa, 1983
Human pheochromocytoma	Radiometric	0.17		Maneckjee and Baylin, 1983

Voltattorni et al., 1983). In summary, the Michaelis constants of AADC for aromatic amino acid substrates are lowest for DOPA, 5-HTP, *o*-tyrosine and *m*-tyrosine. However, in all cases, the calculated Michaelis constant for each amino acid substrate greatly exceeds the endogenous concentration of the aromatic amino acid. Thus, AADC activity appears to be largely governed by the V_{max} for a given substrate and, therefore, the endogenous concentration of each amino acid substrate is probably a very important factor regulating AADC activity in vivo.

3.3. pH Dependency

The rate of in vitro decarboxylation of aromatic amino acids is highly dependent on the pH of the reaction medium. Additionally, different substrates are characterized by different pH optima. As shown in Table 2, L-DOPA is reported to have a pH optimum near 7.0, which is an entire pH unit lower than the optimum for 5-HTP. A number of other studies have also examined the pH dependency of this enzyme with respect to other aromatic amino acids. Awapara et al. (1962) reported a pH optimum of 7.2 for *o*-tyrosine, whereas Lovenberg et al. (1962) reported a pH optimum of 9 (Tris buffer) for all naturally occurring amino acids with the exception of DOPA. Schayer (1957) reported the pH optimum for histidine was 8.0. Christenson et al. (1970) reported the pH optimum for phenylalanine and tryptophan was 8.5. Recently, Bowsher and Henry (1983) reexamined the pH dependency of *p*-tyrosine decarboxylation and concluded that the pH optimum was 8.0 in Tris buffer (37°C). In summary, AADC possesses a broad range of pH optima that are a function of the respective substrate present. The unique pH dependency of AADC probably reflects the complex nature of the enzyme's active site and probably is not because of catechol oxidation or substrate ionization (Bender and Coulson, 1977). Accordingly, AADC is postulated to contain multiple binding sites for substrates that are adjacent to the PLP coenzyme (Bender and Coulson, 1972, 1977; Borri Voltattorni et al., 1983). Thus, the rate of amino acid decarboxylation is a function of both substrate binding and PLP catalysis that are influenced by the pH of the incubation medium.

3.4. Stimulation of AADC Activity by Organic Solvents

Certain organic solvents have been reported to stimulate the in vitro decarboxylation of aromatic amino acids. The serendipitous addition of toluene to decarboxylase reactions, in order to maintain sterility, may have been a major factor that permitted the early

Table 2
pH Dependency for Decarboxylation of DOPA and 5-HTP

Enzyme source	Optimum for DOPA	Optimum for 5-HTP	Reference
Guinea pig kidney	6.8	8.1	Clark et al., 1954
Guinea pig kidney	6.7–7.0	9.0	Lovenberg et al., 1962
Rat liver	6.8	7.8	Bender and Coulson, 1972
Rat brain	6.8	8.1–8.5	Sims et al., 1973
Rat brain	6.9	7.8	Corgier and Pacheco, 1975
Rat liver		8.0	Bouchard and Roberge, 1979
Rat serum	7.2	8.3	Rahman et al., 1981
Hog kidney	7.0		Okuno and Fujisawa, 1983

demonstration of AADC activity by bioassay methodology (Blaschko, 1945). Later studies by Waton (1956a,b) showed benzene, toluene, and chloroform dramatically stimulated the in vitro decarboxylation of histidine by extracts of rabbit kidney, and that aromatic solvents were the most efficient activators. Lovenberg et al. (1962) subsequently showed that benzene activated the decarboxylation of the naturally occurring amino acids *p*-tyrosine, phenylalanine, and tryptophan. Bowsher and Henry (1983) have recently reexamined the stimulation of the *p*-tyrosine decarboxylase activity of AADC by organic solvents and demonstrated a highly specific structure–activity relationship. Benzene and toluene were most effective, whereas cyclohexane and cyclopentane were less effective (Table 3). Slight structural modifications of these solvents eliminated their stimulatory activity, and other solvents with a wide range of polarities were also inactive. The rate enhancement was independent of enzyme purity, therefore suggesting a direct interaction between the protein and organic solvent. Preliminary kinetic studies have suggested this activation results from an increase in the V_{max} for *p*-tyrosine decarboxylation. Benzene dramatically stimulated the *p*-tyrosine decarboxylase activity of both rat and human renal AADC, but failed to increase the rate of 5-HTP decarboxylation. The mechanism for this differential activation of AADC activity by organic solvents remains unclear. However, aromatic solvents may facilitate diffusion of carbon dioxide from the hydrophobic catalytic site during the rate-limiting step of amino acid decarboxylation (Minelli et al., 1979). Owing to the specificity of AADC activation by organic solvents, we have previously suggested that it would be appropri-

Table 3
Stimulation of the *p*-Tyrosine Decarboxylase Activity of AADC by Organic Solvents

Solvent added	% of Control
Benzene	713
Toluene	417
Cyclopentane	258
Cyclohexane	185
Ethylbenzene	111
Control	100
Heptane	96
Hexane	93
Chloroform	82
m-Xylene	76
o-Xylene	63
Methanol	63
p-Xylene	61
5% phenol	55
Acetonitrile	54
Dioxane	53
Tetralin	53
25% Pyridine	53
Methylethylketone	52
n-Propanol	52
Ethyl acetate	47

ate to search for an endogenous activator of this activity (Bowsher and Henry, 1983).

4. Molecular Properties of AADC

4.1. Molecular Weight and Subunit Structure

Aromatic L-amino acid decarboxylase, isolated from a number of mammalian tissues, is an asymmetric molecule with an approximate molecular weight of 100,000 (Table 4). However, there has been disagreement regarding the number and molecular weights of AADC subunits. Christenson et al. (1970) initially reported that hog kidney AADC was composed of three subunits, whereas all subsequent studies have concluded AADC is a dimer (Table 4).

Table 4
Molecular Weight Estimates for AADC

Enzyme source	Native mol wt, kdaltons	Subunit mol wt, kdaltons	Reference
Hog kidney	112	57, 40, 21	Christenson et al., 1970
Hog kidney	90	50, 44	Lancaster and Sourkes, 1972
Rat brain	115	—	Corgier and Pacheco, 1975
Hog kidney	103	—	Borri Voltattorni et al., 1979
Hog kidney	95	—	Rudd and Thanassi, 1981
Hog kidney	93	50, 43	Borri Voltattorni et al., 1982
Human pheochromocytoma	100	50, 50	Maneckjee and Baylin, 1983

Borri Voltattorni et al. (1982) have resolved the apparent nonidentical subunits of hog renal AADC by cation-exchange chromatography and reported molecular weights of 50,000 and 43,000 respectively. Furthermore, using fluorescence spectrometry, they have concluded the coenzyme is bound only to the larger molecular weight subunit. This observation has been independently corroborated using radiolabeled α-monofluoromethyldopa, a specific "suicide inhibitor" of AADC, and enzyme isolated from human pheochromocytoma tissue (Maneckjee and Baylin, 1983). However, these investigators have speculated the smaller subunit may be the result of proteolysis. The presence of a noncatalytic subunit in AADC raises the question regarding its role in decarboxylation. The polypeptide may contribute to the complex active site of AADC and partially explain apparent broad substrate specificity and the rate enhancement seen with some substrates after the addition of certain organic solvents.

4.2. Coenzyme

A catalytic role for PLP in the decarboxylation of aromatic amino acids is well established (Borri, Voltattorni et al., 1971, 1979; Lancaster and Sourkes, 1972; Srinivasan and Awapara, 1978). AADC contains one mole of PLP per mole of protein (Christenson et al., 1970; Borri Voltattorni et al., 1979), which is tightly bound to the ϵ-amino group of a lysine residue (Bossa et al., 1977). Sequencing data involving AADC have also demonstrated considerable primary structure homology with certain other bacterial PLP-dependent decarboxylases. In particular, a histidine residue adjacent to the lysine-PLP residue has been proposed to serve a catalytic function (Bossa et al., 1977).

A catalytic role for PLP is also suggested by the observed stimulation of in vitro decarboxylase activity after the addition of exogenous cofactor (Buzard and Nytch, 1957; Christenson et al., 1970; Lancaster and Sourkes, 1972; Sims et al., 1973; Bouchard and Roberge, 1979; Rahman et al., 1982). However, the degree of enhancement may be dependent on the substrate present and other variables, such as pH, enzyme source, and freeze–thawing (Awapara et al., 1962; Sourkes, 1972; Sims et al., 1973; Srinivasan and Awapara, 1978). For example, numerous studies have reported that exogenous PLP stimulates DOPA decarboxylase activity, whereas 5-HTP decarboxylase activity is stimulated to a lesser degree or not at all (Lancaster and Sourkes, 1972; Sourkes, 1972; Sims et al., 1973; Srinivasan and Awapara, 1978). In particular, exogenous PLP is an obligatory requirement for demonstration of AADC activity in human brain tissue (Lloyd and Hornykiewicz, 1970, 1972). Furthermore, PLP deficiency in mammals leads to reduced AADC activity in vitro that can be overcome by exogenous cofactor (Rahman et al., 1982). In conclusion, the mechanism of the differential effect of exogenous PLP on the rate of aromatic amino acid decarboxylation remains unresolved. However, it seems unlikely that the stimulation is simply the result of cofactor binding to apoenzyme, since apoAADC is moderately unstable (Borri Voltattorni et al., 1971). The in vitro activation of enzyme activity by exogenous PLP may be a consequence of multiple coenzyme binding sites (Christenson et al., 1970; Bender and Coulson, 1972, 1977; Borri Voltattorni et al., 1983). Also, Rudd et al. (1981) have recently proposed that exogenous PLP may stimulate in vitro decarboxylase activity by limiting the inactivation of the enzyme via the decarboxylation-dependent transaminase pathway. This could, therefore, explain the differential effect of PLP on decarboxylase activity because 5-HTP is reported not to be metabolized through the abortive transamination pathway.

4.3. Isoelectric Point

The isoelectric point for rat renal AADC has been estimated to be near pH 5.2 (Bowsher and Henry, 1983), whereas the isoelectric point for rat brain AADC was reported to be 6.5 (Corgier and Pacheco, 1975).

5. Inhibitors of AADC

Since inhibition of AADC has been postulated to be therapeutically useful, a number of studies have been directed towards development of specific inhibitors of this enzyme. The topic of AADC inhibition has been extensively reviewed (Clark, 1959; Sourkes and D'Iorio, 1963; Sourkes, 1966; Porter, 1973). AADC inhibitors can be roughly categorized into one of two groups. First, compounds that interact with the coenzyme, PLP, and second, compounds that are directed against the apodecarboxylase. Each category of inhibitor can further be subclassified as reversible or irreversible. However, since AADC catalysis is mechanistically complex, the resulting in vitro inhibition patterns are frequently complex. Also, the in vitro mechanism of AADC inhibition may not accurately reflect the pharmacology of the inhibitor in vivo.

AADC can be inhibited by compounds that react with PLP at its reactive aldehydic site. AADC is especially vunerable to inhibition by nitrogen-containing compounds. Schott and Clark (1952) have described the nonenzymatic condensation of certain monoamines with PLP. Thus, the sequestering of both coenzyme and substrate or product is a potential source of AADC inhibition. These tetrahydroisoquinoline derivatives have been shown to be inhibitory in vitro (Clark, 1959; Sourkes and D'Iorio, 1963), but more recent evidence suggests that the linear noncyclized adduct is the inhibitory species. Accordingly, Rudd et al. (1979) have reported synthesis of the linear noncyclized PLP and L-DOPA adduct, *N*-(5'-phosphopyridoxal)-DOPA, which is a potent noncompetitive AADC inhibitor in vitro (0.02 m*M* produced 63% inhibition versus control).

Carbonyl trapping reagents, such as hydrazine, are potent nonspecific inhibitors of AADC. Accordingly, several hydrazine analogs of L-DOPA have been developed as potentially selective inhibitors of AADC (Fig. 2). Two such agents are L-α-methylhydrazino-3,4-dihydroxyphenylpropionic acid (MK-486, also known as Carbidopa) and seryl-2,3,4-trihydroxybenzylhydrazine (Ro4-4602, also know as benserazide) (Sourkes, 1966; Porter, 1973). These compounds are currently used as inhibitors of peripheral AADC and are used in conjunction with L-DOPA in the treatment of Parkinson's disease (Calne et al., 1977). Recent evidence indicates benserazide undergoes hydrolysis in vivo, yielding trihydroxybenzylhydrazine, Ro4-5127, which is the probable inhibitory species (Borri Voltattorni et al., 1981). Abundant spectral data have demonstrated that these hydrazino drugs bind tightly to the active

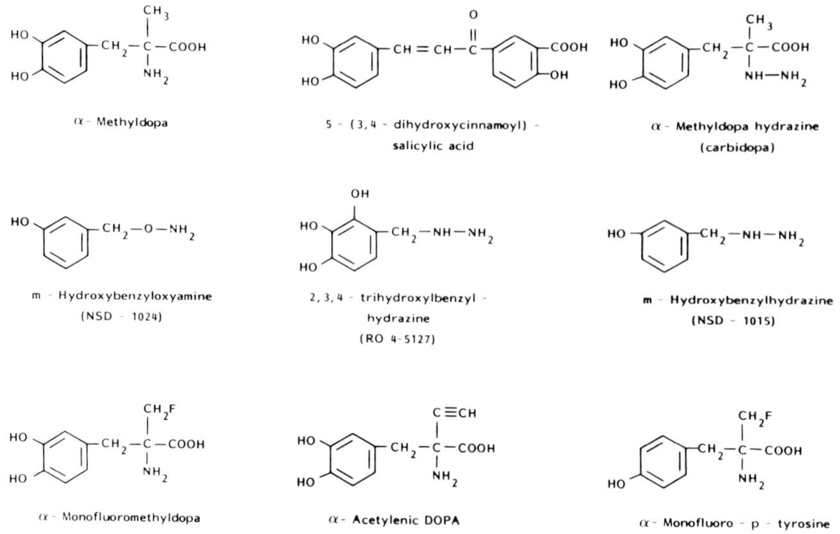

Fig. 2. Inhibitors of aromatic L-amino acid decarboxylase.

site of AADC (Borri Voltattorni et al., 1971; Porter, 1973; Borri Voltattorni et al., 1977a, 1977b, 1981). These drugs do show some specificity for AADC because their corresponding D-isomers are considerably less potent inhibitors in vitro (Porter, 1973). One study has suggested that carbidopa is capable of suppressing the synthesis of noradrenaline in sympathetic neurons (Wurtman and Watkins, 1977). In addition to inhibition of AADC by direct interaction with the holoenzyme, hydrazino drugs may also inhibit AADC indirectly by binding free coenzyme. Thus, hydrazino drugs could decrease the supply of coenzyme available for binding to newly synthesized enzyme and hence depress formation of holoenzyme (Rahman et al., 1982). However, these drugs do not necessarily lead to a generalized pyridoxine deficiency (Bianchine, 1980).

Certain acidic analogs of L-DOPA bind to the active site of AADC and are effective reversible inhibitors of the enzyme in vitro (Barboni et al., 1982). Early work demonstrated that derivatives of phenylpyruvic acid were competitive inhibitors of AADC (Hartman et al., 1955; Clark, 1959; Clark and Pogrund, 1961; Porter, 1973). The most effective compounds were 3-hydroxyphenylpyruvic acid and 3,4-dihydroxyphenylpyruvic acid. Hydroxylated derivatives of cinnamic acid were subsequently shown to be inhibitory. Hartman et al. (1955) systematically extended these investigations and reported their results after examining more than 200 compounds. The most effective in vitro inhibitor was 5-(3,4-dihydroxycinnamoyl)-salicylic acid, which yielded 90% inhibition at a concentration of 1 m*M* (Fig. 2). However, the 3-monohydroxylated derivative was more active in vivo. In general, these inhibitors antagonized the pressor response to iv L-DOPA, but were not effective in blocking noradrenaline synthesis in vivo. The apparent lack of in vivo efficacy was attributed to their rapid metabolism in vivo (Clark and Pogrund, 1961). In summary, acidic analogs of L-DOPA are capable of inhibiting AADC in vitro, but are less effective in vivo. However, one might speculate that acidic metabolites of endogenous monoamines may influence the activity of AADC in vivo and, therefore, represent a potential biosynthetic regulatory mechanism. Various structural analogs of tryptophan, such as β-5-hydroxy-3-benzo[b]thienyl-α-alanine (K_i = 0.06 m*M*), are reversible inhibitors of AADC in vitro (Bosin et al., 1974, 1978).

Various α-methylated derivates of aromatic amino acids are inhibitors of AADC. Sourkes reported that α-methyldopa exerted a biphasic effect on AADC activity in vitro (Sourkes, 1954, 1965). At low concentrations, the compound activated AADC, but at higher concentrations it was a competitive inhibitor. However, later studies showed the type of inhibition was dependent on the in vitro reaction parameters (Lovenberg et al., 1963). These investigators concluded that α-methyldopa bound in a stereospecific manner to the AADC active site and proposed a specific interaction with the coenzyme. α-Methyldopa was also reported to be a substrate for AADC, and the decarboxylase reaction showed an obligatory requirement for exogenous PLP (Weissbach et al., 1960). The observed in vitro inhibition data did not agree well with subsequent in vivo data (Sourkes, 1965). α-Methyldopa, a moderate inhibitor in vitro, produced a temporary decrease in brain dopamine that returned to control levels after several hours. However, the noradrenaline levels, which also were decreased initially, remained suppressed for several days even after α-methyldopa was no longer detectable in brain tissue. Thus, the mechanism of action of α-methylDOPA in vivo was not simply the result of AADC inhibition. Further investigation has revealed α-methyldopa is metabolized to α-methylnoradrenaline, which is capable of displacing endogenous noradrenaline from its storage sites (Carlsson and Lindquist, 1962). In summary, α-methyldopa decreased the concentration of noradrenaline in brain, but the mechanism is thought not to be due to inhibition of AADC, but rather to production of a "false neurotransmitter", α-methylnoradrenaline, which displaces the endogenous neurotransmitter. The resulting hypotension secondary to α-methyldopa administration is currently postulated to be mediated by central alpha adrenergic receptors (Henning, 1975; Frolich, 1980). However, other metabolites of α-methyldopa, such as the *o*-methylated derivatives, have been postulated to be responsible for the hypotensive effect (Zavisca et al., 1979).

One action of chronic α-methyldopa therapy may be to decrease the tissue levels of AADC. Culvenor and Jarrott (1979) have shown by immunotitration that chronic α-methyldopa therapy in rats is associated with a decrease in the amount of soluble AADC protein in adrenals, heart, brain, liver, and kidneys. This decreased activity was solely characterized by a change in the V_{max} of DOPA decarboxylation. The mechanism for this decrease in AADC protein by α-methyldopa therapy remains unclear. However, O'Leary and Baughn (1975, 1977) have postulated that α-methyldopa may decrease AADC activity in vivo by increasing the abortive transamination pathway.

Recent work has culminated in a series of compounds that function as suicide inhibitors of AADC, both in vitro and in vivo (Fig. 2). These inert compounds are derivatives of α-methyldopa that become irreversible inhibitors after activation by AADC. These newly developed compounds are α-vinyldopa (Ribereau-Gayon et al., 1979), α-acetylenicdopa (Ribereau-Gayon et al., 1979). α, α-difluoromethyldopa (Palfreyman et al., 1978; Ribereau-Gayon et al., 1980), and α-monofluoromethyldopa (Maycock et al., 1980; Fozard, 1982). The mono- and difluoro compounds are the most widely studied and their mechanism of action is summarized in the following steps: (1) the drug forms a Schiff base with the coenzyme in the active site of AADC; (2) the carboxyl group of the drug is lost as carbon dioxide; (3) the highly electronegative fluorine causes the α-methyl carbon to become electron-deficient and hence susceptible to nucleophilic attack; (4) a nucleophilic residue (N, O, S) on the protein attacks the methylene carbon, and the fluorine is simultaneously lost as a fluoride ion; and (5) the drug now becomes covalently bound to the protein and renders it catalytically inactive. These new inhibitors are highly specific for AADC because they first require decarboxylation before they can inhibit. In this regard, they may be superior to the currently used hydrazine based inhibitors. Both acetylenicdopa and vinyldopa appear to have a reversible competitive component to their inhibition pattern (Ribereau-Gayon et al., 1979). α-Monofluoromethyldopa is a potent inhibitor of in vivo activity in all tissues, including brain (Jung et al., 1979; Bey et al., 1980; Fozard et al., 1980). Bey et al. (1980) have demonstrated in the mouse that 24 h after a single 100 mg/kg ip injection of α-monofluoromethyldopa, brain concentrations of dopamine, noradrenaline, and serotonin were significantly decreased, whereas L-DOPA levels were increased. Thus, α-monofluoromethyldopa is the first compound that appears capable of blocking monoamine synthesis at the level of AADC.

Recently, Jung et al. (1984) reported that the specificity of AADC inhibition may be enhanced by using α-monofluoro-*p*-tyrosine. This compound first requires metabolic transformation by the neuronal enzyme tyrosine hydroxylase before it can inhibit AADC. Thus, AADC inhibition should be restricted to catecholaminergic neuronal systems. This seems to be the case because the compound had no effect on the levels of serotonin and 5-HTP in rat brain, although efficiently lowering the concentration of noradrenaline.

In summary, a wide variety of compounds has been shown to inhibit AADC in vitro. However, until very recently, most were incapable of inhibiting monoamine synthesis at the level of AADC. Presumably, this was because AADC was not rate-limiting in monoamine synthesis (Levitt et al., 1965; Wang and Perlman, 1981). However, the recent proliferation of irreversible AADC inhibitors contradicts this concept. In conclusion, AADC inhibitors have been, and will continue to be, useful pharmacologic tools and therapeutic agents.

6. Properties and Functional Significance of AADC in the CNS

6.1. Regional Distribution of AADC in Brain

The presence of AADC activity in mammalian brain has been demonstrated by a number of experimental approaches. First, AADC activity has been frequently assayed in extracts of brain, using DOPA and 5-HTP as substrates (Kuntzman et al., 1961; Sims et al., 1973; Rahman and Nagatsu, 1982). Second, polyclonal antisera to purified AADC have been used to document the presence of AADC in mammalian brain by immunologic criteria (Christenson et al., 1972; Hokfelt et al., 1973b; Bowsher and Henry, 1983). Third, the presence of AADC in brain tissue has also been corroborated by studies involving isotopically labeled substrates and inhibitors (Wade and Katzman, 1975; Wooten and Horne, 1982).

The wide and heterogeneous distribution of AADC in mammalian brain parallels tissue monoamine content. Studies with rats (Lamprecht and Coyle, 1972; Sims et al., 1973; Rahman et al., 1981; Bowsher and Henry, 1983), cats (Kuntzman et al., 1961; Roberge and Poirier, 1973), rabbits (McCaman et al., 1965), and monkeys (Goldstein et al., 1969), have revealed essentially the

Table 5
Regional Distribution of AADC in Brain

	DOPA Decarboxylase activity, nmol/h/g wet tissue			
	Rat[a]	Rabbit[b]	Cat[c]	Human[d]
Caudate	6402	1530	2130	682
Hypothalamus	4260	—	1268	396
Midbrain	—	—	1095	68
Colliculi	2442	1240	340	—
Brain stem	1482	550	370	75
Cerebral cortex	882	170	86	5
Cerebellum	756	210	36	—

[a]Rahman et al., 1981
[b]McCaman et al., 1965
[c]Kuntzman et al., 1961
[d]MacKay et al., 1978

same regional distribution, with high levels of decarboxylase activity occurring in the striatum, midbrain, hypothalamus, and brain stem (Table 5). High AADC activity has also been found in the pineal gland (Lloyd and Hornykiewicz, 1972; Rahman et al., 1981). The distribution of AADC in human brain is similar to that described for other mammals (Lloyd and Hornykiewicz, 1970, 1972; MacKay et al., 1978; Rahman and Nagatsu, 1982). However, studies involving human brain have consistently reported that this organ contains a much lower amount of AADC than other species (Vogel et al., 1970; Lloyd and Hornykiewicz, 1972; Nagatsu et al., 1979). Rahman and Nagatsu (1982) have used a new HPLC assay method with electrochemical detection and reported DOPA decarboxylase rates of 39 nm/h/g and 14 nm/h/g for human hypothalamus and caudate, respectively. Also, the level of decarboxylase in human brain is highly variable, yielding large standard errors in experimental data (Lloyd and Hornykiewicz, 1972; Sacks et al., 1979; Rahman and Nagatsu, 1982). These observations have prompted speculation that AADC may be rate-limiting in monoamine synthesis in human brain (Robins et al., 1967; Lloyd and Hornykiewicz, 1970; Sacks et al., 1979). The low level of AADC activity in human brain does not appear to be the consequence of postmortem instability (MacKay et al., 1978; Sacks et al., 1979). Also, the in vitro decarboxylation of DOPA by extracts of human brain has an obligatory requirement for the addition of exogeneous cofactor (Lloyd and Hornykiewicz, 1970, 1972). In summary,

AADC is widely distributed in mammalian brain, but appears to be present in human brain tissue at much lower levels. Consequently, previous investigators have suggested rat brain may be an inappropriate model for monoamine synthesis in human brain (Sacks et al., 1979).

6.2. Subcellular Distribution of AADC

A number of studies using centrifugation techniques have concluded that peripheral AADC is a soluble protein (Blaschko et al., 1955; Laduron and Belpaire, 1968; Christenson et al., 1972). This concept is reinforced by the fact that all published purification procedures have used high-speed supernatant as the starting source for isolation of AADC. In contrast, several studies have reported brain AADC maybe partially particulate. Sims et al. (1973) compared the centrifugational sedimentation of DOPA and 5-HTP decarboxylase activities of rat brain. They concluded that a greater percentage of 5-HTP decarboxylase activity was associated with the synaptosomal fraction, and that a substantial fraction of 5-HTP decarboxylase was not liberated by hypotonic osmotic shock. However, these conclusions have not been corroborated by subsequent studies. Recently, the subcellular distribution of AADC has been studied in mouse brain (Gardner and Richards, 1981). These investigators, using DOPA as the substrate, concluded greater than 90% of AADC activity could be accounted for by soluble and synaptosomal fractions. In summary, evidence suggests that AADC is predominantly a soluble protein in both peripheral tissues and brain. However, it is interesting to speculate that the subcellular environment of AADC may influence its structural conformation and substrate specificity.

6.3. Ontogeny of AADC

The ontogeny of rat brain AADC parallels the development of other monoamine synthetic enzymes, such as tyrosine hydroxylase, dopamine-β-hydroxylase, and tryptophan hydroxylase (Bennett and Giarman, 1965; Lamprecht and Coyle, 1972). Brain AADC activity increases substantially prenatally, and then slightly or not at all after birth (Lamprecht and Coyle, 1972). In contrast, AADC activity in peripheral tissues, which is primarily nonneuronal (Klingman, 1965), increases considerably postnatally (Awapara and Saine, 1975). Studies involving the ontogeny of AADC in human tissues have been complicated by the large variability in decarboxylase activity. However, Vogel et al. (1970) have reported higher AADC activity in fetal peripheral tissues than in adult, whereas fetal brain AADC activity did not differ substantially from adult brain.

6.4. Functional Significance of AADC in the CNS

Neuronal AADC catalyzes the decarboxylation of aromatic amino acids and, therefore, plays an important role in the synthesis of monoamine neurotransmitters. This biosynthetic role in catecholaminergic and serotoninergic neurons is well established and is supported by an abundance of biochemical, pharmacologic, and immunologic data. However, recent evidence also suggests AADC may play a role in the synthesis of other putative monoamine neurotransmitters or neuromodulators, such as *p*-tyramine, *m*-tyramine, *o*-tyramine, phenylethylamine, histamine, and tryptamine. These "trace amines" are normal constitutents of mammalian brain (Axelrod et al., 1976; Boulton, 1978; Karoum et al., 1980), and much biochemical and pharmacologic evidence suggests AADC is responsible for catalyzing their formation (Jones et al., 1983; Bowsher and Henry, 1983; Juorio, 1983; Dyck et al., 1983). Furthermore, Jaeger et al. (1983a, b, 1984), using immunocytochemical methods, have recently found neurons in rat spinal chord, hypothalamus, and brain stem that contain AADC but lack tyrosine hydroxylase or serotonin. Also, these investigators have concluded AADC and vasopressin coexist in some cells of the hypothalamus (Jaeger et al., 1983b). Thus, current evidence suggests the biosynthetic role for AADC in neuronal tissue should be expanded beyond the catecholamines, serotonin, and melatonin, to include the "trace amines."

We currently hypothesize the following factors are important for determining the monoamine profile in AADC-containing neurons. First, the presence of other monoamine synthetic enzymes, such as tyrosine hydroxylase or tryptophan hydroxylase, will define the types of monoamines synthesized in a given neuron. Second, because aromatic amino acids are present in tissues at concentrations much below their respective Michaelis constants for decarboxylation, the Michaelis-Menten rate equation simplifies to $V = V_{max}/K_m \times$ [amino acid] (Segel, 1975; Wang and Perlman, 1981). Thus, the rate of monoamine synthesis should be first order and directly proportional to the concentration of free precursor amino acid in that neuron. This is depicted in Fig. 3, which shows the potential metabolic interactions in a hypothetical dopaminergic neuron. As shown, multiple monoamines could arise intraneuronally through the actions of tyrosine hydroxylase and AADC. Therefore, the mechanisms for delivery and uptake of aromatic amino acids and storage of monoamine products are of

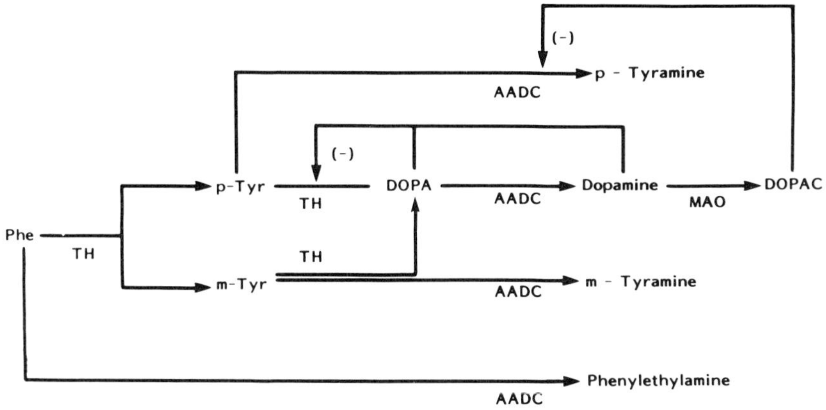

Fig. 3. Potential metabolic interactions in a dopaminergic neuron. Numerous monoamine products can potentially be synthesized in dopaminergic neurons through the independent actions of tyrosine hydroxylase (TH) and AADC. Additionally, acid metabolites that arise through the action of monoamine oxidase (MAO) may also be important as feedback inhibitors of AADC.

pivotal importance in defining the monoamine profile in a given neuron. This concept is supported by experiments that have demonstrated an inverse relationship between the turnover of dopamine and the concentration of *p*-tyramine in the caudate nucleus (Juorio, 1979; Juorio and Jones, 1981; Jones et al., 1983). This hypothesis is also compatible with the concept of DOPA decarboxylation by adjacent striatal serotoninergic neurons or glia during L-DOPA therapy for treatment of Parkinson's disease (Melamed et al., 1981).

AADC has been characterized in a number of CNS structures associated with neuroendocrine function. The pineal gland contains high levels of AADC (Lloyd and Hornykiewicz, 1972; Rahman et al., 1981), which function in the synthesis of melatonin, an indolealkylamine with antigonadotropic properties (Reichlin, 1981). A number of studies have also demonstrated an effect of monoamines, especially dopamine, on the secretion of hormones by the anterior pituitary (Weiner and Ganong, 1978). AADC activity is heterogenously distributed in the pituitary, and recent studies have demonstrated trace levels of AADC in the anterior pituitary, low levels in the neurointermediate pituitary, and substantial levels in the median eminence (Szabo et al., 1977; Reymond and Porter, 1982; Johnston et al., 1984). Thus, these results are compatible with the current view of a tuberohypophyseal origin for monoamines.

AADC activity has been demonstrated in the visual pathway

of the cat and rabbit, including the optic nerve (Kuntzman et al., 1961; McCaman et al., 1965). More recent studies have also shown that the rat retina has a high level of dopamine (DaPrada, 1977; Kato et al., 1980), and that the administration of AADC inhibitors results in a significant accumulation of retinal L-DOPA (Morgan and Kamp, 1982). Furthermore, the retina dopaminergic system appears to be extremely photoresponsive, as indicated by changes in tyrosine hydroxylase activity (Iuvone et al., 1978) and dopamine turnover (Iuvone et al., 1978; Morgan and Kamp, 1980).

A fraction of total brain AADC is localized in a nonneuronal tissue. An early study using histochemical techniques suggested the presence of AADC in brain capillary endothelium of rat and mouse, and proposed the concept of an enzymatic barrier for the penetration of DOPA and 5-HTP into the brain (Bertler et al., 1963). These observations have been extended to a number of mammalian species (Langelier et al., 1972; Hardebo et al., 1980). In contrast, spontaneously hypertensive rats may contain substantially less AADC in cerebral vessels than their normotensive genetic controls, the Kyoto Wistars (Tanaka, 1970). Dihydroxyphenylserine, a substrate for purified AADC (Inagaki and Tanaka, 1978; Bowsher and Henry, 1983), also undergoes decarboxylation within brain capillary endothelium (Bartholini et al., 1971, 1975), therefore further suggesting the presence of AADC in this tissue. Most recent studies have indicated this enzyme barrier is saturable (Wade and Katzman, 1975; Hardebo et al., 1977), and that AADC activity is limited to the wall of brain microvessels, distinct from large brain arteries and the vessels of the choroid plexus (Hardebo, 1980; Hardebo and Owman, 1980). The physiologic significance of AADC as an enzyme component of the blood–brain barrier system is well discussed in a review by Hardebo and Owman (1980).

7. Properties and Functional Significance of AADC in Peripheral Tissues

In the periphery, AADC has been described in a number of neuronal compartments: First, AADC is present in postganglionic neurons of the sympathetic nervous system, where this enzyme plays an unequivocal role in the synthesis of noradrenaline; second, AADC is present in the adrenal medulla, where it functions in the synthesis of adrenaline; third, neurons containing AADC, but lacking dopamine β-hydroxylase (DβH), have been demonstrated using immunohistochemical methods in the sympathetic ganglia of the dog (Muller et al., 1984). Presumably, these neurons are dopaminergic and they may function as interneurons or they may be postganglionic neurons projecting to the kidney and hindlimb. Neurons present in the mammalian myenteric plexus decarboxylate 5-HTP (Gershon and Ross, 1966a,b) and synthesize serotonin in vitro (Dreyfus et al., 1977). Additionally, a large amount of biochemical and pharmacologic evidence suggests serotonin may be a neurotransmitter in the myenteric plexus (Bulbring and Gershon, 1967; Gershon et al., 1976, 1977; Costa and Furness, 1979; Furness et al., 1980). Thus, AADC is present in intestinal neurons where this enzyme may catalyze synthesis of serotonin or a related indoleamine (Furness et al., 1980).

AADC is present in certain types of paraneurons. Paraneurons are cells that traditionally have not been considered neurons, but are closely related to neurons because of their structure, metabolism, and neurosecretory-like function (Fujita, 1977; Fujita and Kobayashi, 1979). In particular, AADC activity has been described in APUD (amine precursor uptake and decarboxylase) cells and SIF (small intensely fluorescent) cells. APUD cells are polypeptide hormone-secreting cells that have the capacity to take up amine precursors, DOPA and 5-HTP, and store monoamines (Oie et al., 1983). The properties and functional significance of APUD cells have been reviewed by Pearse (1969, 1974). SIF cells are small monoamine-containing cells located near the sympathetic ganglion as interneurons (type 1), or in close proximity to blood vessels (type 2) (Neff et al., 1983). Type 1 dopaminergic SIF cells have been suggested to modulate ganglionic transmission by altering the sensitivity of postganglionic neurons to acetylcholine (Libet, 1977; Eranko, 1978). In summary, AADC appears to be present in a number of neuroendocrine-like cells, paraneurons, where it catalyzes formation of monoamines that may modulate various neuroendocrine functions.

In addition to its localization in neurons and paraneurons, peripheral AADC is also compartmentalized nonneuronally. Klingman (1965), using immunosympathectomy, has demonstrated that nonneuronal AADC comprizes the vast majority of the enzyme present in the periphery. AADC is widely distributed in the periphery, with high levels of enzyme activity localized in liver, kidney, and intestine (Clark et al., 1954; Davis and Awapara, 1960; McCaman et al., 1965; Vogel et al., 1970; Bouchard and Roberge, 1979; Rahman et al., 1981, 1982). Based on antigenic criteria, peripheral AADC is the same protein that is present in brain tissue (Christenson et al., 1972; Hokfelt et al., 1973a; Jaeger et al., 1983b; Bowsher and Henry, 1983). However, the functional significance of peripheral nonneuronal AADC remains unclear. Peripheral AADC has been proposed to be of metabolic importance under circumstances associated with elevation of plasma aromatic amino acids (David et al., 1973, 1974). In this regard, nonneuronal AADC may facilitate elimination of aromatic amino acids from plasma. Recently, Van Huysse et al. (1983a) have demonstrated that monoamines derived from aromatic amino acids, especially *p*-tyramine, occur in high concentrations in mammalian urine. Subsequent studies have also shown *p*-tyramine undergoes net renal synthesis and secretion by the rabbit kidney in vivo (Van Huysse et al., 1983b, 1984a). Renal *p*-tyramine clearance was significantly decreased by 80% after iv administration of the specific AADC suicide inhibitor α-monofluoromethyldopa (Van Huysse et al., 1984b). Also, Stier et al. (1984) have recently shown by perfusion studies that rat renal AADC catalyzes decarboxylation of 5-HTP, and this is associated with an increase in renal vascular resistance. AADC activity is localized primarily within cortical tissue in kidney (Beaven et al., 1978; Bowsher and Henry, 1983). Thus, nonneuronal AADC can catalyze formation of monoamines within renal tissue, where these substances could serve a physiologic function.

AADC is present in the gastric mucosa of several mammalian species (Hakanson and Owman, 1966). Mammalian stomach also contains the specific histidine decarboxylase enzyme (Aures et al., 1968b; Bergmark et al., 1976; Beaven et al., 1982). Presumably, one or both of these enzymes catalyze formation of histamine, a mediator of gastric acid secretion. However, substantial species differences exist concerning the cellular localization of these enzymes. In rat gastric mucosa, AADC activity is present primarily in enterochromaffin-like cells (Hakanson and Owman, 1967; Aures et al., 1968b; Soll et al., 1981). Low levels of AADC are also present in rat mast cells (Soll et al., 1981). Both of these cell types contain histidine decarboxylase and store histamine. Rat mast cells also contain serotonin (Soll et al., 1981). In contrast, the gastric mucosa of other species, such as dog and man, contain only mast cells, and these cells lack APUD properties (Aures et al., 1968b) and detectable levels of AADC and serotonin (Beaven et al., 1982). The cellular origin of histamine in these species remains unresolved. In conclusion, the majority of AADC in mammalian gastric mucosal cells appears localized within enterochromaffin-like cells rather than mast cells.

8. AADC Assay Methodology

A multitude of AADC assays have been published. All reported methods, except for the early bioassays, involve quantifying the accumulation of reaction products—either the monoamine or carbon dioxide. Bioassays represent the earliest attempts at estimation of in vitro AADC activity (Blaschko, 1945). The respective aromatic amino acid was incubated with a crude tissue extract for a long period of time and then examined for a predetermined physiologic response, such as a change in blood pressure. Complimentary methods, such as product isolation and pharmacologic manipulations, were used to further corroborate the presence of the monoamine product. These methods were sensitive and fairly specific, but were laborious and time-consuming. Another disadvantage is that bioassays are indirect analytical methods. In other words, the enzymic reaction is monitored by its ability to elicit a physiologic response, rather than by direct chemical analysis.

Two types of decarboxylase assays are based on quantifying the enzymatic evolution of carbon dioxide. Manometric analysis represents the first widely used analytical method for measuring AADC activity. This technique is based on quantification of the volume of carbon dioxide liberated from the amino acid substrate. Though suitable for substrates with a high V_{max}, such as L-DOPA, this technique is generally too insensitive for quantifying the decarboxylation of many aromatic amino acids (Fellman, 1959; Holtz, 1959; Hagen, 1962). Undoubtedly, this methodology helped foster the controversy concerning the substrate specificity of AADC.

The second method, based on measurement of carbon dioxide evolution, is the radiometric assay (Christenson et al., 1970; Beaven et al., 1978; Okuno and Fujisawa, 1983). This technique involves incubation of a [^{14}C-carboxy]amino acid with a tissue extract and chemically trapping the radioactive carbon dioxide in a basic solution, such as hyamine hydroxide. The amount of radioactive carbon dioxide liberated is directly proportional to the amount of decarboxylase activity present in the tissue extract and is quantified by liquid scintillation spectrometry. These methods are sensitive, but are dependent on the availability of [^{14}C-carboxy]-labeled amino acids. Also, L-DOPA may be subject to erroneous results because of the nonenzymatic evolution of carbon dioxide (Mackowiak et al., 1972). However, Okuno and Fujisawa (1983) have recently reported that the combined use of 1 m*M* mercaptoethanol, a reducing agent, and 0.1 m*M* EDTA blocks the nonenzymatic decarboxylation of L-DOPA.

The alternate approach for the quantification of AADC activity is the measurement of the amine product. This requires an obligatory separation step because the amino acid substrate will frequently interfere with analysis. The charge difference between the amino acid substrate and amine product has been frequently exploited as the basis for separation. The various methods used for separating the substrate and product include selective solvent extraction (Kuntzman et al., 1961; Snyder and Axelrod, 1964; Laduron and Belpaire, 1968; Bowsher and Henry, 1983), ion-pair extraction (McCaman et al., 1972; Melamed et al., 1980), ion-exchange chromatography (Bertler and Rosengren, 1959b; Davis and Awapara, 1960; Lancaster and Sourkes, 1972), and adsorptive chromatography (Dietrich, 1953; Lovenberg et al., 1962). The efficiency of the separative procedure is an important factor that can have an impact on the sensitivity of the assay. Accordingly, the lack of adequate separation results in an unsatisfactory assay blank and, therefore, poor assay sensitivity. The recent trend in AADC assay development has been to compliment an initial separative procedure with subsequent higher resolution chromatographic techniques. This increases the manipulative effort but dramatically increases the assay sensitivity and specificity. These methods include paper chromatography (Bender and Coulson, 1972), thin-layer chromatography (Aures et al., 1968a; Bowsher and Henry, 1983), and paper electrophoresis (Somerville, 1964). The optimization of modern chromatographic methods has resulted in a recent proliferation of sensitive HPLC assays for AADC activity. A number of different detection methods have been used to quantify the amine reaction products. These include spectrophotometric (D'Erme et al., 1980), fluorometric (Tabakoff and Black, 1980), and electrochemical (Nagatsu et al., 1979; Rahman et al., 1980; Bowsher and Henry, 1983). The electrochemical detection methods are intrinsically the most sensitive. AADC assays have also been reported that employ chemical derivatization of the reaction product, dopamine, using 2,4,6-trinitrobenzene-1-sulfonic acid (Streffer, 1967; Sherald et al., 1973; Charteris and John, 1975). The derivative is then selectively extracted into benzene, and its absorbance measured at 340 nm. The benzene extraction is necessary to separate 2,4,6-trinitrobenzene-1-dopamine from 2,4,6-trinitrobenzenedopa. This method is basically a solvent extraction approach, but the unique absorption maximum of the dopamine derivative may increase assay specificity. Several radioenzymatic assays have been described for the quantification of in vitro AADC activities (Taylor and Snyder, 1972; Bowsher and Henry, 1983).

One limitation of AADC assays that are based on the quantification of the amine reaction product is the interference associated with endogenous monoamines. This is especially true for spectrophotometric- and fluorometric-based assays. Consequently, these assays are quite satisfactory with purified preparations of AADC or with tissue extracts rich in decarboxylase activity, such as kidney. However, they are not well suited for studies involving crude extracts of tissues, such as brain, which are low in enzyme activity and rich in the amine product (Aures et al., 1970).

AADC assays based on quantifying a radiolabeled amine product represent the potentially most sensitive and specific approach. The sensitivity of this approach is only limited by the availability of radiolabeled amino acid, and most aromatic amino acids are currently available labeled with tritium in high specific activities. Also, tritium-labeled amino acids are considerably less expensive than their corresponding [^{14}C-carboxy]amino acids. Furthermore, the tissue content of endogenous monoamines is of no consequence. Thus, some of the deficits of spectrophotometric and fluorometric methods are overcome by this analytical approach. Assays based on the isolation of a radioactive monoamine product may also be less susceptible to interference by nonenzymatic decarboxylation. Finally, isolation of the radioactive monoamine may be more specific than isolation of radioactive carbon dioxide, because other metabolic pathways can liberate carbon dioxide from the α-carboxyl position of aromatic amino acids (Knox, 1951).

In summary, a large number of AADC assays have been described. Most methods have used either L-DOPA or 5-HTP as the substrate and involve quantification of either the amine or carbon dioxide reaction product. The various AADC assays are characterized by a wide range of sensitivity limits, and all are potentially useful under appropriate conditions. However, careful consideration must be given to the type of assay used for documentation of the substrate specificity of AADC because of the wide range of V_{max} values for decarboxylation of aromatic amino acids. Undoubtedly, this factor has helped foster the controversy concerning the substrate specificity of AADC.

8.1. AADC Assay Protocol

5-Hydroxytryptophan decarboxylase, a marker activity for AADC, was quantified by measuring the conversion of tritiated 5-HTP to tritiated serotonin (Bowsher and Henry, 1983). This assay is similar to other 5-HTP decarboxylase assays that are based on the selective isolation of radioactive serotonin (Snyder and Axelrod, 1964). Prior to the assay, the tritiated substrate was purified by batch ion-exchange chromatography using AG 50 W-X2 (200–400 mesh) in 1 m*M* Tris, pH 8.3. A 100 μL aliquot of tritiated D,L-5-HTP

(5 Ci/mm) was combined with 400 μL of distilled water and then treated with 50 μL of a suspension of the AG 50 W-X2 in Tris buffer (1:1). The tube was vortexed gently, centrifuged, and the supernatent containing the tritiated 5-HTP used directly in the assay. This simple and quick manipulation resulted in substantial blank reduction. Each 100 μL reaction was carried out in a borosilicate culture tube (12 × 75 mm) that contained the following components: 15 μmol of potassium phosphate, pH 7.8; 0.12 μmol of EDTA; 0.025% bovine serum albumin (BSA); and 1 nmol of prepurified D,L-5-hydroxy[G-^{3}H]-tryptophan. The decarboxylase reaction was initiated by the addition of 25 μL of enzyme and then incubated for 15 min at 37°C. The blank tubes received 25 μL of BSA instead of enzyme. The reaction was terminated by the addition of 75 μL of 2.5 *M* potassium borate, pH 11.0 and 1.25 mL of toluene/isopentyl alcohol (3:1). After vortexing and centrifugation at 1500*g* for 5 min, a 1 mL aliquot of the organic phase was transferred to another culture tube containing 0.25 mL of 2*M* formic acid. The tubes were vortexed, centrifuged, and the organic phases removed by aspiration. A 150 μL aliquot of the aqueous phase was then transferred to a scintillation vial containing 1 mL of 1*M* potassium phosphate, pH 7.1 and 10 mL of 2% bis-diethylhexyl hydrogen phosphate—Econofluor. The vials were shaken and then quantified as a two-phase system by liquid scintillation spectrometry (Bowsher and Henry, 1983; Bowsher et al., 1983).

9. AADC Purification

AADC has been partially purified from guinea-pig kidney (Lovenberg et al., 1962; Aures et al., 1970), beef adrenals (Fellman, 1959; Hagen, 1962; Laduron and Belpaire, 1968), rat liver (Awapara et al., 1962; Bender and Coulson, 1972), rabbit kidney (Rosengren, 1960), rat brain (Corgier and Pacheco, 1975), and human pheochromocytoma (Hagen, 1962), and homogeneous protein has been successfully isolated from hog kidney, guinea pig kidney, rat kidney, and human pheochromocytoma (Table 6). AADC activity has been extracted from tissue sources using a variety of solutions, such as water, dilute buffers, and sucrose. There is no clear-cut advantage to the selection of any one of the above homogenization media, except that it is probably desirable to use an isotonic solution because AADC is a soluble protein. Investigators have also tried various additions, such as reducing agents, chelators, PLP, and glycerol in order to stabilize AADC during purification. Any conclusions regarding the usefulness of these agents are difficult to interpret because of the inconstancies in the literature. However, during our studies on the purification of rat renal AADC we consistently observed that buffers containing 10% glycerol stabilized the protein (Bowsher and Henry, 1983). Ammonium sulfate fractionation between 30 and 60% saturation has been widely used as an initial purification step (Christenson et al., 1970; Lancaster and Sourkes, 1972; Srinivasan and Awapara, 1978; Bowsher and Henry, 1983). Also, a heat treatment step (Christenson et al., 1970; Lancaster and Sourkes, 1972) or acid precipitation (Fellman, 1959; Bowsher and Henry, 1983) have been frequently incorporated early into reported protein purification schemes. Anion-exchange chromatography using DEAE has been the most widely used column chromatographic procedure used for isolating AADC (Christenson et al., 1970; Lancaster and Sourkes, 1972; Corgier and Pacheco, 1975; Srinivasan and Awapara, 1978; Borri Voltattorni et al., 1979; Bowsher and Henry, 1983). Other successful chromatographic techniques have included adsorption on calcium phosphate gel (Fellman, 1959), alumina gel chromatography (Clark et al., 1954; Christenson et al., 1970), molecular size exclusion chromatography (Lancaster and Sourkes, 1972; Borri Voltattorni, 1979), and ion-exchange chromatography using QAE-cellulose (Borri Voltattorni, et al., 1979), and CM-cellulose (Corgier and Pacheco, 1975). Hydroxyapatite has been frequently used as a final chromatographic step in the preparation of highly purified AADC (Christenson et al., 1970; Lancaster and Sourkes, 1972; Corgier and Pacheco, 1975). Recently, several studies have reported the use of hydrophobic interaction chromatography during the purification of AADC. Srinivasan and Awapara (1978) used butyl- and hexylamino-agarose for isolation of guinea-pig renal AADC. However, further analysis of their chromatographic parameters suggests the presence of an ion-exchange component to their chromatography procedure (Hofstee, 1975; Pahlman et al., 1977). We (1983) recently reported the use of phenyl-Sepharose for isolation of rat renal AADC. This pure hydrophobic medium binds AADC and can permit a 20-fold purification of the protein at a single step. Additionally, because protein adsorption occurs in the presence of high salt, phenyl-Sepharose chromatography is compatible with ammonium sulfate fractionation. Few studies have reported the use of electrophoretic methods for isolation of AADC at a preparative scale (Cavalli-Sforza et al., 1974; Landon, 1977). However, we (1983) used preparative isoelectric focusing (pH 4–6 ampholine gradient) for isolation of rat renal AADC. In summary, AADC has been purified to an electrophoretic homogeneous state several times from a number of mammalian tissues. A wide variety of

Table 6
Homogeneous Preparations of AADC

Enzyme source	Purification (-fold)	Reference
Hog kidney	300	Christenson et al., 1970
Hog kidney	190	Lancaster and Sourkes, 1972
Guinea pig kidney	330	Srinivasan and Awapara, 1978
Hog kidney	209	Borri Voltattorni, 1979
Rat kidney	506	Bowsher and Henry, 1983
Human pheochromocytoma	428	Maneckjee and Baylin, 1983

protein purification techniques have been successfully employed during AADC purification, however, most currently reported protocols are variants of the initial procedure described by Christenson et al. (1970).

9.1. Protocol for Isolation of Rat AADC

The following purification scheme is a slight modification of the protocol Bowsher and Henry (1983) used for the preparation of homogeneous rat renal AADC. All steps in this procedure were performed at 0–4°C. The following buffers were used: buffer A, 100 m*M* potassium phosphate, 0.1 m*M* EDTA, pH 7.5; buffer B, 1 m*M* potassium phosphate, 0.1 m*M* EDTA, pH 7.5; and buffer C, 25 m*M* bis-Tris HCl, 0.1 m*M* EDTA, 10% (vol/vol) glycerol, pH 6.5; and buffer D, 1 m*M* potassium phosphate, 5% (vol/vol) glycerol, 0.1 m*M* dithiothreitol pH 7.0.

Thirty-five male Wistar rats (200–300 g) were killed by decapitation and their kidneys removed, defatted, and immediately chilled in ice-cold saline. The resulting tissue was minced, homogenized with 3.5 vol of 0.25*M* sucrose using a Brinkmann Polytron, and centrifuged at 40,000*g* for 20 min. The supernatent was then recentrifuged at 220,000*g* for 1 h.

9.1.1. Acid Treatment and Ammonium Sulfate Fractionation

The high-speed supernatent was adjusted to pH 5.0 by the slow addition of cold 2*M* acetic acid. After stirring 20 min at pH 5.0, the suspension was centrifuged at 40,000*g* for 10 min and the precipitate discarded. The supernatent was then titrated to pH 7.0 by the addition of cold 1*M* ammonium hydroxide. Next, the protein preparation was fractionated between 1.2 and 2.2*M* (31–56%

saturation) with solid ammonium sulfate (Wood, 1976). The final precipitate was collected by centrifugation for 10 min at 40,000*g* and resuspended with 30 mL of buffer A containing 0.1 m*M* PLP.

9.1.2. Phenyl-Sepharose Chromatography

The protein was loaded on a Phenyl-Sepharose column (20 × 2.2 cm) that had been previously equilibrated with 500 mL of buffer A. The column was eluted with Buffer A until the 280 nm absorbance had returned to baseline. The column was then step-eluted with buffer B at a flow rate of 1.5 mL/min, and AADC was efficiently eluted at the front of the eluting protein peak. Fractions containing peak enzyme activity, as determined by the previously described assay, were pooled and desalted against buffer C in an Amicon ultrafiltration cell using a PM-10 Diaflo membrane.

9.1.3. DEAE-Sephacel Anion-Exchange Chromatography I

The desalted protein was loaded on a DEAE-Sephacel column (8 × 1.6 cm) that had been previously equilibrated with 300 mL of buffer C. After elution of a small protein peak, the column was eluted with a linear gradient of 0–0.2*M* NaCl in a total volume of 200 mL of buffer C. The column was eluted at 0.75 mL/min, and 7.5 mL fractions were collected. Fractions containing peak decarboxylase activity were pooled.

9.1.4. Preparative Isoelectric Focusing

The protein was desalted against 0.5% (wt/vol) glycine, 5% (vol/vol) glycerol, pH 7.0, 0.1 m*M* dithiothreitol in an Amicon ultrafiltration cell, and then concentrated to 2 mL. Preparative, flat-bed isoelectric focusing was performed on a prefocused granulated gel bed using a pH 4–6 Ampholine gradient. The enzyme preparation was focused for 16 h at 8 W constant power. Protein bands were localized by a print-staining technique using Coomassie brilliant blue R-250 (Radola, 1973). Protein bands were eluted from the gel bed with buffer C containing 0.1 m*M* dithiolthreitol. After localization of AADC activity, the protein preparation was dialyzed against buffer C containing 0.1 m*M* dithiothreitol and concentrated to approximately 5 mL.

Preparative isoelectric focusing routinely yields substantial overall protein purification. However, if the appropriate electrophoretic equipment is unavailable, then this step may be replaced by hydroxylapatite chromatography, but with less than optimal results. Accordingly, the protein is then dialyzed against buffer D and loaded on a hydroxylapatite column (8 × 1.6 cm) that had been previously equilibrated with 250 mL of buffer D. After elution of a small protein peak, the column is developed with a linear 200 mL gradient from 1 to 75 m*M* potassium phosphate containing 5% (vol/vol) glycerol, pH 7.0.

9.1.5. DEAE-Sephacel Anion-Exchange Chromatography II

The dialyzed protein preparation was loaded on a DEAE-Sephacel column (2 × 1.6 cm) that had been previously equilibrated with 200 mL of buffer C containing 0.1 m*M* dithiothreitol. The AADC was eluted with a linear gradient of 0–0.125 m*M* NaCl in a total volume of 200 mL of buffer C. Fractions containing peak AADC activity were pooled and concentrated to 2 ml, yielding approximately 0.5 mg of homogeneous protein. This amount of protein represents a sufficient amount for formation of a polyclonal anti-AADC antiserum in rabbits.

10. Conclusion

A great deal of information about AADC has accumulated since its initial characterization approximately 50 yrs ago. However, despite being the first reported monoaminergic synthetic enzyme, major questions remain regarding the biochemical and functional aspects of this important enzyme. AADC catalyzes the decarboxylation of all naturally occurring aromatic amino acids in vitro and probably is responsible for the synthesis of a number of important monoamines in vivo. Major biochemical aspects of AADC that require more clarification include further characterization of the AADC active site with respect to the number of substrate and PLP binding sites and definition of the mechanism of the decarboxylase rate enhancement seen after the addition of certain organic solvents and its physiologic revelance. Despite its complex kinetics, which suggest that enzyme regulation is possible, no evidence for the control of the AADC reaction has yet been described. However, the following factors suggest the potential for regulation of the enzyme in vivo. First, AADC is unique in that it participates in the in vivo formation of several known neurotransmitters and hormones, such as the catecholamines, serotonin, melatonin, and the trace amines. Second, the ubiquitous distribution of AADC in mammalian neuronal and nonneuronal tissues. The physiologic role of this enzyme in peripheral nonneuronal tissues, such as liver and kidney, remains unclear. Third, the presence of AADC in certain pathophysiological conditions, such as tumors of neural crest origin. However, no pathophysiologic condition has been identified that is directly related to an abnormality or malfunction of AADC. The prevailing concept of low and variable levels of AADC in human brain suggests this enzyme may be rate-limiting in monoamine synthesis in human brain. The successful development of AADC inhibitors should continue with emphasis on increased specificity. The recent synthesis of the catecholamine specific inhibitor α-monofluoromethyl-*p*-tyrosine suggests that α-monofluoromethyltryptophan may be specific for serotonergic systems. It is in these areas that important future advances in the study of AADC will most likely arise.

References

Aures D. and Hakanson R. (1971) Histidine Decarboxylase, in *Methods in Enzymology,* Vol. 18, pt. B (Colowick S. P. and Kaplan N. O., eds.), Academic, New York.

Aures D., Fleming R., and Hakanson R. (1968a). Separation and detection of biogenic amines by thin-layer chromatography *J. Chromatogr.* **33,** 480–493.

Aures D., Hakanson R., and Clark W. G. (1970) Histidine Decarboxylase and Dopa Decarboxylase, in *Handbook of Neurochemistry* (Lajtha A., ed.), pp. 165–196, Plenum, New York.

Aures D., Hakanson R., and Schauer A. (1968b) Histidine decarboxylase and dopa decarboxylase in the rat stomach. Properties and cellular localization. *Eur. J. Pharmacol.* **3,** 217–234.

Awapara J. and Saine S. (1975) Fluctuations in dopa decarboxylase activity with age. *J. Neurochem.* **24,** 817–818.

Awapara J., Perry T. L., Hanly C., and Peck E. (1964) Substrate specificity of dopa decarboxylase. *Clin. Chim. Acta* **10,** 286–289.

Awapara J., Sandman R. P., and Hanly C. (1962) Activation of dopa decarboxylase by pyridoxal phosphate. *Arch. Biochem. Biophys.* **98,** 520–525.

Axelrod J., Saavedra J. and Usdin E. (1976) Trace Amines in the Brain, in *Psychopharmacology* Vol. 1 (Usdin E. and Sandler M., eds.), pp. 1–20. Marcel, New York.

Barboni E., Borri Voltattorni C., D'Erme M., Fiori A., Minelli A., and Rosei A. (1982) Inhibitors binding to AADC. *Life Sci.* **31,** 1519–1524.

Bartholini J., Constantinidis J., Puig M., Tissot R., Pletscher A. (1975) The stereoisomers of 3,4-dihydroxyphenylserine as precursors of norepinephrine. *J. Pharmacol. Exp. Ther.* **193,** 523–532.

Bartholini G., Constantinidis J., Tissot R., and Pletscher A. (1971) Formation of monoamines from various amino acids in the brain after inhibition of extracerebral decarboxylase. *Biochem. Pharmacol.* **20,** 1243–1247.

Beaven M. A., Soll A. H., and Lewin K. J. (1982) Histamine synthesis by intact mast cells from canine fundic mucosa and liver. *Gastroenterology* **82,** 254–262.

Beaven M. A., Wilcox G., and Terpstra G. K. (1978) A microprocedure for the measurement of $^{14}CO_2$ release from [^{14}C]carboxy-labeled amino acids. *Anal. Biochem.* **84,** 638–641.

Bender D. A. and Coulson W. F. (1977) AADC: pH dependence of substrates and inhibitors. *Biochem. Soc. Trans.* **5,** 1353–1356.

Bender D. A. and Coulson W. F. (1972) Variations in AADC activity towards DOPA and 5HTP caused by pH changes and denaturation. *J. Neurochem.* **19,** 2801–2810.

Bennett D. S. and Giarman N. J. (1965) Schedule of appearance of serotonin and associated enzyme in the developing rat brain. *J. Neurochem.* **12,** 911–918.

Bergmark J., Granerus G., and Henningsson S. (1976) Histamine metabolism of the guinea pig gastric mucosa. *J. Physiol.* (Lond.) **257,** 419–431.

Bertler A. and Rosengren E. (1959a) On the distribution in brain monoamines and of enzymes responsible for their formation. *Experientia* **15,** 382–384.

Bertler A. and Rosengren E. (1959b) Occurrence and distribution of catecholamines in brain. *Acta Physiol. Scand.* **47,** 350–361.

Bertler A., Falck B., and Rosengren E. (1963) The direct demonstration of a barrier mechanism in brain capillaries. *Acta Pharmacol. Toxicol.* **20,** 317–321.

Bey P., Jung M. J., Koch-Weser J., Palfreyman M. G., Sjoerdsma A., Wagner J., and Zraika M. (1980) Further studies on the inhibition of monoamine synthesis by monofluoromethyldopa. *Br. J. Pharmacol.* **70,** 571–576.

Bianchine J. R. (1980) Drugs For Parkinson's Disease, in *The Pharmacologic Basis of Therapeutics* (Gilman A. G., Goodman L. S., Gilman A., eds.), 6th ed., pp. 482–483, Macmillan, New York.

Blaschko H. (1939) The specific action of L-dopa decarboxylase. *J. Physiol.* (Lond.) **96,** 50P–51P.

Blaschko H. (1942) The activity of L-dopa decarboxylase. *J. Physiol.* (Lond.) **101,** 337–349.

Blaschko H. (1945) The amino acid decarboxylases of mammalian tissues. *Adv. Enzymol.* **5,** 67–85.

Blaschko H. (1950) Substrate specificity of amino acid decarboxylases. *Biochim. Biophys. Acta* **4,** 130–137.

Blaschko H. and Holton P. (1950) Enzymic formation of *ortho*-tyramine. *J. Physiol.* (Lond.) **110,** 482–487.

Blaschko H., Burn J. H., and Langemann H. (1950) The formation of noradrenaline from dihydroxyphenylserine. *Br. J. Pharmacol.* **5,** 431–437.

Blaschko H., Hagen P., and Welch A. D. (1955) Observations on the intracellular granules of the adrenal medulla. *J. Physiol.* (Lond.) **129,** 27–49.

Blaschko H., Holton P., and Stanley G. H. S. (1949) Enzymic formation of pressor amines. *J. Physiol.* (Lond.) **108,** 427–439.

Borri Voltattorni C., Minelli A., and Borri P. (1977a) Interaction of L-α-methyl-α-hydrazino-3,4-dihydroxyphenylpropionic acid with dopa decarboxylase from pig kidney. *FEBS Lett.* **75,** 277–280.

Borri Voltattorni C., Minelli A., and Borri P. (1977b) Interaction of *N*-(D,L-seryl)*N*'-(2,3,4-trihydroxybenzyl)-hydrazine with dopa decarboxylase from pig kidney. *Experientia* **33,** 158–160.

Borri Voltattorni C., Minelli A., and Borri P. (1981) The interaction of 2,3,4-trihydroxybenzylhydraine with dopa decarboxylase from pig kidney. *Life Sci.* **28,** 103–108.

Borri Voltattorni C., Minelli A., Cirotto C., Barra D., and Turano C. (1982) Subunit structure of dopa decarboxylase from pig kidney. *Arch. Biochem. Biophys.* **217,** 58–64.

Borri Voltattorni C., Minelli A., and Dominici P. (1983) Interaction of aromatic amino acids in D and L forms with DOPA decarboxylase from pig kidney. *Biochemistry* **22,** 2249–2254.

Borri Voltattorni C., Minelli A., and Turano C. (1971) Spectral properties of the coenzyme bound to dopa decarboxylase from pig kidney. *FEBS Lett.* **17,** 231–235.

Borri Voltattorni C., Minelli A., Vecchini P., Fiori A., and Turano C. (1979) Purification and characterization of dopa decarboxylase from pig kidney. *Eur. J. Biochem.* **93,** 181–188.

Bosin T. R., Baldwin J. R., and Maickel R. P. (1978) Inhibition of dopa decarboxylation by analogues of tryptophan. *Biochem. Pharmacol.* **27,** 1289–1291.

Bosin T. R., Buckpitt A. R., and Maickel R. P. (1974) Substrate specificity of AADC. *Life Sci.* **14,** 899–908.

Bossa F., Martini F., Barra D., Borri Voltattorni C., Minelli A., and Turano C. (1977) The chymotryptic phosphopyridoxal peptide of dopa decarboxylase from pig kidney. *Biochem. Biophys. Res. Commun.* **78,** 177–184.

Bouchard S. and Roberge A. G. (1979) Biochemical properties and kinetic parameters of DOPA-5HTP decarboxylase in brain, liver, and adrenals of cat. *Can. J. Biochem.* **57,** 1014–1018.

Boulton A. A. (1978) The tyramines: Functionally significant biogenic amines or metabolic accidents? *Life Sci.* **23,** 659–672.

Bowsher R. R. and Henry D. P. (1983) Decarboxylation of *p*-tyrosine: A potential source of *p*-tyramine in mammalian tissues. *J. Neurochem.* **40,** 992–1002.

Bowsher R. R., Verburg K. M., and Henry D. P. (1983) Rat histamine *N*-methyltransferase: Quantification, tissue distribution, purification, and immunologic properties. *J. Biol. Chem.* **258,** 12215–12220.

Bulbring E. and Gershon M. D. (1967) 5-Hydroxytryptamine participation in the vagal inhibitory innervation of the stomach. *J. Physiol.* (Lond.) **192,** 823–846.

Buzard J. A. and Nytch P. D. (1957) Some characteristics of rat kidney 5HTP decarboxylase. *J. Biol. Chem.* **227,** 225–230.

Calne D. B., Teychenne P. F., and Pfeiffer R. F. (1977) Treatment of parkinsonism. *Adv. Exp. Med. Biol.* **90,** 49–54.

Carlsson A. and Lindquist M. (1962) In vivo decarboxylation of α-methyl-dopa and α-methyl-metatyrosine. *Acta Physiol. Scand.* **54,** 87–94.

Cavalli-Sforza L. L., Santachiara S. A., and Wang L. (1974) Electrophoretic study of 5HTP decarboxylase from brain and liver in several species. *J. Neurochem.* **23,** 629–634.

Charteris A. and John R. (1975) An investigation of the assay of dopamine using trinitrobenzenesulfonic acid. *Anal. Biochem.* **66,** 365–371.

Christenson J. G., Dairman W., and Udenfriend S. (1970) Preparation and properties of a homogeneous aromatic L-amino acid decarboxylase from hog kidney. *Arch. Biochem. Biophys.* **141,** 356–367.

Christenson J. G., Dairman W., and Udenfriend S. (1972) On the identity of DOPA decarboxylase and 5HTP decarboxylase. *Proc. Natl. Acad. Sci. USA* **69,** 343–374.

Clark W. G. (1959) Studies on inhibition of dopa decarboxylase in vitro and in vivo. *Pharmacol. Rev.* **11,** 330–349.

Clark W. G. and Pogrund R. S. (1961) Inhibition of dopa decarboxylase in vitro and in vivo. *Circ. Res.* **9,** 721–732.

Clark C. T., Weissbach H., and Udenfriend S. (1954) 5-Hydroxytryptophan decarboxylase: Preparation and properties. *J. Biol. Chem.* **210,** 139–148.

Corgier M. and Pacheco H. (1975) Purification and properties of aromatic L-amino acid decarboxylase of rat brain. *Biochimie* **57,** 1005–1017.

Costa M. and Furness J. B. (1979) On the possibility that an indoleamine is a neurotransmitter in the gastrointestinal tract. *Biochem. Pharmacol.* **28,** 565–571.

Coulson W. F., Henson G., and Jepson J. B. (1968) The specificity of rat liver aromatic amino acid decarboxylase. *Biochem. J.* **107,** 17P–18P.

Culvenor A. J. and Jarrott B. (1979) Reduction of AADC protein in rats after chronic administration of alphamethyldopa. *Mol. Pharmacol.* **15,** 86–98.

Dairman W., Horst W. D., Marchelle M. E., and Bautz G. (1975) The proportionate loss of DOPA and 5HTP decarboxylating activity in rat central nervous system following intracisternal administration of 5,6-dihydroxytryptamine or 6-hydroxydopamine. *J. Neurochem.* **24,** 619–623.

DaPrada M. (1977) Dopamine Content and Synthesis in Retina and N. Accumbens Septi: Pharmacological and Light-Induced Modifications, in *Advances in Biochemical Psychopharmacology*, Vol. 16 (Costa E. and Gessa G. L., eds.), pp. 311–319, Raven, New York.

David J. C., Dairman W., and Udenfriend S. (1973) On the significance of the large amounts of non-neuronal AADC in kidney and liver. *Pharmacologist* **15,** 246.

David J. C., Dairman W., and Udenfriend S. (1974) On the importance of decarboxylation in the metabolism of phenylalanine, tyrosine, and tryptophan. *Arch. Biochem. Biophys.* **160,** 561–568.

Davis V. E. and Awapara J. (1960) A method for the determination of some amino acid decarboxylases. *J. Biol. Chem.* **235,** 124–127.

D'Erme M., Rosei M. A., Fiori A., and DiStazio G. (1980) Assay of AADC by high performance liquid chromatography. *Anal. Biochem.* **104,** 59–61.

Dietrich L. S. (1953) A rapid method for the determination of dopa decarboxylase in animal tissues. *J. Biol Chem.* **204,** 587–591.

Dreyfus C. F., Bornstein M. B., and Gershon M. D. (1977) Synthesis of serotonin by neurons of the myenteric plexus *in situ* and in organotypic tissue culture. *Brain Res.* **128,** 125–139.

Dyck L. E., Yang C. R., and Boulton A. A. (1983) The biosynthesis of *p*-tyramine, *m*-tyramine, and β-phenylethylamine by rat striatal slices. *J. Neurosci. Res.* **10,** 211–220.

Eranko O. (1978) Small intensely fluorescent (SIF) cells and nervous transmission in sympathetic ganglia. *Ann. Rev. Pharmacol. Toxicol.* **18,** 417–430.

Fellman J. H. (1959) Purification and properties of adrenal L-dopa decarboxylase. *Enzymologia* **20,** 366–376.

Ferrini R. and Glasser A. (1964) In vitro decarboxylation of new phenylalanine derivatives. *Biochem. Pharmacol.* **13,** 798–800.

Florkin M. and Stotz E. H., eds. (1964) *Comprehensive Biochemistry*, Vol. 13, Report of the Commission on Enzymes of the International Union of Biochemistry, pp. 122–123, Elsevier, New York.

Florkin M. and Stotz E. H., eds. (1973) *Comprehensive Biochemistry*, Vol. 13, Enzyme Nomenclature, pp. 272–273, Elsevier, New York.

Fozard J. R. (1982) Highly potent irreversible inhibitors of AADC. *Trends Pharmacol. Sci.* **3,** 429.

Fozard J. R., Spedding M., Palfreyman M. G., Wagner J., Mohring J., and Koch-Weser J. (1980) Depression of sympathetic nervous function by D,L-α-monofluoromethyl dopa, an enzyme-activated, irreversible inhibitor of AADC. *J. Cardiovasc. Pharmacol.* **2,** 229–245.

Frohlich E. D. (1980) Methyldopa: Mechanisms and treatment 25 years later. *Arch. Intern. Med.* **140,** 954–959.

Fujita T. (1977) Concepts of paraneurons. *Arch. Histol. Jpn.* **40,** 1–12.

Fujita T. and Kobayashi S. (1979) Current views on the paraneurone concept. *Trends in Neurosci.* **2,** 27–30.

Furness J. B., Costa M., and Howe P. R. C. (1980) Intrinsic Amine-Handling Neurons in the Intestine, in *Advances in Biochemical Psychopharmacology* (Eranko O., Soinila S. and Paivarinta H., eds.), Vol. 25, Raven, New York.

Gardner C. R. and Richards H. H. (1981) Use of D,L-alpha-monofluoromethyldopa to distinguish subcellular pools of AADC in mouse brain. *Brain Res.* **216,** 291–298.

Gershon M. D. and Ross L. L. (1966a) Radioisotopic studies of the binding exchange and distribution of 5-hydroxytryptamine synthesized from its radioactive precursor. *J. Physiol.* (Lond.) **186,** 451–476.

Gershon M. D. and Ross L. L. (1966b) Location of sites of 5-hydroxytryptamine storage and metabolism by radioautography. *J. Physiol.* (Lond.) **186,** 477–492.

Gershon M. D., Dreyfus C. F., Pickel V. M., Joh T. H., and Reis D. J. (1977) Serotonergic neurons in the peripheral nervous system: Identification in gut by histochemical localization of tryptophan hydroxylase. *Proc. Natl. Acad. Sci. USA* **74,** 3086–3089.

Gershon M. D., Robinson R. G., and Ross L. L. (1976) Serotonin accumulation in the guinea-pig myenteric plexus: Ion dependence, structure–activity relationship and the effect of drugs. *J. Pharmacol. Exp. Ther.* **198,** 548–561.

Goldstein M., Anagnoste B., Battista A. F., Owen W. S., and Nakatani S. (1969) Studies of amines in the striatum in monkeys with nigral lesions. *J. Neurochem.* **16,** 645–653.

Goldstein M., Fuxe K., and Hokfelt T. (1972) Characterization and tissue localization of catecholamine-synthesizing enzymes. *Pharmacol. Rev.* **24,** 293–295.

Gonnard P. and Camier M. (1966) Multiplicity of decarboxylases for cyclic amino acids. *Bull. Soc. Chim. Biol.* **48,** 225–238.

Hagen P. (1962) Observations on the substrate specificity of dopa decarboxylase from ox adrenal medulla, human phaeochromocytoma and human argentaffinoma. *Br. J. Pharmacol.* **18,** 175–182.

Hakanson R. and Owman C. (1966) Distribution and properties of amino acid decarboxylases in gastric mucosa. *Biochem. Pharmacol.* **15,** 489–499.

Hakanson R. and Owman C. (1967) Concomitant histochemical demonstration of histamine and catecholamines in enterochromaffin-like cells of gastric mucosa. *Life Sci.* **6,** 759–766.

Hardebo J. E. and Owman C. (1980) Barrier mechanisms for neurotransmitter monoamines and their precursors at the blood–brain interface. *Ann. Neurol.* **8,** 1–11.

Hardebo J. E., Edvinsson L., Owman Ch., and Rosengren E. (1977) Quantitative evaluation of the blood–brain barrier capacity to form dopamine from circulating L-dopa. *Acta Physiol. Scand.* **99,** 377–384.

Hardebo J. E., Emson P. C., Falck B., Owman Ch., and Rosengren E. (1980) Enzymes related to monoamine transmitter metabolism in brain microvessels. *J. Neurochem.* **35,** 1388–1393.

Hartman W. J., Akawie R. I., and Clark W. G. (1955) Competitive inhibition of dopa decarboxylase *in vitro. J. Biol. Chem.* **216,** 507–529.

Henning M. (1975) Central Action of Alpha-Methyldopa, in *Central Action of Drugs in Blood Pressure Regulation* (Davies D. S. and Reid J. L., eds.), pp. 157–165, University Park, Baltimore.

Hofstee B. H. J. (1975) Fractionation of protein mixtures through differential adsorption on a gradient of substituted agaroses of increasing hydrophobicity. *Prep. Biochem.* **5,** 7–19.

Hokfelt T., Fuxe K., and Goldstein M. (1973a) Immunohistochemical localization of AADC in central dopamine and 5-hydroxytryptamine nerve cell bodies of the rat. *Brain Res.* **53,** 175–180.

Hokfelt T., Fuxe K., and Goldstein M. (1973b) Immunohistochemical studies on monoamine-containing cell systems. *Brain Res.* **62,** 461–469.

Holtz P. (1959) Role of L-dopa decarboxylase in the biosynthesis of catecholamines in nervous tissue and the adrenal medulla. *Pharmacol. Rev.* **11,** 317–329.

Holtz P., Heise R., and Luedtke K. (1938) Enzymic destruction of L-dopa by the kidney. *Arch. Exp. Path. Pharmakol.* **191,** 87–118.

Inagaki C. and Tanaka C. (1978) Characteristics of enzymic decarboxylation of L-threo-3,4-dihydroxyphenylserine using hog renal AADC. *Biochem. Pharmacol.* **27,** 1081–1086.

Iuvone P. M., Galli C. L., Garrison-Gund C. K., and Neff N. H. (1978) Light stimulates tyrosine hydroxylase activity and dopamine synthesis in retinal amacrine neurons. *Science* **202,** 901–902.

Jaeger C. B., Albert V. R., Joh T. H., and Reis D. J. (1983a) AADC in the rat brain: Coexistance with vasopressin in small neurons of the suprachiasmatic nucleus. *Brain Res.* **276,** 362–366.

Jaeger C. B., Ruggiero D. A., Albert V. R., Park D. H., Joh T. H., and Reis D. J. (1984) AADC in the rat brain: Immunocytochemical localization in neurons of the brain stem. *Neuroscience* **11,** 691–713.

Jaeger C. B., Teitelman G., Joh T. H., Albert V. R., Park D. H., and Reis D. J. (1983b) Some neurons of the rat central nervous system contain AADC but not monoamines. *Science* **219,** 1233–1235.

Johnston C. A., Spinedi E., and Negro-Vilar A. (1984) AADC activity in the rat median eminence, neurointermediate lobe and anterior lobe of the pituitary. *Neuroendocrinology* **39,** 54–59.

Jones R. S. G., Juorio A. V., and Boulton A. A. (1983) Changes in levels of dopamine and tyramine in the rat caudate nucleus following alterations of impulse flow in the nigrostriatal pathway. *J. Neurochem.* **40,** 396–401.

Jung M. J., Hornsperger J-M., Gerhart F., and Wagner J. (1984) Inhibition of AADC and depletion of biogenic amines in brain of rats treated with α-monofluoromethyl-*p*-tyrosine: Similitudes and differences with the effects of α-monofluoromethyldopa. *Biochem. Pharmacol.* **33,** 327–330.

Jung M. J., Palfreyman M. G., Wagner J., Bey P., Ribereau-Gayon G., Zraika M., and Koch-Weser J. (1979) Inhibition of monoamine synthesis by irreversible blockage of AADC with α-monofluoromethyldopa. *Life Sci.* **24,** 1037–1042.

Juorio A. V. (1979) Drug-induced changes in the formation, storage, and metabolism of tyramine in the mouse. *Br. J. Pharmacol.* **66,** 377–384.

Juorio A. V. (1983) The effect of some decarboxylase inhibitors on striatal tyramines in the mouse. *Neuropharmacology* **22,** 71–73.

Juorio A. V. and Jones R. S. G. (1981) The effect of mesencephalic lesions on tyramine and dopamine in the caudate nucleus of the rat. *J. Neurochem.* **36,** 1898–1903.

Karoum F., Potkin S. G., and Wyatt R. J. (1980) Quantitation and Metabolism of Phenylethylamine and Tyramines Three Isomers in Humans, in *Noncatecholic Phenylethylamines,* (Mosnain A. D. and Wolf M. E., eds.), Marcel Dekker, New York.

Kato S., Nakamura T., and Negishi K. (1980) Postnatal development of dopaminergic cells in the rat retina. *J. Comp. Neurol.* **191,** 227–236.

Klingman G. I. (1965) Catecholamine levels and dopa-decarboxylase activity in peripheral organs and adrenergic tissues in the rat after immunosympathectomy. *J. Pharmacol. Exp. Ther.* **148,** 14–21.

Knox W. E. (1951) The oxidation in liver of L-tyrosine to acetoacetate through *p*-hydroxyphenylpyruvate and homogentisic acid. *Biochem. J.* (Lond.) **49,** 686–693.

Kuntzman R., Shore P. A., Bogdanski D., and Brodie B. B. (1961) Microanalytical procedures for fluorometric assay of brain dopa-5HTP decarboxylase, norepinephrine and serotonin, and a detailed mapping of decarboxylase activity in brain. *J. Neurochem.* **6,** 226–232.

Laduron P. and Belpaire F. (1968) A rapid assay and partial purification of dopa decarboxylase. *Anal. Biochem.* **26,** 210–218.

Lamprecht F. and Coyle J. T. (1972) Dopa decarboxylase in the developing rat brain. *Brain Res.* **41,** 503–506.

Lancaster G. A. and Sourkes T. L. (1972) Purification and properties of hog-kidney dopa decarboxylase. *Can. J. Biochem.* **50,** 791–797.

Landon M. (1977) A method for the direct detection of AADC in electrophoretic media. *Anal. Biochem.* **77,** 293–297.

Langelier P., Parent A., Poirier L. J. (1972) Decarboxylase activity of the brain capillary walls and parenchyma in the rat, cat and monkey. *Brain Res.* **45,** 622–629.

Langemann H. (1958) 5-Hydroxytryptophan Decarboxylase, in *5-Hydroxytryptamine* (Lewis G. P., ed.), pp. 153–157, Pergamon, Oxford.

Levitt M., Spector S., Sjoerdsma A., and Udenfriend S. (1965) Elucidation of the rate-limiting step in norepinephrine biosynthesis in the perfused guinea-pig heart. *J. Pharmacol. Exp. Ther.* **148,** 1–8.

Libet B. (1977) The Role SIF Cells Play in Ganglionic Transmission, in *Advances in Biochemical Psychopharmacology* (Costa E. and Gessa G. L., eds.) Vol. 16, 541–546, Raven, New York.

Lloyd K. G. and Hornykiewicz O. (1970) Occurrence and distribution of dopa decarboxylase in the human brain. *Brain Res.* **22,** 426–428.

Lloyd K. G. and Hornykiewicz O. (1972) Occurrence and distribution of AADC in the human brain. *J. Neurochem.* **19,** 1549–1559.

Lovenberg W., Barchas J., Weissbach H., and Udenfriend S. (1963) Characteristics of the inhibition of AADC by α-methylamino acids. *Arch. Biochem. Biophys.* **103,** 9–14.

Lovenberg W., Weissbach H., and Udenfriend S. (1962) Aromatic L-amino acid decarboxylase. *J. Biol. Chem.* **237,** 89–93.

MacKay A. V. P., Davies P., Dewar A. J., and Yates C. M. (1978) Regional distribution of enzymes associated with neurotransmission by monoamines, acetylcholine and gaba in the human brain. *J. Neurochem.* **30,** 827–839.

Mackowiak E. D., Hare T. A., and Vogel W. H. (1972) Measurement of AADC—a technical comment. *Biochem. Med.* **6,** 562–567.

Maneckjee R. and Baylin S. B. (1983) Use of radiolabeled monofluoromethyldopa to define the subunit structure of human L-DOPA decarboxylase. *Biochemistry* **22,** 6058–6063.

Maycock A. L., Aster S. D., and Patchett A. A. (1980) Inactivation of 3-(3,4-dihydroxyphenyl)alanine decarboxylase by 2-(fluoromethyl)-3-(3,4-dihydroxyphenyl)alanine. *Biochemistry* **19,** 709–718.

McCaman M. W., McCaman R. E., and Lees G. J. (1972) Liquid cation exchange—a basis for sensitive radiometric assays for AADC. *Anal. Biochem.* **45,** 242–252.

McCaman R. E., McCaman M. W., Hunt J. M., and Smith M. S. (1965) Microdetermination of monoamine oxidase and 5HTP decarboxylase activities in nervous tissues. *J. Neurochem.* **12,** 15–23.

Melamed E., Hefti F., Pettibone D. J., Liebman J., and Wurtman R. J. (1981) AADC in rat corpus striatum: Implications for action of L-dopa in Parkinsonism. *Neurology* **31,** 651–655.

Melamed E., Hefti F., and Wurtman R. J. (1980) DOPA and 5HTP decarboxylase activities in rat striatum: Effect of selective destruction of dopaminergic or serotoninergic input. *J. Neurochem.* **34,** 1753–1756.

Minelli A., Charteris A. T., Borri Voltattorni C., and John R. A. (1979) Reactions of dopa decarboxylase with dopa. *Biochem. J.* **183,** 361–368.

Morgan W. and Kamp C. W. (1980) A GABAergic influence on the light-induced increase in dopamine turnover in the dark-adapted rat retina *in vivo*. *J. Neurochem.* **34,** 1082–1086.

Morgan W. W. and Kamp C. W. (1982) Postnatal development of the light response of the dopaminergic neurons in the rat retina. *J. Neurochem.* **39,** 283–285.

Muller B., Harris T., Borri Voltattorni C., and Bell C. (1984) Distribution of neurons containing dopa decarboxylase and dopamine-β-hydroxylase in some sympathetic ganglia of the dog: A quantitative study. *Neuroscience* **11,** 733–740.

Nagatsu, T. (1973) *Biochemistry of Catecholamines,* pp. 50–60, University Park, Baltimore.

Nagatsu T., Yamamoto T., and Kato T. (1979) A new and highly sensitive voltammetric assay for AADC activity by high-performance liquid chromatography. *Anal. Biochem.* **100,** 160–165.

Neff N. H., Karoum F., and Hadjiconstantinou N. (1983) Dopamine-containing small intensely fluorescent cells and sympathetic ganglion function. *Fed. Proc.* **42,** 3009–3011.

Nomenclature Commission of the International Union of Biochemistry (1979) Enzyme Nomenclature, 1978, pp. 920–923, Academic, New York.

Oie H. K., Gazdar A. F., Minna J. D., Weir G. C., and Baylin S. B. (1983) Clonal analysis of insulin and somatostatin secretion and dopa decarboxylase expression by a rat islet cell tumor. *Endocrinology* **112,** 1070–1075.

Okuno S. and Fujisawa H. (1983) Accurate assay of dopa decarboxylase by preventing nonenzymatic decarboxylation of dopa. *Anal. Biochem.* **129,** 412–415.

O'Leary M. H. and Baughn R. L. (1975) New pathway for metabolism of dopa. *Nature* **253,** 52–53.

O'Leary M. H. and Baughn R. L. (1977) Decarboxylation-dependent transamination catalyzed by mammalian dopa decarboxylase. *J. Biol. Chem.* **252,** 7168–7173.

Pahlman S., Rosengren J., and Hjerten S. (1977) Hydrophobic interaction chromatography on uncharged Sepharose derivatives. J. *Chromatogr.* **131,** 99–108.

Palfreyman M. G., Danzin C., Bey P., Jung M. J., Ribereau-Gayon G., Aubry M., Vevert J. P., and Sjoerdsma A. (1978) α-Difluoromethyl dopa, a new enzyme-activated irreversible inhibitor of AADC. *J. Neurochem.* **31,** 922–932.

Pearse A. G. (1974) The apud cell concept and its implication in pathology. *Pathol. Ann.* **9,** 27–34.

Pearse A. G. E. (1969) The cytochemistry and ultrastructure of polypeptide hormone-producing cells of the apud series and the embryologic, physiologic and pathologic implications of the concept. *J. Histochem. Cytochem.* **17,** 303–313.

Porter C. C. (1973) Inhibitors of AADC-their biochemistry. *Adv. Neurol.* **2,** 37–58.

Radola B. J. (1973) Analytical and preparative isoelectric focusing in gel-stabilized layers. *Ann. NY Acad. Sci.* **209,** 127–143.

Rahman M. K. and Nagatsu T. (1982) Demonstration of AADC activity in human brain with L-DOPA and 5HTP as substrates by high-performance liquid chromatography with electrochemical detection. *Neurochem. Int.* **4,** 1–6.

Rahman M. K., Nagatsu T., and Kato T. (1980) New and highly sensitive assay for 5HTP decarboxylase activity by high-performance liquid chromatography-voltammetry. *J. Chromatogr.* **221,** 265–270.

Rahman M. K., Nagatsu T., and Kato T. (1981) Aromatic L-amino acid decarboxylase activity in central and peripheral tissues and serum of rats with DOPA and 5HTP as substrates. *Biochem. Pharmacol.* **30,** 645–649.

Rahman M. K., Nagatsu T., Sakuri T., Hori S., Abe M., and Matsuda M. (1982) Effect of pyridoxal phosphate deficiency on AADC activity with DOPA and 5HTP as substrates in rats. *Jpn. J. Pharmacol.* **32,** 803–811.

Reichlin S. (1981) Neuroendocrinology, in *Textbook of Endocrinology,* 6th ed. (Williams R. H., ed.), pp. 625–628, Saunders, Philadelphia.

Reymond M. J. and Porter J. C. (1982) Hypothalamic secretion of dopamine after inhibition of AADC activity. *Endocrinology* **111,** 1051–1056.

Ribereau-Gayon G., Danzin C., Palfreyman M. G., Aubry M., Wagner J., Metcalf B. W., and Jung M. J. (1979) In vitro and in vivo effects of α-acetylenic dopa and α-vinyl dopa on AADC. *Biochem. Pharmacol.* **28,** 1331–1335.

Ribereau-Gayon G., Palfreyman M. G., Zraika M., Wagner J., and Jung M. J. (1980) Irreversible inhibition of AADC by α-difluoromethyl-DOPA and metabolism of the inhibitor. *Biochem. Pharmacol.* **29,** 2465–2469.

Roberge A. G. and Poirier L. J. (1973) Effect of chronically administered L-dopa on dopa/5HTP decarboxylase and tyrosine and tryptophan hydroxylases in cat brain. *J. Neural. Trans.* **34,** 171–185.

Robins E., Robins J. M., Croninger A. B., Moses S. G., Spencer J., and Hudgens R. W. (1967) The low level of 5HTP decarboxylase in human brain. *Biochem. Med.* **1,** 240–251.

Rosengren E. (1960) Are dopa decarboxylase and 5-hydroxytryptophan decarboxylase individual enzymes? *Acta Physiol. Scand.* **49,** 364–369.

Rudd E. A., Cunningham W. C., and Thanassi J. W. (1979) Coenzyme-substrate adducts as inhibitors of mouse liver DOPA decarboxylase *J. Med. Chem.* **22,** 233–237.

Rudd E. A. and Thanassi J. W. (1981) Inhibition of AADC by coenzymeamino acid adducts. *Biochemistry* **20,** 7469–7475.

Sacks W., Vogel W. H., Nagatsu T., Lloyd K. G., and Sandler M. (1979) Round Table On: Is There DOPA Decarboxylase in Human Brain?, in *Catecholamines: Basic and Clinical Frontiers,* Vol. 1 (Usdin E., Kopin I. J., and Barchas J., eds.), pp. 127–131, Pergamon, New York.

Schayer R. W. (1957) Histidine decarboxylase of rat stomach and other mammalian tissues. *Am. J. Physiol.* **189,** 533–536.

Schott H. F. and Clark W. G. (1952) Dopa decarboxylase inhibition through the interaction of coenzyme and substrate. *J. Biol. Chem.* **196,** 449–462.

Segel I. H. (1975) *Enzyme Kinetics,* pp. 41–43, Wiley, New York.

Sherald A. F., Sparrow J. C., and Wright T. R. F. (1973) A spectrophotometric assay for drosophilia dopa decarboxylase. *Anal. Biochem.* **56,** 300–305.

Sims K. L. and Bloom F. E. (1973) Rat brain DOPA and 5HTP decarboxylase activities: Differential effect of 6-hydroxydopamine. *Brain Res.* **49,** 165–175.

Sims K. L., Davis G. A., and Bloom F. E. (1973) Activities of DOPA and 5HTP decarboxylases in rat brain: Assay characteristics and distribution. *J. Neurochem.* **20,** 449–464.

Snyder S. H. and Axelrod J. (1964) A sensitive assay for 5HTP decarboxylase. *Biochem. Pharmacol.* **13,** 805–806.

Soll A. H., Lewin K. J., and Beaven M. A. (1981) Isolation of histamine-containing cells from rat gastric mucosa: Biochemical and morphologic differences from mast cells. *Gastroenterology* **80,** 717–727.

Somerville A. R. (1964) The assay of AADC using radioactive substrates. *Biochem. Pharmacol.* **13,** 1681–1683.

Sourkes T. L. (1954) Inhibition of dopa decarboxylase by derivatives of phenylalanine. *Arch. Biochem. Biophys.* **51,** 444–456.

Sourkes T. L. (1965) The action of α-methyldopa in the brain. *Br. Med. Bull.* **20–21,** 66–69.

Sourkes T. L. (1966) Dopa decarboxylase: Substrates, coenzyme, inhibitors. *Pharmacol. Rev.* **18,** 53–60.

Sourkes T. L. (1972) Influence of specific nutrients on catecholamine synthesis and metabolism. *Pharmacol. Rev.* **24,** 349–352.

Sourkes T. L. (1977) Enzymology of Aromatic Amino Acid Decarboxylases, in *Structure and Function of Monoamine Enzymes* (Usdin E., Weiner N., and Youdim M. B. H., eds.), pp. 477–496, Dekker, New York.

Sourkes T. L. and D'Iorio (1963) Inhibitors of Catecholamine Metabolism, in *Metabolic Inhibitors* (Hochster R. M. and Quastel J. H., eds.), pp. 79–87, Academic, New York.

Sourkes T., Heneage P., and Trano Y. (1952) Enzymatic decarboxylation of isomers and derivatives of dihydroxyphenylalanine. *Arch. Biochem. Biophys.* **40,** 185–193.

Srinivasan K. and Awapara J. (1978) Substrate specificity and other properties of dopa decarboxylase from guinea pig kidneys. *Biochim. Biophys. Acta* **526,** 597–604.

Stier C. T., McKendall G., and Itskovitz H. D. (1984) Serotonin formation in nonblood-perfused rat kidneys. *J. Pharmacol. Exp. Ther.* **228,** 53–56.

Streffer C. (1967) A method for determination of AADC activity. *Biochim. Biophys. Acta* **139,** 193–195.

Szabo M., Nakawatase C., Kovathana N., and Frohman L. A. (1977) Effect of the dopa decarboxylase inhibitor MK486 on L-dopa-induced inhibition of prolactin secretion. Evidence for CNS participation in the L-dopa effects. *Neuroendocrinology* **24,** 24–34.

Tabakoff B. and Black R. F. (1980) A high performance liquid chromatography method for measuring brain DOPA levels and dopamine synthesis rates. *J. Neurochem.* **34,** 1707–1711.

Tanaka C. (1972) Low Activity of DOPA Decarboxylase in Cerebral Vessels of Spontaneously Hypertensive Rats, in *Spontaneous Hypertension: Its Pathogenesis and Complications* (Okamoto K., ed.), pp. 62–63, Springer-Verlag, Berlin.

Taylor K. M. and Snyder S. H. (1972) Isotopic microassay of histamine, histidine decarboxylase and histamine methyltransferase in brain tissue. *J. Neurochem.* **19,** 1343–1358.

Tran V. T. and Snyder S. H. (1981) Histidine decarboxylase. *J. Biol. Chem.* **56,** 680–686.

Udenfriend S., Clark C. T., and Titus E. (1953) 5-Hydroxytryptophan decarboxylase: A new route of metabolism of tryptophan. *J. Amer. Chem. Soc.* **75,** 501–502.

Van Huysse J. W., Bowsher R. R., Henry D. P., and Willis L. R. (1984b) In vivo renal synthesis of *para*-tyramine in the rabbit. *Fed. Proc.* **43,** 452.

Van Huysse J. W., Henry D. P., and Willis L. R. (1983a) Amine profile of rabbit and human urine. *Fed. Proc.* **42,** 1132.

Van Huysse J. W., Henry D. P., and Willis L. R. (1983b) In vivo synthesis *para*-tyramine by the mammalian kidney. *Clin. Res.* **31,** 752A.

Van Huysse J. W., Henry D. P., and Willis L. R. (1984a) Renal tubular secretions of *para*-tyramine and norepinephrine in the rabbit. *Pharmacologist* **26,** 132.

Vogel W. H., McFarland H., and Prince L. N. (1970) Decarboxylation of dopa in various human adult and fetal tissues. *Biochem. Pharmacol.* **19,** 618–620.

Wade L. A. and Katzman R. (1975) Rat brain regional uptake and decarboxylation of L-dopa following carotid injection. *Am. J. Physiol.* **228,** 352–359.

Wang F. and Perlman R. L. (1981) Activity of AADC in pheochromocytoma cells. *Biochem. Pharmacol.* **30,** 3355–3359.

Waton N. G. (1956a) The effect of organic solvents on mammalian histidine decarboxylase. *Biochem. J.* **64,** 318–322.

Waton N. G. (1956b) Studies on mammalian histidine decarboxylase. *Br. J. Pharmacol.* **11,** 119–127.

Weiner R. I. and Ganong W. F. (1978) Role of monoamines and histamine in regulation of anterior pituitary secretion. *Physiol. Rev.* **58,** 905–976.

Weissbach H., Lovenberg W., and Udenfriend S. (1960) Enzymatic decarboxylation of α-methyl amino acids. *Biochem. Biophys. Res. Commun.* **3,** 225–227.

Weissbach H., Lovenberg W., and Udenfriend S. (1961) Characteristics of mammalian histidine decarboxylating enzymes. *Biochim. Biophys. Acta* **50,** 177–179.

Westermann E., Balzer H., and Knell J. (1958) Inhibition of serotonin synthesis by α-methyl-dopa. *Arch. Exp. Path. Pharmakol.* **234,** 194–205.

Wood W. I. (1976) Tables for the preparation of ammonium sulfate solutions. *Anal. Biochem.* **73,** 250–257.

Wooten G. F. and Horne M. K. (1982) A new autoradiographic approach for imaging forebrain dopamine distribution. *Ann. Neurol.* **12,** 163–168.

Wurtman R. J. and Watkins C. J. (1977) Suppression of noradrenaline synthesis in sympathetic nerves by carbidopa, an inhibitor of peripheral dopa decarboxylase. *Nature* **265,** 79–80.

Yuwiler A., Geller E., and Eiduson S. (1959) Studies on 5-hydroxytryptophan decarboxylase. I. In vitro inhibition and substrate interaction. *Arch. Biochem. Biophys.* **80,** 162–173.

Yuwiler A., Geller E., and Eiduson S. (1960) Studies on 5-hydroxytryptophan decarboxylase. II. Additional inhibition studies and suggestions on the nature of the enzymic site. *Arch. Biochem. Biophys.* **89,** 143–147.

Zavisca F. G., Breau A. P., and Wurtman R. J. (1979) Mechanism of action of methyldopa in the rat: role of 3-*O*-methylated metabolites. *Circ. Res.* **45,** 684–690.

Chapter 3
Dopamine β-Hydroxylase

Toshiharu Nagatsu

1. Introduction

Dopamine β-hydroxylase [DBH; dopamine β-monooxygenase; EC 1.14.17.1 3,4-dihydroxyphenylethylamine; ascorbate: oxygen oxidoreductase (hydroxylation)] is a copper-containing monooxygenase that catalyzes hydroxylation of a phenylethylamine (mainly dopamine in vivo) at the β-C of the side chain to a phenylethanolamine [noradrenaline (NA) from dopamine], using molecular oxygen and ascorbic acid as an electron donor. DBH localizes in noradrenergic and adrenergic neurons of the brain, in noradrenergic neurons of peripheral sympathetic nerves, and in NA and adrenaline (A) cells of the adrenal medulla, and synthesizes NA and A, which are important neurotransmitters and hormones (Nagatsu, 1973) (Fig. 1).

DBH was first partially purified and characterized from the bovine adrenal medulla in 1960 (Levin et al., 1960) and was homogeneously purified and characterized as a copper-containing monooxygenase in 1965 (Friedman and Kaufman, 1965).

DBH is localized in catecholamine storage vesicles in nerve endings of noradrenergic and adrenergic neurons and in adrenal medulla chromaffin cells, and is secreted, together with NA or A, by a process of exocytosis to appear, finally, in the cerebrospinal fluid (CSF) from the brain and in the serum from the periphreal sympathetic nerves and the adrenal medulla (Axelrod, 1972; Nagatsu, 1977; Kopin, 1978; Weinshilboum, 1979). DBH is considered to be a biochemical index related to the activity of catecholamine-containing cells. Many assay methods have therefore been developed for its detection.

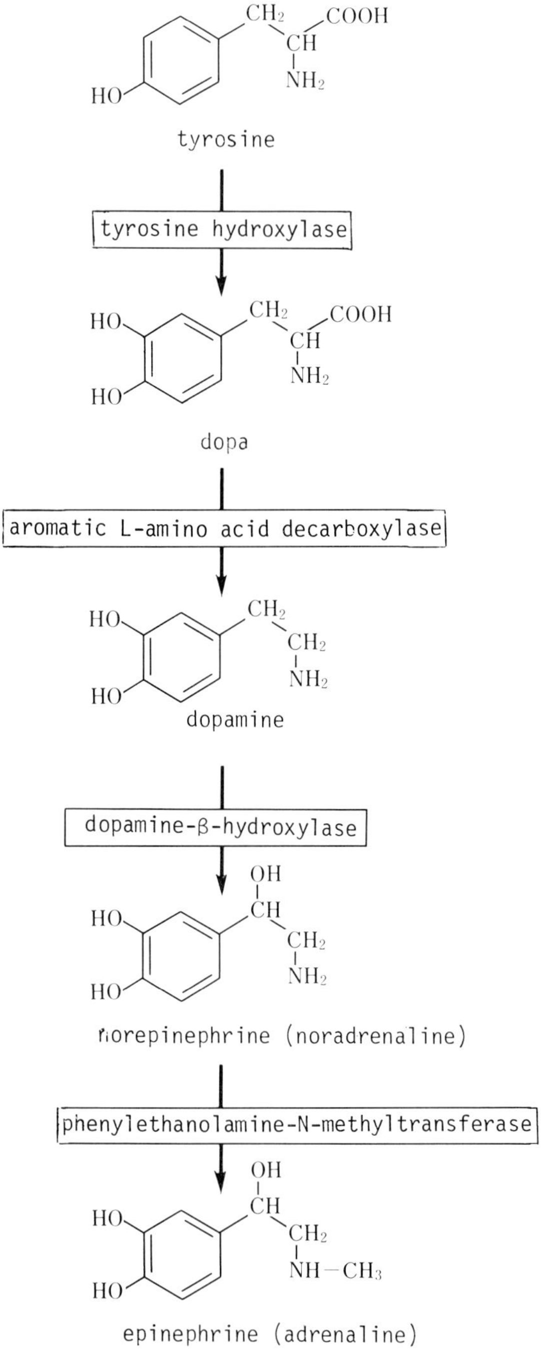

Fig. 1. Biosynthesis of catecholamines.

2. The Reaction

DBH catalyzes the following reaction (Fig. 2):

dopamine + ascorbic acid + $O_2 \rightarrow$ noradrenaline + dehydroascorbic acid + H_2O

The enzyme requires two substrates, dopamine (DA) and molecular oxygen, and an electron donor, ascorbic acid. Dopamine and ascorbic acid are considered to be required in vivo, but they are not specific substrates in vitro. Any phenylethylamine can act as a substrate to form corresponding phenylethanolamines: DA (to NA); phenylethylamine (to phenylethanolamine); tyramine (TA) [to octopamine (OA)]; and epinine (to A). Ascorbate is not specific, and other reducing agents, such as ferrocyanide, can act as electron donors in vitro. Ascorbic acid is considered to be a natural electron donor, because it is most efficient (Levin et al., 1960) and is found in high concentration in chromaffin granules of the adrenal medulla (Terland and Flatmark, 1975).

DBH is activated by various anions, such as fumarate, which stimulates the reaction by 30% at $10^{-2}M$. The binding of anions to a basic group adjacent to the imidazole group at the active site is proposed to explain the mechanism of activation (Levin et al., 1960;

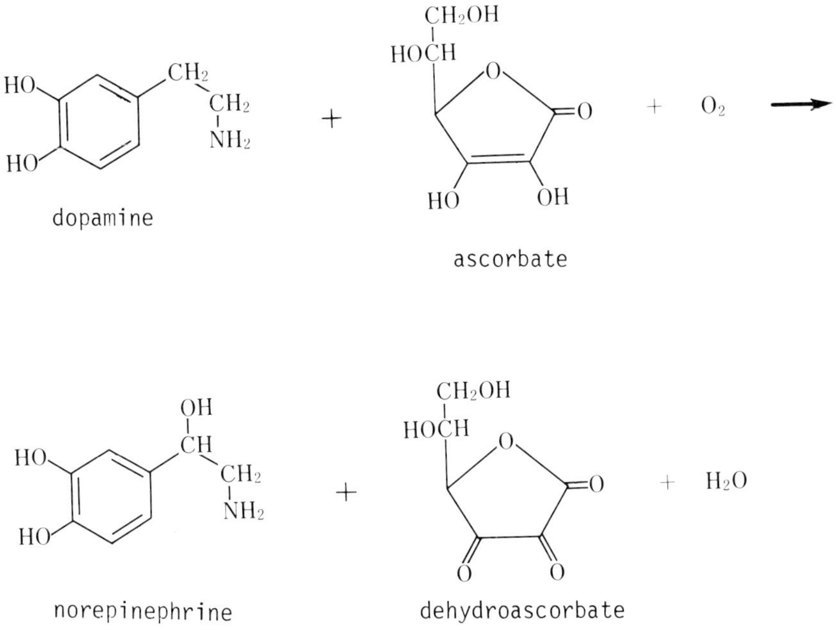

Fig. 2. The reaction of dopamine β-hydroxylase.

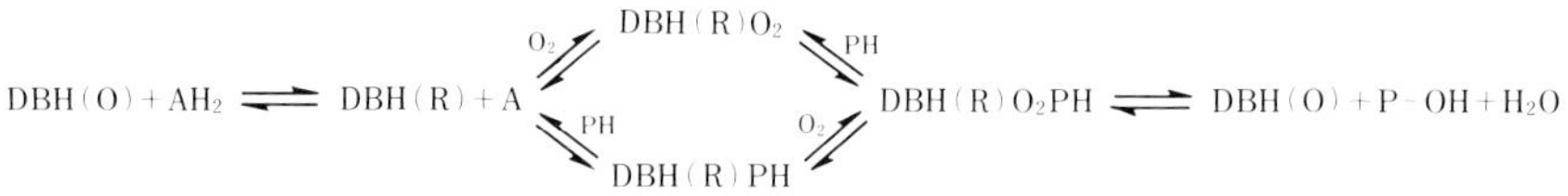

Fig. 3. Kinetics of the dopamine β-hydroxylase reaction.

Craine et al., 1973). The K_m for ascorbic acid is 1 m*M* in the presence of 10 m*M* fumarate and in air (about 0.22 m*M* oxygen), and is independent of pH. The K_m for TA is dependent on pH and is about 1 m*M* at pH 5.0 (Aunis et al., 1974).

The reaction with DBH is stereospecific, and the natural form of D – or R – NA is formed from DA (Levin et al., 1960; Battersby et al., 1976).

A uni-uni-random-bi-uni ping-pong mechanism is proposed for the DBH reaction (Friedman and Kaufman, 1965; Goldstein et al., 1968; Long et al., 1981). As shown in Fig. 3, ascorbate first reacts with oxidized DBH to form reduced DBH and dehydroascorbate, and oxygen or a phenylethylamine substrate then reacts with the reduced DBH to produce a phenylethanolamine and H_2O. Oxygen and phenylethylamine are assumed to form a ternary complex with the reduced DBH.

3. Occurrence and Distribution

DBH is localized in NA- or A-containing cells in central or peripheral neurons and in the adrenal medulla. In the brain, the locus ceruleus, with its noradrenergic cell bodies, and the hypothalamus, with rich noradrenergic nerve terminals, have relatively high DBH activity, and the distribution is parallel with the distribution of NA. In the peripheral tissues, DBH is high in the adrenal medulla and sympathetically innervated tissues, such as the vas deferens, heart, and spleen. The adrenal medulla shows the highest activity; the activity in the sympathetically innervated tissues and in the brain is relatively low. In any tissue, DBH activity can not be measured in crude tissue preparations because of the presence of potent endogenous inhibitors, and DBH activity appears only after partial purification of the enzyme (Nagatsu et al., 1967). The natural inhibitors are mostly sulfhydryl compounds, such as glutathione and cysteine, and DBH activity in the crude enzyme preparations appears by adding *N*-ethylmaleimide or Cu^{2+}, which remove the inhibition by endogenous sulfhydryl compounds (Nagatsu et al., 1967; Duch et al., 1968; Kuzuya and Nagatsu, 1969). DBH also exists in the serum and CSF (Nagatsu, 1977). Man has the highest serum DBH activity among various animals (Axelrod, 1972;

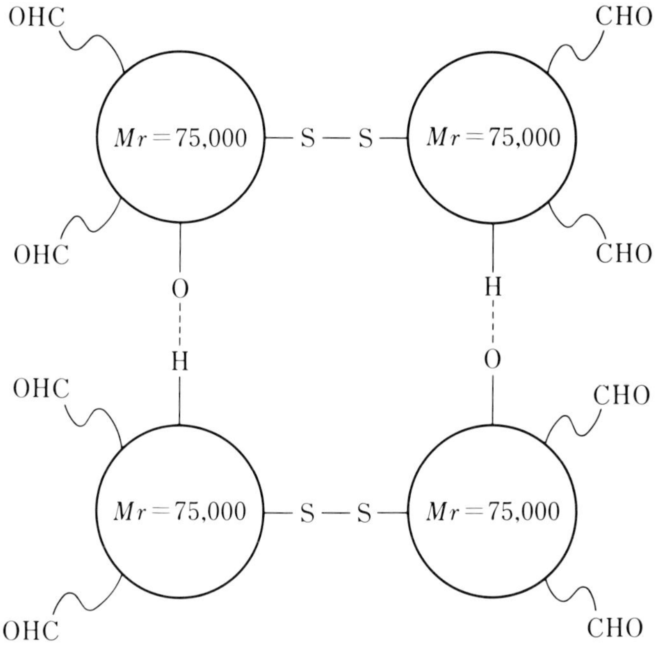

Fig. 4. A schematic model of dopamine β-hydroxylase.

Weinshilboum, 1979). Activity of the enzyme in the CSF is very low; approximately 1/400 of that in the serum (Fujita et al., 1982; Matsui et al., 1981).

4. Enzyme Properties

DBH is a tetrameric enzyme consisting of four identical subunits of 75 kdaltons. As shown in a schematic model in Fig. 4 (Friedman and Kaufman, 1965; Craine et al., 1973; Wallace et al., 1973), two dimers joined by disulfide bonds form the tetramer by hydrogen bonding. The molecular weight of the tetramer is 290,000. The model contains eight carbohydrate side chains.

In human serum, DBH exists primarily as the tetramer, but a small portion exists as the dimer, which has a low but significant activity (Ikeno et al., 1977).

DBH in the soluble matrix of bovine adrenal medullary chromaffin granules contains biantennary oligosaccharides and high-mannose oligosaccharides in a molar ratio of approximately 2:1 (Margolis et al., 1984). It would appear that the four DBH subunits

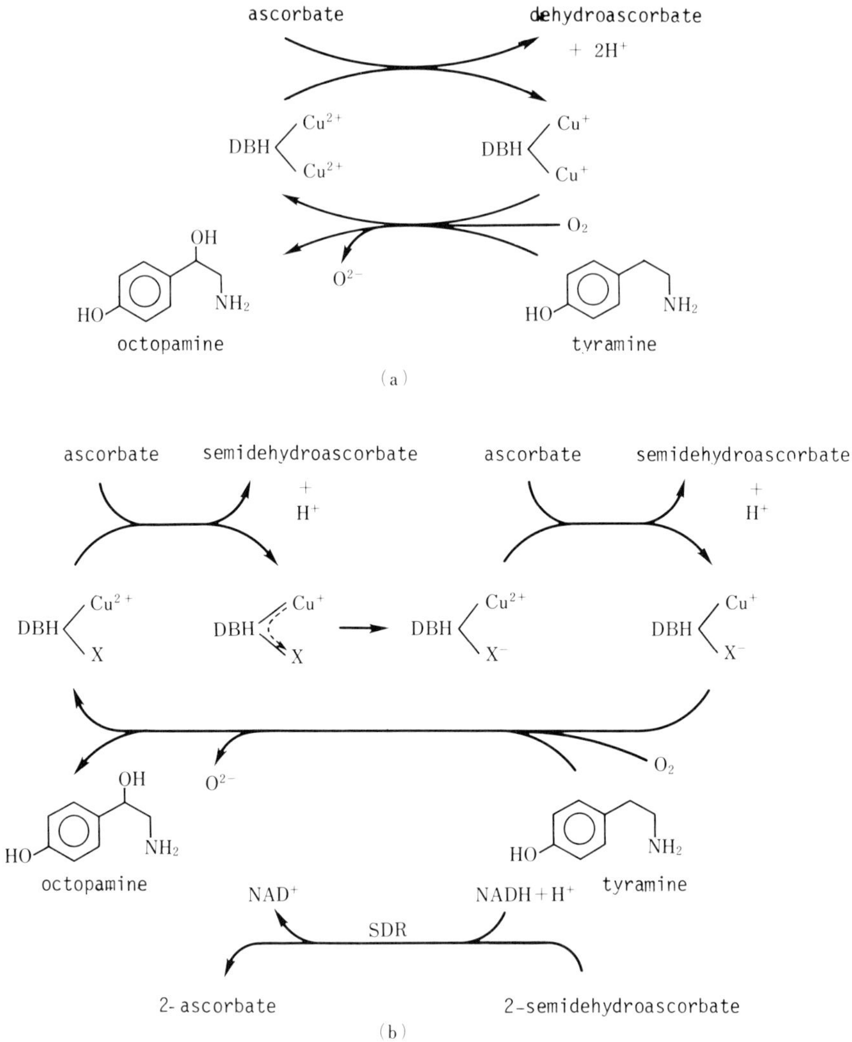

Fig. 5. Mechanisms of action of dopamine β-hydroxylase proposed (a) by Friedman and Kaufman (1965) and (b) by Diliberto and Allen (1981).

of M_r = 75 kdaltons are not identical with respect to their oligosaccharide moieties.

The copper content of DBH is not well established. Four atoms per tetramer are frequently found in pure DBH, but recent studies indicate a model consisting of four identical active subunits, each containing two copper atoms. DBH may specifically bind 8 mol of copper per tetramer and this stoichiometry is required for maximal activity (Ash et al., 1984; Klinman et al., 1984).

Ascorbate may function in enzymatic oxidation as either a one- or two-electron donor (Fig. 5). In two-electron reduction [Fig. 5(a)]

(Friedman and Kaufman, 1965), each ascorbate ion transfers two electrons to the enzyme, giving dehydroascorbate as the product of enzyme catalysis. In one-electron reduction [Fig. 5(b)], each ascorbate ion transfers one electron to the enzyme, and semidehydroascorbate, the unstable free radical, is the product of enzyme catalysis (Skotland and Ljones, 1980; Diliberto and Allen, 1981). In the model proposed by Diliberto and Allen (1981) [Fig. 5(b)], the electron-accepting group in the enzyme is bound Cu^{2+}, and the addition of a second electron follows an intramolecular electron shift. Reoxidation of reduced enzyme with two electrons occurs through an interaction with molecular oxygen and substrate [*see* Fig. 5(a)]. Semidehydroascorbate rapidly dissimilates to one ascorbate and one dehydroascorbate. However, Diliberto and Allen (1981) proposed that dissimilation may not be the major pathway for disappearance of semidehydroascorbate and that semidehydroascorbate dehydrogenase may be physiologically important in the maintenance of active cofactor for β-hydroxylation, possibly through a direct transfer of ascorbate from the mitochondria, where semidehydroascorbate reductase localizes.

DBH is a transmembrane protein that has its active site exclusively on the intragranular face, with a unique portion of its polypeptide chain exposed on the cytoplasmic face of the intact chromaffin granule (Zaremba and Hogue-Angeletti, 1981). Since the active site of DBH in the chromaffin granule is located inside the granule, a mechanism must exist for transferring electrons across the membrane from the cytoplasmic surface to the matrix surface. Cytochrome b-561 is proposed as the likely candidate (Grouselle and Phillips, 1982; Njus et al., 1983).

DBH is present in the bovine adrenal medulla in soluble and membrane-bound forms. Previous isolation procedures for the membranous DBH have resulted in a form of enzyme identical in subunit structure with the soluble form, i.e., tetramer of identical $M_r = 75$ kdaltons subunits. However, recent reports indicate that a membrane-bound form of DBH is structurally different from the soluble form. The isolated membranous form has a large apparent molecular weight on gel filtration, is amphiphilic, and contains bound phospholipid that is predominantly phosphatidylserine. Polyacrylamide gel electrophoresis in the presence of sodium dodecyl sulfate shows that the membranous DBH contains two nonidentical subunits under both reducing and nonreducing conditions. Under reducing conditions, the apparent molecular weights of the two subunits are 70,000 and 75,000 and both contain carbohydrate (Saxena and Fleming, 1983). In the PC12 pheochromocytoma cell line, both soluble and membrane-bound forms of

DBH are synthesized and core glycosylated in the endoplasmic reticulum. DBH is present as two subunit forms (M_r = 77 kdaltons and 73 kdaltons); the M_r = 77 kdaltons form is initially synthesized, and a portion of it is converted to the M_r = 73 kdaltons form. There is probably a precursor–product relationship between the M_r = 77 kdaltons and the M_r = 73 kdaltons subunit forms of DBH (Sabban et al., 1983).

5. Specific Enzyme Inhibitors

Many enzyme inhibitors have been reported, as shown in Table 1. Some of the inhibitors can block NA biosynthesis in vivo, have potent pharmacological effects, such as hypotension and sedation, and are useful chemical tools in the study of catecholamine metabolism.

Substrate analogs, such as benzyloxyamine or benzylhydrazine, are competitive inhibitors with the phenylethylamine substrate (Creveling et al., 1962). Benzyl cyanide analogs are substrates and inhibitors, inactivate DBH during catalysis in the presence of ascorbate or ferrocyanide, and thus, are suicide inactivators (Colombo et al., 1984a,b,c).

All the copper-chelating agents, such as diethyldithiocarbamate (Aunis et al., 1973a), alkylthiourea (Johnson et al., 1969), and troopolon (Goldstein et al., 1964) inhibit DBH. Disulfiram is reduced to diethyldithiocarbamate in vivo and inhibits DBH (Goldstein et al., 1964).

SH-containing compounds, such as cysteine, glutathione, and coenzyme A, inhibit DBH as a result of their chelating action with copper (Nagatsu et al., 1967). Tissues contain potent endogenous inhibitors. Most of the natural inhibitors are sulfhydryl compounds, and the inhibition can be reversed by such SH reagents as *N*-ethylmaleimide and copper (Nagatsu et al., 1967; Duch et al., 1968). These SH reagents are used to measure DBH activity in crude tissue preparations.

Ascorbate-2-sulfate inhibits DBH in competition with ascorbate (Hatanaka et al., 1975).

Umezawa et al. (1972) screened DBH inhibitors of microbial origin, and discovered new natural DBH inhibitors. Fusaric acid (5-butylpicolinic acid) (Hidaka et al., 1969; Nagatsu et al., 1970) is one of the most potent and specific inhibitors and inhibits DBH by 50% at $10^{-7}M$. It has a hypotensive activity (Hidaka et al., 1969). All the 5-alkyl-picolinic acids inhibit DBH; 5-butyl- (fusaric acid) and 5-pentyl-picolinic acid are most potent and have the strongest

Table 1
Inhibitors of Dopamine β-Hydroxylase

Inhibitors	Concentration for 50% inhibition, IC_{50}, *M*	Reference
Aquayamycin	4×10^{-7}	Nagatsu et al. (1968)
Ascorbate-2-sulfate	10^{-3}	Hatanaka et al. (1975)
A-32390 [1,6-di-*O*-(2-isocyano-3-methylcrotonyl)-*D*-mannitol]	10^{-6}	Fuller et al. (1976)
Benzyl cyanide		Colombo et al. (1984a,b,c)
Benzylhydrazine	10^{-5}	Creveling et al. (1962)
Benzyloxyamine	10^{-5}	van der Schoot et al. (1963)
Bleomycin (A_2)	8×10^{-8}	Matsui et al. (1980) Ravatte and Douglas (1983)
BRL-8242 [2-(2-benzylimidazoyl)-amino-2-imidazoline dihydrochloride]	5×10^{-6}	Claxton et al. (1976)
Carbon disulphide	10^{-4}	Catignani and Neal (1975)
Carbon monoxide		Friedman and Kaufman (1965)
Carnosine	2×10^{-3}	Izumi et al. (1975)
Coenzyme A	10^{-4}	Nagatsu et al. (1967)
Cyanide	10^{-6}	Green (1964)

(continued on next page)

Table 1
Inhibitors of Dopamine β-Hydroxylase (continued)

Inhibitors	Concentration for 50% inhibition, IC_{50}, M	Reference
Cysteine	10^{-4}	Nagatsu et al. (1967)
Diethyldithiocarbamate	10^{-6}	Aunis et al. (1973a)
α,α-Dipyridyl	10^{-5}	Green (1964)
Disulfiram		Goldstein et al. (1964)
Dopastin	5×10^{-6}	Iinuma et al. (1972) Iinuma et al. (1974)
Echinochrome A	10^{-5}	Kuzuya et al. (1973)
EDTA	5×10^{-5}	Green (1964) Goldstein et al. (1964)
FLA-63 [bis-(4-methyl)-homopiperazinylthiocarbonyl disulphide]	5×10^{-5}	Florvall and Corrodi (1970)
Glutathione	10^{-5}	Nagatsu et al. (1967)
Histidine	6×10^{-6}	Izumi et al. (1975a,b)
Hydrazine	10^{-5}	Liu et al. (1974)
Hydroquinone	10^{-6}	Aunis et al. (1973a)
6-Hydroxydopamine	10^{-6}	Aunis et al. (1973b)

Mercaptoethylguanidine	6×10^{-5}	Diliberto et al. (1973)
Mimosine	2×10^{-4}	Hashiguchi and Takahashi (1977)
(1-alkyl-2-)mercaptoimidazole		
1-isobutyl	8×10^{-6}	Hanlon and Shuman (1973)
1-isopentyl	2×10^{-6}	Fuller et al. (1976)
o-Phenanthroline	10^{-5}	Goldstein et al. (1964)
1-Phenyl-1-aminomethylethene (suicide inhibition)	10^{-2}	May et al. (1983)
(5-alkyl)Picolinic acid		Hidaka et al. (1969)
5-butyl(fusaric acid)	10^{-7}	Hidaka (1973)
5-pentyl	10^{-8}	Nagatsu et al. (1970)
5-halobutyl	10^{-9}	Nagatsu et al. (1973)
		Suda et al. (1969)
Prazosine	10^{-3}	Frigon et al. (1978)
Propylthiouracil	10^{-4}	Catignani and Neal (1975)
Spinochrome A	10^{-5}	Kuzuya et al. (1973)
(Aromatic and alkyl)Thioureas	10^{-5} - 10^{-6}	Johnson et al. (1969)
Thiazoline-2-carboxylate		Naber et al. (1982)
Tropolon	10^{-5}	Goldstein et al. (1964)

hypotensive action (Hidaka and Takeya, 1972). Fusaric acid is a copper chelator, but its inhibition may not only be a result of its chelating action; fusaric acid inhibits DBH competitively with ascorbate and uncompetitively with TA, whereas copper-chelating inhibitors inhibit DBH noncompetitively with ascorbate and TA. Copper chelating agents may combine with both Cu(I) and Cu(II) enzymes, whereas fusaric acid may combine only with Cu(II) enzyme. Fusaric acid inhibits DBH in vivo, decreasing NA in tissues (Nagatsu et al., 1970).

Hydrogen peroxide (H_2O_2) inhibits DBH. Since the electron donor, ascorbate, produces H_2O_2 by autooxidation, catalase must be included in the DBH reaction mixture (Levin et al., 1960).

6. Clinical Significance of DBH in Relation to Nervous and Mental Diseases

Noradrenaline is a neurotransmitter in the central and peripheral noradrenergic neurons that regulates many important functions, such as blood pressure, sleep, emotion, and memory. DBH catalyzes the formation of NA from DA, and is therefore closely connected to various nervous and mental diseases.

Blood levels of DBH are elevated in patients with pheochromocytoma and decline after removal of the tumor. The role of NA in essential hypertension has been speculated on for many years, but DBH levels in serum seems to be normal in hypertensive patients. In amyloid neuropathy, DBH and NA in the peripheral sympathetic nerve disappear, and thus the patients suffer from severe orthostatic hypotension (Suzuki et al., 1979). In Parkinson's disease, the main biochemical changes are decreases of tyrosine hydroxylase and the cofactor tetrahydrobiopterin in the nigrostriatal dopaminergic regions, but DBH in central noradrenergic neurons and in CSF also decreases (Nagatsu et al., 1977, 1979, 1981, 1982, 1983). Significance of NA and/or serotonin in manic depressive diseases has also been speculated. Dopaminergic overactivity is the main hypothesis in schizophrenia, but involvement of NA has also been proposed (Hornykiewicz, 1982). DBH has been reported to be decreased in the serum of schizophrenic patients (Fujita et al., 1978), although there have also been several reports that failed to show any such significant decrease. This may be because of the type of schizophrenic patients studied and partly because of the assay methods. We have proposed that serum DBH may be decreased in type II (Crow, 1980) schizophrenics with negative symptoms.

7. Assay of DBH

Because of the physiological and clinical importance of DBH, many assay methods for the enzyme have been reported (Nagatsu, 1983). DBH is assayed primarily on the basis of enzyme activity, but the enzyme protein can also be measured by radioimmunoassay, which is not influenced by the presence of endogenous inhibitors, as is the activity assay (Rush et al., 1975a,b). A comparison of various methods of assaying DBH is shown in Table 2.

DBH is high in the adrenal medulla and in tissues containing noradrenergic neurons, such as the noradrenergic brain regions (locus ceruleus, hypothalamus, and so on) and peripheral sympathetically innervated tissues (vas deferens, heart, and so on). DBH is released in the nerve terminals of noradrenergic neurons, together with NA, and appears in blood and CSF (Nagatsu, 1977). Analysis of DBH in blood (serum or plasma) and CSF is frequently required in clinical studies.

Homogenates (0.32*M* sucrose) or high speed (10,000–100,000*g* for 10–60 min) supernatants of the homogenate with a solvent containing 0.1% Triton X-100 or Cutscum are used as crude enzyme preparations.

Assaying DBH activity in crude tissue preparations (homogenates or particulate fractions of tissues, serum, plasma, or CSF) is difficult because of the presence of endogenous inhibitors, mainly sulfhydryl compounds, which can be inactivated by SH-blocking agents, such as *N*-ethylmaleimide (Nagatsu et al., 1967), *p*-chloromercuribenzoate (Duch et al., 1968), or copper (Nagatsu et al., 1967; Duch et al., 1968). When copper is used, a high concentration is inhibitory and the optimum copper concentration must first be determined by titration with each enzyme preparation. It is not necessary to do a titration when *N*-ethylmaleimide (30 m*M*) or *N*-ethylmaleimide (30 m*M*) plus a small amount of copper (5 μ*M*) is used in the presence of a large amount of catalase (500 μg/mL) in the incubation mixture. DBH requires ascorbic acid as cofactor, and ascorbic acid produces H_2O_2 by autooxidation. Because H_2O_2 is a potent inhibitor of DBH, catalase must be present in the incubation mixture to protect the enzyme. The inhibition of DBH at high concentrations of copper can also be prevented by catalase. Therefore, in the presence of a large amount of catalase, a higher concentration of copper can be added to inactivate the endogenous inhibitors.

Since DBH activity is very low, except in the adrenal medulla, a highly sensitive assay method is required for the assay of the

Table 2
Assay of Dopamine β-Hydroxylase

Principle	Substrate and Reaction	Sensitivity	References
Photometry	tyramine → octopamine ↓ $NaIO_4$ *p*-hydroxybenzaldehyde	2 nmol	Pisano et al. (1960) Kuzuya and Nagatsu (1969) Nagatsu and Udenfriend (1972)
		500 pmol (dual-wavelength photometry)	Kato et al. (1974; 1978)
Fluorometry	dopamine → noradrenaline	—	Levin et al. (1960)
	phenylethylamine → phenylethanolamine ↓ $NaIO_4$ benzaldehyde ↓ reaction with 1,2-diaminonaphthalene	800 pmol	Ohkura et al. (1977)
Phosphorimetry	tyramine → octopamine ↓ $NaIO_4$ *p*-hydroxybenzadlehyde (in alkaline medium)	6 pmol	Yamaguchi et al. (1980)
Radioassay Two-step method	phenylethylamine → phenylethanolamine (or tyramine → octopamine) PNMT ↓ $[^{14}C]$-SAM or $[^{3}H]$-SAM *N*-($[^{14}C]$- or $[^{3}H]$-methyl)-phenylethanolamine (or octopamine)	10–100 pmol	Molinoff et al. (1971) Weinshilboum and Axelrod (1971) Goldstein et al. (1971)

One-step method	[2-^{14}C]-tyramine → [2-^{14}C]-octopamine ↓ $NaIO_4$ [^{14}C]-*p*-hydroxybenzaldehyde	10–100 pmol	Nagatsu et al. (1973) Wise (1976) Boulier et al. (1977) Frigon et al. (1978)
	[1-^{14}C]-tyramine → [1-^{14}C]-octopamine ↓ $NaIO_4$ $^{14}CO_2$	2 nmol	Joh et al. (1974)
In vitro	[7-^{3}H]-dopamine → [7-^{3}H]-noradrenaline + ^{3}H $^{3}H + H_2O \rightarrow {}^{3}HHO + H^{+}$		Goldstein et al. (1965) Wilcox and Beaven (1977) Katz (1982)
In vivo			Hoeldtke and Kaufman (1978) Hoeldtke and Stetson (1980)
TLC/ fluorescence detection	tyramine → octopamine ↓ dansyl chloride fluorescent product ↓ TLC	50 pmol	Kopun and Heschel (1978)
HPLC/ fluorescence detection	tyramine → octopamine ↓ *o*-phthalaldehyde fluorescent product	20 pmol	Fujita et al. (1977)
	tyramine → octopamine ↓ $NaIO_4$ *p*-hydroxybenzaldehyde ↓ 2,2'-dithiobis(1-aminonaphthalene) fluorescent product	10 pmol	Nohta et al. (1982a,b)

(continued on next page)

Table 2
Assay of Dopamine β-Hydroxylase (continued)

Principle	Substrate and Reaction	Sensitivity	References
	tyramine → octopamine (native fluorescence)	50 pmol	Flatmark et al. (1978)
HPLC/UV method	tyramine → octopamine	20 pmol	Feilchenfeld et al. (1982)
HPLC/ electrochemical detection	dopamine → noradrenaline	10 pmol	Kissinger et al. (1977) Davis and Kissinger (1979) Sperk et al. (1980) Matsui et al. (1981)
GC	tyramine → octopamine	—	Markianos and Nyström (1974)
GC-MS	dopamine → noradrenaline	—	Hashimoto and Miyazaki (1979)
Radioimmunoassay	DBH protein	100 ng	Rush et al. (1975a) Rush et al. (1975b) Ebstein et al. (1973)

enzyme, especially in the brain, peripheral sympathetically innervated tissues, blood, and CSF.

Tyramine is most frequently used as a substrate, and the product, OA, is measured. Dopamine, which is the main substrate for DBH in vivo, is unstable, and the product, NA, is also unstable and difficult to measure in the presence of a large amount of the substrate DA. Tyramine gives the same V_{max} value as DA (Creveling et al., 1962).

Several sensitive radioisotopic methods have been developed to measure the activity. In 1971, a two-step, coupled radioenzymatic method, using phenylethanolamine *N*-methyltransferase (PNMT), was first reported (Molinoff et al., 1971; Goldstein et al., 1971). In this method, phenylethylamine or TA is used as a substrate, and phenylethanolamine or OA, formed by DBH reaction, is converted to the radioactive *N*-methyl derivative in the second incubation with PNMT and radioactive *S*-adenosylmethionine. This method is highly sensitive and as little as 10 pmol of the DBH reaction product can be measured. However, it has some disadvantages. Since high concentrations of the substrate for DBH inhibit PNMT, a high, saturating substrate concentration cannot be used, and the maximum velocity (V_{max}) cannot be obtained. One-step radioisotopic methods have also been developed using [2-^{14}C]-TA (Nagatsu et al., 1973b; Wise, 1976; Bouclier et al., 1977; Frigon et al., 1978a,b), [1-^{14}C]-TA (Joh et al., 1974), and [7-^{3}H]-DA (Wilcox and Beaen, 1977). These one-step isotopic methods are also highly sensitive, but as in the two-step enzyme-isotopic method, a saturating substrate concentration cannot be used, since addition of nonradioactive substrate dilutes the specific radioactivity of the radioactive substrate to such an extent that the sensitivity is no longer sufficient to measure low DBH activity.

A sensitive spectrophotometric method has been developed, using TA as a substrate; the product, OA, is isolated on a microcolumn of Dowex-50-H^+, chemically converted to *p*-hydroxybenzaldehyde with periodate, and assayed spectrophotometrically at 330 nm (Nagtsu and Udenfriend, 1972). The sensitivity is lower than that in radioassays (about 2 nmol), but can be further increased up to 250 pmol by separation of *p*-hydroxybenzaldehyde and by the use of dual-wavelength spectrophotometry (Kato et al., 1974, 1978). The advantage of the spectrophotometric method is that the maximum velocity can be determined under saturating substrate (20 m*M* TA) and cofactor (10 m*M* ascorbic acid) concentrations, and the method is therefore sensitive enough to measure DBH activity in various crude tissues of mammals.

High-pressure liquid chromatography (HPLC) with fluorescence (Fujita et al., 1977; Flatmark et al., 1978), electrochemical detection (Kissinger et al., 1977; Sperk et al., 1980; Matsui et al., 1981), thin layer chromatography (Kopun and Herschel, 1978), or gas chromatography combined with chemical-ionization mass fragmentography (Hashimoto and Miyazaki, 1979) have been applied to the assay of DBH activity.

A fluorometric method uses phenylethylamine as substrate, and the product, phenylethanolamine is then oxidized to benzaldehyde and determined by reaction with 1,2-diaminonaphthalene (Ohkura et al., 1977). A phosphorometric assay of *p*-hydroxybenzaldehyde derived from enzymatically formed OA can give a high sensitivity (limit of detection of OA, 60 pmol) and can measure DBH activity in rat serum (Yamaguchi et al., 1980).

Among many assay procedures for DBH activity, the most convenient methods may be the spectrophotometric method (Nagatsu and Udenfriend, 1972; Kato et al., 1974, 1978), the coupled enzyme-isotopic method (Molinoff et al., 1971), the [2-^{14}C]-TA method (Nagatsu et al., 1973b), and the HPLC method (Fujita et al., 1977; Matsui et al., 1981).

7.1. Photometric Method

7.1.1. General Principles

Pisano et al. (1960) first described the photometric assay of DBH activity based on the oxidative formation of *p*-hydroxybenzaldehyde from enzymatically formed OA.

$$p\text{-Tyramine} \xrightarrow{\text{DBH}} \text{Octopamine}$$

$$\text{Octopamine} \xrightarrow{\text{Periodate oxidation}} p\text{-Hydroxybenzaldehyde}$$

When the inhibitory effect of endogenous tissue inhibitors is removed by *N*-ethylmaleimide (Nagatsu et al., 1967), the photometric assay is sensitive enough to measure DBH activity in tissues with relatively high activity, such as adrenal glands, vas deferens, human serum (plasma), or hypothalamus of the brain (Kuzuya and Nagatsu, 1969; Nagatsu and Udenfriend, 1972). The sensitivity of the assay is further increased by the use of dual-wavelength spectrophotometry (Kato et al., 1974, 1978).

7.1.2. Procedures

7.1.2.1. Simple Photometric Method. This method (Nagatsu and Udenfriend, 1972) is less sensitive (limit of sensitivity, 2 nmol of OA formed), but very simple and inexpensive, and can be carried out in any laboratory.

The standard incubation mixture (total volume = 1.0 mL, final concentrations in parentheses) contains: an enzyme preparation (in case of human serum or plasma, 2–50 μL) and water to dilute to 500 μL; 0.2*M* freshly prepared ascorbic acid, 50 μL (10 m*M*); 20 m*M* pargyline HCl (a monoamine oxidase inhibitor), 50 μL (1 m*M*); and 0.2*M* freshly prepared *N*-ethylmaleimide, 150 μL (30 m*M*). The various reagents used to make up the final reaction mixture can be premixed into one reagent (500 μL total volume) and added to the enzyme (diluted to 500 μL). As blank, 2 m*M* fusaric acid, 50 μL (0.1 m*M*) is added to the incubation to inhibit DBH, or a sample of boiled (95°C for 5 min) enzyme preparation is added instead of active enzyme. The reaction mixture is incubated at 37°C for 40–60 min in a waterbath with continual shaking; the reaction contents are exposed to air. The incubation is stopped by adding 0.2 mL of 3*M* trichloroacetic acid to the reaction mixture in an icebath, the mixture is immediately centrifuged at 2000 rpm for 10 min, and the supernatant fluid is immediately transferred to a small column of Dowex-50-X4-H^+ (200–400 mesh, packed volume = 0.2 mL) that has been prepared in a disposable Pasteur pipet (0.5 × 10 cm). The tube and precipitate are washed with 1 mL of water, and the washings are also transferred to the column. This column procedure should be carried out rapidly, since the blank value is increased gradually in trichloroacetic acid solutions. The column is washed two more times with 2 mL of water and the adsorbed amines (TA and OA) are eluted with 1.0 mL of 4*M* NH_4OH. Octopamine in the eluate is converted to *p*-hydroxybenzaldehyde by adding 0.10 mL of $NaIO_4$ solution (20 g/L). Then 0.1 mL of $Na_2S_2O_5$ solution (100 g/L) is immediately added to reduce excess periodate. The time between the additions of $NaIO_4$ and $Na_2S_2O_5$ is not critical. Absorbance is measured against water at 330 nm in a microcuvet with a 1-cm lightpath by use of a spectrophotometer. When 20 nmol of OA in 1.0 mL of 4*M* NH_4OH is carried through the oxidation procedure as a standard, an absorbance of 0.43 is obtained. Absorbance is linear, with OA concentration from 5 to 140 nmol; the latter gives an absorbance of about 3.10. The blank (boiled enzyme) absorbance is about 0.05.

The microcolumn of Dowex-50 can be regenerated after the assay by washing with 1 mL of 4*M* NH_4OH, 2 mL of water, and then with 1 mL of 6*M* HC1 and 2 mL of water two more times.

7.1.2.2. Dual-Wavelength Spectrophotometry. The use of dual-wavelength spectrophotometry (Kato et al., 1974, 1978) at 333 nm and 360 nm can increase the assay sensitivity by about 10-fold. The effect of endogenous inhibitors, which are sulfhydryl compounds (Nagatsu et al., 1967), in bovine adrenal medulla homogenate or human serum can be completely removed by the addition of enough (10–30 m*M*) *N*-ethylmaleimide alone. However, in tissues with very low DBH activity, in which a large amount of tissue is required as a source of enzyme for the assay, addition of Cu^{2+} (1–10 μ*M*), together with *N*-ethylmaleimide (30 m*M*), is necessary to completely inactivate the endogenous inhibitors. Addition of Cu^{2+} at concentrations higher than 1 μ*M* inhibits DBH activity, but the inhibition is reversed by an excess amount (500 μg, 25,000 U) of catalase. Addition of 1–10 μ*M* Cu^{2+} plus 30 m*M* *N*-ethylmaleimide in the presence of 500 μg catalase/mL can completely remove the effect of endogenous inhibitors, even when using a large amount (about 30 mg) of any crude tissue as enzyme.

The incubation mixture (total volume = 1.0 mL; the final concentration in parenthesis) is prepared as follows: The enzyme solution is added to a 15-mL centrifuge tube. Water is added to make up 500 μL of an enzyme preparation. Fifty μL of 2 m*M* fusaric acid (100 μ*M*) is included in another enzyme preparation as a blank incubation. Twenty μL of an internal standard solution containing 2.00 nmol of OA is added to another blank incubation mixture. Then 100 μL of 2*M* sodium acetate buffer, pH 5.0 (0.2*M*), 150 μL of 0.2*M* *N*-ethylmaleimide (30 m*M*), 50 μL of 100 μ*M* $CuSO_4$ (5 μ*M*), 25 μL of 20 mg/mL aqueous solution of catalase (25,000 U, 500 μg), and 25 μL of 40 m*M* pargyline HCl (1 m*M*) are added, and the solution is mixed to inactivate the endogenous inhibitors and monoamine oxidase in the enzyme preparation. Then the following reaction mixture is added: 50 μL of 0.2*M* ascorbic acid (10 m*M*), 50 μL of 0.2*M* sodium fumarate (10 m*M*), and 50 μL of 0.4*M* tyramine HCl (20 m*M*). The reaction mixtures are incubated at 37°C for 45 min in air with continual shaking. The incubation is stopped by adding 0.2 mL of 3*M* trichloroacetic acid in an icebath and the mixture is immediately centrifuged at 2500 rpm for 10 min (when the mixture is left at room temperature, the blank value is slowly but continuously increased). The supernatant fluid is immediately transferred to a small glass column (0.5 × 10 cm) of Dowex 50W-X4-H^+ (200–400 mesh, packed volume = 0.2 mL). The tube and precipitate are washed with 1 mL of water, centrifuged again, and the supernatant fluid transferred to the column. The column is washed two more times with 2 mL of water and the adsorbed amines are eluted with 1.0 mL of 3*M* NH_4OH. Octopamine in the eluate is converted to *p*-hydroxybenzaldehyde by adding 10 μL of 20 g/L $NaIO_4$ solution. Excess periodate is then reduced by adding 10 μL of 100 g/L $Na_2S_2O_5$ solution. The solution is then acidified by adding 0.5 mL of 6*M* HCl. The tube is kept in an icebath during this acidification procedure. *p*-Hydroxybenzaldehyde in the solution is extracted into 5.0 mL of ethyl ether by vigorous shaking on a Vortex mixer for 30 s. Four milliliters of the ether phase are transferred to another 15-mL glass-stoppered centrifuge tube, and extracted with 1.0 mL of 3*M* NH_4OH by vigorous shaking on a Vortex mixer for 30 s. The ether layer is removed by aspiration and the NH_4OH layer is incubated at 37°C for 30 min and transferred into a cuvet with 1 cm lightpath. The difference between the absorbances, δA, at wavelengths 333 and 360 nm, is measured with a dual-wavelength double-beam recording spectrophotometer. When a dual-wavelength spectrophotometer is not available, it is also possible to use a regular single-wavelength spectrophotometer. The absorption spectra of the samples from the experimental, blank, and internal standard incubations are recorded from 300 to 360 nm, and $\delta A = A_{333\ nm} - A_{360\ nm}$ is measured on the spectrum.

The absolute amount of OA formed is calculated from the following equation:

$$\text{Octopamine formed (nmol)} = \frac{(\delta A \text{ of experimental}) - (\delta A \text{ of blank})}{(\delta A \text{ of internal standard}) - (\delta A \text{ of blank})} \times 2.00$$

7.2. Two-Step, Coupled Radioenzymatic Method Using Phenylethanolamine N-Methyltransferase (PNMT)

7.2.1. General Principles

Phenylethylamine or TA is used as a substrate, and phenylethanolamine or OA formed by DBH reaction is converted to the radioactive *N*-methyl derivative in the second incubation with PNMT (Molinoff et al., 1971; Weinshilboum and Axelrod, 1971; Goldstein et al., 1971).

$RCH_2CH_2NH_2 + O_2 +$ ascorbic acid $\xrightarrow{DBH}$ $RCHOHCH_2NH_2 + H_2O$ + dehydoascorbic acid

$RCHOHCH_2NH_2$ + S-([^3H]-methyl or [^{14}C]-methyl)adenosyl-methionine $\xrightarrow{PNMT}$

$RCOHCH_2NH$-([^3H]-methyl or [^{14}C]-methyl) + S-adenosyl-homocysteine

The radioactive *N*-methyl derivative is isolated by solvent extraction and measured. Either copper (Molinoff et al., 1971; Weinshilboum and Axelrod, 1971) or *N*-ethylmaleimide (Goldstein et al., 1971) is used to inactivate endogenous inhibitors. Since copper or *N*-ethylmaleimide inhibits the second PNMT reaction, it must be removed by adding cysteine for *N*-ethylmaleimide or EDTA for copper. This method is highly sensitive and as little as 10 pmol of the DBH reaction product can be measured. However, it has one disadvantage in that, since high concentrations of the substrate for DBH, phenylethylamine, or tyramine inhibit PNMT, a saturating substrate concentration cannot be used. Therefore, the maximum velocity (V_{max}) cannot be measured.

7.2.2. Procedure

Tissues are homogenized in 25–400 volume of 0.005*M* Tris buffer, pH 7.5, containing 0.1% Triton X-100 (Molinoff et al., 1971). Homogenates are centrifuged at 10,000*g* for 10 min. Two hundred μL of the supernatant fluid is added to a 15 mL glass-stoppered centrifuge tube containing a reaction mixture of 25 μL of 0.048*M* ascorbic acid, pH 6; 25 μL of 0.5*M* sodium fumarate, pH 6; 20 μL of 0.006*M* pargyline (a monoamine oxidase inhibitor); 1500 U of catalase in 10 μL of 0.005*M* Tris buffer, pH 7.4; 10 μL of 0.03*M* phenylethylamine; 10 μL of 1*M* Tris buffer at pH 6; and 10 μL of $CuSO_4$ in appropriate concentration (final concentration 10–100 μ*M* to give an optimum activity for the tissue being assayed. The optimum concentration of $CuSO_4$ should be examined preliminarily with various amounts of the homogenate. The reaction mixture (total volume 310 μL) is incubated at 37°C for 20 min. The DBH portion of the assay is then stopped, and the PNMT reaction initiated by adding to each tube 100 μL of a mixture of 10 μL of purified PNMT, 10 μL of *S*-([^{3}H]- or [^{14}C]-methyl)adenosylmethionine (1 nmol), and 80 μL of 1*M* Tris buffer, pH 8.6. At pH 8.6, DBH has less than 5% of the activity at pH 6. The PNMT reaction is allowed to proceed for 30 min at 37°C and is then stopped by the addition of borate buffer, pH 10. The *N*-([^{3}H]- or [^{14}C]-methyl)phenylethanolamine is extracted into 6 mL of toluene containing 3% isoamyl alcohol by vigorous shaking on a Vortex mixer for 15 s. After centrifugation at low speed, 4 mL of the organic phase is transferred to a counting vial with 10 mL of phosphor for radioactive determination. Blanks consisting of tissue homogenate heated at 95°C for 5 min are run in all experiments, as is an internal standard consisting of 100 ng of phenylethanolamine added to the entire reaction mixture. The internal standard is used in the calculation of absolute amounts of DBH activity and corrects for any inhibition of PNMT.

When TA is used as the substrate, the assay is carried out in a similar way, except that 10 μL of 0.003*M* TA is used as substrate and 40 ng of OA HCl is used as the internal standard. After the reaction is stopped with borate buffer, the *N*-([^{3}H]- or [^{14}C]-methyl)-OA formed is extracted into a mixture of toluene and isoamyl alcohol, 3:2 (v/v). After centrifugation, 4 mL of the organic phase is transferred to counting vials containing an additional 2 mL of toluene-isoamyl alcohol, 3:2. The samples are then dried in a chromatography oven at 80°C. This drying step is necessary to remove volatile radioactive contaminants that are extracted into the toluene-isoamyl alcohol. After drying, the residue is dissolved in 1 mL of ethanol, and the radioactivity determined after the addition of 10 mL of phosphor. Although the use of TA can increase the sensitivity of the assay, the drawbacks are high blank values and possible interference from endogenous tissue stores of OA.

7.3. One-Step Radioenzymatic Method Using [2-^{14}C]-TA

7.3.1. General Principles

This radioassay (Nagatsu et al., 1973) is based on the enzymatic conversion of [2-^{14}C]-TA to OA, which is then oxidized to *p*-[^{14}C]-hydroxybenzaldehyde. The latter compound is separated from the amine by ether extraction and its radioactivity determined in a liquid scintillation counter.

7.3.2. Reagents

[2-^{14}C]-Tyramine (50 mCi/mmol) (Schwartz-Mann) is purified by paper chromatography (ascending method) using *n*-butanol:acetic acid:water (4:1:2, v/v/v) as solvent. Immediately after chromatography, the position of [2-^{14}C]-TA on the chromatogram is located by autoradiography, using Kodak X-ray film BB-54 and a 3-h exposure. The band of radioactivity in the chromatogram is traced from the film, cut out, and eluted with 0.01*M* HCl. The TA concentration in the eluate is measured fluorometrically in a small aliquot by the fluorescamine method (Udenfriend et al., 1972) and the specific radioactivity determined. The purified TA is evaporated to dryness under nitrogen, dissolved in a small amount of methanol, and stored at –20°C. Since the sensitivity of any radioassay is limited by the blank value, this previous purification of [2-^{14}C]-TA is essential. The lowest blank value obtained after purification by paper chromatography is 130 cpm from 1.0×10^6 cpm of [2-^{14}C]-TA added to an incubation flask. Wise (1976) purified [2-^{14}C]-TA by periodate oxidation and washing with ether.

7.3.3. Procedure

Before assay, an aliquot of the purified [2-^{14}C]-TA solution, sufficient for all the incubation tubes to be set up (1.0×10^6 cpm/incubation tube), is transferred to a small test tube, evaporated to dryness under nitrogen, and dissolved in a small volume of 0.01*M* HCl. This acid solution is washed three times with 5 volumes of ethyl ether (previously washed with 5% aqueous $FeSO_4$ solution and water) using a Vortex mixer, and the aqueous layer evaporated under nitrogen. The [2-^{14}C]-TA is redissolved in the reagent mixture to give a final concentration of 1 m*M*. The reagent mixture contains the following in a final volume of 6 μL: sodium acetate buffer, pH 5.0, 2.0 μmol; disodium fumarate, 0.1 μmol; ascorbic acid, 0.1 μmol; catalase (Sigma), 0.50 μg (25 U); pargyline HCl, 0.01 μmol; *N*-ethylmaleimide, 0.3 μmol; and $CuSO_4$, 0.04 μmol. Four microliters of enzyme solution (for example, 40-fold diluted human serum or twofold diluted rat serum) is added to 6 μL of the reagent mixture to give a final volume of 10 μL. These additions are pipeted into 10 × 75 mm tubes in an icebath, each tube is sealed with parafilm, and the mixture is collected at the bottom by a brief centrifugation in the cold. Incubation is then carried out at 37°C for 45 min in air. A sample of a heated enzyme preparation (95°C for 5 min) serves as blank. After incubation, the mixture is immediately placed in an icebath and 0.10 mL of 10 m*M* OA is added as carrier (1.0 μmol), followed by 0.90 mL of water and 0.10 mL of 3*M* trichloroacetic acid. After mixing and centrifuging at 2000 rpm for 10 min, the supernatant fluid is immediately transferred to a small column of Dowex-50-H^+ (200–400 mesh, packed volumes = 0.2 mL) that has been prepared in a disposable Pasteur pipet. The test tube and the precipitate are washed with 1 mL of water, and the washings are also transferred to the column. After the solution has passed through, the column is washed three more times with 2 mL of water, and the adsorbed amines eluted with 1.0 mL of 4*M* NH_4OH into a 20 mL glass-stoppered centrifuge tube. The tube is then placed in an icebath and the OA in the eluate converted to *p*-hydroxybenzaldehyde by addition of 50 μL of 2% $NaIO_4$. Excess $NaIO_4$ is destroyed by the addition of 50 μL of 10% $Na_2S_2O_5$. The time between the addition of $NaIO_4$ and $Na_2S_2O_5$ does not affect the assay values over the range of 0.5–4 min. After this, 1.0 mL of 5*M* HCl and 10 mL of ethyl ether (previously washed with 5% aqueous $FeSO_4$ and water) are added, and the *p*-hydroxybenzaldehyde is extracted into the ether by mixing for 15 s on a Vortex mixer. An 8.0 mL aliquot of the ether layer is transferred to a counting vial and evaporated to dryness under nitrogen. The residue is dissolved in 0.80 mL of water. Fifty microliters of the solution are transferred to a test tube containing 1.05 mL of 4*M* NH_4OH, to determine the recovery of the carrier OA. Fifty nanomoles of OA in 1.0 mL of 4*M* NH_4OH, oxidized by adding 50 μL of 2% $NaIO_4$ followed by 50 μL of 10% $Na_2S_2O_5$, serve as a standard.

The mean recovery of carrier OA is 80%. Because the recoveries are relatively constant, it is not necessary to determine the recovery for each sample. To the remaining 0.75 mL of the solution in the counting vial, 10 mL of a scintillation cocktail is added, and the radioactivity measured in a liquid scintillation counter.

When an enzyme preparation with relatively high DBH activity, such as human serum is used, the reaction can be carried out at saturating concentrations of TA (10 m*M*). However, addition of nonradioactive TA dilutes the specific radioactivity of the radioactive TA to an extent that the sensitivity is no longer sufficient to measure low enzyme activity, such as that in rat serum. It is necessary, therefore, to dilute [^{14}C]-TA to 1 m*M*. Values must, therefore, be compared with standards prepared by using identical concentrations of substrate.

7.4. HPLC Method With Fluorescence Detection

7.4.1. General Principles

Tyramine is used as a substrate and is incubated under optimal conditions (Fujita et al., 1977, 1978, 1982). Endogenous inhibitors in the enzyme preparation are completely inactivated by addition of *N*-ethylmaleimide and a small amount of Cu^{2+}. After isolating the amines on a small Dowex-50 column, the eluate is analyzed by HPLC with fluorescence detection, using *o*-phthaladlehyde as a fluorescence reagent. The product, OA, is eluted and completely separated from TA, and the height of the fluorescence peak gives the amount of enzymatically formed OA. This method is highly sensitive (limit of sensitivity, 30 pmol of enzymatically formed OA), and it is possible to measure DBH activity in 0.25 μL of human serum (Fujita et al., 1977) and 500 μL of CSF (Fujita et al., 1977; 1978; 1982).

7.4.2. Procedures

The incubation mixture (total volume = 1.0 mL, final concentration in parenthesis) for the DBH assay contains in a 15 mL centrifuge tube: 500 μL of diluted enzyme solution; 100 μL of 2*M* sodium acetate buffer, pH 5.0 (0.2*M*); 150 μL of 0.02*M* *N*-ethylmaleimide (30 m*M*); 50 μL of 100 μ*M* $CuSO_4$ (5 μ*M*); 25 μL of 20 mg/mL aqueous solution of catalase (25,000 U, 500 μg); 25 μL of 40 m*M* pargyline HCl (1 m*M*); 50 μL of 0.2*M* ascorbic acid (10 m*M*); 50 μL of 0.2*M* sodium fumarate (10 m*M*); and 50 μL of 0.4*M* TA HCl (20 m*M*). Incubated in the blank is 50 μL of 2 m*M* fusaric acid (0.1 m*M*), a specific DBH inhibitor (Hidaka et al., 1969; Nagatsu et al., 1970). Incubation is carried out at 37°C for 45 min. The reaction is stopped by adding 0.6 mL of ice-cold 3*M* trichloroacetic acid. The mixture is centrifuged immediately at 2500 rpm for 10 min at 4°C. The supernatant solution is transferred to a column (0.9 × 10.5 cm) of Dowex-50W-X8-H^+ (200–400 mesh). The tube and the precipitate are washed with 0.6 mL of 0.5*M* trichloroacetic acid, centrifuged again, and the supernatant fluid transferred to the column. The column is washed four times with 5 mL of 0.1*M* sodium phosphate buffer, pH 8.0, containing 0.05*M* NaCl, and four times with 5 mL of water, and the adsorbed amines eluted into a test tube (12 × 75 mm) with 3.0 mL of 3.5% NH_4OH. The eluate is evaporated on a Vortex evaporator (Buchler Instruments) at 60°C, and the residue dissolved in 200 μL of the buffer for HPLC. The buffer is prepared by dissolving 5.25 g of citric acid, 4.1 g of sodium acetate, 2.9 g of sodium hydroxide, 23.2 g of sodium chloride, and 1.05 mL of acetic acid in 1L of water purified through Milli Q system (Millipore Corp.). The buffer, which has a pH of 5.10 ± 0.02, is passed through a Millipore filter (0.45 ± 0.02 μm, diameter 47 mm, HAWPO 4700) and evacuated for 15 min under vacuum. The HPLC (Waters Associates Inc.) is equipped with two columns of μBondapak C18 (1/4 inch × 1 ft), three M-6000A pumps, two Model U6K injectors, a fluorescence detector (an Aminco-Bowman spectrophotofluorometer with a flow cell of 10 μL, excitation at 340 nm, fluorescence at 455 nm), and a recorder (JEOL Unicorder U-125 M). *o*-Phthalaldehyde solution is prepared as follows. Boric acid (25 g) is dissolved in water to a total volume of 950 mL, the pH adjusted to 10.40 ± 0.01 with a 45% KOH aqueous solution, and the buffer solution filtered through the above-mentioned Millipore filter. To this buffer, 30% Brij 35 aqueous solution (0.9 g dissolved with 3.0 mL of water by mixing for 40 min with a magnetic stirrer) and 2.0 mL of 2-mercaptoethanol are added. To this solution, an *o*-phthalaldehyde solution (500 mg dissolved in 10 mL of methanol, stored at 4°C in the dark) is added. During chromatography, nitrogen gas is bubbled through the *o*-phthalaldehyde solution. The flow rate of the buffer and *o*-phthalaldehyde solution is 1.0 mL/min, and the chart speed is 0.5 cm/min. Two samples can be injected at intervals of 8 min into each injector during a 30-min period. The peak height of OA is measured and converted to nanomoles from the standard curve of OA in the same column. DBH activity is expressed as μmol/min (U) at 37°C.

7.5. HPLC Method With Electrochemical Detection

7.5.1. General Principles

Dopamine is used as a substrate, and NA, enzymatically formed, is separated by HPLC with electrochemical detection. This method was first introduced by Kissinger et al. (1977), and was improved by Sperk et al. (1980) and Matsui et al. (1981). The method used by Matsui is highly sensitive and can measure DBH activity in human CSF.

7.5.2. Procedures

Dopamine is used as a substrate and is incubated under optimal conditions (Matsui et al., 1981). Noradrenaline, formed enzymatically from DA, is isolated by a double-column procedure, the first column of Dowex-50-H^+ and the second column of alumina. Noradrenaline is adsorbed on the second alumina column, eluted with 0.5*M* HCl, and assayed by HPLC with electrochemical detection. Adrenaline is added to each incubation mixture as an internal standard.

The incubation mixture contains (total volume = 1.0 mL, final concentration in parenthesis): 500 μL of a diluted enzyme solution (in the case of CSF, 500 μL), 100 μL of 2*M* sodium acetate buffer, pH 5.0 (0.2*M*), 150 μL of 0.2*M* *N*-ethylmaleimide (30 m*M*), 50 μL of 100 μ*M* $CuSO_4$ (5 μ*M*), 25 μL of 20 mg/mL aqueous solution of catalase (25,000 U, 500 μg), 25 μL of 40 m*M* pargyline HCl (1 m*M*), 50 μL of 0.2*M* ascorbic acid (10 m*M*), 50 μL of 0.2*M* sodium fumarate (10 m*M*), and 50 μL of 0.4*M* DA HCl (20 m*M*). The reaction mixture is preincubated at 37°C for 5 min, and the reaction initiated by addition of DA. Acetate buffer + *N*-ethylmaleimide + $CuSO_4$ (cocktail A), and catalase + pargyline + ascorbic acid + fumarate (cocktail B), can be mixed together; 500 μL of an enzyme preparation, 300 μL of cocktail A and 150 μL of cocktail B, are mixed in this order, and the reaction started by adding 50 μL of the DA solution. The reaction is carried out at 37°C for 45 min in air, and stopped by adding 1 mL of ethanol containing 1 m*M* fusaric acid and 500 pmol of A (internal standard); 100 μL of 10% $Na_2S_2O_5$ and 100 μL of 0.1*M* disodium EDTA are then added. The mixture is left at 0°C for at least 30 min and then centrifuged at 15,000 rpm for 10 min. The supernatant is transferred to a column (4 × 55 mm) of Dowex-50W-X4-H^+ (100–400 mesh). The column is washed with 2 mL of water three times and once with 3 mL of 1*M* HCl. Then NA and A are eluted with 2 mL of 1*M* HCl. Dopamine is retained on the column. The eluate containing NA and A is adjusted to pH 8.4 by addition of 8 mL of 3*M* Tris-HCl buffer, pH 8.6, and transferred to an aluminum oxide column (0.4 cm internal diameter) containing 100 mg of activated alumina (dry powder). The column is washed three times with 4 mL of water and 100 μL of 0.5*M* HCl. Noradrenaline and A (internal standard) are eluted with 200 μL of 0.5*M* HCl. The mean recovery of NA from the Dowex column is 73%, and that from the aluminum oxide column is 63%.

A 10–20-μL aliquot of the eluate from the aluminum oxide column is injected into a high-performance liquid chromatograph (Yanaco L-2000) with a Yanaco VMD-101 electrochemical detector and a Yanapak ODS reversed-phase column (particle size 10 μm, 25 × 0.4 cm id) (Yanaco Ltd., Kyoto, Japan). The mobile phase is 0.1*M* potassium phosphate buffer, pH 3.0, containing pentansulfonic acid (20 mg/100 mL buffer). A flowrate of 0.6 mL/min is used. The detector potential is set at 0.8 V versus the Ag/AgCl electrode. The peak height of NA is measured and converted to pmol from the peak height of A added as an internal standard. The retention times under these conditions are: DA, 12.0 min; NA, 4.4 min; and A, 7.2 min.

8. Purification of Dopamine β-Hydroxylase

DBH was homogeneously purified by Friedman and Kaufman (1965) from the particulate fraction of bovine adrenal glands as follows: solubilization with a nonionic detergent Cutscum, ammonium sulfate fractionation, calcium phosphate gel adsorption, ethanol fractionation, DEAE-cellulose column chromatography, and Sephadex G-200 column chromatography.

Foldes et al. (1972) reported a rapid purification procedure from bovine adrenal medulla. The principle of the purification is based on the isolation of pure chromaffin granules from fresh adrenal medulla using a sucrose density gradient and subsequent solubilization and purification by DEAE-cellulose and Sephadex G-200 chromatographies. Frozen adrenal medulla cannot be used because of damage to chromaffin granules during freezing and thawing.

All purification procedures are carried out at 4°C. Fresh bovine or sheep adrenal glands (about 500 g) are defatted and adrenal medulla dissected out. Fresh adrenal medulla is homogenized in 5 volumes of ice-cold 0.32*M* sucrose, and the homogenate centrifuged at 1000*g* for 15 min. The supernatant is filtered through two layers of cheese cloth and centrifuged at 10,000*g* for 30 min. The precipitate (crude mitochondrial fraction) is suspended in 0.32*M* sucrose (1 mL/2 g of original adrenal medulla). Thirty mL of the crude mitochondrial suspension are layered on 40 mL of 1.6*M* sucrose in a centrifuge tube and centrifuged at 100,000*g* for 60 min in an angle rotor. The pink precipitate on the bottom (chromaffin granules) is carefully separated; the supernatant decanted carefully, the inside of the centrifuge tube above the sediment cleaned, and the light brown loosely adhering layer over the pink sediment of chromaffin granules resuspended by gently shaking the tube with 1.6*M* sucrose, and decanted. The pink chromaffin granule sediment on the bottom is resuspended in 10 m*M* potassium phosphate buffer, pH 7.2, by using 1 mL of buffer for the granules obtained from 10 g of original medulla. The chromaffin granule preparation can be stored frozen.

The granule suspension is thawed and a 10% (v/v) Triton X-100 solution added to a final concentration of 0.1%. The suspension is then stirred for 45 min at 4°C, centrifuged at 1000*g* for 10 min, and the supernatant stored in ice. The sediment is resuspended in a small volume of 10 m*M* phosphate buffer and treated with Triton X-100 once more. The combined supernatants are centrifuged at 100,000*g* for 10 min and the supernatant dialyzed against 3- or 4-h changes of 50 volumes of 10 m*M* phosphate buffer, pH 7.2, containing 10 μ*M* $CuSO_4$, and concentrated by positive pressure dialysis on a Diaflo ultrafilter XM-100 A (Amicon Corp., Lexington, MA). The dialyzed and concentrated supernatant is adsorbed on a DEAE-cellulose column previously equilibrated with 10 m*M* phosphate buffer, pH 7.2 (1.0 × 40 cm). The nonadsorbed protein is washed out with 10 m*M* phosphate buffer until absorbance at 280 nm becomes less than 0.05, and the adsorbed proteins are eluted with a linear salt concentration gradient (0–0.8*M* NaCl in 10 m*M* phosphate buffer, pH 7.2). Each 10-mL fraction is collected. The active fractions are combined and concentrated by positive-pressure dialysis on a Diaflo ultrafilter XM-100 A to less than 4 mL. The enzyme solution is applied to a column of Sephadex G-200 (2.5 × 90 cm) previously equilibrated with 10 m*M* phosphate buffer containing 0.1*M* NaCl, pH 7.2, and the protein eluted with the same buffer. The active fractions containing pure DBH are pooled and concentrated as described above, and stored at 4°C.

Use of a concanavalin A-Sepharose column (Rush et al., 1975a,b), or affinity chromatography using tyramine-Sepharose (Aunis et al., 1974) is also an effective purification procedure. Biogel-A 1.5m (2.5 × 100 cm) may also be used instead of Sephadex G-200 (Hogue-Angeletti, 1977). Human plasma DBH can be purified to homogeneity by Con A-Sepharose, Ocytyl-Sepharose, Bio-Gel A-0.5m, and isoelectric focusing (Frigon and Stone, 1978). Rat pheochromocytoma DBH has been purified by Con A-Sepharose 4B, Sepharose 4B, and sucrose density gradient centrifugation (Fong et al., 1980). Purification of DBH in avian adrenal chromaffin granules by hydroxyapatite and DEAE-Bio-Gel A chromatographies has also been reported (Long et al., 1981).

References

Ash D. E., Papadopoulos N. J., Colombo G., and Villafranca J. J. (1984) Kinetic and spectroscopic studies of the interaction of copper with dopamine β-hydroxylase. *J. Biol. Chem.* **259,** 3395–3398.

Aunis D., Miras-Portugal M. T., and Mandel P. (1973a) Bovine adrenal medullary dopamine β-hydroxylase: Purification by affinity chromatography, kinetic studies, and presence of essential histidyl residues. *Biochim. Biophys. Acta* **327,** 313–327.

Aunis D., Miras-Portugal M. T., and Mandel P. (1973b) Inhibition of dopamine-β-hydroxylase by 6-hydroxydopamine. *Biochem. Pharmacol.* **22,** 2581–2589.

Aunis D., Miras-Portugal M. T., and Mandel P. (1974) pH-Dependent deuterium solvent isotope effects on bovine adrenal medullary dopamine-β-hydroxylase. *Biochem. Biophys. Res. Commun.* **57,** 1192–1199.

Axelrod J. (1972) Dopamine-β-hydroxylase: Regulation of its synthesis and release from nerve terminals. *Pharmacol. Rev.* **24,** 233–243.

Battersby A. R., Sheldrake P. W., Saunton J., and Williams D. C. (1976) Studies of enzyme-mediated reactions. Part VII. Stereospecific synthesis of tritium-labelled (2R)- and (2S)-dopamines: Stereochemical course of hydroxylation of dopamine by dopamine β-hydroxylase (EC1.14.17.1). *J. Chem. Soc. Perkin* **I,** 1056–1062.

Boulier M., Mandel P., and Aunis D. (1977) Microdetermination of dopamine-β-hydroxylase activities using a simplified, single step radioenzymatic assay. *Pharmacol. Res. Commun.* **9,** 743–754.

Catignani G. L. and Neal R. A. (1975) Studies of the inhibition of dopamine-β-hydroxylase by thiono-sulfur-containing compounds. *Life Sci.* **16,** 1915–1922.

Claxton I. M., Palfreyman M. G., Poyser R. H., and Whiting R. L. (1976) BRL-8242 (2-[2-bensimidazolyl]-amino-2-imidazoline dihydrochloride), a new inhibitor of dopamine-β-hydroxylase with antihypertensive activity. *Eur. J. Pharmacol.* **37,** 179–188.

Colombo G., Rajashekhar B., Giedroc D. P., and Villafranca J. J. (1984a) Alternative substrates of dopamine β-hydroxylase. I. Kinetic investigations of benzyl cyanides as substrates and inhibitors. *J. Biol. Chem.* **259,** 1593–1600.

Colombo G., Giedroc D. P., Rajashekhar B., and Villafranca J. J. (1984b) Alternative substrates of dopamine β-hydroxylase. II. Inhibition by benzyl cyanides and reactivation of inhibited enzyme. *J. Biol. Chem.* **259,** 1601–1606.

Colombo G., Rajashekhar B., Ash D. E., and Villafranca J. J. (1984c) III. Stoichiometry of the inactivation reaction with benzyl cyanides and spectroscopic investigations. *J. Biol. Chem.* **259,** 1607–1615.

Craine J. E., Daniels G. H., and Kaufman S. (1973) Dopamine β-hydroxylase. The subunit structure and anion activation of the bovine adrenal enzyme. *J. Biol. Chem.* **248,** 7838–7844.

Creveling C. R., Daly J. W., Witkop B., and Udenfriend S. (1962) Substrates and inhibitors of dopamine-β-hydroxylase. *Biochim. Biophys. Acta* **64,** 125–134.

Crow T. J. (1980) Molecular pathology of schizophrenia: More than one disease process? *Br. Med. J.* 66–68.

Davis G. C. and Kissinger P. T. (1979) Determination of serum dopamine-β-hydroxylase activity by reverse-phase liquid chromatography with column switching. *Anal. Chem.* **51,** 1960–1965.

Diliberto E. J. and Allen P. L. (1981) Mechanism of dopamine-β-hydroxylation. Semidehydroascorbate as the enzymic oxidation of ascorbate. *J. Biol. Chem.* **256,** 3385–3393.

Diliberto E. J., Disefano V., and Smith T. L. (1973) Mechanism and kinetics of the inhibition of dopamine-β-hydroxylase by 2-mercaptoethylguanidine. *Biochem. Pharmacol.* **22,** 2961–2972.

Duch D. S., Viveros O. H., and Kirshner N. (1968) Endogenous inhibitor(s) in adrenal medulla of dopamine-β-hydroxylase. *Biochem. Pharmacol.* **17,** 255–264.

Ebstein R. P., Park D. H., Freedman L. S., Levitz S. M., Ohuchi T., and Goldstein M. (1973) A radioimmunoassay for human circulatory dopamine-β-hydroxylase. *Life Sci.* **13,** 769–774.

Feilchenfeld N. B., Richter H. W., and Waddell W. H. (1982) A time-resolved assay of dopamine β-hydroxylase activity utilizing high-pressure liquid chromatography. *Anal. Biochem.* **122,** 124–128.

Flatmark T., Skotland T., Ljones T., and Ingebretsen O. C. (1978) Fluorimetric detection of octopamine in high-performance liquid chromatography and its application to the assay of dopamine β-monooxygenase in human serum. *J. Chromatogr.* **146,** 433–438.

Florvall L. and Corrodi H. (1970) Preparation and dopamine-β-hydroxylase inhibitory activity of some compounds related to dithiocarbamic acid and thiuram disulfide. *Acta Pharm. Suec.* **7,** 7–22.

Foldes A., Jeffrey P., Preston B. N., and Austin L. (1972) Dopamine-β-hydroxylase of bovine adrenal medulla. A rapid purification procedure. *Biochem. J.* **126,** 1209–1217.

Fong J. C., Shenkman L., and Goldstein M. (1980) Purification and characterization of rat pheochromocytoma dopamine β-hydroxylase. *J. Neurochem.* **34,** 346–350.

Friedman S. and Kaufman S. (1965) 3,4-Dihydroxyphenylethylamine β-hydroxylase. Physical properties, copper content, and role of copper in the catalytic activity. *J. Biol. Chem.* **240,** 4763–4773.

Frigon R. P., Converse S. L., and Stone R. A. (1978a) Plasma dopamine β-hydroxylase species dependence and the in vitro influence of NEM, copper, and pH. *Biochem. Med.* **19,** 1–15.

Frigon R. P., Samples J. R., and Stone R. A. (1978b) Inhibition of dopamine-β-hydroxylase: A proposed mechanism for prazosin hydrochloride. *Clin. Res.* **26,** 363A.

Frigon R. P. and Stone R. A. (1978) Human plasma dopamine β-hydroxylase. Purification and properties. *J. Biol. Chem.* **263,** 6780–6786.

Fujita K., Maruta K., Teradaira R., Beppu H., Ikegame M., Kawai K., and Nagatsu T. (1982) Dopamine β-hydroxylase activity in human cerebrospinal fluid from various age groups. *Clin. Chem.* **28,** 1403–1404.

Fujita K., Ito T., Maruta K., Teradaira R., Beppu H., Nakagami Y., Kato T., Nagatsu T., and Kato T. (1978) Serum dopamine-β-hydroxylase in schizophrenic patients. *J. Neurochem.* **30,** 1569–1572.

Fujita K., Nagatsu T., Maruta K., Tetraira R., Beppu H., Tsuji Y., and Kato T. (1977) Fluorescence assay for dopamine β-hydroxylase activity in human serum by high-performance liquid chromatography. *Anal. Biochem.* **82,** 130–140.

Fuller R. W., Ho P. P. K., Matsumoto C., and Clemens J. A. (1976) New inhibitors of dopamine-β-hydroxylase. *Adv. Enzyme Regul.* **15,** 267–281.

Goldstein M., Freedman L. S., and Bonnay M. (1971) An assay for dopamine β-hydroxylase activity in tissues and serum. *Experientia* **27,** 632–633.

Goldstein M., Joh T. H., and Garvey III T. Q. (1968) Kinetic studies of the enzymatic dopamine β-hydroxylation reaction. *Biochemistry* **7,** 2724–2730.

Goldstein M., Lauber E., and McKereghan M. R. (1964) Inhibition of dopamine-β-hydroxylase by tropolone and other chelating agents. *Biochem. Pharmacol.* **13,** 1103–1106.

Goldstein M., Prochoroff N., and Sirlin S. (1965) A radioassay for dopamine-beta-hydroxylase activity. *Experientia* **21,** 592–593.

Green A. L. (1964) The inhibition of dopamine-β-hydroxylase by chelating agents. *Biochim. Biophys. Acta* **81,** 394–397.

Grouselle M. and Phillips J. H. (1982) Reduction of membrane-bound dopamine β-hydroxylase from the cytoplasmic surface of the chromaffin-granule membrane. *Biochem. J.* **202,** 759–770.

Hanlon D. P. and Shuman S. (1973) Inhibition of brain dopamine-β-hydroxylase activity by methimazole. *Life Sci.* **12,** 417–423.

Hashiguchi H. and Takahashi H. (1977) Inhibition of two copper-containing enzymes, tyrosinase and dopamine β-hydroxylase, by L-mimosine. *Mol. Pharmacol.* **13,** 362–367.

Hashimoto Y. and Miyazaki H. (1979) Simultaneous determination of endogenous norepinephrine and dopamine-β-hydroxylase activity in biological materials by chemical ionization mass fragmentography. *J. Chromatogr.* **168,** 59–68.

Hatanaka H., Egami F., Kato T., and Nagatsu T. (1975) Ascorbate-2-sulfate inhibits dopamine β-hydroxylase but not ascorbate oxidase. *J. Biochem.* **78,** 821–824.

Hidaka H. (1973) Picolinic Acid Derivatives as Inhibitors of Dopamine β-Hydroxylase In Vivo; Their Effects on Blood Pressure and Stress Ulcer, in *Frontiers in Catecholamine Research* (Usdin E. and Snyder S., eds.), pp. 87–90, Pergamon, New York.

Hidaka H. and Takeya K. (1972) Blood pressure decrease by halobutyric acid, an inhibitor of dopamine β-hydroxylase. *Nature* **239,** 334–336.

Hidaka H., Nagatsu T., Takeya K., Takeuchi T., Suda H., Kojiri K., Matsuzaki M., and Umezawa H. (1969) Fusaric acid, a hypotensive agent produced by fungi. *J. Antibiot.* **22,** 228–230.

Hoeldtke R. and Kaufman S. (1978) A tritium-release in vivo assay of dopamine-beta-hydroxylase activity in sympathetic neurons. *Biochem. Pharmacol.* **27,** 2499–2506.

Hoeldtke R. and Stetson P. L. (1980) An in vivo tritium-release assay of human dopamine β-hydroxylase. *J. Clin. Endocrinol. Metab.* **51,** 810–815.

Hogue-Angeletti R. A. (1977) Nonidentity of chromogramin A and dopamine β-monooxygenase. *Arch. Biochem. Biophys.* **184,** 364–372.

Hornykiewicz O. (1982) Brain catecholamines in schizophrenia—a good case for noradrenaline. *Nature* (Lond.) **299,** 484–486.

Iinuma H., Takeuchi T., Kondo S., Matsuzaki M., Umezawa H., and Ohno M. (1972) Dopastin, an inhibitor of dopamine β-hydroxylase. *J. Antibiot.* **25,** 497–500.

Iinuma H., Matsuzaki M., Nagatsu T., Takeuchi T., and Umezawa H. (1974) Biochemical and biological studies on dopastin, an inhibitor of dopamine β-hydroxylase. *Agr. Biol. Chem.* **38,** 2107–2111.

Ikeno T., Hashimoto S., Kuzuya H., and Nagatsu T. (1977) Purification and properties of human serum dopamine-β-hydroxylase. *Mol. Cell. Biochem.* **18,** 117–123.

Izumi H., Hayakari M., Kondo Y., and Taketomo T. (1975a) Inhibition of dopamine-β-hydroxylase by carnosine. *Tohoku J. Exp. Med.* **117,** 101–102.

Izumi H., Hayakari M., Kondo Y., and Takemoto T. (1975b) Inhibition of dopamine-β-hydroxylase by histidine. *Hoppe-Seyler's Z. Physiol. Chem.* **356,** 1831–1833.

Joh T. H., Ross R. A., and Reis D. J. (1974) A simple and sensitive assay for dopamine-β-hydroxylase. *Anal. Biochem.* **62,** 248–254.

Johnson G. A., Boukma S. J., and Kim E. G. (1969) Inhibition of dopamine-β-hydroxylase by aromatic and alkyl thioureas. *J. Pharmacol. Exp. Ther.* **168,** 229–237.

Kato T., Kuzuya H., and Nagatsu T. (1974) A simple and sensitive assay for dopamine-β-hydroxylase activity by dual-wavelength spectrophotometry. *Biochem. Med.* **10,** 320–328.

Kato T., Wakui Y., Nagatsu T., and Ohnishi T. (1978) An improved dual-wavelength spectrophotometric assay for dopamine-β-hydroxylase. *Biochem. Pharmacol.* **27,** 829–831.

Katz I. R. (1982) Oxygen dependence of dopamine-β-hydroxylase activity and lactate metabolism in synaptosomes from rat brain. *Brain Res.* **231,** 399–409.

Kissinger P. T., Bruntlett C. S., Davis G. C., Felice L. J., Riggin R. M., and Shoup R. E. (1977) Recent development in the clinical assessment of the metabolism of aromatics by high-performance, reversed-phase chromatography with amperometric detection. *Clin. Chem.* **23,** 1449–1455.

Klinman J. P., Krueger M., Brenner M., and Edmondson D. E. (1984) Evidence for two copper atoms/subunit in dopamine β-monooxygenase catalysis. *J. Biol. Chem.* **259,** 3399–3402.

Kopin I. J., Kaufman S., Viverous H., Jacobowicz D., Lake R., Ziegler M. G., Lovenberg W., and Goodwin F. K. (1976) Dopamine β-hydroxylase: Basic and clinical studies. *Ann. Int. Med.* **85,** 211–223.

Kopun M. and Hershel M. (1978) Determination of dopamine-β-hydroxylase activity by means of chromatogram-spectrofluorometric measurement in remission. *Anal. Biochem.* **85,** 556–563.

Kuzuya H. and Nagatsu T. (1969) A simple assay of dopamine β-hydroxylase activity in the homogenate of the adrenal medulla. *Enzymologia* **36,** 31–38.

Kuzuya H., Ikuta K., and Nagatsu T. (1973) Inhibition of dopamine-β-hydroxylase by spinochrome A and echinochrome A, naphthoquinon pigments of echinoids. *Biochem. Pharmacol.* **22,** 2772–2774.

Levin E. Y., Levenberg B., and Kaufman S. (1960) The enzymatic conversion of 3,4-dihydroxyphenylethylamine to norepinephrine. *J. Biol. Chem.* **235,** 2080–2086.

Liu T. Z., Shen J. T., and Loken H. F. (1974) Inhibition of dopamine-β-hydroxylase by hydrazine. *Proc. Soc. Exp. Biol. Med.* **145,** 294–297.

Long R. A., Weppelman R. M., Taylor J. E., Tolman R. L., and Olson G. (1981) Purification and characterization of avian dopamine β-hydroxylase. *Biochemistry* **20,** 7423–7431.

Margolis R. K., Finne J., Krusius T., and Margolis R. U. (1984) Structural studies on glycoprotein oligosaccharides of chromaffin granule membranes and dopamine β-hydroxylase. *Arch. Biochem. Biophys.* **228,** 443–449.

Markianos V. E. and Nyström I. (1974) Studie über die Dopamin-β-hydroxylase in Serum. *Arzneimittelforsch* **24,** 1021–1023.

Matsui H., Kato T., Yamamoto C., Fujita K., and Nagatsu T. (1981) Highly sensitive assay for dopamine-β-hydroxylase activity in human cerebrospinal fluid by high performance liquid chromatograph-electrochemical detection: Properties of the enzyme. *J. Neurochem.* **37,** 289–296.

Matsui H., Kato T., Yamamoto C., Takita T., Takeuchi T., Umezawa H., and Nagatsu T. (1981) Inhibition of dopamine-β-hydroxylase, a copper enzyme, by bleomycin. *J. Antibiotics* **33,** 435–440.

May S. W., Mueller P. W., Padgette S. R., Herman H. H., and Phillips R. S. (1983) Dopamine-β-hydroxylase: Suicide inhibition by the novel olefinic substrate, 1-phenyl-1-aminomethylethene. *Biochem. Biophys. Res. Commun.* **110,** 161–168.

Molinoff P. B., Weinshilboum R., and Axelrod J. (1971) A sensitive enzymatic assay for dopamine-β-hydroxylase. *J. Pharmacol. Exp. Ther.* **178,** 425–431.

Naber N., Venkatesan P. P., and Hamilton G. A. (1982) Inhibition of dopamine β-hydroxylase by thiazoline-2-carboxylate, a suspected physiological product of D-amino acid oxidase. *Biochem. Biophys. Commun.* **107,** 374–380.

Nagatsu T. (1973) Biochemistry of Catecholamines. University of Tokyo and University Park, Tokyo and Baltimore. pp. 60–179.

Nagatsu T. (1977) Dopamine β-hydroxylase in blood and cerebrospinal fluid. *Trends Biochem. Sci.* **2,** 217–219.

Nagatsu T. (1983) Analysis of Monooxygenases in Catecholamine Biosynthesis: Tyrosine Hydroxylase and Dopamine β-Hydroxylase, in *Methods in Biogenic Amine Research* (Parvez S., Nagatsu T., Nagatsu I., and Pravez H., eds.), pp. 329–357. Elseveir, Amsterdam.

Nagatsu T. and Udenfriend S. (1972) Photometric assay of dopamine-β-hydroxylase activity in human blood. *Clin. Chem.* **18,** 980–983.

Nagatsu T., Ayukawa S., and Umezawa H. (1968) Inhibition of dopamine-β-hydroxylase by aquayamycin. *J. Antibiotics* **21,** 354–357.

Nagatsu T., Hidaka H., Kuzuya H., Takeya K., Umezawa H., Takeuchi T., and Suda H. (1970) Inhibition of dopamine β-hydroxylase by fusaric acid (5-butylpicolinic acid) in vitro and in vivo. *Biochem. Pharmacol.* **19,** 35–44.

Nagatsu T., Kato T., Kuzuya H., Umezawa H., and Takeuchi T. (1973a) New Inhibitors of Microbial Origin for Dopamine-β-Hydroxylase, in *Frontiers in Catecholamine Research* (Usdin E. and Snyder S., eds.), pp. 83–86, Pergamon, New York.

Nagatsu T., Thomas P., Rush R., and Udenfriend S. (1973b) A radioassay for dopamine β-hydroxylase in rat serum. *Anal. Biochem.* **55,** 615–619.

Nagatsu T., Kato T., Numata(Sudo) Y., Ikuta K., Sano M., Nagatsu I., Kondo Y., Inagaki S., Iizuka R., Hori A., and Narabayashi H. (1977) Phenylethanolamine *N*-methyltransferase and other enzymes of catecholamine metabolism in human brain. *Clin. Chim. Acta* **75,** 221–232.

Nagatsu T., Kuzuya H., and Hidaka H. (1967) Inhibition of dopamine β-hydroxylase by sulfhydryl compounds and the nature of the natural inhibitors. *Biochim. Biophys. Acta* **139,** 319–327.

Nagatsu T., Kato T., Nagatsu I., Kondo Y., Inagaki S., Iizuka R., and Narabayashi H. (1979) Catecholamine-Related Enzymes in the Brain of Patients With Parkinsonism and Wilson's Disease, in *Advances in Neurology* (Poirier L. J., Sourkes T. L., and Bédard P. J., eds.), pp. 283–292, Raven, New York.

Nagatsu T., Wakui Y., Kato T., Fujita K., Kondo T., Yokochi F., and Narabayashi H. (1982) Dopamine beta-hydroxylase activity in cerebrospinal fluid of parkinsonian patients. *Biomed. Res.* **3,** 95–98.

Nagatsu T., Yamaguchi T., Kato T., Sugimoto T., Matsuura S., Akino M., Nagatsu I., Iizuka R., and Narabayashi H. (1981) Biopterin in human brain and urine from controls and parkinsonian patients: Application of a new radioimmunoassay. *Clin. Chim. Acta* **109,** 305–311.

Nagatsu T., Yamaguchi T., Rahman M. K., Trocewicz J., Oka K., Hirata Y., Nagatsu I., Narabayashi H., Kondo T., and Iizuka R. (1983) Catecholamine-Related Enzymes and the Biopterin Cofactor in Parkinson's Disease and Related Extrapyramidal Diseases, in *Advances in Neurology* (Hassler R. G. and Christ J. F., eds.), pp. 467–473, Raven, New York.

Njus D., Knoth J., Cook C., and Kelley P. M. (1983) Electron transfer across the chromaffin granule membrane. *J. Biol. Chem.* **258,** 27–30.

Nohta H., Ohtsubo K., Zaitsu K., and Ohkura Y. (1982a) Assay for dopamine β-hydroxylase in rat serum and adrenal medulla by high-performance liquid chromatography with fluorescence detection. *J. Chromatogr.* **227,** 415–422.

Nohta H., Yamaguchi M., Zaitsu K., and Ohkura Y. (1982b) Assay of dopamine β-hydroxylase in human serum as a modification of the assay for the enzyme in rat serum by high-performance liquid chromatography. *J. Chromatogr.* **233,** 324–327.

Ohkura Y., Ohtsubo K., Kohashi K., and Zaitsu K. (1977) A novel fluormetric assay method of dopamine-β-hydroxylase activity in human serum. *Chem. Pharm. Bull.* **25,** 748–753.

Pisano J. J., Creveling C. R., and Udenfriend S. (1960) Enzymatic conversion of *p*-tyramine *to* p-hydroxyphenylethanolamine (norsynephrine). *Biochim. Biophys. Acta* **43,** 566–568.

Rush R. A., Kindler S. H., and Udenfriend S. (1975a) Solid phase radioimmunoassay on polystyrene beads and the application to dopamine-β-hydroxylase. *Clin. Chem.* **21,** 148–150.

Rush R. A., Thomas P. E., and Udenfriend S. (1975b) Measurement of human dopamine-β-hydroxylase in serum by homologous radioimmunoassay. *Proc. Natl. Acad. Sci. USA* **72,** 750–752.

Ratvatte H. A. M. and Douglas K. T. (1983) Photolabelling of dopamine-β-hydroxylase by bleomycin. *Biochem. Biophys. Res. Commun.* **112,** 273–278.

Sabban E. L., Greene L. A., and Goldstein M. (1983) Mechanism of biosynthesis of soluble and membrane-bound forms of dopamine β-hydroxylase in PC12 pheochromocytoma cells. *J. Biol. Chem.* **258,** 7812–7818.

Saxena A. and Fleming P. J. (1983) Isolation and reconstitution of the membrane-bound form of dopamine β-hydroxylase. *J. Biol. Chem.* **258,** 4147–4152.

Skotland T. and Ljones T. (1980) Direct spectrophotometric detection of ascorbate free radical formed by dopamine β-monooxygenase and by ascorbate oxidase. *Biochim. Biophys. Acta* **630,** 30–35.

Sperk G., Galhaup I., Schlörgl E., Hötnagl H., and Hornykiewicz O. (1980) A sensitive and reliable assay for dopamine β-hydroxylase in tissue. *J. Neurochem.* **35,** 972–976.

Suda H., Takeuchi T., Nagatsu T., Matsuzaki M., Matsumoto I., and Umezawa H. (1969) Inhibition of dopamine-β-hydroxylase by 5-alkylpicolinic acids and their hypotensive effects. *Chem. Pharm. Bull.* **17,** 2377–2380.

Suzuki T., Tsuge I., Higa S., Hayashi A., Yamamura Y., Takaba Y., and Nakajima A. (1979) Catecholamine metabolism in familial amyloid polyneuropathy. *Clin. Genet.* **16,** 117–124.

Terland O. and Flatmark T. (1975) Ascorbate as a natural constituent of chromaffin granules from the bovine adrenal medulla. *FEBS Lett.* **59,** 52–56.

Udenfriend S., Stein S., Böhlen P., Dairman W., Leimgruber W., and Weigele M. (1972) Fluorescamine: A reagent for assay of amino acids, peptides, proteins, and primary amines in the picomole range. *Science* **178,** 871–872.

Umezawa H. (1972) *Enzyme Inhibitors of Microbial Origin.* Univ. Tokyo, Tokyo. pp. 57–106.

van der Schoot J. B., Creveling C. R., Nagatsu T., and Udenfriend S. (1963) On the mechanism of inhibition of dopamine-β-hydroxylase by benzyloxyamines. *J. Pharmacol. Exp. Ther.* **141,** 74–78.

Wallace E. F., Krantz M. J., and Lovenberg W. (1973) Dopamine β-hydroxylase: A tetrameric glycoprotein. *Proc. Natl. Acad. Sci. USA* **70,** 2253–2255.

Weinshilboum R. M. (1979) Serum dopamine β-hydroxylase. *Pharmacol. Rev.* **30,** 133–166.

Weinshilboum R. and Axelrod J. (1971) Serum dopamine-beta-hydroxylase activity. *Circ. Res.* **28,** 307–315.

Wilcox G. and Beaven M. A. (1977) A sensitive and specific tritium-release assay for dopamine-β-hydroxylase (DBH) in serum. *Anal. Biochem.* **75,** 484–497.

Wise C. D. (1976) A sensitive assay for dopamine-β-hydroxylase. *J. Neurochem.* **27,** 883–888.

Yamaguchi M., Yutani Y., Miyamoto S., and Ohkura Y. (1980) Phosphorimetric assay for dopamine β-hydroxylase in rat plasma. *Anal. Biochem.* **104,** 402–406.

Zaremba S. and Hogue-Angeletti R. A. (1981) Transmembrane nature of chromaffin granule dopamine β-monooxygenase. *J. Biol. Chem.* **256,** 12310–12315.

Chapter 4
Monoamine Methyltransferases

Phenylethanolamine *N*-Methyltransferase and Catechol *O*-Methyltransferase

Dong H. Park

1. Phenylethanolamine *N*-Methyltransferase

1.1. Introduction

Phenylethanolamine *N*-methyltransferase (PNMT) (EC 2.1.1.28, *S*-adenosyl-L-methionine: phenylethanolamine *N*-methyltransferase) is the enzyme that catalyzes the biosynthesis of norepinephrine [noradrenaline; (NA)] to epinephrine [adrenaline; (A)] in the presence of *S*-adenosyl-L-methionine (SAM) as the methyl donor. The enzyme is localized mainly in the adrenal medulla and in small amounts in discrete brain regions containing A cell bodies and nerve terminals. Adrenal PNMT has been extensively purified and well characterized, but the brain enzyme has not been well studied, although many similarities between them have been observed.

This chapter is divided into two parts—phenylethanolamine *N*-methyltransferase and catechol *O*-methyltransferase. The first part comprehensively describes an assay method, purification procedures, and important characteristics of PNMT. The second part covers similar studies on catechol *O*-methyltransferase.

1.2. Occurrence and Distribution

PNMT activity is found primarily in the adrenal medulla of rabbit, rat, monkey, cow, guinea pig, cat, frog, man, mouse, dog, and snake (Axelrod, 1962; Axelrod and Vesell, 1970; Kirshner and

Goodall, 1957; Kitabchi and Williams, 1969; Wurtman et al., 1968). The ability of brain tissue to form A in vitro was demonstrated for the first time by McGeer and McGeer (1964), then by Ciaranello et al. (1969) and Pohorecky et al. (1969). The highest regional activity in brain is found in the C_1 area, C_2 area, and hypothalamus, although there is considerable species variation in the level of enzyme activity (Saavedra et al., 1974; Lew et al., 1977; Vogel et al., 1976). Among those species examined, brainstem PNMT activity is the highest in frog, turtle, and rat, whereas hypothalamic enzyme activity is the highest in frog, cat, and rat, in decreasing order (Fuller and Hemrick-Luecke, 1983). The detailed immunohistochemical mapping of PNMT-containing neurons in brain has been reported by Hokfelt et al. (1974) using antibodies to bovine adrenal PNMT.

1.3. Assay Method

The assay is based on the use of radiolabeled *S*-adenosyl-L-methionine as a methyl donor and phenylethanolamine as a methyl acceptor, followed by organic solvent extraction of radiolabeled *N*-methylphenylethanolamine. This assay procedure is a minor modification (Park et al., 1982) of the method described by Axelrod (1962).

1.3.1. Reagents

1*M* Tris-Cl buffer, pH 8.6
$2.5 \times 10^{-3}M$ D,L-phenylethanolamine
S-(Methyl-^{14}C)-adenosyl-L-methionine from New England Nuclear (57.3 mCi/mmol; 50 μCi/0.35 mg/2500 μL; 5.7 μL = 2 nmol)
S-(Methyl-^{3}H)-adensoyl-L-methionine from New England Nuclear (80.6 Ci/nmol; 1 mCi/4.9 μg/2.2 mL; 2.5 μL = 0.014 nmol) plus 0.986 nmol of unlabeled *S*-adenosyl-L-methionine (Sigma Chemical Co.)
Bovine serum albumin (Sigma Chemical Co.) at 10 μg/assay
0.5*M* Sodium borate buffer, pH 10
Isoamyl alcohol-toluene mixture (3:97 v/v)
Aquassure scintillation counting solution (New England Nuclear)

1.3.2. Procedure

The incubation mixture contains $5 \times 10^{-4}M$ phenylethanolamine; 20 μ*M* *S*-(methyl-^{14}C)-adenosyl-L-methionine; 150 m*M*

Tris-Cl buffer pH 8.6; and enzyme solution containing 10 μg bovine serum albumin in a final volume of 100 μL. For a blank, water is substituted for the enzyme solution. After 10–15 min incubation at 37°C, the reaction is stopped by adding 0.5 mL of 0.5*M* sodium borate buffer, pH 10. For extraction (Saavedra et al., 1974) of the radiolabeled methylated product, 6 mL of isoamyl alcohol-toluene mixture is added, and the mixture vortexed for 10 s and centrifuged at 1000 rpm for 5 min. A 4-mL aliquot of the upper organic layer is transferred into a glass counting vial and dried at 40°C under a stream of air. The radioactivity is counted with 1 mL of absolute ethanol and 10 mL of Aquassure in a scintillation counter. For a more sensitive assay, 10 μ*M* *S*-(methyl-^{3}H)-adenosyl-L-methionine is used instead of 20 μ*M* *S*-(methyl-^{14}C)-adenosyl-L-methionine.

1.4. Purification Procedure

PNMT has been purified extensively from adrenal glands of various species (Connett and Kirshner, 1970; Joh and Goldstein, 1973; Lee et al., 1978; Park et al., 1982; Pohorecky and Baliga, 1973). The following procedure is commonly used for purification of adrenal PNMT.

Throughout the purification, potassium phosphate (K-P) buffer, pH 7.0, is used unless otherwise specified, and protein solutions are concentrated by freeze drying. All purification steps are carried out at 0–4°C.

Step 1. Homogenization

Frozen rat adrenal glands (about 15 g, Pel-freeze) or bovine adrenal medullae are homogenized with 5 volumes/g tissue of 10 m*M* K-P buffer in a Waring blender for 1 min and centrifuged at 105,000*g* for 1 h. The supernatant is filtered through a layer of fine glass wool, and the filtrate concentrated and dialyzed in the same buffer.

Step 2. Sepharose 4B Column Chromatography

The dialyzed protein solution (0.5–1 g protein) is applied to a Sepharose 4B column (5 × 86 cm), preequilibrated with 10 m*M* K-P buffer, and the protein eluted with the same buffer. Fractions containing the highest enzyme activity are pooled, concentrated, and dialyzed in the buffer.

Step 3. DEAE-Cellulose Column Chromatography I

A DE 52 (Whatman) column (2.6 × 20 cm) is preequilibrated with 10 m*M* buffer, and the protein solution (200–500 mg protein) from the previous step is loaded. The column is washed with 10-m*M* buffer until the A_{280} nm reading becomes almost zero. The protein is eluted with a linear 10–300 m*M* buffer gradient (10 times the bed volume). Fractions with the highest specific activity are pooled, concentrated, and dialyzed in 10 m*M* buffer.

Step 4. DEAE-Cellulose Column Chromatography II

The protein solution from step 3 is loaded on a second DE 52 column (0.9 × 8 cm), previously equilibrated with 10 m*M* buffer, and eluted with a linear 10–400 m*M* buffer gradient. The highest specific activity is eluted in fractions around 180 m*M* buffer concentration for rat adrenal enzyme and 100 m*M* for bovine adrenal enzyme. The active fractions are pooled, concentrated, and dialyzed in 10 m*M* buffer.

Step 5. Sephadex G-100 Column Chromatography

A Sephadex G-100 column (1.6 × 88 cm) is equilibrated with 10 m*M* buffer, and the protein solution from step 4 is applied to the column. The enzyme is eluted with 10 m*M* buffer, and the most active fractions concentrated and dialyzed in the same buffer.

1.5. Substrate Specificity

The substrate specificity of PNMT has been studied extensively by various investigators (Axelrod, 1962; Fuller et al., 1970, 1977; Grunewald et al., 1975; Hoffman et al., 1975; Kirshner and Goodall, 1957; Rafferty and Grunewald, 1982). SAM has been shown to be the methyl donor in the *N*-methylation reaction catalyzed by PNMT (Kirshner and Goodall, 1957). The enzyme *N*-methylates a wide variety of physiological and nonphysiological phenylethanolamine derivatives, but not phenylethylamine derivatives, with the exception of trifluoromethyl-substituted derivatives of exo-2-aminobenzobicyclo (2.2.1.)heptene with the substituent attached to position 5 or 6 on the aromatic ring (Rafferty and Grunewald, 1982). Therefore, the enzyme has an absolute requirement for the hydroxyl group at the β-position of the ethyl side chain of amines as the methyl acceptor. Naturally occurring phenylethanolamine derivatives methylated by PNMT include NA, normetanephrine, octopamine, synephrine, A, and metanephrine. Primary amines serve as better substrates than do *N*-methylated secondary amines. Some substrate activity can be retained when the β-hydroxyl group of a phenylethanolamine is replaced by an amino group, or a single chloro group occupies the *ortho, meta,* or *para* positions of the ring (Fuller et al., 1977).

Various nonaromatic analogs of phenylethanolamine are also *N*-methylated. Nonaromatic analogs in which the phenyl ring is replaced by a cyclohex-3-enyl, cyclohexyl, or cyclooctyl group are as good as phenylethanolamine itself as a substrate for PNMT (Grunewald et al., 1975).

1.6. Inhibitors

PNMT is strongly inhibited by its own substrates, NA and SAM, as well as its own products, A and *S*-adenosylhomocysteine, above certain concentrations (Connett and Kirshner, 1970; Fuller and Hunt, 1967; Pendleton et al., 1976). Sulfhydryl reagents, such as *p*-chloromecuribenzoic acid, 2,5-dimethyl-1-aminobenzimidazole, *p*-chloromercuriphenylsulfonate, *N*-ethylmaleimide, and mercury, are also in vitro inhibitors of PNMT because the enzyme possesses an essential sulfhydryl group (Axelrod, 1962; Connett and Kirshner, 1970; Pohorecky and Baliga, 1973). In vitro, lead irreversibly inhibits PNMT (Caspers, 1982). Phenylethylamines can combine with PNMT, but are not methylated. Thus, certain aromatic and nonaromatic analogs of phenylethylamines, as well as phenylethanolamines, are potent inhibitors of PNMT (Fuller et al., 1971; Fuller and Molloy, 1977; Grunewald et al., 1975; Krakoff and Axelrod, 1967).

SKF 64139 (7,8-dichloro-1,2,3,4-tetrahydroisoquinoline) is a widely used potent inhibitor, both in vitro and in vivo, of both adrenal and brain PNMT (Pendleton et al., 1976). SKF 29661 (1,2,3,4-tetrahydroisoquinoline-7-sulfonamide) is a potent inhibitor of brainstem PNMT in vitro only because it does not cross the blood–brain barrier (Pendleton et al., 1979). These inhibitors are competitive with phenylethanolamine and noncompetitive with SAM.

Several phenylethylamines, benzylamines, and conformational analogs of these compounds, such as 2,3-dichloro-α-methylbenzylamine (DCMB or LY 78335), 8,9-dichloro-2,3,4,5-tetrahydro-1H-2-benzazepine (LY 134046), 2,3,4,5-tetrahydro-1H-indeno(1,2C)-pyridine (LY 87130), and 2-cyclooctyl-2-hydroxyethylamine, are also widely used in vitro and in vivo inhibitors of PNMT (Fuller, 1982; Grunewald et al., 1975; Lew et al., 1976; Liang et al., 1982).

1.7. Different Forms of the Enzyme

PNMT is highly localized in the soluble fraction of adrenal medulla, and adrenal PNMT has been well characterized.

When adrenal PNMT from rat, mouse, cat, man, dog, cow, and frog is subjected to starch block electrophoresis, electrophoretic mobilities are found to be quite different among various species (Axelrod and Vesell, 1970; Pohorecky and Baliga, 1973; Wurtman et al., 1968). These observations indicate that PNMT from various species differs in charge.

In addition to the presence of different forms of PNMT among species, some have their own multiple forms of the enzyme (Connett and Kirshner, 1970; Joh and Goldstein, 1973; Lee et al., 1978; Park et al., 1982; Pohorecky et al., 1973). Lee et al. isolated five charge isozymes of PNMT from rabbit adrenals and found the isoelectric points to range from 4.4 to 5.9. Pohorecky et al. reported two peaks of PNMT activity in dog adrenal medulla by DEAE-cellulose column chromatography. According to our observations (Park and Joh, 1985), rat adrenal enzyme has only one isoelectric point, pH 4.8, as determined by chromatofocusing, in contrast to the two charge isozymes reported previously (Pohorecky and Baliga, 1973; Yu, 1978). Bovine adrenal enzyme, on the other hand, has five different charge isozymes, ranging in isoelectric point from 5.4 to 6.2 (Joh and Goldstein, 1973; Park et al., 1982; Park and Joh, 1985). We have also found that PNMT from rat and bovine brain have isoelectric points similar to those of the corresponding adrenal enzyme (Park and Joh, 1985).

Although the molecular weight of adrenal PNMT from various species has been reported to range from 35,000 to 40,000 (Connett and Kirshner, 1970; Joh and Goldstein, 1973; Lee et al., 1978; Yu, 1978), our results (Park et al., 1982) have shown that the molecular weight of both rat and bovine adrenal PNMT, as determined by SDS-polyacrylamide gel electrophoresis, is estimated to be 31,000. We have also demonstrated that antibodies directed against either bovine or rat PNMT immunoprecipitate a protein of molecular weight 31,000 from in vitro translation products of bovine poly(A^+)mRNA (Park et al., 1982; Baetge et al., 1983).

1.8. Activation by Various Anions

We have recently demonstrated that adrenal PNMT activity is greatly elevated by a variety of anions (Park and Joh, 1984). PNMT activity in a crude bovine adrenal enzyme preparation is enhanced by phosphate in a concentration-dependent manner, as shown in Fig. 1. The activity increased 680% over the control at 190 m*M* phosphate. In contrast, phosphate activation of the rat adrenal enzyme is minimal, increasing 77% over the control at 30 m*M* phosphate, and gradually decreasing at higher concentrations (Fig. 1). The degree of activation is less when purified preparations of either bovine or rat PNMT are used, although the activation patterns for crude and purified enzyme preparations are similar. The activity of the purified enzyme increases 390% for bovine enzyme and 43% for rat enzyme.

After pretreatment of purified bovine PNMT with neuramini-

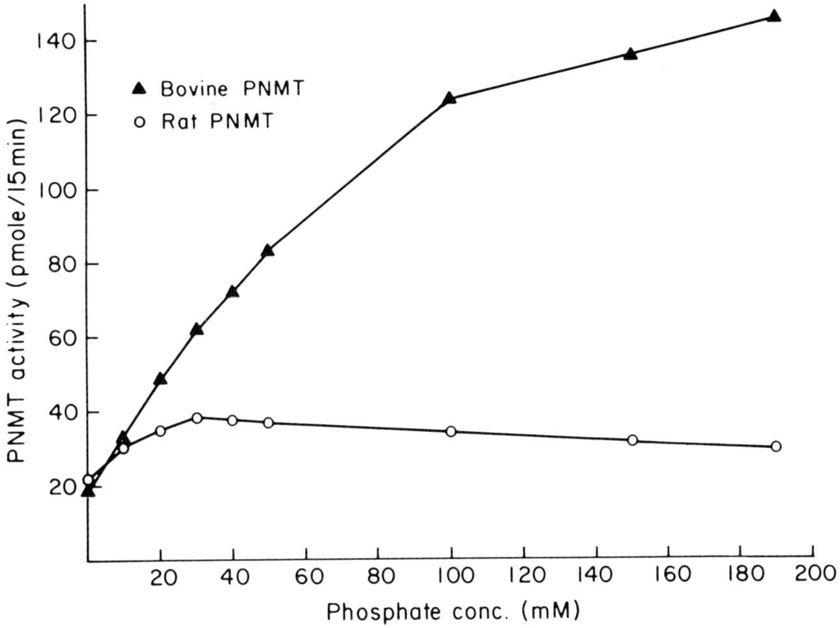

Fig. 1. Phosphate activation of phenylethanolamine *N*-methyltransferase in crude tissue extract. Crude enzyme preparations (105,000 g supernatant) were assayed as described in section 1.3 in the presence of increasing amounts of potassium phosphate. The amount of adrenal enzymes used are as follows: bovine enzyme, 8.7 μg; rat enzyme, 8.5 μg. Each point is the average of duplicate determinations.

dase, enzyme activity is gradually enhanced by phosphate in a concentration-dependent manner and reaches a maximum of only 261% of the control at 200 m*M*. However, neuraminidase pretreatment of purified rat enzyme prevents activation of the enzyme by phosphate.

The activity of purified bovine enzyme is also increased in a concentration-dependent fashion by other anions, including chloride and sulfate. The maximum enhancements of activity are somewhat greater than twice the control levels in the presence of 200 m*M* anion. In contrast, the addition of positive ions, such as Tris, to the assay mixture does not affect the activity of purified PNMT. The activity of purified rat adrenal PNMT is increased by neither the presence of cations nor anions.

Our studies (Park and Joh, 1984) have shown that only bovine PNMT is glycosylated, and that a marked activation of the enzyme by phosphate ion occurs only when the glycosylated form of PNMT is present. These results suggest that the presence or absence of carbohydrate moieties on the enzyme may be largely responsible for the species differences observed. The degree of activation by phosphate is much greater with crude enzyme preparations than with purified enzyme. This may be because of a partial removal of carbohydrate moieties from the enzyme during purification, since our qualitative estimates indicate that only 20% of the purified bovine PNMT is glycosylated.

1.9. Immunochemical Crossreactivity

Antibodies have been raised in rabbit against either purified bovine or rat adrenal PNMT (Joh and Goldstein, 1973; Kondo et al., 1982; Lew et al., 1977; Park et al., 1982; Pohorecky and Baliga, 1973). Immunochemical titration is carried out by adding antibodies to bovine adrenal PNMT to approximately matched activities of adrenal or brainstem PNMT of cow and rat. As shown in Fig. 2, antibodies directed against bovine adrenal PNMT immunoprecipitate both cow and rat PNMT equally well from adrenal and brainstem, although a small portion of the rat enzyme appears to resist inhibition by the antibodies. Antibodies to bovine adrenal PNMT also partially react with enzyme from human brain (Armstrong, Park, and Joh, unpublished observation), mouse, frog (Joh and Goldstein, 1973; Kondo et al., 1982), and chicken (Teitelman et al., 1984). In contrast, antibodies raised against rat adrenal PNMT react strongly only with the rat enzyme and very weakly with PNMT from cow and monkey (Lew et al., 1977; Park et al., 1982; Pohorecky and Baliga, 1973).

In general, antibodies raised against adrenal PNMT from a given species always inhibit or immunoprecipitate the enzyme from the same species more efficiently than from other species. It seems that antibodies do not discriminate between PNMT from adrenal vs brainstem within a given species.

1.10. Other Properties

1.10.1. Thermal Stability

As crude or purified adrenal PNMT from various species is heated at 45–48°C for 5 min in the presence of buffer, the enzyme activity from all species examined is markedly decreased. However, the enzyme from rat, dog, and cow appear to be more heat-resistant than that of mouse, cat, frog, and human (Axelrod and Vesell, 1970; Connett and Kirshner, 1970; Pohorecky and Baliga, 1973; Wurtman et al., 1968). SAM, dithiothreitol, and glycerol seem to have some protective effect on the heat inactivation of the

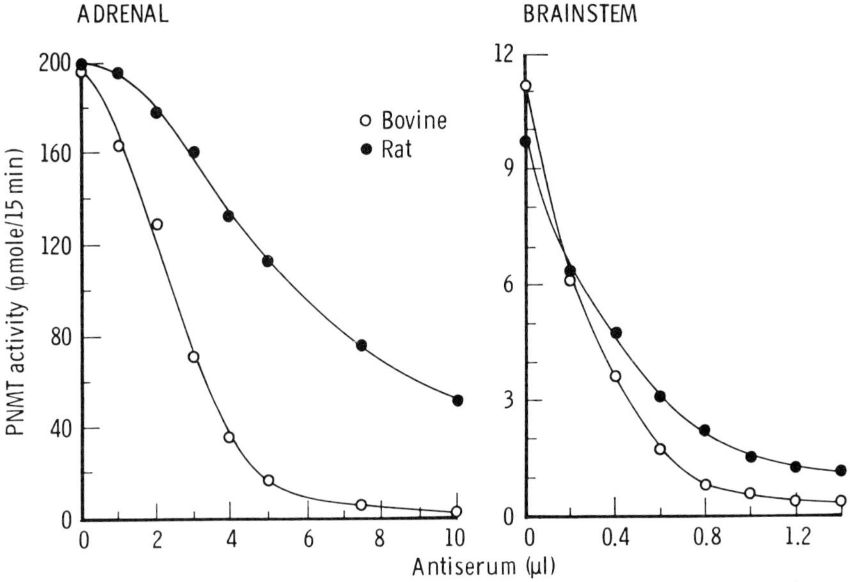

Fig. 2. Immunochemical titration curves for adrenal and brainstem PNMT from rat and cow, using antibodies to bovine adrenal PNMT. Immunochemical titration was carried out by adding increasing amounts of antibodies to bovine adrenal PNMT to 50 µL of crude bovine or rat adrenal PNMT preparation, or bovine or rat brainstem PNMT preparation. The total vol was adjusted to 70 µL with preimmune serum. The mixture was allowed to stand for 1 h at room temperature with occasional shaking and then for 16 h at 4°C. Antigen-antibody precipitate was removed by centrifugation at 9750*g* for 10 min, and a 50-µL aliquot of the supernatant was assayed for PNMT activity.

enzyme (Connett and Kirshner, 1970; Pohorecky and Baliga, 1973). Amines, such as phenylethanolamine, phenylethylamine, and octopamine, are also effective in protecting the enzyme from heat denaturation (Masover et al., 1979). The stabilization effect is greatest when the *meta* and *para* positions of the phenylethylamine nucleus are unsubstituted. The most effective stabilizing substance reported is phenylethylamine, and the least effective is NA. Adrenaline is completely ineffective in stabilizing the enzyme against thermal denaturation.

1.10.2. pH Optimum

The optimal pH for activity reported in the literature varies considerably depending on the buffers, substrates, source of enzyme, and purity of enzyme used. An optimum pH of approximately 8, previously reported with phosphate buffer, may be partly a result of the instability of some substrates used in alkaline pH (Axelrod, 1962; Connett and Kirshner, 1970; Lee et al., 1974; Pohorecky and Baliga, 1973; Wurtman et al., 1968). When phenylethanolamine is used as a substrate, the pH optimum for purified PNMT from rat or cow is approximately 8.8, using either Tris buffer (Diaz-Borges et al., 1978; Park and Joh, unpublished observation; Yu, 1978) or phosphate buffer (Park and Joh, unpublished observation; Yu, 1978).

1.10.3. Kinetics

There are numerous conflicting reports (Axelrod, 1962; Diaz-Borges et al., 1978; Fuller and Hemrick, 1979; Hoffman et al., 1975; Lee et al., 1974; Lew et al., 1977; Masover et al., 1979; Yu, 1978) dealing with the kinetics of PNMT, and it is very difficult to compare one to another because each investigator uses different substrates, enzyme with varying degree of purity and from different species, and a wide variety of assay conditions. In addition, different values for the K_m of the methyl acceptor for both adrenal and brain enzymes within a given species have been reported. The available data suggests that the K_m for the methyl acceptor and the methyl donor for PNMT differs among species. In contrast, our data (Park and Joh, unpublished observation) indicate that the K_m for both phenylethanolamine and SAM for purified rat and bovine adrenal PNMT are similar, as shown in Figs. 3a and b, respectively. Our kinetic studies are carried out under assay conditions of 5 min, thereby ensuring linearity. Phenylethanolamine concentration is varied, from $1 \times 10^{-5}M$ to $1 \times 10^{-3}M$, with SAM present at 20 μM when the K_m for phenylethanolamine is determined. The K_m for SAM is determined at $5 \times 10^{-4}M$ phenylethanolamine, with variable concentrations of SAM ranging from 3.5×10^{-6} to $2 \times 10^{-5}M$. Linear plots are obtained under these conditions with Lineweaver-Burke analysis of initial velocity for both phenylethanolamine (Fig. 3a) and SAM (Fig. 3b). The K_m levels for phenylethanolamine and SAM obtained in this manner are $1.5 \times 10^{-4}M$ and $4.9 \times 10^{-6}M$, respectively.

1.11. Regulation

Wurtman and Axelrod have shown that the activity of rat adrenal PNMT falls markedly following hypophysectomy (Pohorecky and Wurtman, 1971; Wurtman an Axelrod, 1966). This decrease in enzyme level is the result of a decrease in the number of enzyme molecules, judging from the immunochemical titration data (Ciaranello, 1978a,b). Enzyme activity can be restored to normal levels by injecting either adrenocorticotrophic hormone (ACTH) or dexamethasone, a potent synthetic glucocorticoid. The effect of ACTH on PNMT has been shown to be indirect and to be mediated by the release of glucocorticoids from the adrenal cortex. However, PNMT levels cannot be elevated above normal levels by these treatments in either hypophysectomized or normal, adult rats. It has been reported that the decline in PNMT activity following hypophysectomy is from increased degradation of the enzyme through proteolysis (Berenbeim et al., 1979; Ciaranello, 1978a,b; Ciaranello et al., 1978).

Both insulin and glucagon have been reported to stimulate PNMT activity in rat adrenals (Kitabchi et al., 1968). The stimulation is evident within 6 h after a single dose of either hormone.

Adrenal PNMT is also under neuronal control (Ciaranello, 1978b; Molinoff and Axelrod, 1971). Its activity increases markedly following catecholamine depletion with reserpine (Molinoff et al., 1970) and after chemical sympathectomy with 6-hydroxydopamine (Thoenen et al., 1970). The increase after sympathectomy is blocked by adrenal denervation. This process is thought to take place through an increase in *de novo* synthesis of PNMT (Ciaranello, 1978b).

Thus, it seems that the neuronal increase in enzyme synthesis and the hormonal control of enzyme proteolysis are regulatory mechanisms that control PNMT levels in vivo (Ciaranello, 1978b).

1.12. Summary

PNMT is localized in discrete brain regions, as well as in the adrenal medulla. Its estimated molecular weight is 31,000, its pH optimum is 8.8, and K_m for phenylethanolamine and SAM are $1.5 \times 10^{-4}M$ and $4.9 \times 10^{-6}M$, respectively. The enzyme has a broad substrate specificity and can be inhibited by substrate and product analogs, in addition to sulfhydryl reagents. PNMT from certain species has more than one charge isozyme. Antibodies raised against the enzyme of a given species inhibit the enzyme from the same species more efficiently than do ones from other species. PNMT is under hormonal and neuronal control.

2. Catechol *O*-Methyltransferase

2.1. Introduction

Catechol *O*-methyltransferase (COMT) (EC 2.1.1.6, *S*-adenosyl-L-methionine: catechol *O*-methyltransferase) is the en-

Step 5. DEAE-Cellulose Column Chromatography

The dialyzed protein solution is applied to DEAE-cellulose column, preequilibrated with 1 m*M* phosphate buffer, pH 7.5, and the protein eluted with a linear 1–100 m*M* phosphate buffer gradient (10 times the bed volume). Fractions with the highest specific activity are pooled, concentrated, and dialyzed in 10 m*M* phosphate buffer.

Affinity column chromatography (Borchardt et al., 1975; Gulliver and Tipton, 1978b), using either 3,4-dimethoxy-5-hydroxyphenylethylamine-agarose conjugate or 2,6-dimethoxyphenol-azo-phenylmethylene-anilino-agarose, could be used for further purification, if necessary.

2.5. Substrate Specificity

COMT has a broad substrate specificity (Axelrod and Tomchick, 1958; Guldberg and Marsden, 1975) and can *O*-methylate, using SAM as the methyl donor in the presence of Mg^{2+} ion, a wide variety of catechols, regardless of the other substituents on the aromatic nucleus. Monohydroxy derivatives of phenyethylamine are not *O*-methylated. Physiological substrates *O*-methylated by COMT include DOPA, DA, NA, A, 3,4-dihydroxyphenylacetic acid, 3,4-dihydroxymandelic acid, 3,4-dihydroxybenzoic acid, 3,4-dihydroxyphenylglycol, 3,4-dihydroxyphenylethanol, *N*-acetyldopamine, *N*-acetylnorepinephrine, ascorbic acid, hydroxyestradiol, hydroxyestrone, and hydroxyestriol (Axelrod and Tomchick, 1958; Creveling et al., 1972; Knuppen et al., 1969). Exogenously administered catechols, such as 3,4-dihydroxyamphetamine, polyphenols, and many substituted catechols, can also be *O*-methylated by the enzyme (Anderson and D'Ioro, 1968; Axelrod and Tomchick, 1958; Bakke, 1970). This enzyme shows stereospecificity toward the naturally occurring L-isomer, which serves as a better substrate than does the D-form (Gordonsmith et al., 1982). The enzyme *O*-methylates substrates in vivo mainly in the *meta* position, whereas in vitro *O*-methylation occurs in both *meta* and *para* positions, depending on the nature of the aromatic substrate and the pH of the reaction mixture (Creveling et al., 1972; Senoh et al., 1959). The *meta*-methylated isomer predominates with substrates containing a high polar substituent, and for substrates with nonpolar substituents the ratio of the *meta*- and *para*-methylation is nearly one (Creveling et al., 1972).

2.6. Inhibitors

Inhibitors of COMT are more or less structurely related to either substrates, products, metal effector, chelating agents, or sulfhydryl binding agents (Axelrod and Laroche, 1959; Axelrod and Tomchick, 1958; Borchardt et al., 1974; Flohe, 1974; Guldberg and Marsden, 1975; Molinoff and Axelrod, 1971).

Phenolic and polyhydroxy aromatic compounds, such as pyrogallol and its derivatives, e.g., 3-hydroxyestradiol, serve as alternative substrates of the enzyme, and thus inhibit the methylation of substrate mainly as potent competitive inhibitors, but are less effective inhibitors in vivo (Anderson and D'Ioro, 1968; Borchardt et al., 1974, 1982; Fahn et al., 1979; Flohe, 1974; Guldberg and Marsden, 1975; Knuppen et al., 1969). SAM, *S*-adenosylhomocysteine, and their derivatives also strongly inhibit COMT (Borchardt and Cheng, 1978; Coward et al., 1973; Deguchi and Barchas, 1971; Flohe, 1974). Calcium ions act as noncompetitive inhibitors (Flohe, 1974). Chelating agents, such as the seven-membered ring compound tropolone (structurally similar to catechol) and its derivatives, are also effective inhibitors of COMT in vitro and in vivo, with a short duration of action (Flohe, 1974; Belleau and Burba, 1963; Guldberg and Masden, 1975).

2.7. Different Forms of the Enzyme

There are still conflicting reports (Guldberg and Marsden, 1975) in the literature concerning the existence of multiple forms of soluble COMT, in addition to membrane-bound COMT, among different species as well as in various tissues of a given species.

Axelrod and Vesell (1970) have reported that when liver COMT from rat, cat, dog, and man is subjected to starch block electrophoresis, two species-specific forms of COMT are observed. However, when the electrophoretic patterns from various rat tissues are compared, the two corresponding isozymes have the same mobility, regardless of their tissue origins. Huh and Friedhoff (1979) also observed two COMT forms from rat liver by either gel filtration, ion exchange chromatography, or sucrose density gradient centrigfugation. However, in this case, the molecular weights determined indicate that two forms have a monomer and dimer relationship. In contrast to the above findings, most other investigators (Anderson and D'Ioro, 1968; Assicot and Bohuon, 1970; Borchardt and Cheng, 1978; Borchardt et al., 1975; Gulliver and Tipton, 1978b, 1979; Raxworthy et al., 1982; White and Wu, 1975) have observed, predominantly, a single soluble molecular form of COMT in various tissues and species. Its molecular weight has been estimated to be around 24,000 by SDS-gel electrophoresis, gel filtration, or sucrose density gradient centrifugation. They have also reported that the membrane-bound form represents only a small percentage of the total activity. In contrast, recent findings of

Roth and coworkers (Roth, 1980; Rivett et al., 1983a,b) indicate that membrane-bound COMT represents more than 50% of the total activity in brain of certain species, and has approximately a 100-fold lower K_m value for catecholamines than does soluble COMT. After injections of kainic acid into rat striatum, membrane-bound COMT activity is decreased in the lesioned striatum and soluble COMT activity is elevated. From these observations, they suggest that membrane-bound enzyme may reside in neurons, whereas the soluble form is localized in glial cells.

2.8. Immunochemical Crossreactivity

Antibodies have been raised in rabbits by injecting either purified soluble rat liver COMT (Assicot and Bohuon, 1969; Creveling et al., 1973) or soluble pig liver COMT (Gulliver and Tipton, 1979). The inhibitory potency of the antibodies produced is measured by immunochemical titrations that are performed by adding an increasing amount of antibodies to a fixed amount of COMT, or vice versa.

Antibodies to rat liver soluble COMT inhibit by almost 100% the enzyme activity from liver and other tissues of rat (Assicot and Bohuon, 1969; Borchardt and Cheng, 1978; Creveling et al., 1973; Guldberg and Marsden, 1975; Kaplan et al., 1980). However, inhibitory potency is quite different when COMT from various species is compared. Inhibition of COMT from liver of man, cat, and guinea pig is around 50%, whereas no inhibition of COMT from bovine and rabbit liver is observed with antibodies to rat liver COMT (Assicot and Bohuon, 1969). The antibodies also inhibit soluble or membrane-bound enzyme almost equally well (Borchardt and Cheng, 1978; Creveling et al., 1973), although a small percentage of membrane-bound activity appears to resist the inhibition. On the other hand, Assicot and Bohuon (1969) have reported that antibodies to soluble enzyme do not inhibit membrane-bound COMT. Antibodies to soluble pig liver COMT also show similar inhibition of brain and liver enzyme of pig (Gulliver and Tipton, 1979).

2.9. Other Properties

2.9.1. Thermal Stability

Heat stability of liver COMT seems to differ among various species examined (Assicot and Bohoun, 1970; Axelrod and Vesell,

1970). When partially purified liver enzyme is subjected to heat treatment in the presence of 50 m*M* phosphate buffer, pH 7.9, at 51°C for 3 min, enzyme activity from all species examined is drastically reduced. The enzyme from dog and cat is more heat-sensitive than that of man and rat. Enzyme activity is completely lost at 60°C within 5 min.

2.9.2. pH Optimum

Depending on enzyme source and purity, substrate, and assay conditions, minor differences exist, but the pH optimum for COMT is generally between pH 7.3 and 8.5 (Assicot and Bohuon, 1970; Axelrod, 1962; Axelrod and Tomchick, 1958; Bade et al., 1976; Borchardt, 1981; Borchardt et al., 1974; Gulliver and Tipton, 1978b). Purified pig liver COMT shows a bell-shaped activity profile, with a pH optimum between 7.1 and 7.4 (Gulliver and Tipton, 1978b). On the other hand, purified rat liver soluble COMT has optimal activity between pH 8.2 and 8.5, and purified rat liver membrane-bound enzyme exhibits a pH optimum of 7.8–8.0 (Assicot and Bohuon, 1970; Borchardt et al., 1974). Bade et al. (1976) reported that crude enzyme from rat brain, erythrocytes, and liver has a pH optimum of 8.

2.9.3. Kinetics

A wide variety of kinetic data on COMT has been reported in the literature, with use of various substrates, crude and purified soluble or membrane-bound enzyme, enzyme from different tissues or species, and various assay conditions (Axelrod and Vesell, 1970; Borchardt and Cheng, 1978; Borchardt et al., 1974; Gordonsmith et al., 1982; Gulliver and Tipton, 1978b; Rivett and Roth, 1982; Roth, 1980; Tunnicliff and Ngo, 1983; White and Wu, 1975).

K_m values for SAM, reported by different workers, range from 3 to 69 μ*M* (Borchardt and Cheng, 1978; Borchardt et al., 1974; Gulliver and Tipton, 1978b; Rivett and Roth, 1982; Roth, 1980; Tunnicliff and Ngo, 1983; White and Wu, 1975). When 3,4-dihydroxybenzoic acid is used as substrate, the K_m of COMT ranges between 100 and 740 μ*M*, depending upon enzyme sources (Borchardt and Cheng, 1978; Borchardt et al., 1974; Gulliver and Tipton, 1978b; Tunnicliff and Ngo, 1983). There are also other kinetic data available in the literature resulting from the use of different substrates, such as DA and NA (Axelrod and Vesell, 1970; Borchardt and Cheng, 1978; Borchardt et al., 1974; Gordonsmith et al., 1982; Rivett and Roth, 1982; Roth, 1980; White and Wu, 1975).

2.10. Psychiatric Diseases

The availability and balance of brain catecholamines are thought to play an important role, directly or indirectly, in a variety of psychiatric disorders. Therefore, COMT, one of the catecholamine-metabolizing enzymes, may influence the availability of brain catecholamines at key sites. Erythrocyte COMT has been measured in patients with various mental illnesses, and its relationship to these disorders has been reported. However, studies of erythrocyte COMT levels in patients with particular types of mental illness have yielded conflicting results (Briggs and Briggs, 1973; Cohn et al., 1970; Davidson et al., 1979; Dunner, 1979; Dunner et al., 1971, 1977; Gershon and Jonas, 1975; Gershon et al., 1979, 1980; Guldberg and Marsden, 1975; White et al., 1976).

The initial studies suggest reduced mean erythrocyte COMT activity in women with primary affective disorder (history of mania, depression, or both), as compared with controls of men with the same illness (Cohn et al., 1970; Dunner et al., 1971). It was also observed that erythrocyte COMT activity is diminished in unipolar depressed women, compared to women with bipolar depressive states (depression with a history of mania or hypomania) and normal women (Briggs and Briggs, 1973; Cohn et al., 1970, Dunner et al., 1971). In contrast, Gershon and Jonas (1975) observed that erythrocyte COMT activity in patients with primary affective disorder is elevated instead of decreased. Several other studies demonstrated that there is no significant difference in the enzyme levels in patients with primary affective disorder and in controls (Davidson et al., 1979; Dunner, 1979; Dunner et al., 1977; Gershon et al., 1979; White et al., 1976). These inconclusive findings may be partly because of differences in criteria of patient selection and sample sizes.

The transmethylation hypothesis in schizophrenia suggests that a defect in the methylation of catecholamines might elicit some of the schizophrenic symptoms (Kety, 1967). Therefore, the erythrocyte COMT activity from schizophrenic patients has been examined by a number of investigators. These studies have, however, produced conflicting results. The majority of these studies show no significant differences in erythrocyte COMT activity found in schizophrenic patients and controls (Baron et al., 1984; Briggs and Briggs, 1973; Cohn et al., 1970; Dunner et al., 1971; Ebstein et al., 1976; Floderus et al., 1981; Groshong et al., 1978; Lewander et al., 1981). However, some studies (Matthysse and Baldessarini, 1972; Poitou et al., 1974; Shopsin et al., 1973; White et al., 1976) have reported significantly higher erythrocyte COMT activity in schizophrenic patients than in controls.

2.11. Summary

COMT is localized in brain, as well as in various peripheral tissues. Its estimated molecular weight is around 24,000, and its optimal pH is between 7.3 and 8.5. The enzyme has a broad substrate specificity and its inhibitors are structurally similar to either substrate or product. Its antibodies inhibit the enzyme of a given species more efficiently than one from another species. Erythrocyte COMT levels have been implicated in various psychiatric disorders, but conflicting results have been reported.

Acknowledgments

I am greatly indebted to Drs. Tong H. Joh and Donald J. Reis for their guidance and support. I would also like to express my deep gratitude to my wife, Min Jung, and my children, Henry and Bernard, who gave me encouragement and continuous moral support.

References

Anderson P. J. and D'Ioro A. (1968) Purification and properties of catechol-*O*-methyltransferase. *Biochem. Pharmacol.* **17,** 1943–1949.

Assicot M. and Bohuon C. (1969) Production of antibodies to catechol-*O*-methyltransferase (EC 2.1.1.6.) of rat liver. *Biochem. Pharmacol.* **18,** 1893–1898.

Assicot M. and Bohuon C. (1970) Purification and studies of catechol-*O*-methyltransferase of rat liver. *Eur. J. Biochem.* **12,** 490–495.

Axelrod J. (1962) Purification and properties of phenylethanolamine *N*-methyltransferase. *J. Biol. Chem.* **237,** 1657–1660.

Axelrod J. and Laroche M. J. (1959) Inhibitor of *O*-methylation of epinephrine and norepinephrine in vitro and in vivo. *Science* **130,** 800.

Axelrod J. and Tomchick R. (1958) Enzymatic *O*-methylation of epinephrine and other catechols. *J. Biol. Chem.* **233,** 702–705.

Axelrod J. and Vesell E. S. (1970) Heterogeneity of *N*- and *O*-methyltransferase. *Mol. Pharmacol.* **6,** 78–84.

Axelrod J., Albers W., and Clemente C. D. (1959) Distribution of catechol-*O*-methyltransferase in the nervous system and other tissues. *J. Neurochem.* **5,** 68–72.

Bade P., Christ W., Rakow D., and Coper H. (1976) Comparison of catechol-*O*-methyltransferase from rat brain, erythrocytes and liver. *Life Sci.* **19,** 1833–1844.

Baetge E. E., Moon H. M., Kaplan B. B., Park D. H., Reis D. J., and Joh T. H. (1983) Identification of clones containing DNA complementary to phenylethanolamine *N*-methyltransferase mRNA. *Neurochem. Intl.* **5,** 611–617.

Bakke O. M. (1970) *O*-methylation of simple phenols in the rat. *Acta Pharmacol. Toxicol.* **28,** 28–38.

Ball P., Knuppen R., and Breuer H. (1971) Purification and properties of a catechol-*O*-methyltransferase of human liver. *Eur. J. Biochem.* **21,** 517–525.

Baron M., Gruen R., Levitt M., Hunter C., and Asnis L. (1984) Erythrocyte catechol *O*-methyltransferase activity in schizophrenia: Analysis of family data. *Am. J. Psychiatry* **141,** 29–32.

Belleau B. and Burba J. (1963) Occupancy of adrenergic receptors and inhibition of catechol *O*-methyltransferase by tropolones. *J. Med. Chem.* **6,** 755–759.

Berenbeim D. M., Wong D. L., Masover S. J., and Ciaranello R. D. (1979) Regulation of synthesis and degradation of rat adrenal phenylethanolamine *N*-methyltransferase. III. Stabilization of PNMT against thermal and tryptic degradation by *S*-adenosylmethionine. *Mol. Pharmacol.* **16,** 482–490.

Borchardt R. T. (1981) Catechol *O*-Methyltransferase in *Methods in Enzymology* (Jakoby W. B., ed.), vol. 77, pp. 267–272, Academic, New York.

Borchardt R. T. and Cheng C. F. (1978) Purification and characterization of rat heart and brain catechol methyltransferase. *Biochim. Biophys. Acta* **522,** 49–62.

Borchardt R. T., Cheng C. F., Cooke P. H., and Creveling C. R. (1974) The purification and kinetic properties of liver microsomal-catechol *O*-methyltransferase. *Life Sci.* **14,** 1089–1100.

Borchardt R. T., Cheng C. F., and Thakker D. R. (1975) Purification of catechol *O*-methyltransferase by affinity chromatography. *Biochem. Biophys. Res. Commun.* **63,** 69–77.

Borchardt R. T., Huber J. A., and Houston M. (1982) Catechol-*O*-methyltransferase. 10. 5-substituted 3-hydroxy-4-methoxybenzoic acid (isovanillic acids) and 5-substituted 3-hydroxy-4-methoxybenzaldehydes (isovanillins) as potential inhibitors. *J. Med. Chem.* **25,** 258–271.

Briggs M. H. and Briggs M. (1973) Hormonal Influences on erythrocyte catechol-*O*-methyl transferase activity in humans. *Experientia* **29,** 278–280.

Caspers M. L. (1982) Inhibition by lead of phenylethanolamine *N*-methyltransferase. *Biochem. Pharmacol.* **31,** 1985–1988.

Chen C. H., Klein D. C., and Robinson J. C. (1974) Catechol-*O*-methyltransferase in rat placenta, human placenta and choriocarcinoma grown in culture. *J. Reprod. Fertil.* **39,** 407–410.

Ciaranello R. D. (1978a) Regulation of phenylethanolamine *N*-methyltransferase synthesis and degradation. I. Regulation by rat adrenal glucocorticoids. *Mol. Pharmacol.* **14,** 478–489.

Ciaranello R. D. (1978b) Regulation of phenylethanolamine *N*-methyltransferase. *Biochem. Pharmacol.* **27,** 1895–1897.

Ciaranello R. D., Barchas R. E., Byers G. S., Stemmle D. W., and Barchas J. D. (1969) Enzymatic synthesis of adrenaline in mammalian brain. *Nature* **221,** 368–369.

Ciaranello R. D., Wong D. L., and Berenbeim D. M. (1978) Regulation of phenylethanolamine *N*-methyltransferase synthesis and degradation. II. Control of the thermal stability of the enzyme by an endogenous stabilizing factor. *Mol. Pharmacol.* **14,** 490–501.

Cohn C. K., Dunner D. L., and Axelrod J. (1970) Reduced catechol-*O*-methyltransferase activity in red blood cells of women with primary affective disorder. *Science* **170,** 1323–1324.

Connett R. J. and Kirshner N. (1970) Purification and properties of bovine phenylethanolamine *N*-methyltransferase. *J. Biol. Chem.* **245,** 329–334.

Coward J. K., Slisz E. P., and Wu F. Y. H. (1973) Kinetic studies on catechol *O*-methyltransferase. Product inhibition and the nature of the catechol binding site. *Biochemistry,* **12,** 2291–2297.

Creveling C. R., Morris N., Shimizu H., Ong H. H., and Daly J. (1972) Catechol *O*-methyltransferase. IV. Factors affecting *m*- and *p*-methylation of substituted catechols. *Mol. Pharmacol.* **8,** 398–409.

Creveling C. R., Borchardt R. T., and Isersky C. (1973) Immunological Characterization of Catechol-*O*-Methyltransferase, in *Frontiers in Catecholamine Research* (Usdin E. and Snyder S. H., eds.), pp. 117–119, Pergamon, New York.

Davidson J. R. T., McLeod M. N., Turnbull C. D., White H. L., and Feuer E. J. (1979) Catechol-*O*-methyltransferase activity and classification of depression. *Biol. Psychiatry* **14,** 937–942.

Deguchi T. and Barchas J. (1971) Inhibition of transmethylations of biogenic amines by *S*-adenosylhomocysteine: Enhancement of transmethylation by adenosylhomocysteinase. *J. Biol. Chem.* **246,** 3175–3181.

Diaz-Borges J. M., Urbina M., and Drujan B. D. (1978) Some properties of phenylethanolamine *N*-methyltransferase of rat brain. *Neurochem. Res.* **3,** 15–26.

Dunner D. L. (1979) The use of erythrocyte catechol-*O*-methyltransferase activity as a possible marker for vulnerability of affective disorder. *Psychopharmacol. Bull.* **15,** 32.

Dunner D. L., Cohn C. K., Gershon E. S., and Goodwin F. K. (1971) Differential catechol-*O*-methyltransferase activity in unipolar and bipolar affective illness. *Arch. Gen. Psychiatry* **25,** 348–353.

Dunner D. L., Levitt M., Kumbaraci T., and Fieve R. R. (1977) Erythrocyte catechol-*O*-methyltransferase in primary affective disorder. *Biol. Psychiatry* **12,** 237–244.

Ebstein R., Belmaker R. H., Benbenisty D., and Rimon R. (1976) Electrophoretic pattern of red blood cell catechol-*O*-methyltransferase in schizophrenia and manic-depressive illness. *Biol. Psychiatry* **11,** 613–623.

Fahn S., Comi R., Snider S. R., and Prasad A. L. N. (1979) Effect of a catechol-*O*-methyltransferase inhibitor, U-0521, with levo-DOPA administration. *Biochem. Pharmacol.* **28,** 1221–1225.

Floderus Y., Book J. A., and Wetterberg L. (1981) Erythrocyte catechol-*O*-methyltransferase activity in related families with schizophrenia. *Clin. Genet.* **19,** 379–385.

Flohe L. (1974) Catechol-*O*-methyltransferase. *Int. Pharmacopsychiatry* **9,** 52–60.

Flohe L. and Schwabe K. P. (1970) Kinetics of purified catechol-*O*-methyltransferase. *Biochim. Biophys. Acta* **220,** 469–476.

Fuller R. W. (1982) Pharmacology of brain epinephrine neurons. *Ann. Rev. Pharmacol. Toxicol.* **22,** 31–55.

Fuller R. W. and Hemrick S. K. (1979) Properties of norepinephrine *N*-methyltransferase from rat, cat, and chicken brain. *Comp. Biochem. Physiol.* **62C,** 243–245.

Fuller R. W. and Hemrick-Luecke S. K. (1983) Species differences in epinephrine concentration and norepinephrine *N*-methyltransferase activity in hypothalamus and brain stem. *Comp. Biochem. Physiol.* **74C,** 47–49.

Fuller R. W. and Hunt J. M. (1967) Inhibition of phenylethanolamine *N*-methyltransferase by its product, epinephrine. *Life Sci.* **6,** 1107–1112.

Fuller R. W. and Molloy B. B. (1977) Inhibition in vitro of norepinephrine *N*-methyltransferase by 2-aminotetralins, analogs of phenylethylamines with rigid conformation. *Biochem. Pharmacol.* **26,** 446–447.

Fuller R. W., Warren B. J., and Molloy B. B. (1970) Substrate specificity of phenylethanolamine *N*-methyltransferase from rabbit adrenal. *Biochim. Biophys. Acta* **222,** 210–212.

Fuller R. W., Mills J., and Marsh M. M. (1971) Inhibition of phenethanolamine *N*-methyltransferase by ring-substituted α-methyl-phenethylamines (amphetamines). *J. Med. Chem.* **14,** 322–325.

Fuller R. W., Hemrick, S. K., and Molloy B. B. (1977) Position of a chloro substituent on phenylethanolamine as a determinant of methylation by rabbit adrenal norepinephrine *N*-methyltransferase. *Res. Commun. Chem. Pathol. Pharmacol.* **18,** 577–580.

Gershon E. S. and Jonas W. Z. (1975) Erythrocyte soluble catechol-*O*-methyltransferase activity in primary affective disorder: A clinical and genetic study. *Arch. Gen. Psychiatry* **32,** 1351–1356.

Gershon E. S., Lake C. R., and Leckman J. F. (1979) Reduced Erythrocyte

COMT and Plasma DBH Activities not Associated With Affective Disorders in Patients and Relatives, in *Catecholamines: Basic and Clinical Frontiers* (Usdin E., Kopin I. J., and Barchas J., eds.), pp. 1934–1936, Pergamon, New York.

Gershon E. S., Goldin L. R., Lake C. R., Murphy D. L., and Guroff J. J. (1980) Genetics of Plasma Dopamine-β-Hydroxylase (DBH), Erythrocyte Catechol-*O*-Methyltransferase (COMT), and Platelet Monoamine Oxidase (MAO) in Pedigrees of Patients With Affective Disorders, in *Enzymes and Neurotransmitters in Mental Disease* (Usdin E., Sourkes T. L., and Youdim M. B. H., eds.), pp. 281–299, John Wiley, New York.

Gordonsmith R. H., Raxworthy M. J., and Gulliver P. A. (1982) Substrate stereospecificity and selectivity of catechol-*O*-methyltransferase for DOPA, DOPA derivatives and α-substituted catecholamines. *Biochem. Pharmacol.* **31,** 433–437.

Groshong R., Baldessarini R. J., Gibson D. A., Lipinski J. F., Axelrod D., and Pope A. (1978) Activities of type A and B MAO and catechol-*O*-methyltransferase in blood cells and skin fibroblasts of normal and chronic schizophrenic subjects. *Arch. Gen. Psychiatry* **35,** 1198–1205.

Grunewald G. L., Grindel J. M., Vincek W. C., and Borchardt R. T. (1975) Importance of the aromatic ring in adrenergic amines: Nonaromatic analogues of phenylethanolamine as substrates for phenylethanolamine *N*-methyltransferase. *Mol. Pharmacol.* **11,** 694–699.

Guldberg H. C. and Marsden C. A. (1975) Catechol-*O*-methyl transferase: Pharmacological aspects and physiological role. *Pharmacol. Rev.* **27,** 135–206.

Gulliver P. A. and Tipton K. F. (1978a) Direct extraction radioassay for catechol-*O*-methyl-transferase activity. *Biochem. Pharmacol.* **27,** 773–775.

Gulliver P. A. and Tipton K. F. (1978b) The purification and properties of pig-liver catechol-*O*-methyl transferase. *Eur. J. Biochem.* **88,** 439–444.

Gulliver P. A. and Tipton K. F. (1979) The purification and properties of pig brain catechol-*O*-methyltransferase. *J. Neurochem.* **32,** 1525–1529.

Hoffman A. R., Ciaranello R. D., and Axelrod J. (1975) Substrate and inhibitor kinetics of bovine phenylethanolamine *N*-methyltransferase. *Biochem. Pharmacol.* **24,** 544–546.

Hokfelt T., Fuxe K., Goldstein M., and Johansson O. (1974) Immunohistochemical evidence for the existence of adrenaline neurons in the rat brain. *Brain Res.* **66,** 235–251.

Huh M. M. O. and Friedhoff A. J. (1979) Multiple molecular forms of catechol-*O*-methyltransferase: Evidence for two distinct forms, and their purification and physical characterization. *J. Biol. Chem.* **254,** 299–308.

Joh T. H. and Goldstein M. (1973) Isolation and characterization of multiple forms of phenylethanolamine *N*-methyltransferase. *Mol. Pharmacol.* **9,** 117–129.

Kaplan G. P., Hartman B. K., and Creveling C. R. (1980) Anti-catechol-*O*-methyltransferase: Demonstration of specificity and immunological cross-reactivity with the enzyme from rat liver, kidney, brain and choroid plexuses. *Neurochem. Res.* **5,** 869–877.

Kety S. S. (1967) Current biochemical approaches to schizophrenia. *N. Engl. J. Med.* **276,** 325–331.

Kirshner N. and Goodall McC. (1957) The formation of adrenaline from noradrenaline. *Biochim. Biophys. Acta* **24,** 658–659.

Kitabchi A. E. and Williams R. H. (1969) Phenylethanolamine *N*-methyltransferase in human adrenal gland. *Biochim. Biophys. Acta* **178,** 181–184.

Kitabchi A. E., Solomon S. S., and Williams R. H. (1968) Stimulatory effects of insulin and glucagon on phenylethanolamine *N*-methyltransferase of rat adrenal. *Proc. Soc. Exp. Biol. Med.* **127,** 296–300.

Knuppen V. R., Holler M., Tilmann D., and Breuer H. (1969) Interactions between oestrogens and catecholamines. II. Effect of oestrogen on the degradation and methylation of adrenaline in mouse under in vivo conditions. *Hoppe-Seylers Z. Physiol. Chem.* **350,** 1301–1309.

Kondo Y., Nagatsu I., Karasawa N., Kojima H., and Nagatsu T. (1982) Immunohistocytochemical and biochemical studies on catecholamine-synthesizing enzymes of the lumbar sympathetic ganglia and sciatic nerves of bullfrogs. *Biomed. Res.* **3,** 197–206.

Krakoff L. R. and Axelrod J. (1967) Inhibition of phenylethanolamine *N*-methyltransferase. *Biochem. Pharmacol.* **16,** 1384–1386.

Lee H. S., Schulz A. R., and Fuller R. W. (1974) Isolation of multiple forms of rabbit adrenal norepinephrine *N*-methyltransferase. *Arch. Biochem. Biophys.* **163,** 735–741.

Lee H. S., Schulz A. R., and Fuller R. W. (1978) Isolation and purification of rabbit adrenal norepinephrine *N*-methyltransferase isozyme. *Arch. Biochem. Biophys.* **185,** 222–227.

Lew J. Y., Miyamoto T., and Goldstein M. (1976) Inhibition of PNMT activity in the adrenal glands and brainstem of rats. *Biochem. Pharmacol.* **25,** 1433–1434.

Lew J. Y., Matsumoto Y., Pearson J., Goldstein M., Hokfelt T., and Fuxe K. (1977) Localization and characterization of phenylethanolamine *N*-methyltransferase in the brain of various mammalian species. *Brain Res.* **119,** 199–210.

Lewander T., Pongracz G. V., Backstrom M., and Wetterberg L. (1981) Dopamine metabolism in red blood cells in schizophrenia. *Clin. Genet.* **19,** 410–413.

Liang N. Y., Tessel R. E., Grunewald G. L., and Borchardt R. T. (1982) Inhibitors of phenylethanolamine *N*-methyltransferase. 1. Effects of 2-cyclooctyl-2-hydroxyethylamine on rat brain and adrenal catecholamine content and blood pressure. *J. Pharmacol. Exp. Ther.* **223,** 375–381.

Masover S. J., Berenbeim D. M., and Ciaranello R. D. (1979) Regulation of synthesis and degradation of rat adrenal phenylethanolamine *N*-methyltransferase. IV. Synergistic stabilization of the enzyme against thermal and tryptic degration by *S*-adenosylmethionine and biogenic amine substrates. *Mol. Pharmacol.* **16,** 491–503.

Matthysse S. and Baldessarini R. J. (1972) *S*-adenosylmethionine and catechol-*O*-methyltransferase in schizophrenia. *Am. J. Psychiatry* **128,** 1310–1312.

McGeer P. L. and McGeer E. G. (1964) Formation of adrenaline by brain tissue. *Biochem. Biophys. Res. Commun.* **17,** 502–507.

Molinoff P. B. and Axelrod J. (1971) Biochemistry of catecholamines. *Ann. Rev. Biochem.* **40,** 465–499.

Molinoff P. B., Brimijoin S., Weinshilboum R., and Axelrod J. (1970) Neurally mediated increase in dopamine-β-hydroxylase activity. *Proc. Natl. Acad. Sci. USA* **66,** 453–458.

Park D. H. and Joh T. H. (1985) Species-specific charge forms of phenylethanolamine *N*-methyltransferase. *Brain Res.,* in press.

Park D. H. and Joh T. H. (1984) Activation of adrenal phenylethanolamine *N*-methyltransferase by phosphate. *Biochem. Pharmacol.* **33,** 2911–2913.

Park D. H., Baetge E. E., Kaplan B. B., Albert V. R., Reis D. J., and Joh T. H. (1982) Different forms of adrenal phenylethanolamine *N*-methyltransferase: Species-specific posttranslational modification. *J. Neurochem.* **38,** 410–414.

Pendleton R. G., Kaiser C., and Gessner G. (1976) Studies on adrenal phenylethanolamine *N*-methyltransferase (PNMT) with SKF 64139, a selective inhibitor. *J. Pharmacol. Exp. Ther.* **197,** 623–632.

Pendleton R. G., Gessner G., Weiner G., Jenkins B., Sawyer J., Bondinell W., and Intoccia A. (1979) Studies on SKF 29661, an organ-specific inhibitor of phenylethanolamine *N*-methyltransferase. *J. Pharmacol. Exp. Ther.* **208,** 24–30.

Pohorecky L. A. and Baliga B. S. (1973) Purification and properties of rat adrenal phenylethanolamine *N*-methyl transferase. *Arch. Biochem. Biophys.* **156,** 703–711.

Pohorecky L. A. and Wurtman R. J. (1971) Adrenocortical control of epinephrine synthesis. *Pharmacol. Rev.* **23,** 1–35.

Pohorecky L. A., Zigmond M., Karten H., and Wurtman R. J. (1969) Enzymatic conversion of norepinephrine to epinephrine by the brain. *J. Pharmacol. Exp. Ther.* **165,** 190–195.

Pohorecky L. A., Baliga B. S., Wurtman R. J., and Bartter F. C. (1973) Adrenocortical control of catecholamine metabolism in the dog adrenal medulla: Relationship to protein synthesis. *Endocrinology* **93,** 566–574.

Poitou P., Assicot M., and Bohuon C. (1974) Soluble and Membrane catechol-*O*-methyltransferase in red blood cells of schizophrenic patients. *Biomedicine* **21,** 91–93.

Rafferty M. F. and Grunewald G. L. (1982) The remarkable substrate activity for phenylethanolamine *N*-methyltransferase of some conformationally defined phenylethylamines lacking a side-chain hydroxyl group. *Mol. Pharmacol.* **22,** 127–132.

Raxworthy M. J., Gulliver P. A., and Hughes P. J. (1982) The cellular location of catechol-*O*-methyltransferase in rat liver. *Naunyn Schmiedebergs Arch. Pharmacol.* **320,** 182–188.

Rivett A. J. and Roth J. A. (1982) Kinetic studies on the *O*-methylation of dopamine by human brain membrane-bound catechol *O*-methyltransferase. *Biochemistry* **21,** 1740–1742.

Rivett A. J., Francis A., and Roth J. A. (1983a) Distinct cellular localization of membrane-bound and soluble forms of catechol-*O*-methyltransferase in brain. *J. Neurochem.* **40,** 215–219.

Rivett A. J., Francis A., and Roth J. A. (1983b) Localization of membrane-bound catechol-*O*-methyltransferase. *J. Neurochem.* **40,** 1494–1496.

Roth J. A. (1980) Presence of membrane-bound catechol-*O*-methyltransferase in human brain. *Biochem. Pharmacol.* **29,** 3119–3122.

Saavedra J. M., Palkovits M., Brownstein M. J., and Axelrod J. (1974) Localization of phenylethanolamine *N*-methyltransferase in the rat brain nuclei. *Nature* **248,** 695–696.

Saavedra J. M., Brownstein M. J., and Palkovits M. (1976) Distribution of catechol-*O*-methyltransferase, histamine *N*-methyltransferase and monoamine oxidase in specific areas of the rat brain. *Brain Res.* **118,** 152–156.

Senoh S., Daly J., Axelrod J., and Witkop B. (1959) Enzymatic *p*-*O*-methylation by catechol *O*-methyl transferase. *J. Am. Chem. Soc.* **81,** 6240–6245.

Shopsin B., Wilk S., Gershon S., Roffman M., and Goldstein M. (1973) Collaborative Psychopharmacologic Studies Exploring Catecholamine Metabolism in Psychiatric Disorders, in *Frontiers in Catecholamine Research* (Usdin E. and Snyder S. H., eds.), Pergamon, New York.

Teitelman G., Skaper S., Baker H., Park D. H., Joh T. H., and Adler R. (1984) Expression of phenylethanolamine *N*-methyltransferase in sympathetic neurons and extraadrenal chromaffin tissue of chick embryos in vivo and in vitro. *Dev. Brain Res.* **13,** 283–291.

Thoenen H., Mueller R. A., and Axelrod J. (1970) Neuronally dependent induction of phenylethanolamine *N*-methyltransferase by 6-hydroxydopamine. *Biochem. Pharmacol.* **19,** 669–673.

Tunnicliff G. and Ngo T. T. (1983) Kinetics of rat brain soluble catechol-*O*-methyltransferase and its inhibition by substrate analogues. *Int. J. Biochem.* **15,** 733–738.

Vogel W. H., Lewis L. E., and Boehme D. H. (1976) Phenylethanolamine *N*-methyltransferase activity in various areas of human brain, tissues and fluids. *Brain Res.* **115,** 357–359.

White H. L. and Wu J. C. (1975) Properties of catechol *O*-methyltransferases from brain and liver of rat and human. *Biochem. J.* **145,** 135–143.

White H. L., McLeod M. N., and Davidson J. R. T. (1976) Catechol *O*-methyltransferase in red blood cells of schizophrenic, depressed and normal human subjects. *Br. J. Psychiatry* **128,** 184–187.

Wurtman R. J. and Axelrod J. (1966) Control of enzymatic synthesis of adrenaline in the adrenal medulla by adrenal cortical steroids. *J. Biol. Chem.* **241,** 2301–2305.

Wurtman R. J., Axelrod J., Vesell E. S., and Ross G. T. (1968) Species differences in inducibility of phenylethanolamine *N*-methyltransferase. *Endocrinology* **82,** 584–590.

Yu P. H. (1978) Phenylethanolamine *N*-methyltransferase from the brain and adrenal medulla of the rat: A comparison of their properties. *Neurochem. Res.* **3,** 755–762.

Zurcher G. and Da Prada M. (1982) Rapid and sensitive single-step radiochemical assay for catechol-*O*-methyltransferase. *J. Neurochem.* **38,** 191–195.

Chapter 5
Histamine *N*-Methyltransferase
Enzymology and Functional Aspects

Kenneth M. Verburg and David P. Henry

1. Introduction

Histamine *N*-methyltransferase (HNMT; EC 2.1.1.8) was discovered a little over 25 yr ago. Since that time, HNMT has been the subject of numerous investigations, and although these studies have provided a fundamental understanding of this enzyme, HNMT has not been the subject of a comprehensive review or analysis. This is surprising since HNMT and diamine oxidase (DAO; EC 1.3.4.6) are the only known enzymes that terminate the physiologic activity of histamine (HA) in mammals. However, to a large extent the physiologic significance of HNMT remains a conundrum, and the importance of this enzyme may have been overshadowed by the lack of pharmacologic or pathologic conditions that are characterized by abnormalities in HNMT activity.

The preponderance of HNMT research has centered on the characteristics of the HNMT reaction, enzyme properties, and pharmacologic inhibitors of HNMT activity. Information on the distribution, localization, and function of HNMT is more limited. In most instances, it has been difficult to project the importance or significance of in vitro findings to the in vivo state. Additionally, it is easy to envision that the enclosed nature of HNMT research may have caused the selection of inadequate or improper experimental procedures and could account for the limited success of studies designed to evaluate the functional significance of HNMT in vivo.

Recently, major advancements have occurred in the enzymologic study of HNMT. For example, within the past 5 yr, HNMT was purified to a homogeneous state, an antibody to HNMT was produced, a specific pharmacologic inhibitor of HNMT that is active both in vitro and in vivo was identified, and substantial improvements in HNMT assay procedures have taken place. The progress of HNMT research in these areas may well precipitate a renewed effort and expedite future analysis of the function and importance of HNMT in vivo.

This chapter will review enzymologic and functional aspects of HNMT. This discussion is organized to address characteristics of the HNMT reaction, inhibitors of HNMT, and the distribution and functional significance of HNMT in the CNS and the periphery. The final section reviews the properties of HNMT and the methodologic protocols that have been established to study this important enzyme.

2. Historical Perspective

The nomenclature of imidazole compounds is unnecessarily confusing. The ambiguity of imidazole terminology stems from the use of nonstandardized nomenclature with respect to substitutions on the imidazole ring. The nomenclature used in the present discussion will be that recommended by Black and Ganellin (1974). The major pathways for HA catabolism in mammals are depicted in Fig. 1. However, other enzymatic conversions of HA are possible (Schayer, 1959), and the products of these potential interconversions, as well as some synthetic imidazole compounds, are shown in Fig. 2.

Imidazoleacetic acid was the first metabolite of HA to be identified from urine following the administration of [^{14}C]-HA (Schayer, 1952; Mehler et al., 1952). These studies provided the first evidence to indicate the importance of DAO in the catabolism of HA in vivo. From in vitro studies, it was already established that HA could be oxidized by DAO and an aldehyde dehydrogenase to imidazoleacetic acid (Zeller, 1951). Later, Karjala et al. (1956) found that another major product of HA catabolism in vivo was not a result of the action of DAO. This compound was identified as *tele*-methylimidazole acetic acid, and suggested that a potential precursor of this metabolite could be *tele*-methylhistamine (Karjala and Turnequest, 1955). Later evidence demonstrated that *tele*-methylhistamine was, in fact, derived from HA in vivo and was present in urine, although most was sequentially oxidized by monoamine oxidase type B and an aldehyde dehydrogenase to

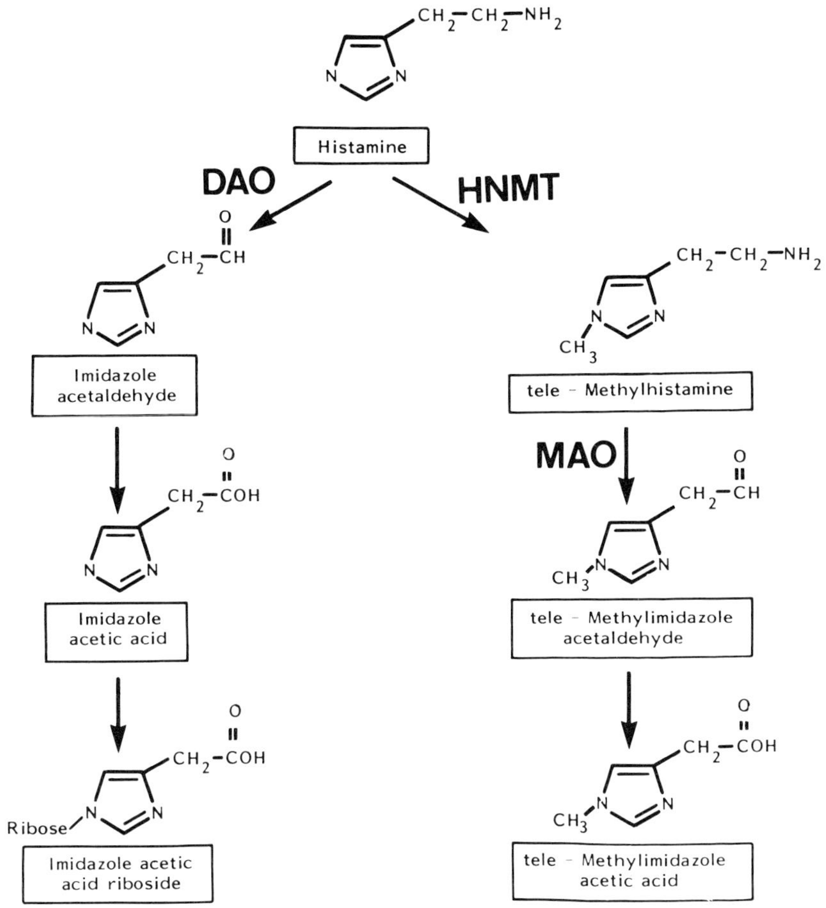

Fig. 1. The major catabolic pathways for histamine in mammals are shown. Histamine catabolism occurs via methylation and oxidative deamination, and is very similar, therefore, to monoamine and indoleamine metabolism. The relative contribution of each pathway to the inactivation of histamine is species- and tissue-dependent.

tele-methylimidazoleacetic acid (Schayer and Karjala, 1956; Rothschild and Schayer, 1958).

The discovery of *tele*-methylhistamine was the impetus for the initial characterization of HNMT. Lindahl (1958) demonstrated the formation of *tele*-methylhistamine when HA was incubated with mouse liver homogenates and the cosubstrate *S*-adenosyl-L-methionine (SAME). Brown et al. (1959a,b) developed the initial method for quantifying HNMT activity from mammalian tissues and then reported on the partial purification, tissue distribution,

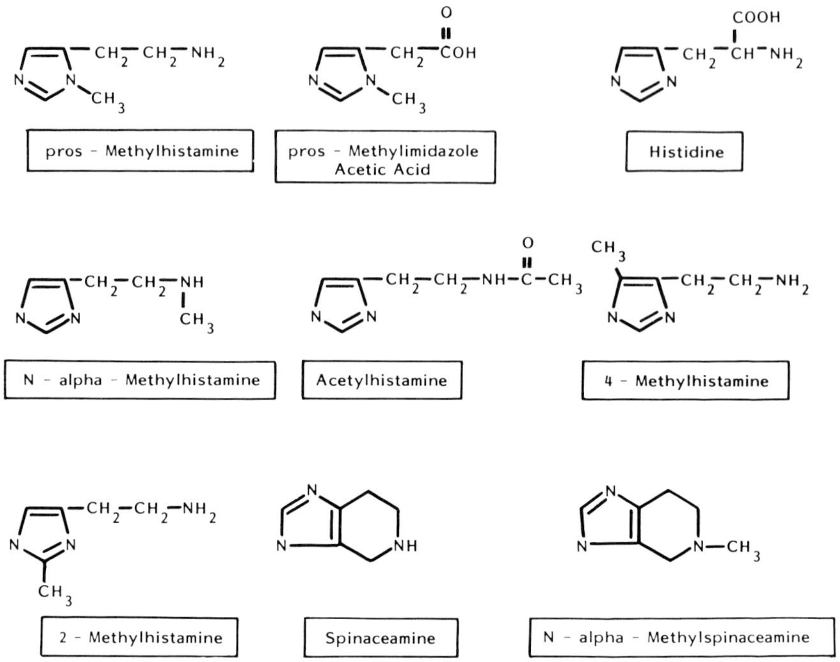

Fig. 2. The structures of other important imidazole compounds are shown. Histidine is the precursor for histamine synthesis and acetylhistamine is an inactive metabolite of histamine formed in bacteria. *Pros*-methyhistamine and *pros*-methylimidazoleacetic acid are also inactive metabolites that may be derived from histamine in mammalian species. *N*-alpha-methylhistamine, spinaceamine, and *N*-alpha-methylspinaceamine are substrates for HNMT, although the synthesis and physiologic significance of these compounds remains unresolved. 2-Methylhistamine and 4-methylhistamine are synthetic histamine receptor agonists.

and characterization of HNMT. Shortly thereafter, Lindahl (1960) confirmed and extended the initial characterization of HNMT.

3. Characteristics of the HNMT Reaction

HNMT exhibits a high methyl-acceptor substrate specificity when compared with other biogenic amine methyltransferases, such as catechol *O*-methyltransferase (COMT) or phenylethanolamine *N*-methyltransferase (PNMT) (Lindahl, 1960; Gustafsson and Forshell, 1964; Guldberg and Marsden, 1975; Fuller et al., 1970). Histamine is most certainly the kinetically favored substrate for HNMT. However, questions remain as to whether it is the only endogenous or physiological substrate present in mammalian tissues.

The methylation of HA by HNMT occurs exclusively at the *tele*-nitrogen position of the imidazole ring (Fig. 1). A prerequisite for methylation of HA congeners by HNMT is an intact imidazole ring, unsubstituted at the *tele-*, *pros-*, and the two-position. However, certain HA analogs, distinguished by modifications of the ethylamine side chain or, to a lesser extent, by substitutions at the four-position of the imidazole ring, may be substrates for HNMT.

HNMT is not the only mammalian enzyme that will methylate HA in vitro. Highly purified indoleamine *N*-methyltransferase (INMT; EC 2.1.1.49) isolated from rabbit lung also has HA methylating activity (Herman and Henry, 1984). The product of the INMT reaction is not *tele*-methylhistamine, however. Instead, INMT catalyzes the formation of both *pros*-methylhistamine and *N*-alpha-methylhistamine from HA in vitro (Fig. 2). The formation of *pros*-methylhistamine from HA appears to be favored since this product is formed at three times the rate of *N*-alpha-methylhistamine formation. The importance of INMT activity in the metabolism of HA in vivo is not defined at the present time.

3.1. Methyl Acceptors

3.1.1. Synthetic Substrates

A limited number of synthetic histamine congeners are substrates for HNMT. Homohistamine, an imidazole compound with a propylamine side chain, is a substrate for HNMT and has a K_m value similar to HA (Dent et al., 1982). Alpha-methylhistamine is composed of an imidazole ring substituted with an isopropylamine moiety. The K_m of this substrate for HNMT is 10 times higher than the K_m of HA (Hough et al., 1981). 2-Amino-ethylmercapto-methylimidazole is also a substrate for HNMT, but the methylation of this compound is minimal in comparison to HA (Barth et al., 1980).

These studies point out that very few modifications of the ethylamine side chain of HA are allowed before activity as a methyl acceptor for HNMT is reduced. The side chain must contain a basic group or center of positive charge within a defined distance from the imidazole ring. Substitutions on the alkyl side chain will reduce the activity of the compound as a substrate for HNMT. Imidazole compounds with a nonpolar or negatively charged side chain show no activity as substrates or, for that matter, as inhibitors of HNMT (Barth and Lorenz, 1978; Barth et al., 1980).

3.1.2. Naturally Occurring and Endogenous Substrates

Despite the restricted substrate specificity of HNMT, HA is not the only naturally occurring substrate for HNMT. These substrates are most easily classified as HA congeners that have alkyl substitutions on the alpha-nitrogen of the ethylamine side chain. For example, *N*-alpha-methylhistamine and *N,N*-alpha-dimethylhistamine are substrates for HNMT (Gustafsson and Forshell, 1964; Barth et al., 1980). The importance of this observation lies in the fact that both *N,N*-alpha-methylhistamine and *N*-alpha-dimethylhistamine are potent HA receptor agonists and could represent alternate endogenous ligands for this receptor system (Lin et al., 1962). Both compounds have been tentatively identified in human urine (Kapeller-Adler and Iggo, 1957) and mammalian gastric mucosal extracts (Navert et al., 1967; Kovacs and Heisler, 1970; Code et al., 1976). However, more recent studies have not substantiated these findings (Maslinski et al., 1977; Verburg and Henry, 1984).

Erspamer et al. (1963) found high concentrations of spinaceamine (4,5,6,7-tetrahydroimidazo-[5,4-*c*]pyridine) and *N*-alpha-methylspinaceamine (6-methyl-4,5,6,7-tetrahydroimidazo-[5,4-*c*]pyridine) in the skin of some South American amphibian species. Spinaceamine and *N*-alpha-methylspinaceamine are structurally similar to *N*-alpha-methylhistamine and *N,N*-alpha-dimethylhistamine, except that the side chain is cyclized into a six-membered ring by attachment of an *N*-alpha-methylene group to the four-position of the imidazole ring (Fig. 2). Spinaceamine and *N*-alpha-methylspinaceamine are actually better substrates for HNMT than are the noncyclized structural isomers (Barth et al., 1980; Verburg and Henry, 1984). High concentrations of spinaceamine and *N*-alpha-methylspinaceamine are found in rat urine; smaller quantities are also present in human and dog urine (Verburg and Henry, 1984). More studies are required to validate these findings and to elucidate whether these compounds originate from dietary intake or are endogenously synthesized in some mammalian species. These cyclic HA derivatives have little or no H-1 or H-2 receptor agonist or antagonist properties (Bertaccini and Vitale, 1964). However, this does not preclude the possibility that these compounds are of physiologic importance.

Benzimidazole and purine compounds are similar to spinaceamine in that both are bicyclic structures comprised of an imidazole ring, although in each case the six-membered ring is aromatic. Purine and adenine are weak substrates for HNMT, whereas guanine, xanthine, hypoxanthine, and uric acid are not methylated to any degree. Surprisingly, benzimidazole and 5,6-dimethylbenzimidazole are also substrates for HNMT (Verburg and Henry, 1984). Apparently, the electron-withdrawing characteristics of the benzene ring are close enough to an ethylaminoalkyl side chain to allow methylation of benzimidazole by HNMT. Benzimidazole compounds are not usually considered to be naturally occurring compounds. However, 5,6-dimethylbenzimidazole is a structural component of vitamin B-12 (Harper et al., 1979) and could be liberated by vitamin B-12 catabolism. It is also interesting to note that some H-1 antagonists, such as clemazole, contain the benzimidazole moiety and inhibit HNMT activity in vitro (Netter and Bodenschatz, 1967).

The list of substrates for HNMT has continued to expand and now includes many HA congeners as well as other imidazole-containing compounds that are not necessarily derived from, or related to, HA metabolism. At present, it is difficult to speculate on the importance of these findings. Compounds that contain an imidazole ring and are known to be important in mammalian biochemistry, such as adenine or purine, are very poor substrates for HNMT. On the other hand, the existence or role of kinetically favored HNMT substrates, such as spinaceamine or *N*-alpha-methylspinaceamine in mammals, remains a mystery.

3.2. Methyl Donors

Lindahl (1958) was the first to recognize that the methyl donor in the HNMT reaction is SAME. *S*-Adenosyl-D-methionine is inactive as a methyl donor, although this compound does bind to HNMT (Zappia et al., 1969). Thus, the stereoisomeric configuration of the methyl donor is an important factor for the transfer of a methyl group. Other synthetic structural derivatives of SAME can serve as the methyl donor in the HNMT reaction (Borchardt, 1980). However, any modification of the base, amino acid, or sugar portion of the SAME molecule results in a substantially higher K_m value for HNMT, but has little impact on V_{max}. Thus, structural modifications of the SAME molecule appear to influence binding of the methyl donor to the HNMT catalytic site, but not the ability to transfer the methyl group.

3.3. Kinetic Characterization of HNMT Substrates

The K_m values of SAME and HA for HNMT have been reported from several laboratories. The kinetic constants for HNMT are actually quite similar despite the uncertainty surrounding factors that may affect these determinations, such as the ionic strength and pH of the incubation medium, the influence of endogenous substrate concentrations or the concentration of various endogenous inhibitors, and differences in the enzyme source and purity or the possibility of HNMT isozymes.

Guinea pig brain contains a high level of HNMT activity (Brown et al., 1959b). Accordingly, HNMT from this tissue has been used for numerous kinetic studies (Table 1). Although the activity of HNMT in mouse and rat brain is substantially less than in guinea pig brain (Brown et al., 1959b), these tissues have also been utilized in kinetic studies. The K_m values for both rat and mouse brain HNMT are similar in magnitude to guinea pig brain HNMT (Taylor and Snyder, 1972c; Orr and Quay, 1978). The K_m values of both substrates for HNMT obtained from various peripheral tissues have also been reported (Table 2). Once again, the K_m values for peripheral tissue HNMT are similar despite differences in enzyme purification and other experimental parameters used for K_m determinations.

Reaction temperature is an important consideration when estimating the K_m of HA for HNMT. The K_m value of HA for highly purified rat kidney HNMT falls precipitously as the reaction temperature is reduced from 37 to 2°C (Verburg et al., 1984). The K_m value of HA for HNMT at a fixed SAME concentration was reduced from 13.7 μM at 37°C to 0.9 μM at 2°C. A low incubation temperature does not diminish K_m values for any other HNMT substrate or

Table 1
K_m Comparison for Guinea Pig Brain HNMT

K_m for histamine, μM	K_m for SAME, μM	Reference
380	35	Brown et al. (1959b)
37	9.4	Suriychan and Thithapanda (1972)
8	6	Baudry et al. (1973)
43	9.7	Thithapanda and Cohn (1978)
13	—	Dent et al. (1982)
13.6	6.1	Matuszewska and Borchardt (1983)

Table 2
K_m Determinations for HNMT From Various Peripheral Tissues

Tissue	K_m for histamine, μM	K_m for SAME, μM	Reference
Rat kidney	6–9		Taylor and Snyder (1972c)
	2.9		Hough et al. (1981)
	7		Beaven and Shaff (1979a)
	13.7		Verburg et al. (1984a)
Cat kidney	26	13	Suriychan and Thithapanda (1972)
Cat intestine	27	11	
Rabbit intestine	12	8.7	
Pig liver	22	16	Gustafsson and Forshell (1964)
Pig gastric mucosa	23	43	Lorenz et al. (1970)
Human skin	4.2	1.8	Francis et al. (1980)

for substrates of other methyltransferase activities. This phenomenon has proved to be of great utility in optimizing specificity of the HNMT-based radioenzymatic assay for HA (Verburg et al., 1984). Additionally, this observation suggests that HNMT activity is maintained in tissues, such as skin, at temperatures less than 37°C, and that postmortem reductions in tissue HA levels may occur even at low temperatures as a result of the continued activity of HNMT.

3.4. Substrate Inhibition

High concentrations of HA are found to inhibit HNMT activity in vitro. Taylor and Snyder (1972c) observed that HA concentrations in excess of 10 μM inhibited HNMT activity from various mouse and rat tissues. Substrate inhibition of HNMT by HA is independent of the SAME concentration that is utilized. Furthermore, purification of HNMT does not eliminate substrate inhibition. Homogeneous HNMT activity from guinea pig brain is inhibited by high concentrations of HA (Matuszewska and Borchardt, 1983). Similar findings were reported for HNMT purified from the gastric mucosa of the pig antrum. Substrate inhibition began at HA concentrations of 120 μ*M* (Lorenz et al., 1970; Barth et al., 1973b). Histamine in concentrations of 25–100 μ*M* inhibited HNMT purified from human skin uncompetitively with respect to SAME. Additionally, HNMT activity was not inhibited by high concentrations of SAME (Francis et al., 1980).

The observations on substrate inhibition do not uniformly agree as to the HA concentration that defines the apex of the HNMT velocity curve. This may well be a function of the endogenous tissue level of HA and purity of the HNMT preparation. A reasonable estimation of the HA concentration that will first elicit substrate inhibition of HNMT activity places this value somewhere between 10 and 100 μ*M*.

The significance of substrate inhibition on HNMT activity in vivo is unknown. To an extent, it is disconcerting to rationalize the importance of substrate inhibition in view of certain conditions, such as anaphylactic shock, when free HA concentrations are dramatically elevated and could approach levels that are found to inhibit HNMT activity in vitro. Maximum HNMT activity would seem to be a pivitol compensatory mechanism in this situation. However, enzyme and substrate compartmentalization and other interacting factors must be clarified before the significance of substrate inhibition on the activity of HNMT in vivo can be evaluated.

3.5. Product Inhibition

Tele-methylhistamine and *S*-adenosyl-L-homocysteine (SAH) are formed by the HNMT reaction, and both products are known to inhibit HNMT activity in vitro. The inhibition of HNMT by *tele*-methylhistamine was first recognized in the initial characterization of this enzyme (Brown et al., 1959a,b; Lindahl, 1960). Later, Zappia et al. (1969) found that SAH also inhibited HNMT. The K_i of inhibition and the type of inhibition that each product exerts with respect to each substrate is of interest for two reasons. First, the pattern of product inhibition in relation to each substrate provides evidence to support a kinetic mechanism for HNMT. Second, a low in vitro K_i of inhibition coupled with a comparison of endogenous concentrations of SAH and *tele*-methylhistamine could provide a clue to the effect of product inhibition on HNMT activity in vivo.

Baudry et al. (1973) demonstrated that SAH inhibited guinea pig brain HNMT competitively with respect to SAME and noncompetitively with respect to HA. The K_i of inhibition was approximately 5 μ*M*. Orr and Quay (1978) confirmed these findings and extended this evaluation by characterizing *tele*-methylhistamine

Table 3
Product Inhibition of HNMT[a]

Inhibitor	Type of inhibition vs SAME	Type of inhibition vs histamine
tele-Methylhistamine	Noncompetitive	Noncompetitive or uncompetitive
S-Adenosyl-L-homocysteine	Competitive	Noncompetitive

[a]From Baudry et al. (1973); Orr and Quay (1978); Francis et al. (1980).

inhibition of rat brain HNMT (Table 3). The K_i of inhibition was approximately 9 μM. The concentrations of SAH and *tele*-methylhistamine in rat brain are approximately 10 times lower than the K_i values of these products as inhibitors of HNMT activity (Hough and Domino, 1979; Gharib et al., 1982). Francis et al. (1980) performed product inhibition studies on HNMT isolated from human skin. The pattern of inhibition observed for both products was nearly identical to previous studies.

In summary, the K_i values of inhibition for both HNMT products are in the range of K_m values reported for HA and SAME. The endogenous levels of *tele*-methylhistamine and SAH in rat brain are lower than their respective K_i values, but these concentrations may still be important factors in the modulation of HNMT activity in vivo. Again however, the compartmentalization of HNMT in relation to substrates and products is an undetermined variable. The kinetic mechanism for HNMT suggested by the pattern of product inhibition is discussed in the following section.

3.6. Kinetic Mechanism of HNMT

A comprehensive appraisal of the kinetic mechanism for highly purified guinea pig brain HNMT was performed by Thithapanda and Cohn (1978). The evidence from this study was consistent with a ping-pong mechanism. First, the concentration of the fixed cosubstrate was found to affect the K_m determination for the other substrate. Second, the pattern of *tele*-methylhistamine inhibition in relation to both substrates was consistent with a ping-pong mechanism.

A ping-pong mechanism implies that SAME is the first substrate to bind to HNMT. This interaction results in the formation of a methylated enzyme and the release of SAH. Histamine then binds to the methylated enzyme and the methyl group is transferred to this substrate, forming *tele*-methylhistamine and regenerated HNMT. However, the existence of a methylated enzyme could not be demonstrated by Thithapanda and Cohn (1978), confirming the earlier report of Gustafsson and Forshell (1964).

Recent evidence indicates that the enzymatic transfer of a methyl group from SAME to HA, catalyzed by HNMT, occurs with configurational inversion of the methyl group hydrogens (Asano et al., 1984). A reasonable interpretation of this observation is that the HNMT reaction involves only a single or direct transfer of the methyl group from one substrate to another. These results essentially exclude a ping-pong mechanism since the transfer of a methyl group to HNMT and then to HA would lead to a net retention of configuration.

The findings of other studies are also inconsistent with a ping-pong mechanism, and most evidence indicates that an ordered bi-bi mechanism best describes the HNMT reaction. Taylor and Snyder (1972c) found that varying the concentration of the fixed cosubstrate had no influence on the K_m for the other substrate. Also, the inhibition pattern for SAH and *tele*-methylhistamine, as described by others, differs from the predicted pattern for a ping-pong mechanism and is consistent with an ordered bi-bi mechanism (Orr and Quay, 1978; Baudry et al., 1973). Francis et al. (1980) also examined, in detail, the kinetic mechanism of HNMT from human skin and found a bi-bi sequential mechanism best characterized their observations on HNMT kinetics. In this scheme, SAME is still the first substrate to bind to HNMT, but SAH is not released until after HA binds to HNMT and *tele*-methylhistamine has been formed and released.

4. Inhibition of HNMT

The pharmacologic inhibition of HNMT activity has attracted a great deal of attention, and this subject is dealt with extensively in the HNMT literature. At first glance, it may be difficult to avoid questioning whether the intensive study of HNMT inhibition is commensurate with the importance of this topic. Upon closer examination, however, this viewpoint is most certainly premature.

Five general conclusions can be drawn from the subject of HNMT inhibition. First, in contrast to the restricted substrate specificity of HNMT, a diverse group of synthetic pharmacologic agents and naturally occurring compounds inhibit HNMT activity in vitro. Certain structural features that are common to most

HNMT inhibitors may account for this observation. Second, the potency of HNMT inhibitors in vitro is not necessarily an index of in vivo potency, as evidenced by antimalarial drugs and H-1 antagonists. Third, most competitive inhibitors of HNMT are found to stimulate or inhibit HNMT activity in vitro in a manner dependent on both the HA and inhibitor concentration. The occurrence or significance of this type of modulation on HNMT activity in vivo has not been elucidated. However, this observation may explain the lack of in vivo activity of competitive HNMT inhibitors. Fourth, some compounds have been shown to inhibit HNMT activity in vivo, although the utilization or effectiveness of these agents in vivo is generally limited by low potency and nonspecificity. SKF compound 91488 is an important exception to this rule, although experience with this compound is limited. Finally, many endogenous compounds, such as serotonin and dopamine, are potent inhibitors of HNMT in vitro. The importance of endogenous inhibitors on the modulation of HNMT activity in vivo remains unresolved. However, the presence of endogenous inhibitors is likely to affect the quantification of HNMT activity in tissue homogenates. In light of the preceding points of emphasis, this section reviews HNMT inhibitors by pharmacologic class and will discuss common structural characteristics, kinetics of inhibition, and in vivo activity of these compounds.

4.1. Phenothiazines

Phenothiazines possess a wide range of pharmacologic properties, including antagonism of H-1 receptors. Phenothiazines also inhibit HNMT activity in vitro (Brown et al., 1959b; Gustafsson and Forshell, 1963). Although phenothiazines are less potent than other antihistamines for antagonism of H-1 receptors, they are about equipotent with other H-1 antagonists as inhibitors of HNMT (Merril et al., 1966; Taylor and Snyder, 1972c; Barth et al., 1973a).

Phenothiazines appear to have minimal impact on HNMT activity in vivo. For example, chlorpromazine produces only a small decrease in the methylation of [^{14}C]-HA administered intraperitoneally to mice (Snyder and Axelrod, 1964). Chlorpromazine was shown to elevate endogenous HA levels in brain (Green and Erickson, 1964, 1967; Adam and Hye, 1966). However, Taylor and Snyder (1971, 1972a) found no evidence to support the idea that chlorpromazine inhibited HNMT activity in vivo. In contrast to previous studies, chlorpromazine failed to elevate brain HA levels.

4.2. Antimalarial Drugs

Antimalarial drugs may represent the most potent class of HNMT inhibitors in vitro. Cohn (1965) presented the initial observations on antimalarial inhibition of HNMT and reported that amodiaquine and quinacrine are potent inhibitors of HNMT. Quinacrine has a K_i value of 0.1 μM for mouse brain HNMT (Taylor and Snyder, 1972c). The K_i of amodiaquine is 0.019 μM for rat kidney HNMT (Beaven and Shaff, 1979b) and 0.012 μM for guinea pig brain HNMT (Thithapanda and Cohn, 1978). Both of these antimalarial agents contain a diethylaminoalkyl moiety that appears to be important in the recognition of the HNMT active site and is similar to the dimethylaminoalkyl group that many H-1 antagonists possess (Fig. 3). The inhibition of HNMT by quinacrine and amodiaquine is competitive with respect to HA (Taylor and Snyder, 1972c; Thithapanda and Cohn, 1978). However, these compounds are unique with respect to other competitive HNMT inhibitors (*see* H-1 antagonists, section 4.3).

Attempts to inhibit HNMT in vivo by amodiaquine and quinacrine have met with limited success in a broad spectrum of experiments. This was especially true when endogenous HA levels were used to evaluate HNMT inhibition. For example, Taylor and Snyder (1971, 1972a) observed no elevation in brain HA levels following quinacrine administration. Amodiaquine was also ineffective in altering endogenous HA concentrations in various tissues, although the methylation of [^{3}H]-HA was reduced (Andre and Schwartz, 1973; Schayer and Reilly, 1974, 1975a; Schwartz et al., 1974).

There is some indirect evidence that inhibition of HNMT by antimalarial agents reduces the methylation of HA or elevates endogenous HA levels. For example, amodiaquine and quinacrine potentiate the response of aortic strips to HA (Kalsner, 1970). Additionally, amodiaquine substantially enhances HA-stimulated gastric acid secretion (Troidl et al., 1973; Barth et al., 1975b). Both of these observations could result from HNMT inhibition and prevention of HA inactivation. Finally, the urinary excretion of HA in the rat is increased following amodiaquine administration (Kim and Cohn, 1966).

In general, the efficacy of quinacrine and amodiaquine as inhibitors of HNMT activity in vivo is much less than anticipated, based on in vitro findings. However, these agents may have limited access to HNMT. For example, antimalarial agents are known to be significantly bound to plasma proteins (Goldstein et al., 1974).

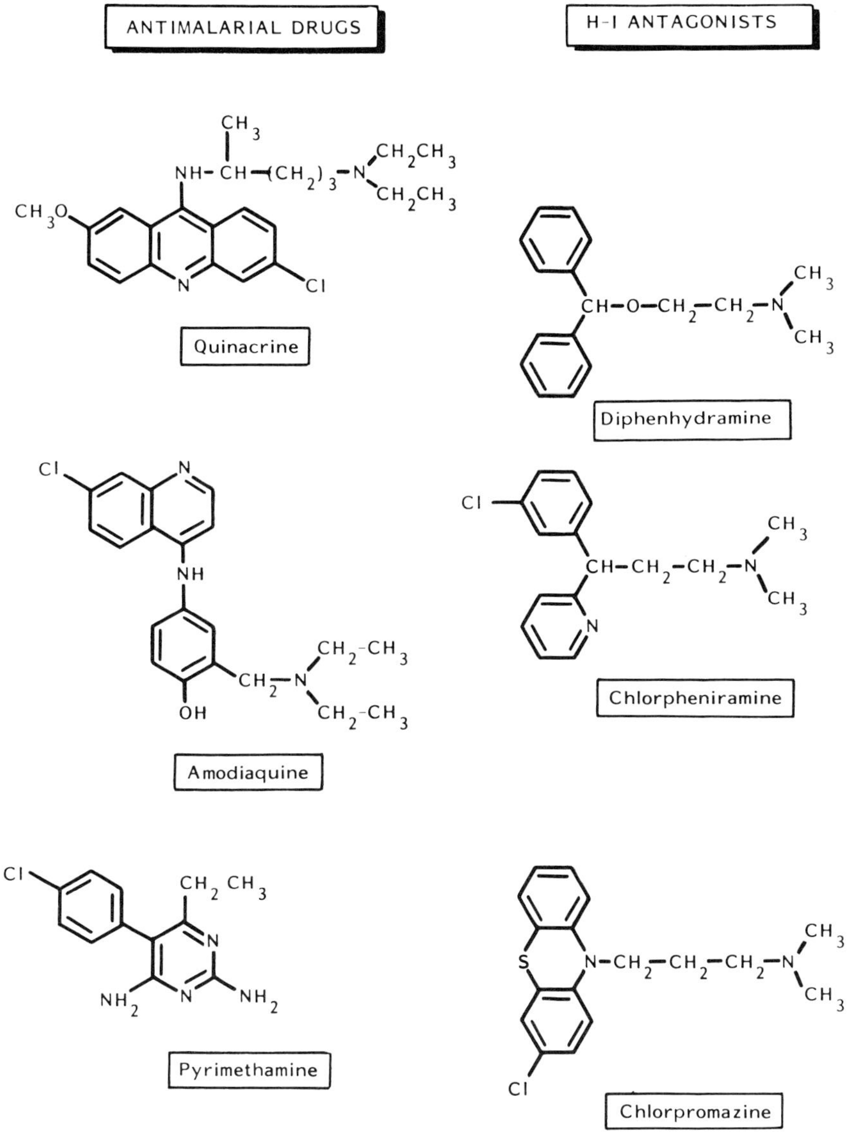

Fig. 3. Antimalarial drugs and H-1 antagonists inhibit HNMT activity in vitro. Representative compounds from both pharmacologic classes are shown above. There is a striking resemblance between the diethylaminoalkyl moiety of antimalarial agents and the dimethylaminoalkyl group that H-1 antagonists possess. These structural configurations appear to confer the ability of these compounds to bind to the HNMT active site and inhibit enzyme activity.

4.3. H-1 Antagonists

The studies demonstrating that chlorpromazine, an H-1 antagonist, inhibits HNMT activity in vitro led Netter and Bodenschatz (1967) to examine whether other antihistamines had the same effect. They found that compounds from the benzimidazole and ethylenediamine classes of H-1 antagonists inhibit rat kidney HNMT. Taylor and Snyder (1972c) continued the investigation of HNMT inhibition by H-1 antagonists. They examined a series of H-1 antagonists as inhibitors of HNMT purified sixfold from mouse brain. Chlorpheniramine and brompheniramine, among the most potent H-1 antagonists, are also the most potent inhibitors of HNMT. The K_i values determined for chlorpheniramine and brompheniramine are 1.5 and 1.9 μM, respectively. An interesting corollary to this observation is the finding that the D-isomers of these alkylamine antihistamines are more potent HNMT inhibitors than the L-isomers (the same is true for H-1 receptor activity).

Although exceptions do exist, the correlation between potency of HNMT inhibition and antagonist activity at the H-1 receptor suggests a similarity in the two binding sites. Structure–activity studies have shown that an important structural requirement for H-1 receptor recognition is the terminal portion of the side chain, specifically a dimethylaminoalkyl moiety (Fig. 3). The activity of a given ligand for the H-1 receptor is governed by the aromatic portion of the molecule. An imidazole ring confers agonist properties, whereas a nitrogen or carbon substituted with two six-membered aromatic rings confers antagonistic properties (Casey, 1978). A similar trend is observed for inhibition of HNMT. Although recognition of the HNMT active site is not exclusively limited to this type of structural moiety, the diethylaminoalkyl group contained within some antimalarial drugs and the dimethylaminoalkyl group contained within the structure of H-1 antagonists appear to be one important factor in the recognition of HNMT active site.

The kinetic mechanism of HNMT inhibition by H-1 antagonists is complex. When mouse brain HNMT activity was determined in the presence of 100 μM HA, varying the drug concentration stimulated HNMT activity at low drug concentrations and inhibited enzyme activity at high drug concentrations (Taylor and Snyder, 1972c). Adjusting the SAME concentration had no effect on this pattern. Similar findings were reported for purified HNMT from porcine antral gastric mucosa (Barth et al., 1973a). As previously discussed, HNMT is subject to substrate inhibition by high concentrations of HA in vitro. At high HA concentrations, H-1 antagonists stimulate HNMT activity by a mechanism proposed to be the attenuation of substrate inhibition since V_{max} values are not actually increased, and consequently, true stimulation of HNMT activity does not occur (Taylor and Snyder, 1972c; Barth et al., 1973a).

The inhibition of HNMT by H-1 antagonists is competitive with respect to HA, and a correlation exists between the relative potencies of H-1 antagonists to inhibit and stimulate HNMT activity. Not all competitive inhibitors exert this effect on HNMT. Quinacrine inhibition of HNMT is also competitive; however, no stimulation of HNMT activity occurs (Taylor and Snyder, 1972c; Thithapanda and Cohn, 1978).

Several studies have evaluated the inhibition of HNMT activity in vivo by H-1 antagonists. Brompheniramine administration elevated HA levels in whole mouse brain, possibly inhibiting HNMT activity in vivo (Taylor and Snyder, 1972a). Other H-1 antagonists have no effect on endogenous HA levels (Green and Ericksson, 1964, 1967; Taylor, 1973; Netter and Bodenschatz, 1967) or the peripheral metabolism of [^{14}C]-HA (Reilly and Schayer, 1971).

4.4. H-2 Antagonists

H-2 antagonists also inhibit HNMT activity in vitro, although the potency of these compounds for inhibition of enzyme activity is much lower than with H-1 antagonists. For example, burimamide has an IC_{50} value of 160 μM for inhibition of gastric mucosal HNMT (Barth et al., 1973a). The K_i of burimamide for rat kidney HNMT is 300 μM (Beaven and Shaff, 1979a) and 240–280 μM for HNMT derived from various guinea pig tissues (Taylor, 1973). Burimamide does not stimulate gastric mucosal HNMT activity in the presence of high HA concentrations and inhibits HNMT by a mixed uncompetitive and noncompetitive mechanism (Barth et al., 1973a). Burimamide inhibits rat kidney HNMT noncompetitively with respect to HA (Beaven and Schaff, 1979a).

Metiamide, a more potent H-2 antagonist, also inhibits gastric mucosal HNMT, but is approximately one order of magnitude weaker than burimamide (Barth et al., 1973a). However, metiamide is more potent than burimamide for inhibition of rat kidney HNMT, exhibiting a K_i value of 60 μM (Beaven and Shaff, 1979a). The mechanism of metiamide inhibition of HNMT is competitive with respect to HA and, in a fashion similar to H-1 antagonists, metiamide both stimulates and inhibits HNMT activity. The mechanism of 4-methylburimamide inhibition of HNMT is also competi-

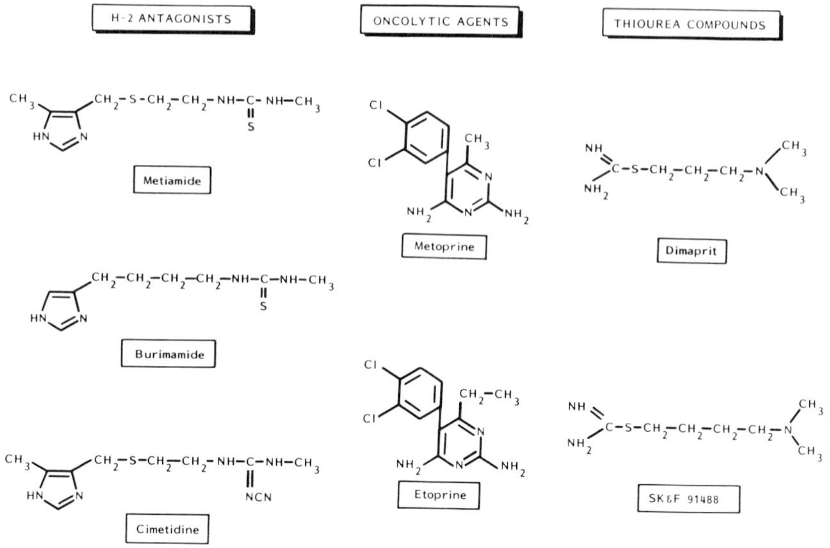

Fig. 4. HNMT inhibitors from three other pharmacologic classes are shown. H-2 antagonists that contain an imidazole ring are weak inhibitors of HNMT. Metoprine and etoprine are structurally similar to pyrimethamine (Fig. 3) and are potent inhibitors of HNMT. Thiourea compounds have an isothiourea group that undergoes prototopic tautomerism and a dimethylaminoalkyl moiety. Both dimaprit and SKF 91488 are potent inhibitors of HNMT, although only SKF 91488 is without potent H-2 agonist properties.

tive (Barth et al., 1973a). Consequently, differences in the mechanism of HNMT inhibition by burimamide and metiamide reside in the methyl group substitution at the four-position of the imidazole ring (Fig. 4). This structure–activity relationship was extended to include other compounds in a subsequent study. For example, cimetidine contains a 4-methylimidazole ring, and, although less potent, this compound exhibits a mechanism of inhibition identical to metiamide and 4-methylburimamide (Barth and Lorenz, 1978).

Studies examining the structure–activity relationships of the H-2 antagonists for blockade of H-2 receptors have demonstrated that a methylene-imidazole moiety, unsubstituted or substituted with a methyl group at the four-position, is important in the recognition of the H-2 receptor site, whereas the side chain confers agonist or antagonist properties (Ganellin, 1978). Although this structure–activity relationship is the direct opposite of H-1 receptor requirements, it does point out that recognition of the HNMT active site can be accomplished by either component necessary for binding to HA receptors.

H-2 antagonists exhibit very little activity as inhibitors of HNMT activity in vivo. Taylor (1973) found that burimamide did not alter endogenous HA levels in various mouse tissues. The results of burimamide administration on the catabolism of [^{3}H]-HA are contradictory. Burimamide prevents the methylation of [^{3}H]-HA by isolated guinea pig atrium, but has no effect on the uptake of [^{3}H]-HA. In contrast, burimamide reduces the uptake of [^{3}H]-HA into neoplastic mouse mast cells, but has no effect on HNMT activity (Moroni et al., 1973).

4.5. Oncolytic Agents

Duch et al. (1978) observed HA-like CNS, cutaneous, and gastrointestinal toxicities during clinical trials of the dihydrofolate reductase inhibitors metoprine and trizinate. These investigators recognized that metoprine (Fig. 4) was a structural analog of pyrimethamine, an antimalarial compound (Fig. 3) previously shown to be an inhibitor of HNMT in vitro (Cohn, 1965). This observation led to the characterization of HNMT inhibition by metoprine, etoprine, and trizinate. All three drugs inhibited bovine cerebral cortex HNMT competitively, with K_i values of 1.0, 7.6, and 6.0 μM, respectively. Metoprine is a remarkably active inhibitor of HNMT in vivo. This drug is lipid soluble, and 5 h after the oral administration of metoprine, endogenous HA levels were elevated fivefold in rat brain (Duch et al., 1979). The greatest increase in HA levels occurred in the kidney,—the endogenous HA concentration in this tissue increased 40-fold.

4.6. Thiourea Compounds

Thiourea compounds that inhibit HNMT activity contain a dimethylaminoalkyl group identical to H-1 antagonists, and the isothiourea moiety present in these compounds undergoes prototopic tautomerism similar to the case with the imidazole ring (Fig. 4). The potency of HNMT inhibition by thiourea compounds is dependent on the length of the alkyl chain separating the isothiourea moiety and the dimethylamino group. The highly selective H-2 agonist dimaprit (*S*-[3-(dimethylaminopropyl)]isothiourea), is a potent noncompetitive inhibitor of HNMT in vitro (Shaff and Beaven, 1977). The K_i for dimaprit inhibition of HNMT is 8 μM. This observation led to the examination of compounds structurally related to dimaprit. SKF compound 91488 (*S*-[4-(dimethylaminobutyl)]isothiourea) contains one extra methylene group compared to dimaprit and exhibits no intrinsic HA receptor activity (Beaven and Shaff, 1979a). SKF 91488 is a potent noncompetitive inhibitor of HNMT, with K_i values of 0.9–1.6 μM for rat kidney HNMT and 3.0 μM for guinea pig brain HNMT.

SKF 91488 is also an effective inhibitor of HNMT in vivo. This compound prevents the methylation of [^{3}H]-HA in mice, as evidenced by elevations in the urinary excretion of [^{3}H]-HA, coupled with a simultaneous decrease in [^{3}H]-*tele*-methylhistamine excretion (Beaven and Shaff, 1979b). SKF 91488 does not cross the blood–brain barrier following peripheral administration (Beaven and Shaff, 1979b). However, the intracerebroventricular administration of this compound was effective in elevating endogenous hypothalamic HA levels (Klein and Gertner, 1981). In conclusion, SKF 91488 appears to be the most potent and selective inhibitor of HNMT in vivo that is available at the present time.

4.7. Ring-Methylated HA Analogs

In addition to *tele*-methylhistamine, other ring-methylated HA derivatives are also inhibitors of HNMT. A comparison of these compounds reveals that 2-methylhistamine, a specific H-1 agonist, is the most potent inhibitor of HNMT (Barth and Lorenz, 1978; Beven and Shaff, 1979a). *Tele*-methylhistamine, 2-methylhistamine, and *pros*-methylhistamine inhibit HNMT by a mixed noncompetitive and uncompetitive mechanism (Barth and Lorenz, 1978). None of these compounds stimulate HNMT activity in vitro at high concentrations of HA, in contrast to the pattern of inhibition observed for 4-methylhistamine.

Inhibition of porcine gastric mucosal HNMT by 4-methylhistamine is competitive with respect to HA. This mechanism is analogous to that of H-2 antagonists that contain a 4-methylimidazole ring (Barth and Lorenz, 1978). However, this observation was not supported by a later study in which 4-methylhistamine inhibited rat kidney HNMT by an uncompetitive mechanism (Beaven and Shaff, 1979a). This conflict could be the result of two factors. First, 4-methylhistamine may be a substrate for HNMT (Barth et al., 1980) and could distort inhibition studies of HNMT. Second, rat kidney appears to be unique with respect to inhibition by this series of compounds. For example, *tele*-methylhistamine is a poor inhibitor of HNMT derived from rat kidney, as evidenced by a K_i value of 300 μM (Beaven and Shaff, 1979a); the K_i for rat brain HNMT was reported to be 9.6 μM (Orr and Quay, 1978). In another study, no inhibition of rat kidney HNMT occurred in the presence of 100 μM *tele*-methylhistamine, whereas 2-methylhistamine was an effective inhibitor at the same concentration. In contrast, mouse brain HNMT was inhibited by both compounds to almost the same degree (Schayer and Reilly, 1975b). A direct comparison of the effects of *tele-*, 2-, and 4-methylhistamine on the fate of [^{3}H]-HA in mice demonstrated that all three were effective in preventing HA methylation in vivo (Reilly and Schayer, 1970; Schayer and Reilly, 1975b). As observed in vitro, 2-methylhistamine was the most effective inhibitor of HNMT activity in vivo.

4.8. S-Adenosyl-L-Homocysteine Derivatives

A product of the HNMT reaction, *S*-adenosyl-L-homocysteine (L-SAH), is an inhibitor of HNMT and may be important in the regulation of HNMT activity in vivo. Certain synthetic derivatives of L-SAH are also effective, and in some instances, irreversible, inhibitors of HNMT. In comparison to L-SAH, *S*-adenosyl-D-homocysteine (D-SAH) is a more potent inhibitor of HNMT, but is less effective than L-SAH in the inhibition of other methyltransferases (Borchardt, 1977). The mechanism of HNMT inhibition by L-SAH and D-SAH is competitive with respect to SAME, and K_i values of 18.1 and 10.5 μM, respectively, have been reported.

Many base-modified analogs of SAH that lack a six-amino group are ineffective HNMT inhibitors, as are many amino acid-modified SAH derivatives (Borchardt, 1977). For example, replacement of homocysteine with cysteine substantially reduces the capability to inhibit HNMT (Barth et al., 1974). One modification of the ribose moiety resulted in the development of compounds that exhibit HNMT inhibitory properties. Treatment of SAH with periodic acid oxidizes the *cis*-diol of the ribose molecule to the corresponding 2',3'-dialdehyde. L-SAH dialdehyde, D-SAH dialdehyde, and L-SAME-dialdehyde produce an irreversible inhibition of HNMT (Borchardt et al., 1978). Since the inactivation of HNMT by dialdehydes is completely irreversible, these compounds may be useful as HNMT affinity labeling reagents. Interestingly, the inhibitory activity of the dialdehyde derivatives is apparently limited to HNMT because they are inactive as inhibitors of PNMT and COMT.

4.9. Monoamine and Indoleamine Compounds

Serotonin (5-hydroxytryptamine, 5-HT) and tryptamine were shown to inhibit HNMT in vitro during the initial characterization of HNMT (Brown et al., 1959b; Gustafsson and Forshell, 1963). Later, the K_i values of tryptamine and serotonin for guinea pig brain HNMT were reported to be 4.5 and 33 μM, respectively (Merril et al., 1966). Other methylated indoleamines were also found to competitively inhibit HNMT purified from mouse and rat brain (Sellinger et al., 1978). Similar to most competitive inhibitors, the modulation of HNMT by these compounds consisted of stimulation and inhibition in a manner that was dependent on the HA concentration.

Substantial inhibition of rat and mouse cerebral HNMT occurs in the presence of dopamine, tyramine, octopamine, 3-methoxydopamine, and 4-methoxyphenethylamine (Sellinger et al., 1978). Dopamine is a noncompetitive inhibitor of HNMT with a K_i value of 0.045 μM, and, following the pattern of other noncompetitive inhibitors of HNMT, dopamine does not stimulate HNMT activity in the presence of high HA concentrations.

5. Distribution and Function of HNMT in Brain

Histamine is considered a neurotransmitter in mammalian brain. Support for a neurotransmitter function and the potential roles of HA in the CNS have been the subjects of several reviews (Snyder and Taylor, 1972; Schwartz, 1977; Liebowitz, 1979; Roberts and Calcutt, 1983). Previous studies indicate that HNMT plays a dominant role in the metabolic regulation of HA levels in brain. It follows, then, that HNMT may be a central factor in the modulation of HA activity in the CNS. Support for this concept comes from several lines of evidence.

In contrast to peripheral tissues, mammalian brain contains little or no DAO activity (Burkhard et al., 1963). The conversion of HA to imidazoleacetic acid also seems unlikely because of the central pharmacologic effects of imidazoleacetic acid mediated through the GABAergic system (Green et al., 1978). The same argument could apply to *tele*-methylhistamine since this metabolite may not be completely inactive in the CNS (Haas and Wolf, 1977), although *tele*-methylhistamine has no HA receptor activity (Black et al., 1972). Nevertheless, [^{3}H]-*tele*-methylhistamine and [^{3}H]-*tele*-methylimidazoleacetic acid are the only products formed in vivo by mouse, rat, and guinea pig brain following the administration of [^{3}H]-histamine (Schwartz et al., 1971; Schayer and Reilly, 1973). Furthermore, *tele*-methylhistamine is readily oxidized by monamine oxidase type B and an aldehyde dehydrogenase to *tele*-methylimidazoleacetic acid (Waldmeier et al., 1977), and the regional distribution of *tele*-methylhistamine and *tele*-methylimidazoleacetic acid in brain is similar to that of HA (Hough and Domino, 1979; Khandelwal et al., 1984).

Previous studies have been unable to demonstrate a high-affinity uptake system for HA in brain (Green et al., 1978; Schwartz et al., 1980). This indicates that synaptic reuptake of HA may not be a major regulatory mechanism. Furthermore, the turnover of HA in brain has been estimated by a variety of methods, and all determinations indicate the half-life of HA in rat brain is less than 1 h (Pollard et al., 1974; Dismukes and Snyder, 1974; Hough et al., 1981; Oishi et al., 1984). This value is somewhat lower than that of all other CNS neurotransmitters except for acetylcholine (Green et al., 1978; Schwartz et al., 1980), and suggests that methylation is an extremely efficient process in the CNS. This concept is corroborated by HNMT inhibition studies. Inhibition of HNMT activity produced a fivefold elevation in the endogenous concentration of HA in brain (Duch et al., 1979). Additionally, cardiovascular changes were observed following an elevation of hypothalamic HA caused by the inhibition of HNMT (Klein and Gertner, 1981). Thus, convincing evidence has been obtained to propose a major role for HNMT in the CNS and provides the conceptual backdrop for the discussion on distribution and regulation of brain HNMT activity.

5.1. Regional Distribution of HNMT in Brain

HNMT is heterogenously distributed in mammalian brain, with the highest activity normally present in the hypothalamus. Highest HNMT activity in monkey brain is found in the hypothalamus, and is closely followed by areas located in the limbic system, such as the septum, amygdala, and hippocampus (Axelrod et al., 1961). The regional distribution of HNMT in goat brain (Tuomisto and Eriksson, 1982) and rat brain (Taylor and Snyder, 1972b) is similar to that in the monkey. Rat hypothalamus contains the highest HNMT activity, whereas lowest enzyme activity is found in the cerebellum. Except for the hypothalamus, HNMT activity correlates with HA levels (Taylor and Snyder, 1972b). Mast cells are prevalent in or near the hypothalamus (Edvinsson et al., 1977) and may be an important consideration in the correlation of HNMT activity with the concentration of HA in brain since mast cell HA turns over at a very slow rate. In general, HNMT activity in mammalian brain is higher in forebrain regions than areas of the hindbrain, and the cerebellum normally contains the lowest HNMT activity of all regions in the CNS.

Although HNMT activity is highest in rat hypothalamus, specific nuclei or areas within the limbic system and brainstem contain as much or more HNMT activity than specific nuclei or subregions of the hypothalamus (Saavedra et al., 1976). A similar distribution has been observed for HA (Pollard et al., 1976, 1978; Correa and Saavedra, 1981). These results suggest that HA could serve as a neurotransmitter in many brain regions outside of the hypothalamus.

The distribution of HNMT in guinea pig brain appears to be different from that in other mammalian species (Tuomisto, 1977). In the adult guinea pig, the regional distribution of HNMT in the CNS is nearly homogeneous, with highest activity located in the cerebellum, closely followed by forebrain, hypothalamic, and hindbrain regions. However, the regional distribution of HA is identical to that in the rat (Tuomisto, 1977). Thus, guinea pig cerebellum contains the highest HNMT activity and lowest concentration of HA. Interestingly, guinea pig cerebellum has the highest concentration of H-1 receptors of all brain regions, as evidenced by in vitro [^{3}H]-mepyramine binding studies (Hill et al., 1978). In contrast, rat cerebellum exhibits the lowest degree of [^{3}H]-mepyramine binding, whereas the highest density of H-1 receptors is found in the hypothalamus (Chang et al., 1978). The significance of these differences with regard to guinea pig brain remains undetermined.

5.2. Subcellular Localization of HNMT

Snyder et al. (1974) found that 80% of the total HNMT activity present in the hypothalamus was localized in the soluble fraction and 14% was located in the crude mitochondrial fraction (P_2). Further purification of the mitochondrial fraction revealed that 75% of the HNMT activity located in the P_2 fraction was located in the supernatant, indicating that most of the HNMT within synapatosomes is cytosolic. The other 25% of HNMT activity present in the P_2 fraction may have also been cytosolic and located in unbroken synaptosomes.

5.3. Cellular Localization of HNMT

Only indirect evidence is available concerning the cellular localization of HNMT in brain. HNMT is present in two types of glial cell cultures, as well as a neuroblastoma cell culture (Garbarg et al., 1975). In all cell lines, the magnitude of HNMT activity was similar to that in rat brain. This evidence suggests HNMT is localized in both glial and neuronal elements in brain. The ontogeny of HNMT activity indicates a potential compartmentalization of HNMT in neuronal elements since enzyme activity increases during synaptogenesis to a degree reminiscent of exclusively neuronal enzymes (Kouvelas et al., 1976; Hough et al., 1981; Tuomisto, 1977).

Lesion studies have also provided some evidence on the cellular compartmentalization of HNMT in brain. Garbarg et al. (1974) demonstrated that medial forebrain bundle (MFB) lesions reduced histidine decarboxylase (HDC) and HA in selected forebrain regions. In a similar study, HDC and HNMT activity in the rat striatum, cerebral cortex, hippocampus, and anterior hypothalamus were evaluated following ipsilateral MFB lesioning (Bischoff and Korf, 1978). HDC activity was reduced in all four areas ipsilateral to the lesion and HNMT activity remained unchanged. These findings could be interpreted to support either a nonneuronal compartmentalization of HNMT, or a disassociation of HDC- and HNMT-containing neurons.

Sperk et al. (1981) examined HNMT activity in the rat striatum following the local injection of kainic acid to destroy intrinsic striatal neurons indiscriminately, or of 6-hydroxydopamine to destroy specifically dopamine-containing neurons. Ipsilateral electrolytic lesions of the MFB were also utilized. All treatments caused a significant decrease in HNMT activity, although 6-hydroxydopamine was less effective than kainic acid or MFB lesions in reducing HNMT activity. The combination of kainic acid and MFB lesioning caused a 50% decrease in HNMT activity, suggesting that HNMT has a major neuronal localization in the rat striatum. The results of this later investigation indicate that HNMT is partially located in striatal intrinsic neurons and not located exclusively in HA- or HDC-containing neurons. However, the reason for the discrepancy in the effect of MFB lesions in this study and the previous study is not readily apparent.

In summary, the accumulated evidence suggests, but does not prove, that HNMT activity is located in both neurons and glial elements in brain, and that neuronal HNMT is not necessarily localized in histaminergic neurons. The production of an HNMT antibody (Bowsher et al., 1983) and utilization in immunocytochemical studies may be useful in the further elucidation of HNMT cellular compartmentalization in brain.

5.4. Neurohypophyseal HNMT

HNMT is present in the neurohypophysis, but the amount of activity is dependent on the species examined. In monkey, HNMT activity in the neural lobe exceeds levels of HNMT found in the hypothalamus (Axelrod et al., 1961). Rat neurohypophyseal HNMT activity is also high, but slightly below whole brain HNMT activity (Bowsher et al., 1983). HNMT appears to be localized in descending neurons since stalk transection alters neural lobe HNMT activity (Verburg et al., 1983b).

5.5. Regulation of Brain HNMT Activity

The regulation of HNMT and the factors that influence enzyme activity have been investigated in a limited number of studies. However, there is no evidence at present to suggest that alterations of HNMT activity are responsible for, or the result of, any CNS disease state. There is speculation that HNMT activity, as well as other brain methyltransferase activities, may be regulated by endogenous levels of SAME and SAH (Deguchi and Barchas, 1971). If so, methyltransferase activities may be linked to each other through the availability of a common substrate, SAME, and the generation of a common product, since SAH is a potent inhibitor of most methyltransferases (Borchardt, 1977). This may be particularly true for HNMT and COMT because the regional, cellular, and subcellular distribution of these enzymes in brain is remarkably similar (Garbarg et al., 1975).

One experimental approach designed to evaluate the impact of SAH on brain HNMT activity in vivo consisted of loading animals with COMT substrates to elevate endogenous SAH concentrations, and evaluating the effect of this treatment on the methylation of [^{3}H]-HA and the endogenous HA concentration (Baudry et al., 1973; Schwartz et al., 1974). Pyrogallol and L-DOPA reduced [^{3}H]-HA methylation, although no significant alterations in endogenous HA concentrations were observed. The interpretation of these studies is complicated by other effects elicited by COMT substrate loading. For example, COMT substrates may have a direct inhibitory effect on HNMT activity in vivo. Furthermore, endogenous SAME levels are reduced by these agents, and it is impossible to establish whether alterations in the concentration of SAME, SAH, or both, were responsible for the reduction in exogenous HA methylation.

SAH hydrolase catalyzes the hydrolysis of SAH to adenosine and homocysteine (Broch and Ueland, 1980), and neither product influences HNMT activity (Borchardt et al., 1978). Therefore, SAH hydrolase activity may be important in modulating HNMT activity. Intracisternal administration of SAH had no effect on the methylation of HA by HNMT in vivo (Baudry et al., 1973; Schwartz et al., 1974). Schatz et al. (1981) also found SAH administration to be ineffective. However, the simultaneous administration of adenosine and homocysteine elevated endogenous brain SAH levels since the thermodynamic equilibrium of SAH hydrolase actually favors SAH synthesis. Importantly, brain SAME levels are unchanged by this treatment. Adenosine and homocysteine administration inhibits HNMT as determined in vitro, and full enzyme activity is restored following dialysis. This treatment also inhibits HNMT in vivo, as shown by the reduced methylation of [^{3}H]-HA and elevations in the endogenous concentration of HA in brain.

Schatz and Sellinger (1975b) have proposed that increased transmethylation occurs following the administration of the convulsant L-methionine-D,L-sulfoxamine (MSO). Although SAME synthesis is unaffected, MSO reduces endogenous brain SAME concentrations in a region-specific fashion (Schatz and Sellinger, 1975a). Reduced SAME concentrations correlate with observed increases in HNMT activity in most brain regions (Schatz and Sellinger, 1975b). However, MSO was also found to decrease cerebral SAH concentrations (Schatz et al., 1977). The effect of MSO on the methylation of [^{3}H]-HA in vivo has also been examined (Schatz et al., 1978). [^{3}H]-*tele*-Methylhistamine levels are significantly elevated by MSO, whereas levels of [^{3}H]-HA and [^{3}H]-*tele*-methylimidazoleacetic acid are unaffected. Administration of a monamine oxidase inhibitor with MSO further increased the levels of [^{3}H]-*tele*-methylhistamine and reduced the concentration of [^{3}H]-*tele*-methylimidazole acetic acid. It is not clear from these studies if alterations of SAME and SAH levels in brain are the cause of MSO-induced stimulation of HNMT activity, or whether the levels just reflect the result of increased transmethylation caused by an entirely different mechanism of action of MSO.

In conclusion, HNMT activity in brain appears to be dynamically regulated, although the factors responsible for the modulation have not been completely elucidated. Current evidence does suggest that SAME and SAH levels may be important determinants for the metabolism of HA in brain, although more studies are needed to rigorously define this relationship.

5.6. Ontogeny of Cerebral HNMT Activity

The ontogenesis of HNMT activity in brain has been examined in several species. HNMT activity in mouse brain was observed to increase substantially after birth, reaching adult levels 40 d after parturition and paralleling the development of the neuronal enzyme DOPA decarboxylase (Laduron et al., 1975). HNMT activity in guinea pig brain increases sharply from the 28th d of gestation until birth, although enzyme activity continues to increase after parturition until adult levels are reached in 20-d-old animals. No regional differences in HNMT development have been observed (Tuomisto, 1977).

Development of HNMT activity in rat brain also begins *in utero,* but the largest increase in HNMT activity occurs after parturition until adult levels are reached in 7–10 d (Kouvelas et al., 1976; Hough et al., 1982). Martres et al. (1975) did not find this degree of postnatal HNMT development, but did report a large and region-specific increase in HDC activity following parturition. However, Maura et al. (1977) found that the development of HNMT in the rat cerebral cortex and hypothalamus occurred at a faster rate than did HDC in these regions. There is a large increase in the concentration of *tele*-methylhistamine in rat brain following birth, which parallels the development of HNMT activity (Hough et al., 1982). There is also a large increase in the number of H-1 receptors in the CNS during this period (Subramanian et al., 1981). So even though slight differences may occur between the development of HNMT and other strictly neuronal enzymes (Martres et al., 1975), the ontogeny of HNMT in mammalian brain seems to occur during synaptogenesis in general, and specifically during the development of the histaminergic system.

5.7. Diurnal Variation of Hypothalamic HNMT

Two studies have shown that hypothalamic HNMT undergoes a diurnal variation (Orr and Quay, 1975; Tuomisto and Tuomisto, 1982). In both studies, peak HNMT activity occurred midway through a 12-h dark phase. Variations in HNMT activity do not correlate with observed changes in HDC and HA levels. The diurnal variation of HNMT is much smaller or nonexistent in other brain regions. It is not known if the diurnal variation in hypothalamic HNMT is restricted to certain nuclei or subregions. Furthermore, the functional significance of the diurnal variation in HNMT activity has not been studied.

6. Distribution and Function of HNMT in Peripheral Tissues

Most mammalian tissues contain at least a low level of HNMT activity, but the amount of enzyme activity is dependent on the tissue and species investigated (Table 4). Some mammalian tissues exhibit the capability not only to catabolize HA that originates locally within those tissues, but also to take up and catabolize HA from the blood delivered to them. Thus, HNMT activity in a given tissue provides a means to regulate the local HA concentration, as well as participating in the regulation of HA levels for the entire organism.

It is perhaps best to view the functional significance of periph-

Table 4
Distribution of Histamine *N*-Methyltransferase in Rat and Guinea Pig Tissues

Tissue	Rat HNMT, Units/g tissue[a]	Guinea Pig HNMT, Units/g tissue[a]
Kidney	1370 ± 90.7	99.9 ± 9.4
Mid-jejunum	284.5 ± 7.7	101.0 ± 9.9
Duodenum	99.7 ± 24.0	154.0 ± 15.6
Ileum	98.4 ± 13.3	56.9 ± 5.3
Brain	22.0 ± 0.3	79.4 ± 3.9
Spleen	0.060 ± 0.023	273.0 ± 8.8
Heart	0.055 ± 0.014	26.0 ± 1.3
Lung	0.051 ± 0.014	59.0 ± 6.3
Stomach	0.035 ± 0.015	112.3 ± 9.7
Liver	0.033 ± 0.017	30.0 ± 1.9

[a]An enzyme Unit catalyzes the formation of 1 nmol of *tele*-methylhistamine/h, $\bar{x}$ ± SEM (n = 6).

eral HNMT at two distinct levels of perspective. The evidence supplied from both HNMT and HA distribution studies, as well as physiologic evidence, implies that protection of the cardiovascular system from excessive HA concentrations is of primary importance (Brown et al., 1959b; Vugman and Rocha e Silva, 1966; Beaven, 1978; Ind et al., 1982; Bowsher et al., 1983). At the organism level, then, it appears that an important function of HNMT is to participate in the homeostatic compartmentalization of HA most compatible with normal cardiovascular function.

The functional significance of HNMT may also be viewed in terms of the metabolic compartmentalization of HA, although this analysis is difficult to generalize from tissue to tissue (Beaven, 1978; Marshall, 1981; Rocklin and Beer, 1983). In those tissues in which HA functions as a modulator or messenger of a physiologic process, it is logical to assume that HNMT also functions as a metabolic mechanism to regulate the concentration of HA at a local level. For example, HNMT activity localized to the gastric mucosa may be important in the physiologic control of gastric acid secretion by regulating the concentration of HA present at the parietal cell.

The presence of DAO, another major catabolic enzyme for HA, complicates the analysis of peripheral HNMT function, regardless of the perspective that is taken. The relative contribution of both enzyme activities to the catabolism of HA must be assessed before any conclusive statements can be made with respect to the importance of either enzyme. The following discussion will focus on the distribution and function of HNMT in selected peripheral tissues that are meant to be representative of the potential importance of HNMT activity in the function of peripheral tissues or organs in general.

6.1. Stomach

Histamine is a potent gastric acid secretogogue, and most evidence points to an important physiologic role for HA in the stimulation of gastric acid secretion (Ivy and Bachrach, 1966; Johnson, 1978; Code, 1982). The events leading to gastric acid secretion and the mechanism for termination of secretion are not completely understood. The inactivation of extracellular HA in the gastric mucosa could occur by several different mechanisms, including local enzymatic catabolism. Except for the rat, little or no DAO is present in the mammalian gastric mucosa (Kim et al., 1969). The absence of DAO activity suggests that enzymatic inactivation of HA in the gastric mucosa is primarily through methylation by HNMT. In the first tissue distribution study of HNMT, Brown et al. (1959b) found HNMT activity in the stomach of the guinea pig, cat, rabbit, and mouse. No HNMT activity was present in the rat stomach. Kim et al. (1969) reported a similar species variation in HNMT from stomach, with the following descending order; dog > guinea pig > cat > rabbit > mouse > rat.

Lorenz et al. (1970) examined the distribution of HNMT in the gastric mucosa of pig, dog, cow, and human. Highest HNMT activity was located in porcine stomach, and lowest activity was observed in the bovine stomach. The values for human and dog were very similar. In all species, the antral gastric mucosa contained the highest level of HNMT activity, followed by the fundus and corpus. No HNMT activity was present in the musculature of the fundus and corpus of the pig or dog. The high level of HNMT activity in the fundus implies that methylation could represent an important mechanism for catabolism of HA and inactivation of a potent stimulus for gastric acid secretion. The significance of high HNMT activity in the antral gastric mucosa remains unknown and essentially uninvestigated. HNMT activity appears to be located, at least in part, in gastric mucosal parietal cells. Isolated dog and guinea pig parietal cells have been shown to take up HA by a Na^{+}-dependent process and to convert this compound to *tele*-methylhistamine (Albinus and Sewing, 1981).

Inhibition of gastric mucosal HNMT by amodiaquine augments acid secretion induced by exogenous HA in the dog (Troidl et al., 1973; Barth et al., 1975b). Additionally, HNMT activity is reduced in the gastric mucosa of duodenal ulcer patients. Thus, in two distinct situations, reduced HNMT activity correlates with increased gastric secretion (Barth et al., 1977b). Furthermore, alterations in HNMT activity of guinea pig gastric mucosa occur following the application of several gastric acid secretogogues (Bunce et al., 1976; Barth et al., 1977a).

In summary, high levels of HNMT are present in the mammalian gastric mucosa, and enzyme activity is localized, in part, to parietal cells. HNMT appears to be regulated by endogenous factors, and reduced enzyme activity correlates with increased gastric acid secretion. Further physiologic studies are required, however, to clarify the importance of HNMT in the regulation of gastric acid secretion induced by endogenous HA.

6.2. Small Intestine

The level of HNMT activity in the small intestine is similar to that found in stomach. Unlike the stomach, however, the small intestine also contains a large amount of DAO (Kim et al., 1969; Backus and Kim, 1970; Suriychan and Thithapanda, 1972). Thus, the small intestine contains a substantial capacity to enzymatically catabolize HA. Kim et al. (1969) found a species variation in the amount of HNMT activity present in the small intestine, with the following descending order: cat > guinea pig > rabbit > rat > dog > mouse. The distribution of DAO activity is inversely related to the species variation of HNMT. HNMT activity is unevenly distributed throughout the small intestine. The level of HNMT in the duodenum and jejunum greatly exceeds the level of enzyme activity in the ileum (Kim et al., 1969; Bowsher et al., 1983). The small intestine participates in the degradation of [^{14}C]-HA administered intravenously (Reilly and Schayer, 1971; Schayer and Reilly, 1974) and could also participate in the catabolism of HA in foodstuffs or HA synthesized by intestinal bacteria (Granerus, 1968; Taylor and Lieber, 1979).

6.3. Kidney

The kidney is one of the richest tissue sources of HNMT activity. The amount of HNMT activity in the rat kidney far exceeds that of any other rat tissue (Bowsher et al., 1983). A regional dissection of rat kidney tissue reveals that HNMT is highly localized to one region, although HNMT activity is present in all areas

Table 5
Regional Distribution of HNMT in Rat Kidney[a]

Tissue	Pelvis	Cortex	Subcortex	Medulla	Papilla
Artery	Interlobe	Interlobular	Arcuate	Interlobular	None
HNMT activity[b]	940 ± 14	302 ± 12	2752 ± 195	823 ± 153	102 ± 8

[a]From Henry et al., 1983.
[b]nmol/hr/g tissue; $\bar{x} \pm$ SEM (n = 6)

(Henry et al., 1983). As shown in Table 5, the highest level of HNMT is located in the subcortical region. The subcortex contains the majority of arcuate arteries (Fourman and Moffat, 1971), and immunofluorescent histochemical studies indicate that HNMT is localized in the intima and muscularis of these blood vessels (Henry et al., 1983). The renal pelvis, cortex, and medulla also contain substantial amounts of HNMT activity that may be localized to arteries of the interlobe or interlobular class. No immunofluorescence is seen in arterioles, glomeruli, venous structures, or tubules. A similar distribution has been reported for feline kidney (Backus and Kim, 1970). Almost twice as much HNMT activity is localized to the inner and outer cortex when compared to other regions of the kidney.

In light of high HNMT activity located in the rat kidney arcuate arteries, a substantial portion of circulating HA that reaches the kidney could be taken up and methylated. Few studies have focused on the metabolic significance of HNMT with respect to potential physiologic or pathophysiologic functions of HA in the kidney. When radiolabeled HA is infused into the renal artery of dogs, a large percentage is converted to *tele*-methylhistamine (Lindell and Schayer, 1958). There is a marked gender difference in the amount of HNMT activity in the rat kidney (Snyder and Axelrod, 1965). The male rat kidney contains twice the amount of HNMT activity present in the female kidney, and this may explain why the urinary excretion of HA by the female rat is greater than that of the male. Inhibition of HNMT activity in the kidney causes a large elevation in the concentration of renal HA (Duch et al., 1979). Obstructive hydronephrosis also produces a large elevation of renal HA, and is followed by induction of renal HNMT activity (Barth et al., 1975a).

The kidney can also synthesize HA, as evidenced by the presence of HDC in renal tissue (Heald and Hollis, 1976). However, HDC activity appears to be predominantly located in glomeruli. Thus, HA synthesized in the kidney is essentially inaccessible to the majority of renal HNMT activity. It seems likely, therefore, that a portion of the HA found in urine is derived directly from the kidney. Finally, HA receptors are located in the renal microvasculature and glomeruli (Banks et al., 1978; Chansel et al., 1982). The results from micropuncture studies are compatible with the concept that HA changes glomerular surface area, but does not affect glomerular permeability characteristics (Ichikawa and Brenner, 1979). In summary, current evidence suggests that HA may play a role in normal renal homeostasis, and that the strategic location of HNMT indicates that pharmacologic or physiologic modulation of renal HNMT activity could significantly enhance the activity of circulating HA on blood vessels and glomerular HA receptors.

6.4. Cardiovascular System

High concentrations of HA are present in mammalian heart and blood vessels (Ryan and Brody, 1972; Howland and Spector, 1972). For the most part, the localization and function of HA in cardiovascular tissue remains unresolved (Kahlson and Rosengren, 1968; Mannaioni, 1972; El-Ackad and Brody, 1975). Some have suggested that HA has a functional role in the regulation of the cardiovascular system. For example, HA may be an intrinsic regulator of the microcirculation (Schayer, 1965) and a mediator of the active component of reflex vasodilation (Heitz and Brody, 1975; Brody, 1978).

HNMT activity is present in myocardial tissue of most mammalian species examined (Brown et al., 1959b). Additionally, in some species, the uptake and metabolism of HA by the heart and certain blood vessels has been demonstrated (Mannaioni and Moroni, 1973; Adams and Hudgins, 1976; Moroni et al., 1977). HNMT activity is also present in blood vessels. The rat renal artery contains the highest enzyme activity of all rat blood vessels examined and the fifth highest among all rat tissues (Bowsher et al., 1983). The renal vein also contains HNMT activity, as does the rat thoracic aorta and vena cava. Furthermore, inhibition of HNMT and DAO activity elevates endogenous HA levels in the rat aorta, mesenteric artery, and mesenteric vein (Howland and Spector, 1972).

Robinson-White and Beaven (1982) examined HNMT activity in microvascular endothelial cells from rat and guinea pig tissues. Perirenal fat pad, myocardium, and brain tissue were investigated. All microvascular preparations from guinea pig tissue contained high HNMT activity, but those isolated from rat tissue had little or none. Conversely, rat microvessels contained more DAO activity than did guinea pig microvessels. These results parallel the observation that the predominant route of HA catabolism is methylation in the guinea pig and oxidative deamination in the rat (Schayer, 1959). Because rat endothelial cells contain little HNMT activity, they may catabolize HA less efficiently than guinea pig cells and could be a contributing factor to high plasma HA concentrations that circulate in the rat (Almeida et al., 1980). In summary, HNMT activity is present in blood vessels and appears to be regionally distributed, species-dependent, and could represent a substantial portion of the total HNMT activity present in a given tissue. Additionally, the strategic location of HNMT implies that this enzyme participates in the homeostatic regulation of plasma HA concentrations.

6.5. *Liver*

Liver was used as the tissue source for studies initially characterizing HNMT (Lindahl, 1960). Gustafsson and Forshell (1963) used porcine liver as the source of HNMT for purification and characterization. Brown et al. (1959b) established that liver tissue from the cat, rabbit, and mouse contained approximately 50% of the HNMT activity present in guinea pig brain. Guinea pig liver contained 25% of the HNMT activity present in brain tissue, and rat liver had none.

Catabolism of [^{14}C]-HA by liver in vivo parallels in vitro findings. Mouse liver appears to have a large capacity to extract and metabolize HA introduced into the circulation. Guinea pig liver was found to have a moderate capacity, whereas this phenomena was negligible for rat liver (Reilly and Schayer, 1971; Schayer and Reilly, 1974, 1975a). The clearance of HA across the canine liver was estimated to be 65% in contrast to the kidney or femoral vein, where the clearance of HA was nearly complete (Beaven et al., 1982). The significance of hepatic HA catabolism in man has not been elucidated. The urinary excretion rate of [^{14}C]-HA metabolites is somewhat slower in certain patients with liver dysfunction (Beall and VanArsdel, 1960). However, the urinary excretion of HA is within normal limits in patients with jaundice or cirrhosis (Mitchell et al., 1954). Thus, hepatic HNMT may contribute to, but is not solely responsible for, the enzymatic catabolism of HA in most mammalian species, including man.

6.6. Lung

HNMT activity is present in mammalian lung, and in some species the amount of enzyme activity is substantial (Brown et al., 1959b). Many questions about the role of HA in the respiratory system remain unanswered. Thus, the importance of regulating endogenous pulmonary HA, and the knowledge of whether alterations of HA inactivation can result in pathologic conditions, are unclear. The rat lung contains little HNMT activity (Bowsher et al., 1983), but is also insensitive to HA-induced bronchoconstriction. However, in most other species, inhalation of HA causes bronchoconstriction (Eyre and Chand, 1982). There is a strong correlation between increased plasma HA concentrations and increased airway resistance in allergic obstructive airway disease (Ulmer et al., 1982). Additionally, patients with chronic obstructive bronchitis have elevated sputum HA concentrations (Thomas and Simmons, 1969). Direct inhibition of rat lung HNMT results in a more than twofold increase in the concentration of endogenous HA (Duch et al., 1979). However, at the present time, there is no evidence to establish that inhibition of HNMT in the lung results in any type of acute or chronic physiologic changes.

7. Properties of HNMT

7.1. HNMT Assay: General Principles

The determination of HNMT activity is based on the assumption that only this enzyme catalyzes the transfer of a methyl group to HA in the presence of SAME. No other cofactors are necessary for this reaction to take place (Gustafsson and Forshell, 1964; Suriyachan and Thithapanda, 1972). The methodology for determining HNMT activity can be subdivided into two types. The first type of assay is based on the disappearance of HA. A known amount of HA and SAME are used to initiate the reaction. The amount of HA remaining after the enzyme reaction is isolated from other contaminating material, derivatized with *o*-phthalaldehyde, and assayed spectrofluorometrically (Cohn, 1965; Lorenz et al., 1970). *o*-Phthalaldehyde derivatization of HA and HPLC separation with fluorescence detection have also been used to quantify HNMT activity (Tsuruta et al., 1981). The second type of HNMT assay is composed of radiometric procedures incorporating radiolabeled HA (Brown et al., 1959a,b; Lindahl, 1960; Gustafsson and Forshell, 1964) or, more commonly at the present time, radiolabeled SAME (Snyder and Axelrod, 1965; Taylor and Snyder, 1972b; Tuomisto, 1977; Beaven and Horakova, 1978; Bowsher et al., 1983) to form radiolabeled *tele*-methylhistamine. This compound is then isolated and quantified by liquid scintillation spectrometry.

With some modifications, two product-isolation procedures have been primarily used to extract and isolate radiolabeled *tele*-methylhistamine following the enzyme reaction. In the first, 5*M* NaOH is used to terminate the reaction and permit extraction of *tele*-methylhistamine into chloroform or toluene:isopentanol (3:2, v:v) (Brown et al., 1959a; Beaven and Horakova, 1978; Thithapanda and Cohn, 1978). The organic layer is then quantified directly or evaporated in a scintillation vial. In a second product-isolation procedure, the reaction is terminated with 0.5*M* sodium borate buffer, pH 10–11, and *tele*-methylhistamine is then extracted into toluene:isopentanol (1:1) and once again is either quantified directly, or the organic solvent is evaporated in a scintillation vial (Snyder and Axelrod, 1965; Taylor and Snyder, 1972b). As shown in Table 6, both product-isolation procedures result in substantial HA- and tissue-dependent blanks when [^{3}H]-SAME is utilized. Recently, a product-isolation procedure was developed that substantially reduced both blank values (Table 6). A 2.5*M* potassium borate buffer, pH 11, is utilized to terminate the enzyme reaction and permit extraction of radiolabeled *tele*-methylhistamine into toluene:isopentanol (3:2). The product is then back-extracted into 0.5*M* HCl, and finally, an ion-pair extraction is executed in the scintillation vial (Bowsher et al., 1983).

HNMT is considered to be a cytosolic enzyme, and for most tissues HNMT activity can be determined in a 40,000*g* supernatant. Blank reduction results in improved assay sensitivity, permitting detection of lower amounts of HNMT activity. A sensitive HNMT assay is necessary since many endogenous compounds inhibit HNMT activity (Sellinger et al., 1978; Taylor and Lieber, 1979) and dilution of tissue extracts could minimize this effect and permit accurate determinations of HNMT activity without extensive dialysis or purification of the enzyme preparation. The importance of including tissue-dependent blanks is also shown in Table 6, and inappropriate blank selection may explain why discrepancies in the amount of HNMT activity have been reported for certain tissues, such as rat liver (Brown et al., 1959b; Snyder and Axelrod, 1965; Duch et al., 1979; Bowsher et al., 1983).

It is important that sufficient attention is given to the kinetics of the enzyme reaction in the determination of HNMT activity. The concentration of both substrates must be saturating with respect to the enzyme concentration. No more than 5% product conversion

Table 6
Comparison of Product Isolation Procedures for Existing HNMT Assays

Product isolation procedure used	Incubation parameters	Enzyme activity,[a] cpm	
		Liver	Whole brain
Borate and toluene-isopentyl alcohol (1:1)	No enzyme	5832	5832
	No histamine	59168	26231
	Enzyme and histamine	56967	138762
NaOH and toluene-isopentyl alcohol (3:2)	No enzyme	35891	35891
	No histamine	192153	63359
	Enzyme and histamine	196376	226343
New method	No enzyme	633	633
	No histamine	4221	4363
	Enzyme and histamine	4058	119823

[a]Each reaction containing enzyme was incubated with 40,000*g* supernatant of liver or whole brain (5 mg tissue/tube) using the assay conditions described in the text. Each value represents the mean of duplicate determinations.

should occur since initial velocities are being determined, and both reaction products inhibit HNMT activity (Baudry et al., 1973; Orr and Quay, 1978). The linear range of HNMT velocity vs both incubation time and the amount of tissue must be determined. The optimal concentrations of SAME and HA depend on a number of factors, such as tissue selection and preparation, incubation medium, ionic strength, and pH. Routinely, sodium or potassium phosphate buffers, pH 7.4–7.8, ranging in molarity from 40–125 m*M*, have been utilized in the enzyme reaction (Lorenz et al., 1970; Orr and Quay, 1978; Bowsher et al. 1983).

Another important consideration in the selection of the appropriate HA concentration is that substrate inhibition occurs at high concentrations (Taylor and Snyder, 1972c; Barth et al., 1973b). The HA concentration should always be kept below the concentration that results in substrate inhibition and will depend on tissue selection and preparation. The range of HA concentrations used in previous HNMT assays was from 9 μM to 2m*M* (Snyder and Axelrod, 1965; Cohn, 1965; Taylor and Snyder, 1972b; Thithapanda and Cohn, 1978; Bowsher et al., 1983). The radiolabeled SAME concentration should be adjusted to fulfill the criteria previously mentioned and provide an adequate enzyme velocity as determined by a sample-to-blank cpm ratio. The range of SAME concentrations used in previous assays was from 1.25 to 125 μ*M*. The inclusion of aminoguanidine in the incubation mixture may be required to inhibit DAO activity. The effect of aminoguanidine should be examined in those instances when HNMT activity is being determined in a tissue that contains a large amount of DAO activity.

7.2. HNMT Assay Protocol

Each 100 μL reaction is carried out in a borosilicate culture tube (12 × 75 mm) that contains the following components: 12.5 μmol of potassium phosphate, pH 7.8; 0.1 μmol of EDTA; 0.9 nmol of HA; 1.25 μCi of *S*-adenosyl-L-[methyl-^{3}H]-methionine (5–15 Ci/mmol); and 0.025% bovine serum albumin. The reaction is initiated by the addition of 25 μL of a tissue homogenate and then incubated for 30 min at 37°C. The reaction is terminated by the addition of 75 μL of 2.5*M* potassium borate, pH 11.0, and 1.25 mL of toluene:isopentanol (3:2). After vortexing and centrifugation at 1500*g* for 5 min, a 1-mL aliquot of the organic phase is transferred to another culture tube containing 250 μL of 0.5*M* HCl. The tubes are vortexed, centrifuged, and the organic phase removed by aspiration. A 200 μL aliquot of the aqueous phase is then transferred to a scintillation vial containing 1 mL of 1*M* potassium phosphate, pH 7.1, and 10 mL of Econofluor with 2% bis-diethylhexylhydrogen phosphate. The vials are shaken and then quantified as a two-phase system by liquid scintillation spectrometry.

7.3. HNMT Purification: General Principles

HNMT has been successfully purified from various tissue sources, including guinea pig brain (Thithapanda and Cohn, 1978; Matuszewska and Borchardt, 1983), porcine gastric mucosa (Lorenz et al., 1972), liver (Gustafsson and Forshell, 1963), rat and mouse brain (Sellinger et al., 1978), and rat kidney (Bowsher et al., 1983). Guinea pig brain was initially used for purification of HNMT. Brown et al. (1959b) utilized a purification scheme that included 50,000*g* centrifugation, 45–75% ammonium sulfate fractionation, calcium phosphate negative adsorption, and alumina gel adsorption followed by dialysis. This procedure accomplished a 30-fold purification. Lindahl (1960) also found that a similar ammonium sulfate fractionation was a successful procedure for mouse liver HNMT purification. Subsequently, ammonium sulfate fractionation has been included in other purification schemes (Sellinger et al., 1978; Bowsher et al., 1983).

Gustafsson and Forshell (1963) obtained a 260-fold purification of HNMT from porcine liver, although only 7% of enzyme activity was recovered. They incorporated a diethylaminoethyl (DEAE) cellulose anion-exchange chromatographic procedure and achieved a 25-fold purification of HNMT with this technique alone. DEAE anion-exchange chromatography has been used in all subsequent HNMT purification schemes, although the elution parameters have been modified extensively.

Lorenz et al. (1972) obtained a 120-fold purification of HNMT from porcine antral gastric mucosa using previously discussed techniques. Thithapanda and Cohn (1978) purified HNMT 199-fold from crude guinea pig brain homogenate with an overall recovery of 18%. A distinct procedural change from previous HNMT purification schemes was to extend protein separation by centrifugation techniques. A 225,000*g* ultracentrifugation, lasting for 12 h, was utilized to provide a more complete separation of the cytosolic fraction contining HNMT from cellular particulate matter. The combined centrifugation techniques resulted in a 11.2-fold purification, but only 52% of HNMT activity was recovered. Low recovery could have been the result of protease-catalyzed degradation of HNMT. Once again, the most successful purification technique was DEAE-cellulose chromatography. Further attempts to purify HNMT proved to be unsuccessful, although the preparation was not homogeneous.

Matuszewska and Borchardt (1983), also using guinea pig brain, were successful in purification of HNMT to homogeneity. A purification scheme identical to that of Thithapanda and Cohn (1978) was utilized through the DEAE-cellulose chromatography procedure, and further purification was obtained by affinity chromatography. The affinity media consisted of SAH covalently linked through the homocysteinyl amino group to agarose. Following the elution of nonadsorbed protein, HNMT was eluted with a buffer containing both high salt (0.2*M* NaCl) and 0.1 m*M* SAME. The HNMT eluate was then applied to a Sephadex G-25 molecular size exclusion column for desalting purposes. The recovery through the two chromatographic procedures was 54%, with a 28-fold purification. Overall purification of HNMT was 4400-fold with respect to crude homogenate, with a 12% recovery of enzyme activity.

HNMT was also recently purified to homogeneity from rat

kidney (Bowsher et al., 1983). Ultracentrifugation, acid precipitation, and 54–82% ammonium sulfate fractionation were included to remove as much protein as possible, and maximize HNMT recovery. DEAE-Sephacel anion exchange chromatography was an excellent initial chromatographic procedure since most protein did not adsorb to the column in the presence of 100 m*M* potassium phosphate, pH 7.5. Elution of HNMT activity was achieved with a linear 0.1–1.0*M* potassium phosphate, pH 7.5, gradient. This chromatographic procedure resulted in a 12-fold purification with a 92% recovery of HNMT activity. An additional sevenfold purification was obtained with Sephadex G-50 molecular size exclusion chromatography. The effectiveness of molecular size exclusion chromatography appears to be dependent on the total protein concentration and may be influenced by protein–protein interactions. HNMT migrated as a larger protein and no purification of HNMT was achieved when Sephadex G-50 chromatography preceded DEAE-Sephacel anion exchange chromatography.

Although the HNMT preparation was not homogeneous following Sephadex G-50 chromatography, further attempts at purification with hydroxylapatite or anion exchange chromatography proved unsuccessful. Therefore, plans were made to develop an affinity media. HNMT is inhibited by a wide range of pharmacologic agents (Taylor and Snyder, 1972c; Thithapanda and Cohn, 1978; Barth and Lorenz, 1978). However, a structural comparison of these compounds revealed that most HNMT inhibitors contain a dialkylaminoalkyl moiety, and that compounds containing a diethylaminoalkyl group, such as quinacrine and amodiaquine, are the most potent inhibitors of HNMT in vitro (Cohn, 1965). Thus, it was postulated that binding of HNMT to the diethylaminoethyl moiety of DEAE-Sephacel might be an active site-directed phenomenon and more specific than just an ionic interaction. HNMT was displaced from DEAE-Sephacel only at phosphate concentrations of greater than 0.5*M*. However, when absorbed in the presence of 0.1*M* potassium phosphate, pH 7.5, HNMT was efficiently eluted by the addition of 1 m*M* HA (Bowsher et al., 1983). The protein obtained from this purification was homogeneous, as determined by polyacrylamide disc gel electrophoresis. Based on the previous discussion, HNMT purification on DEAE-Sephacel by HA elution may be considered an affinity chromatography procedure. Alternatively, adsorption of divalent HA to HNMT may alter the charge of the HA-enzyme complex, permitting elution from DEAE-Sephacel with lower ionic strength buffers.

7.4. Stability of HNMT During Purification

The stability of HNMT activity during purification is a cause of concern. All HNMT purification procedures have been performed at 0–10°C, and recent purification schemes have included glycerol and dithiothreitol, a reducing agent, (Matuszewska and Borchardt, 1983) or helium-saturated buffers (Bowsher et al., 1983) to protect enzyme stability. It is apparent that HNMT activity is substantially reduced in a short period of time if stored in high ionic strength buffers at 10°C or lost completely if stored at –20°C. All HNMT purification procedures incorporating DEAE-cellulose or DEAE-Sephacel chromatography elute HNMT from this column with high ionic strength or low pH buffers. Therefore, it is necessary to reduce the ionic strength of the protein preparation as rapidly as possible. Purified HNMT stored in a low molarity buffer (25–100 m*M*) and neutral pH is stable at –20°C for a prolonged period of time. However, repeated freeze–thawing of purified HNMT is not recommended since HNMT activity has been found to be reduced by this process (Verburg et al., 1983a).

7.5. Specific Protocol for HNMT Purification

All steps in the enzyme purification scheme are performed at 0–10°C. Fifty male Wistar rats (200–300 g) are sacrificed by decapitation and their kidneys removed, defatted, and immediately chilled in ice-cold saline. The tissue (90 g) is minced and homogenized with 3 vol of 0.25*M* sucrose and centrifuged at 40,000*g* for 20 min. The supernatant is then recentrifuged at 220,000*g* for 60 min. All HNMT activity is present in the supernatant.

7.5.1. Acid Treatment and Ammonium Sulfate Fractionation

The supernatant is adjusted to pH 5.0 by the slow addition of cold 2*M* acetic acid. After stirring 10 min at pH 5.0, the suspension is centrifuged at 40,000*g* for 10 min. The supernatant is titrated to pH 7.0 by the addition of cold 1*M* ammonium hydroxide, and a sufficient vol of 100 m*M* dipotassium EDTA, pH 7.0, is then added to yield a final concentration of 0.1 m*M*. Next, the protein preparation is fractionated between 2.0 and 3.2*M* (51–82% saturation) with solid ammonium sulfate. The final precipitate is collected by centrifugation for 10 min at 40,000*g*, resuspended with 30 mL of 25 m*M* potassium phosphate, 0.2 m*M* EDTA, pH 7.5 (Buffer A), and then dialyzed for 16 h against 2 × 4 L of Buffer A containing 1 m*M* dithiothreitol.

7.5.2. DEAE-Sephacel Chromatography

All chromatographic procedures are performed with helium-saturated buffers. The dialyzed protein is loaded on a DEAE-Sephacel column (5 × 1.6 cm) that is previously equilibrated with 300 mL of Buffer A. The column is eluted with Buffer A until the 280 nm absorbance begins to return to baseline. The column is then eluted with 0.1*M* potassium phosphate and 0.2 m*M* EDTA, pH 7.5 (Buffer B). After the 280 nm absorbance returns to baseline, the column is eluted with a 150-mL linear gradient from 0.1 to 1.0*M* potassium phosphate, 0.2 m*M* EDTA, pH 7.5. HNMT activity elutes at an approximate phosphate concentration of 0.5*M*. Fractions containing peak enzyme activity are pooled and concentrated to 20 mL in an Amicon ultrafiltration cell using a PM-10 Diaflo membrane.

7.5.3. Sephadex G-50 Chromatography

Because of the potential instability of HNMT, the concentrated protein from the DEAE-Sephacel chromatographic procedure is immediately loaded on a Sephadex G-50 (fine) column (90 × 5 cm) that is equilibrated with Buffer A. The column is eluted at a flow rate of 30 mL/h, and 6-mL fractions are collected. Fractions containing peak HNMT activity are pooled and concentrated to 20 mL in an ultrafiltration cell using a PM-10 membrane.

7.5.4. DEAE-Sephacel HA Elution Chromatography

The concentrated protein from Sephadex G-50 chromatography is loaded on a DEAE-Sephacel column (4 × 1 cm) that is equilibrated with 100 mL of Buffer B. After elution of a small protein peak, HNMT activity is eluted with Buffer B containing 1 m*M* HA. After elution of the HNMT activity peak, the column is eluted with 1.0*M* potassium phosphate and 0.2 m*M* EDTA, pH 7.5. Fractions containing peak HNMT activity are pooled and concentrated to 1.25 mL in an ultrafiltration cell using a PM-10 membrane.

7.6. Molecular Weight Determinations

Recent determinations of the molecular weight of HNMT from mouse, rat, and guinea pig brain, as well as rat kidney, are in close agreement. Sellinger et al. (1978) reported the molecular weight of rat and mouse brain HNMT to be 29,000 when estimated by Sephadex G-100 chromatography with known standards. Similarly, Matuszewska and Borchardt (1983) found the molecular weight of guinea pig brain HNMT to also be 29,000 by both gel filtration and polyacrylamide gel electrophoretic techniques. The molecular weight of rat kidney HNMT was found to be 33,400 (Bowsher et al., 1983). Thus, recent estimates of the molecular weight of HNMT are similar to the values obtained for other small molecule methyltransferase enzymes, such as COMT and PNMT (Guldberg and Marsden, 1975; Joh and Goldstein, 1973).

7.7. HNMT Isozymes

The heterogeneity of HNMT may be of biological significance, and the occurrence of isozymes or at least functional states of HNMT in some tissue preparations must be considered. There is speculation that HNMT may exist in multiple forms. Evidence supporting this concept was first presented by Axelrod and Vesell (1970). HNMT was heterogenous as determined by starch gel electrophoresis and was further distinguished by differences in heat stability and kinetic properties. Three distinct forms of HNMT were found, and the multiple forms of this enzyme occurred in tissues within a given species, as well as among species. Suriychan and Thithapanda (1972) reached the same conclusions after they examined the kinetic and physiochemical properties of HNMT derived from cat kidney, cat intestine, guinea pig brain, and rabbit intestine. They detected species differences in heat stability, sensitivity to pharmacologic inhibitors, and energy of activation, although kinetic parameters were nearly identical. Watanabe (1981) was also able to distinguish differences in HNMT derived from various rat and guinea pig tissues based on sensitivity to quinacrine inhibition. In contrast to those differences observed in the physiochemical properties of HNMT and sensitivity to pharmacologic inhibition, the antigenic properties of HNMT appear similar (Fig. 5).

7.8. Production of HNMT Antibody

Previously, the lack of homogeneous HNMT precluded the development of an antibody to HNMT. Recently, a specific antiserum was prepared to rat kidney HNMT (Bowsher et al., 1983). When immunoadsorbed with rat kidney supernatant, this antiserum yielded a single precipitin line on Ouchterlony immunodiffusion. The antibody precipitated HNMT of rat kidney, rat brain, rat intestine, and guinea pig brain (Fig. 5). These data suggest that HNMT from these four tissue sources share common antigenic properties and could be identical molecular species. This HNMT antibody has been used as the basis for fluorescence his-

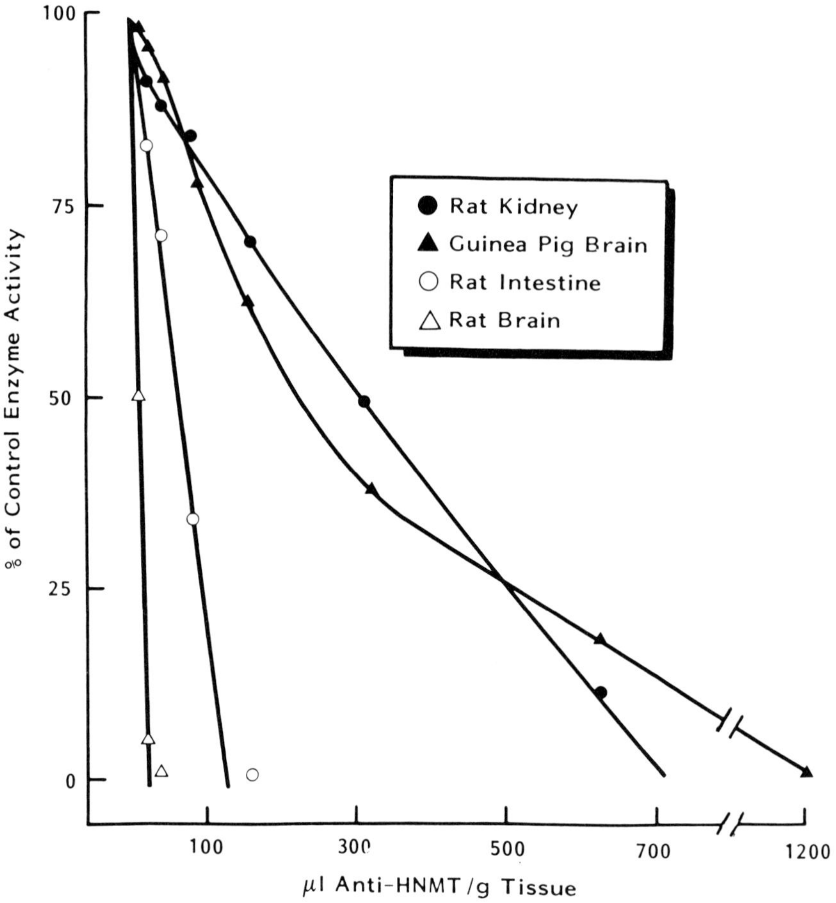

Fig. 5. HNMT activity in the 40,000*g* supernatants of rat kidney, rat intestine, rat brain, and guinea pig brain immunoprecipitated with anti-rat renal HNMT antisera. The antibody titer for each rat tissue reflected the level of HNMT activity present in that tissue and suggests that HNMT in all rat tissues is antigenically identical and could be the same molecular species. Guinea pig brain HNMT activity was also immunoprecipitated by antirat renal HNMT antisera, although a curvilinear titration curve was obtained. This could suggest that two antigenically distinct forms of HNMT are present in guinea pig brain or the antigenic properties of guinea pig brain HNMT are similar to, but not identical with, rat kidney HNMT.

tochemical localization of HNMT in rat kidney (Henry et al., 1983) and may be a useful methodologic tool for other localization studies on HNMT.

7.9. HA Radioenzymatic Assays

Radioenzymatic assays for HA are established and widely used analytic techniques for the quantification of HA from biologic samples (Beaven and Horakova, 1978). HNMT is a primary reagent in this procedure and is used along with radiolabeled SAME to convert HA to radiolabeled *tele*-methylhistamine. This product is then isolated and quantified by liquid scintillation spectrometry. Recent studies have shown that purification of HNMT for use in the HA radioenzymatic assay, as well as adjustment of enzyme reaction conditions based on the kinetic analysis of the HNMT reaction, provide substantial improvements in the sensitivity and specificity of this technique (Verburg et al., 1983a, 1984a).

8. Summary

After 25 yr., there is little doubt that HNMT is an integral component of HA metabolism. However, this fact is greatly overshadowed by the uncertainty that remains concerning the importance of HA or HA metabolism in mammalian physiology. In this regard, the preceding discussion could only emphasize the positive and accentuate the possibilities. Certainly, major advancements have taken place in the enzymologic study of HNMT. For example, there is now a greater appreciation for the complexity of the HNMT reaction as evidenced by modulation of enzyme activity by both substrate and product inhibition. These observations could represent the mechanism for acute regulation of HNMT activity in vivo. Histamine can no longer be considered the sole endogenous methyl acceptor for HNMT and, in turn, HNMT cannot be considered the only enzyme capable of methylating HA. These findings could, in fact, lead to a change in the current simplistic perception of HA metabolism. Although other substrates for HNMT have been identified, the substrate specificity of HNMT must still be considered as restricted in comparison to the susceptibility of HNMT to pharmacologic inhibition in vitro. More often than not, however, the activity of HNMT inhibitors in vitro does not reflect the activity of these compounds in vivo. There are various explanations for this phenomenon, but, above all else, the limited success of inhibiting HNMT activity in vivo points out that actually very little is known about the function and distribution of HNMT activity in mammals.

Recent strides in the development of HNMT methodology should permit a more comprehensive appraisal of the distribution and functional characteristics of this enzyme. For example, observations from HNMT inhibitor studies proved to be invaluable in the preparation of homogeneous HNMT that in turn was used to produce a monospecific antibody. An antibody to HNMT may prove to be useful in HNMT localization studies. Additionally, substantial improvements in the sensitivity and specificity of the HNMT assay should allow greater flexibility in the experimental design of studies aimed at the evaluation of the distribution and regulation of HNMT activity.

References

Adam H. M. and Hye H. K. A. (1966) Concentration of histamine in different parts of brain and hypophysis of cat and its modification by drugs. *Br. J. Pharmacol. Chemother.* **28,** 137–152.

Adams M. D. and Hudgins P. M. (1976) Characteristics of histamine disposition in vascular smooth muscle. *Pharmacology* **14,** 330–338.

Albinus M. and Sewing K. Fr. (1981) Histamine uptake and metabolism in intact isolated parietal cells. *Agents Actions* **11,** 223–227.

Almeida A. P., Flye W., Deveraux D., Horakova Z., and Beaven M. A. (1980) Distribution of histamine and histaminase (diamine oxidase) in blood of various species. *Comp. Biochem. Physiol.* **67C,** 187–190.

Andre J. M. and Schwartz J. C. (1973) Studies on the inhibition of histamine methylation in mice. *Agents Actions* **3,** 172–173.

Asano Y., Woodward R. W., Houck D. R., and Floss H. G. (1984) Stereochemical course of the transmethylation catalyzed by histamine *N*-methyltransferase. *Arch. Biochem. Biophys.* **231,** 253–256.

Axelrod J. and Vesell E. S. (1970) Heterogeneity of *N*- and *O*-methyltransferases. *Mol. Pharmacol.* **6,** 78–84.

Axelrod J., MacLean P. D., Albers R. W., and Weissbach H. (1961) Regional Distribution of Methyltransferase Enzymes in the Nervous System and Glandular Tissues, in *Regional Neurochemistry* (Kety S. S. and Elkes J., eds.) pp. 307–311. Pergamon, New York.

Backus B. and Kim K. S. (1970) Subcellular distribution of diamine oxidase. *Comp. Gen. Pharmacol.* **1,** 196–200.

Banki R. O., Fondacaro J. D., Schwaiger M. M., and Jacobson E. D. (1978) Renal histamine H-1 and H-2 receptors: Characterization and functional significance. *Am. J. Physiol.* **235,** F570–F575.

Barth H. and Lorenz W. (1978) Structural requirements of imidazole compounds to be inhibitors and activators of histamine methyltransferase: Investigation of histamine analogues and H-2 receptor antagonists. *Agents Actions* **8,** 359–365.

Barth H., Niemeyer I., and Lorenz W. (1973a) Studies on the mode of action of histamine H-1 and H-2 receptor antagonists on gastric histamine methyltransferase. *Agents Actions* **3,** 138–147.

Barth H., Lorenz W., and Niemeyer I. (1973b) Inhibition and activation of histamine methyltransferase by methylated histamines. *Hoppe Seylers Z. Physiol. Chem.* **354,** 1021–1026.

Barth H., Niemeyer I., and Lorenz W. (1974) Speculations about the binding sites of *S*-adenosyl-L-homocysteine and some of its synthetic analogues to histamine methyltransferase. *Agents Actions* **4,** 186–188.

Barth H., Naber K., Barthel K. E., Niemeyer I., and Lorenz W. (1975a) Induction of histamine methyltransferase during experimental hydronephrosis: A mechanism of adaptation for histamine homeostasis. *Agents Actions* **5,** 442–443.

Barth H., Lorenz W., and Troidl H. (1975b) Effect of amodiaquine on gastric histamine methyltransferase and on histamine-stimulated gastric secretion. *Br. J. Pharmacol.* **55,** 321–327.

Barth H., Lorenz W., and Stenner W. (1977a) Alterations of histamine methyl transferase activity after feeding or application of gastric secretory stimulants. *Naunyn Schmiedbergs Arch. Pharmacol.* **297** (Suppl. II), R43.

Barth H., Troidl H., Lorenz W., Rohde H., and Glass R. (1977b) Histamine and peptic ulcer disease: Histamine methyltransferase activity in gastric mucosa of control subjects and duodenal ulcer patients before and after surgical treatment. *Agents Actions* **7,** 75–79.

Barth H., Crombach M., Schunack W., and Lorenz W. (1980) Evidence for a less high acceptor substrate specificity of gastric histamine methyltransferase: Methylation of imidazole compounds. *Biochem. Pharmacol.* **29,** 1399–1407.

Baudry M., Chast F., and Schwartz J. C. (1973) Studies on *S*-adenosylhomocysteine inhibition of histamine transmethylation in brain. *J. Neurochem.* **20,** 13–21.

Beall G. N. and VanArsdel P. P. (1960) Histamine metabolism in human disease. *J. Clin. Invest.* **39,** 676–683.

Beaven M. A. (1978) Histamine: Its Role in Physiological and Pathological Processes, in *Monographs in Allergy* Vol. 13 S. Karger, Basel.

Beaven M. A. and Horakova Z. (1978) The Enzymatic Isotopic Assay of Histamine, in *Handbook of Experimental Pharmacology* Vol. XVIII/2 (Rocha e Silva M., ed.) pp. 151–173. Springer-Verlag, Berlin.

Beaven M. A. and Shaff R. E. (1979a) New inhibitors of histamine-*N*-methyltransferase. *Biochem. Pharmacol.* **28,** 183–188.

Beaven M. A. and Shaff R. E. (1979b) Inhibition of histamine methylation *in vivo* by the dimaprit analog, SKF compound 91488. *Agents Actions* **9,** 455–459.

Beaven M. A., Robinson-White A., Roderick N. B., and Kauffman G. L. (1982) The demonstration of histamine release in clinical conditions: A review of past and present assay procedures. *Klin. Wochenschr.* **60,** 873–881.

Bertaccini G. and Vitali T. (1964) Pharmacological properties of some imidazole derivatives occurring in nature. *J. Pharm. Pharmacol.* **16,** 441–449.

Bischoff S. and Korf J. (1978) Different localization of histidine decarboxylase and histamine-*N*-methyltransferase in the rat brain. *Brain Res.* **141,** 375–379.

Black J. W. and Ganellin C. R. (1974) Naming of substituted histamines. *Experientia* **30,** 111–113.

Black J. W., Duncan W. A. M., Durant C. J., Ganellin C. R., and Parsons E. M. (1972) Definition and antagonism of histamine H-2 receptors. *Nature* **236,** 385–390.

Borchardt R. T. (1977) Synthesis and Biological Activity of Analogs of Adenosylhomocysteine as Inhibitors of Methyltransferases, in *The Biochemistry of Adenosylmethionine* (Salvatore F., Borek E., Zappia V., Williams-Ashman H. G., and Schlenk F., eds.), pp. 151–171. Columbia University, New York.

Borchardt R. T. (1980) *N*- and *O*-methylation, in *Enzymatic Basis of Detoxification,* Vol. 2 (Jakoby W. B., ed.), pp. 43–62. Academic, New York.

Borchardt R. T., Wu Y. S., and Wu B. S. (1978) Affinity labeling of histamine *N*-methyltransferase by 2',3'-dialdehyde derivatives of *S*-adenosylhomocysteine and *S*-adenosylmethionine. Kinetics of inactivation. *Biochemistry* **17,** 4145–4153.

Bowsher R. R., Verburg K. M., and Henry D. P. (1983) Rat histamine *N*-methyltransferase: Quantification, tissue distribution, purification and immunological properties. *J. Biol. Chem.* **258,** 12215–12220.

Broch O. J. and Ueland P. M. (1980) Regional and subcellular distribution of *S*-adenosylhomocysteine hydrolase in the adult rat brain. *J. Neurochem.* **35,** 484–488.

Brody M. J. (1978) Histaminergic and Cholinergic Vasodilator Systems, in *Mechanisms of Vasodilation* (Vanhoutte P. M. and Leusen I., eds.), pp. 266–277. S. Karger, Basel.

Brown D. D., Axelrod J., and Tomchick R. (1959a) Enzymatic *N*-methylation of histamine. *Nature* **183,** 680.

Brown D. D., Tomchick R., and Axelrod J. (1959b) The distribution and properties of a histamine-methylating enzyme. *J. Biol. Chem.* **234,** 2948–2950.

Bunce K. T., Golightly L. C., and Lake H. J. (1976) The effect of pentagastrin-stimulated acid secretion on histamine catabolic enzymes in the gastric mucosa of the cat and guinea pig. *J. Physiol.* (Lond.) **263,** 220–221.

Burkhard W., Gey K. F., and Pletscher A. (1963) Diamine oxidase in brain of vertebrates. *J. Neurochem.* **10,** 183–186.

Casy A. F. (1978) Chemistry of Anti-H_1 Histamine Antagonists, in *Handbook of Experimental Pharmacology* Vol. XVIII/2 (Rocha e Silva M., ed.), pp. 175–214. Springer-Verlag, Berlin.

Chang R. S. L., Tran V. T., and Snyder S. H. (1978) Histamine H-1 receptors in brain labeled with ^{3}H-mepyramine. *Eur. J. Pharmacol.* **48,** 463–464.

Chansel D., Oudinet J. P., Nivez M. P., and Ardaillou R. (1982) Histamine H-2 receptors in rat renal glomeruli. *Biochem. Pharmacol.* **31,** 367–375.

Code C. F. (1982) Histamine Receptors and Gastric Secretion, in *Pharmacology of Histamine Receptors* (Ganellin C. R. and Parsons M. E., eds.), pp. 217–235. Wright PSG, Bristol.

Code C. F., Green W. E. R., Kennedy J. C., Ritchie H. D., and Schlegel J. F. (1976) Metabolism of histamine in secreting intact and isolated canine stomach. *Am. J. Physiol.* **30,** 219–227.

Cohn V. H. (1965) Inhibition of histamine methylation by antimalarial drugs. *Biochem. Pharmacol.* **14,** 1686–1688.

Correa F. M. A. and Saavedra J. M. (1981) Increase in histamine concentrations in discrete hypothalmic nuclei of spontaneously hypertensive rats. *Brain Res.* **205,** 445–451.

Deguchi T. and Barchas J. (1971) Inhibition of transmethylation of biogenic amines by *S*-adenosylhomocysteine. *J. Biol. Chem.* **244,** 107–111.

Dent C., Nilam F., and Smith I. R. (1982) Sidechain-modified histamine analogues as substrates for histamine *N*-methyltransferase. *Biochem. Pharmacol.* **32,** 2297–2300.

Dismukes K. and Snyder S. H. (1974) Histamine turnover in rat brain. *Brain Res.* **78,** 467–481.

Duch D. S., Bowers S. W., and Nichol C. A. (1978) Elevation of brain histamine levels by diaminopyrimidine inhibitors of histamine *N*-methyl transferase. *Biochem. Pharmacol.* **27,** 1507–1509.

Duch D. S., Bowers S. W., Edelstein M., and Nichol C. A. (1979) Histamine: Elevation of Brain Levels by Inhibition of Histamine *N*-Methyl Transferase, in *Transmethylation* (Usdin E., Borchardt R. T., and Creveling C. R., eds.), pp. 287–295. Elsevier/North-Holland, New York.

Edvinsson L., Cervos-Navarro J., Larsson L. I., Owman C. H., and Ronnberg A. L. (1977) Regional distribution of mast cells containing histamine, dopamine, or 5-hydroxytryptamine in mammalian brain. *Neurology* **27,** 878–883.

El-Ackad T. M. and Brody M. J. (1975) Evidence for non-mast cell histamine in the vascular wall. *Blood Vessels* **12,** 181–191.

Erspamer V., Vitali T., Roseghini M., and Cei J. M. (1963) Occurrence of new imidazolealkylamines (spinaceamine and 6-methylspinaceamine) in skin extracts of *Leptodactylus Pentadactylus Labyrinthicus. Experientia* **19,** 346–347.

Eyre P. and Chand N. (1982) Histamine Receptor Mechanisms of the Lung, in *Pharmacology of Histamine Receptors* (Ganellin C. R. and Parsons M. E., eds.), pp. 298–321. Wright PSG, Bristol.

Fourman J. and Moffat D. B. (1971) in *The Blood Vessels of the Kidney,* pp. 26–38. Blackwell Scientific, Oxford.

Francis D. M., Thompson M. F., and Greaves M. W. (1980) The kinetic properties and reaction mechanism of histamine methyltransferase from human skin. *Biochem. J.* **187,** 819–828.

Fuller R. W., Warren B. J., and Molloy B. B. (1970) Substrate specificity of phenylethanolamine *N*-methyltransferase from rabbit adrenal. *Biochim. Biophys. Acta* **222,** 21–212.

Ganellin C. R. (1978) Chemistry and Structure-Activity Relationships of H-2 Receptor Antagonists, in *Handbook of Experimental Pharmacology,* Vol. XVIII/2 (Rocha e Silva M., ed.), pp. 251–294. Springer-Verlag, Berlin.

Garbarg M., Barbin G., Feger J., and Schwartz, J. C. (1974) Histaminergic pathway in rat brain evidenced by the lesions of the MFB. *Science* **186,** 833–835.

Garbarg M., Baudry M., Benda P., and Schwartz J. C. (1975) Simultaneous presence of histamine-*N*-methyltransferase and catechol-*O*-methyltransferase in neuronal and glial cells in culture. *Brain Res.* **83,** 538–541.

Gharib A., Sarda N., Chabannes B., Cronenberger L., and Pacheco H. (1982) The regional concentrations of *S*-adenosyl-L-methionine, *S*-adenosyl-L-homocysteine, and adenosine in rat brain. *J. Neurochem.* **38,** 810–815.

Goldstein A., Aronow L., and Kalman S. N. (1974) in *Principles of Drug Action: The Basis of Pharmacology.* 2nd Edit., pp. 154–210. John Wiley & Sons, New York.

Granerus G. (1968) Effects of oral histamine, histidine and diet on urinary excretion of histamine, methylhistamine and 1-methyl-4-imidazole acetic acid in man. *Scand. J. Clin. Lab. Invest.* **22,** Suppl. 104, 49–58.

Green H. and Erickson R. W. (1964) Effect of some drugs upon rat brain histamine content. *Int. J. Neuropharmacol.* **3,** 315–320.

Green H. and Erickson R. W. (1967) Effect of some drugs upon the histamine concentration of guinea pig brain. *Arch. Int. Pharmacodyn. Ther.* **166,** 121–126.

Green J. P., Johnson C. L., and Weinstein H. (1978) Histamine as a Neurotransmitter, in *Psychopharmacology: A Generation of Progress* (Lipton M. A., DiMascio A., and Killam K. F., eds.), pp. 319–332. Raven, New York.

Guldberg H. C. and Marsden C. A. (1975) Catechol *O*-methyltransferase:

Pharmacological aspects and physiological role. *Pharmacol. Rev.* **27,** 135–206.

Gustafsson A. and Forshell G. P. (1963) Purification of a *N*-methyltransferase. *Acta Chem. Scand.* **17,** 541–542.

Gustafsson A. and Forshell G. P. (1964) Studies on the enzymatic methylation of histamine. *Acta Chem. Scand.* **18,** 2069–2076.

Haas H. L. and Wolf P. (1977) Central actions of histamine: Microelectrophoretic studies. *Brain Res.* **122,** 269–279.

Harper H. A., Rodwell V. W., and Mayes, P. A. (1979) in *Review of Physiological Chemistry,* 17th Ed., pp. 178–184. Lange Medical, Los Altos, California.

Heald J. I. and Hollis T. M. (1976) Histidine decarboxylase-mediated histamine synthesis in glomeruli from rat kidneys. *Am. J. Physiol.* **230,** 1349–1353.

Heitz D. C. and Brody M. J. (1975) Possible mechanism of histamine release during active vasodilation. *Am. J. Physiol.* **228,** 1351–1357.

Henry D. P., Bowsher R. R., Myers M. J., Zerbe R. L., and Peterson B. H. (1983) Immunocytochemical localization of histamine *N*-methyltransferase in rat kidney. *Fed. Proc.* **42,** 709.

Herman K. S. and Henry D. P. (1984) A new enzymatic pathway for the formation of *pros*-methylhistamine and *N*-alpha-methylhistamine *in vitro. Fed. Proc.* **43,** 457.

Hill S. J., Emson P. C., and Young J. M. (1978) The binding of [^{3}H]mepyramine to histamine H-1 receptors in guinea pig brain. *J. Neurochem.* **31,** 997–1004.

Hough L. B. and Domino E. F. (1979) *Tele*-methylhistamine distribution in rat brain. *J. Neurochem.* **32,** 1865–1866.

Hough L. B., Khandelwal J. K., and Mittag T. W. (1981) Alphamethylhistamine methylation by histamine methyltransferase. *Agents Actions* **11,** 425–428.

Hough L. B., Khandelwal J. K., and Green J. P. (1982) Ontogeny and subcellular distribution of rat brain *tele*-methylhistamine. *J. Neurochem.* **38,** 1593–1599.

Howland R. D. and Spector S. (1972) Disposition of histamine in mammalian blood vessels. *J. Pharmacol. Exp. Ther.* **182,** 239–245.

Ichikawa I. and Brenner B. M. (1979) Mechanisms of action of histamine and histamine antagonists on the glomerular microcirculation in the rat. *Circ. Res.* **45,** 737–745.

Ind P. W., Brown M. J., Lhoste F. J. M., Macquin I., and Dollery C. T. (1982) Concentration-effect relationships of infused histamine in normal volunteers. *Agents Actions* **12,** 12–16.

Ivy A. C. and Bachrach W. H. (1966) Physiological Significance of the Effect of Histamine on Gastric Secretion, in *Handbook of Experimental*

Pharmacology, Vol. XVIII/1 (Rocha e Silva M., ed.), pp. 810–882. Springer-Verlag, Berlin.

Joh T. H. and Goldstein M. (1973) Isolation and characterization of multiple forms of phenylethanolamine *N*-methyltransferase. *Mol. Pharmacol.* **9,** 117–129.

Johnson L. R. (1978) Histamine and Gastric Secretion, in *Handbook of Experimental Pharmacology,* Vol. XVIII/2 (Rocha e Silva M., ed.), pp. 41–56. Springer-Verlag, Berlin.

Kahlson G. and Rosengren E. (1968) New approaches to the physiology of histamine. *Physiol. Rev.* **48,** 155–196.

Kalsner S. (1970) Potentiation of responses to histamine in aortic strips by steroids. *Can. J. Physiol. Pharmacol.* **46,** 719–730.

Kapeller-Adler R. and Iggo B. (1957) Histamine and its derivatives in human urine. *Biochim. Biophys. Acta* **25,** 394–402.

Karjala S. A. and Turnquest B. W. (1955) The characterization of two methylimidazoleacetic acids as radioactive histamine metabolites. *J. Am. Chem. Soc.* **77,** 6358.

Karjala S. A., Turnquest B., and Schayer R. W. (1956) Urinary metabolites of radioactive histamine *J. Biol. Chem.* **219,** 9–12.

Khandelwal J. K., Hough L. B., and Green J. P. (1984) Regional distribution of the histamine metabolite, *tele*-methylimidazoleacetic acid in rat brain: Effects of pargyline and probenecid. *J. Neurochem.* **42,** 519–522.

Kim K. S. and Cohn V. H. (1966) Inhibition of histamine metabolism. *The Pharmacologist* **8,** 185.

Kim K. S., Backus B., Harris M., and Rourke P. (1969) Distribution of diamine oxidase and imidazole-*N*-methyltransferase along the gastrointestinal tract. *Comp. Biochem. Physiol.* **31,** 137–145.

Klein M. C. and Gertner S. B. (1981) Evidence for a role of endogenous histamine in central cardiovascular regulation: Inhibition of histamine-*N*-methyltransferase by SKF 91488. *J. Pharmacol. Exp. Ther.* **216,** 315–320.

Kouvelas E. D., Savakis Ch. E., Tzebelikos E. Th., Bonatsos G., and Mitrossilis S. (1976) Developmental characteristics of histamine methyltransferase and phenylethanolamine-*N*-methyltransferase. *Experientia* **32,** 1136–1137.

Kovacs E. M. and Heisler S. (1970) Detection and possible role of a chain methylated histamine metabolite(s) in guinea pig stomach. *Eur. J. Pharmacol.* **11,** 349–360.

Laduron P., Janssen P. F. M., Aerts G., and Leysen J. (1975) Ontogenetic development of a 5-methyltetrahydrofolate-requiring enzyme in rats and mice. *Naunyn Schmiedbergs Arch. Pharmacol.* **286,** 379–386.

Leibowitz S. F. (1979) Histamine: Modification of Behavioral and Physiological Components of Body Fluid Homeostasis, in *Histamine Receptors,* (Yellin T. O., ed.), pp. 219–253. Spectrum, New York.

Lin T. M., Alphin R. S., Henderson F. G., Benslay D. N., and Chen K. K. (1962) The role of histamine in gastric hydrochloric acid secretion. *Ann. N.Y. Acad. Sci.* **99,** 30–44.

Lindahl K. M. (1958) Methylation of histamine by mouse liver *in vitro. Acta Physiol. Scand.* **43,** 254–261.

Lindahl K. M. (1960) The histamine methylating enzyme system in liver. *Acta Physiol. Scand.* **49,** 114–138.

Lindell S. E. and Schayer R. W. (1958) The renal removal of injected [^{14}C]histamine from blood in dogs. *Br. J. Pharmacol.* **13,** 44–51.

Lorenz W., Barth H., and Werle E. (1970) Histamine and histamine methyltransferase in the gastric mucosa of man, pig, dog and cow. *Naunyn Schmiedbergs Arch. Pharmacol.* **267,** 421–432.

Lorenz W., Reimann H. J., Barth H., Kusche J., Meyer R., Doenicke A., and Hutzel M. (1972) A sensitive and specific method for the determination of histamine in human whole blood and plasma. *Hoppe Seylers Z. Physiol. Chem.* **353,** 911–920.

Mannaioni P. F. (1972) Physiology and pharmacology of cardiac histamine. *Arch. Int. Pharmacodyn.* Supplementum **196,** 64–78.

Mannaioni P. F. and Moroni F. (1973) Uptake, disposition and metabolism of histamine in isolated heart preparations. *Br. J. Pharmacol.* **49,** 457–465.

Marshall I. (1981) Histamine modulation of neurotransmission in the sympathetic nervous system. *J. Auton. Pharmacol.* **1,** 235–250.

Martres M. P., Baudry M., and Schwartz J. C. (1975) Histamine synthesis in developing rat brain: Evidence for multiple compartmentation. *Brain Res.* **83,** 261–275.

Maslinski S., Schippert B., Kovar K. A., and Sewing K. Fr. (1977) Methylation of histamine in the gastric mucosa. *Digestion* **15,** 497–505.

Matuszewska B. and Borchardt R. T. (1983) Guinea pig brain histamine *N*-methyltransferase: Purification and partial characterization. *J. Neurochem.* **41,** 113–118.

Maura G., Vaccari A., and Timiras P. S. (1977) Effects of chronic stress on the development of histamine enzymes. *Agents Actions* **7,** 177–181.

Mehler A. H., Tabor H., and Bauer H. (1952) The oxidation of histamine to imidazoleacetic acid *in vivo. J. Biol. Chem.* **197,** 475–480.

Merril C. R., Snyder S. H., and Bradley D. F. (1966) Inhibition of histamine methyltransferase by serotonin and chlorpromazine derivatives: Electronic aspects. *Biochim. Biophys. Acta* **118,** 316–324.

Mitchell R. G., Butt H. R., and Code C. F. (1954) Histamine metabolism in diseases of the liver. *J. Clin. Invest.* **33,** 1199–1203.

Moroni F., Fantozzi R., and Mannaioni P. F. (1973) The influence of burimamide on ^{14}C-histamine uptake and metabolism by guinea pig atrium and mouse neoplastic mast cells, *in vitro. Agents Actions* **3,** 187–189.

Moroni F., Fantozzi R., Masini E., and Mannaioni P. F. (1977) The influence of catecholamines and serotonin on histamine uptake and metabolism by guinea pig atrium. *Eur. J. Pharmacol.* **41,** 59–63.

Navert H., Flock E. V., Tyce G. M., and Code C. F. (1967) The fate of ^{14}C histamine during gastric secretion. *Physiologist* **10,** 261.

Netter K. J. and Bodenschatz K. (1967) Inhibition of histamine-*N*-methylation by some antihistamines. *Biochem. Pharmacol.* **16,** 1627–1631.

Oishi R., Nishibori M., and Saeki K. (1984) Regional differences in the turnover of neuronal histamine in the rat brain. *Life Sci.* **34,** 691–699.

Orr E. L. and Quay W. B. (1975) Hypothalamic 24-hour rhythms in histamine, histidine decarboxylase and histamine-*N*-methyltransferase. *Endocrinology* **96,** 941–945.

Orr E. L. and Quay W. B. (1978) Product inhibition of rat brain histamine-*N*-methyltransferase. *J. Neurochem.* **30,** 1539–1542.

Pollard H., Bischoff S., and Schwartz J. C. (1974) Turnover of histamine in rat brain and its decrease under barbiturate anesthesia. *J. Pharmacol. Exp. Ther.* **190,** 88–99.

Pollard H., Bischoff S., Llorens-Cortes C., and Schwartz J. C. (1976) Histidine decarboxylase and histamine in discrete nuclei of rat hypothalamus and the evidence for mast cells in the median eminence. *Brain Res.* **118,** 509–513.

Pollard H., Llorens-Cortes C., Barbin G., Garbarg M., and Schwartz J. C. (1978) Histamine and histidine decarboxylase in brain stem nuclei: Distribution and decrease after lesions. *Brain Res.* **157,** 178–181.

Reilly M. A. and Schayer R. W. (1970) *In vivo* studies on histamine catabolism and its inhibition. *Br. J. Pharmacol.* **38,** 478–489.

Reilly M. A. and Schayer R. W. (1971) Further studies on histamine catabolism *in vivo. Br. J. Pharmacol.* **43,** 349–358.

Roberts F. and Calcutt C. R. (1983) Histamine and the hypothalamus. *Neuroscience* **9,** 721–739.

Robinson-White A. and Beaven M. A. (1982) Presence of histamine and histamine-metabolism enzyme in rat and guinea pig microvascular endothelial cells. *J. Pharmacol. Exp. Ther.* **223,** 440–445.

Rocklin R. E. and Beer D. J. (1983) Histamine and immune modulation. *Adv. Int. Med.* **28,** 225–251.

Rothschild Z. and Schayer R. W. (1958) Synthesis and metabolism of a histamine metabolite, 1-methyl-4-(beta-aminoethyl)-imidazole. *Biochim. Biophys. Acta* **30,** 23–27.

Ryan M. J. and Brody M. J. (1972) Neurogenic and vascular stores of histamine in the dog. *J. Pharmacol. Exp. Ther.* **181,** 83–91.

Saavedra J. M., Brownstein M. J., and Palkovits M. (1976) Distribution of catechol-*O*-methyltransferase, histamine *N*-methyltransferase and monoamine oxidase in specific areas of the rat brain. *Brain Res.* **118,** 152–156.

Schatz R. A. and Sellinger O. Z. (1975a) Effect of methionine and methionine sulphoximine on rat brain *S*-adenosylmethionine levels. *J. Neurochem.* **24,** 63–66.

Schatz R. A. and Sellinger O. Z. (1975b) The elevation of cerebral histamine-*N*- and catechol-*O*-methyl transferase activities by L-methionine-D,L-sulfoximine. *J. Neurochem.* **25,** 73–78.

Schatz R. A., Vunnam C. R., and Sellinger O. Z. (1977) *S*-adenosylhomocysteine in brain: Regional concentrations, catabolism, and the effects of methionine sulfoximine. *Neurochem. Res.* **2,** 27–38.

Schatz R. A., Frye K., and Sellinger O. Z. (1978) Increased *in vivo* methylation of [^{3}H]histamine in the methionine sulfoximine epileptogenic mouse brain. *J. Pharmacol. Exp. Ther.* **207,** 794–800.

Schatz R. A., Wilens T. E., and Sellinger O. Z. (1981) Decreased transmethylation of biogenic amines after *in vivo* elevation of brain *S*-adenosyl-L-homocysteine. *J. Neurochem.* **36,** 1739–1748.

Schayer R. W. (1952) The metabolism of ring-labeled histamine. *J. Biol. Chem.* **196,** 469–475.

Schayer R. W. (1959) Catabolism of physiological quantities of histamine *in vivo*. *Physiol. Rev.* **39,** 116–126.

Schayer R. W. (1965) Histamine and circulatory homeostasis. *Fed. Proc.* **24,** 1295–1297.

Schayer R. W. and Karjala S. A. (1956) Ring *N*-methylation; A major route of histamine metabolism. *J. Biol. Chem.* **221,** 307–313.

Schayer R. W. and Reilly M. A. (1973) Metabolism of ^{14}C-histamine in brain. *J. Pharmacol. Exp. Ther.* **187,** 34–39.

Schayer R. W. and Reilly M. A. (1974) Histamine catabolism in guinea pig, rats and mice. *Eur. J. Pharmacol.* **25,** 101–107.

Schayer R. W. and Reilly M. A. (1975a) Comparison of *in vitro* and *in vivo* histamine methylation by tissues. *Agents Actions* **5,** 119–124.

Schayer R. W. and Reilly M. A. (1975b) Methyl derivatives of histamine; Interaction with histamine metabolism. *Agents Actions* **5,** 231–235.

Schwartz J. C. (1977) Histaminergic mechanisms in brain. *Ann. Rev. Pharmacol. Toxicol.* **17,** 325–339.

Schwartz J. C., Pollard H., and Quach T. T. (1980) Histamine as a neurotransmitter in mammalian brain: Neurochemical evidence. *J. Neurochem.* **35,** 26–33.

Schwartz J. C., Pollard H., Bischoff S., Rehault M. C., and Verdiere-Sahuque M. (1971) Catabolism of ^{3}H-histamine in the rat brain after intracisternal administration. *Eur. J. Pharmacol.* **16,** 326–335.

Schwartz J. C., Baudry M., Chast F., Pollard H., Bischoff S., and Krishnamoorthy M. S. (1974) Histamine in the Brain: Importance of Transmethylation Processes and Their Regulation, in *Central Nervous System: Studies on Metabolic Regulation and Function* (Genazzani E. and Herken H., eds.), pp. 172–184. Springer-Verlag, New York.

Sellinger O. Z., Schatz R. A., and Ohlsson W. G. (1978) Rat and mouse brain histamine *N*-methyltransferase: Modulation by methylated indoleamines. *J. Neurochem.* **30,** 437–445.

Shaff R. E. and Beaven M. A. (1977) Inhibition of histamine-*N*-methyltransferase and histaminase (diamine oxidase) by a new H-2 receptor agonist, dimaprit. *Biochem. Pharmacol.* **26,** 2075–2078.

Snyder S. H. and Axelrod J. (1964) Inhibition of histamine methylation *in vivo* by drugs. *Biochem. Pharmacol.* **13,** 536–537.

Snyder S. H. and Axelrod J. (1965) Sex differences and hormonal control of histamine methyltransferase activity. *Biochim. Biophys. Acta* **111,** 416–421.

Snyder S. H. and Taylor K. M. (1972) Histamine in the Brain: A Neurotransmitter? in *Perspectives in Neuropharmacology: A Tribute to Julius Axelrod.* (Snyder S. H., ed.), pp. 43–73. Oxford Univ. Press, New York.

Snyder S. H., Brown B., and Kuhar M. J. (1974) The subsynaptosomal localization of histamine. Histidine decarboxylase and histamine methyltransferase in rat hypothalamus. *J. Neurochem.* **23,** 37–45.

Sperk G., Hörtnagl H., Reither H., and Hornykiewitz O. (1981) Evidence for neuronal localization of histamine-*N*-methyltransferase. *J. Neurochem.* **37,** 525–526.

Subramanian N., Whitmore W. L., Seidler F. J., and Slotkin T. A. (1981) Ontogeny of histaminergic neurotransmission in the rat brain: Concomitant development of neuronal histamine, H-1 receptors, and H-1 receptor-mediated stimulation of phospholipid turnover. *J. Neurochem.* **36,** 1137–1141.

Suriyachan D. and Thithapanda A. (1972) Distribution and heterogeneity of histamine *N*-methyltransferase. *Biochem. Biophys. Res. Commun.* **48,** 1199–1207.

Taylor K. M. (1973) Comparison of the antihistaminic effect of burimamide and mepyramine and their effect on the activity of histamine methyltransferase. *Biochem. Pharmacol.* **22,** 2775–2776.

Taylor K. M. and Snyder S. H. (1971) Histamine in rat brain: Sensitive assay of endogenous levels, formation *in vivo* and lowering by inhibitors of histidine decarboxylase. *J. Pharmacol. Exp. Ther.* **179,** 619–633.

Taylor K. M. and Snyder S. H. (1972a) Dynamics of the regulation of histamine levels in mouse brain. *J. Neurochem.* **19,** 341–354.

Taylor K. M. and Snyder S. H. (1972b) Isotopic microassay of histamine, histidine, histidine decarboxylase and histamine methyltransferase in brain tissue. *J. Neurochem.* **19,** 1343–1358.

Taylor K. M. and Snyder S. H. (1972c) Histamine methyltransferase: Inhibition and potentiation by antihistamines. *Mol. Pharmacol.* **8,** 300–310.

Taylor S. L. and Lieber E. R. (1979) *In vitro* inhibition of rat intestinal histamine-metabolizing enzymes. *Food Cosmet. Toxicol.* **17,** 237–240.

Thithapanda A. and Cohn V. H. (1978) Brain histamine *N*-methyltransferase purification, mechanism of action, and inhibition by drugs. *Biochem. Pharmacol.* **27,** 263–271.

Thomas H. V. and Simmons E. (1969) Histamine content in sputum from allergic and nonallergic individuals. *J. Appl. Physiol.* **26,** 793–797.

Troidl H., Lorenz W., Barth H., Rohde H., Feifel G., Schmal A., Goecke K., Reimann-Huhnd A., and Seidel W. (1973) Augmentation of pentagastrin-stimulated gastric secretion in the heidenhain pouch dog by amodiaquine: Inhibition of histamine methyltransferase *in vivo? Agents Actions* **3,** 157–167.

Tsuruta Y., Ishida S., Kohashi K., and Ohkura Y. (1981) Fluorometric assay of histamine *N*-methyltransferase by high performance liquid chromatography. *Chem. Pharm. Bull.* **29,** 3398–3400.

Tuomisto L. (1977) Ontogenesis and regional distribution of histamine and histamine-*N*-methyltransferase in the guinea pig brain. *J. Neurochem.* **28,** 271–276.

Tuomisto L. and Tuomisto J. (1982) Diurnal variations in brain and pituitary histamine and histamine-*N*-methyltransferase in the rat and guinea pig. *Med. Biol.* **60,** 204–209.

Tuomisto L. and Eriksson L. (1982) Histamine and histamine-*N*-methyltransferase in the goat brain. *Agents Actions* **12,** 142–145.

Ulmer W. T., Zimmerman I., Bugalho de Almeida A. A., and Park H. S. (1982) Effects of exogenous and endogenous histamine on the respiratory system. *Klin. Wochenschr.* **60,** 991–996.

Verburg K. M. and Henry D. P. (1984) Studies on the substrate specificity of histamine *N*-methyltransferase and characterization of substrates present in mammalian urine. *Pharmacologist* **26,** 222.

Verburg K. M., Bowsher R. R., and Henry D. P. (1983a) A new radioenzymatic assay for histamine using purified histamine *N*-methyltransferase. *Life Sci.* **32,** 2855–2867.

Verburg K. M., Bowsher R. R., Zerbe R. L., and Henry D. P. (1983b) Characterization of neurohypophyseal histamine: Effect of saline-induced dehydration and stalk-sectioning. *Soc. Neurosci. Abstr.* **9,** 705.

Verburg K. M., Bowsher R. R., and Henry D. P. (1984) Kinetic analysis of the histamine *N*-methyltransferase reaction as used in the histamine radioenzymatic assay: Optimization of assay specificity. *Life Sci.* **35,** 241–251.

Vugman I. and Rocha e Silva M. (1966) Biological Determination of Histamine in Living Tissues and Body Fluids, in *Handbook of Experimental Pharmacology.* Vol. XVIII/1 (Rocha e Silva M., ed.), pp. 81–115. Springer-Verlag, Berlin.

Waldmeier P. C., Feldtrauer J. J., and Maitre L. (1977) Methylhistamine: Evidence for selective deamination by MAO B in the rat brain *in vivo*. *J. Neurochem.* **29,** 785–790.

Watanabe T. (1981) Heterogeneity of histamine *N*-methyltransferase. *Chem. Pharm. Bull.* **29,** 2379–2382.

Zappia V., Zydek-Cwick C. R., and Schlenk F. (1969) The specificity of *S*-adenosylmethionine derivatives in methyl transfer reactions. *J. Biol. Chem.* **244,** 4499–4509.

Zeller E. A. (1951) Oxidation of Amines, in *The Enzymes*, Vol. 2, (Sumner J. B. and Myrback K., eds.), pp. 536–558. Academic, New York.

Chapter 6
L-Histidine Decarboxylase
A Marker of Histaminergic Neurons

Hélène Pollard and Jean-Charles Schwartz

1. Introduction

The role of histamine (HA) as a neurotransmitter in the CNS, although proposed a decade ago (Schwartz, 1975), has been only progressively recognized (reviewed by Schwartz et al., 1980, 1982, 1984; Hough and Green, 1984). One of the key pieces of experimental evidence has been the demonstration that synthesis of the amine in CNS depends on a specific enzyme, L-histidine decarboxylase (HDC) (Schwartz et al., 1970), localized in a specific set of neurons, therefore constituting a reliable marker.

Because HA levels in brain are relatively low (about one tenth those of catecholamines, for instance), presumably reflecting a low number of histaminergic neurons, it has long been difficult to measure and localize HDC activity in the CNS. However, several methods are now available for reaching these aims. This chapter will discuss their respective advantages and describe in detail some other selected methods developed and currently utilized in our laboratory.

2. Characterization of the HA-Synthesizing Enzyme in Brain

Formation of HA is a simple, one-step process, but two different enzymes from mammalian tissue are apparently able to catalyze the decarboxylation of the natural amino acid L-histidine (L-His): the aromatic L-amino acid (or DOPA-5-HTP) decarboxylase

(AADC) (EC 4.1.1.28) and the "specific" L-histidine decarboxylase (specific HDC) (EC 4.1.1.22). If cerebral HA was synthesized by the former enzyme, abundant in catecholaminergic and serotoninergic neurons, this could rule out the presence of HA in a specific neuronal system. This is not the case because the HA-synthesizing enzyme in peripheral tissue has been separated from DOPA decarboxylase in the course of purification processes (*see* section 8), and its catalytic properties seem similar in brain and peripheral tissue, differing in both cases from those of DOPA decarboxylase (*see* section 5).

In addition, many pieces of evidence, such as the effects of selective inhibitors, regional distribution, and effects of lesions, clearly show that HA biosynthesis in brain depends on the action of the specific HDC and not on that of AADC. Hence, in homogenates of rat hypothalamus, the region with the highest activity, kinetic data for the HA-forming enzyme (Schwartz et al., 1970) are very similar to those reported for the specific HDC found, for instance, in the rat stomach (Håkanson and Owman, 1966), i.e., the pH optimum of the enzyme is inversely related to the substrate concentration (with a value near 7.0 for $10^{-4}M$ L-His), and its affinity is relatively high (K_m about $10^{-4}M$). On the other hand, the AADC exhibits a much lower affinity for L-His (Lovenberg et al., 1962). The differential effects of decarboxylase inhibitors, both in vitro and in vivo, also indicate that a specific HDC is responsible for cerebral HA biosynthesis (*see* section 6).

Whereas L-His is the main substrate for HDC, the natural amino acid L-3-methylhistidine is also decarboxylated, with formation of *tele*-methylhistamine (Schwartz et al., 1973). *Tele*-methylhistamine is a substance with little biological activity that is predominantly produced in brain by transmethylation of HA. Since the affinity of 3-methylhistidine for HDC is at least 10-times lower than that of L-His, and since its tissue levels are also probably much lower, it can be assumed that formation of *tele*-methylhistamine through this pathway is normally negligible. However, accumulation of [^{3}H]-3-methylhistamine in the brain of mice treated with [^{3}H]-3-methylhistidine has been detected (Schwartz et al., 1973), and operation of this pathway could, at least in part, explain why endogenous *tele*-methylhistamine is found in the subcellular particles of guinea pig brain in which HA is encountered (Green, 1970).

3. Localization of HDC in the CNS

Until the eighties, and in spite of several attempts, antibodies selectively recognizing the HA-synthesizing enzyme had not been

Table 1
Tissue Distribution of L-Histidine Decarboxylase Activity, as Compared to Histamine Levels, in Brain and Various Organs of the Mouse[a]

Brain region or organs	Histidine decarboxylase, dpm/h/μg protein	Histamine, ng/g
Hypothalamus	35.5 ± 3.0	139 ± 9
Striatum	13.4 ± 0.8	32 ± 2
Midbrain	15.9 ± 1.0	82 ± 5
Cerebral cortex	10.5 ± 1.4	35 ± 3
Pons-medulla	8.3 ± 0.8	21 ± 1
Hippocampus	7.4 ± 0.8	36 ± 4
Cerebellum	2.0 ± 0.4	16 ± 1
Liver	0.5 ± 0.2	61 ± 11
Spleen	5.4 ± 0.8	553 ± 80
Heart	1.0 ± 0.4	159 ± 15
Stomach (glandular portion)	95.0 ± 18.0	6800 ± 380

[a]A group of 10 mice was killed and various brain areas or organs rapidly dissected out. HDC activity was measured, as described in section 7, and HA levels by the radioisotopic microassay of Taylor and Snyder (1972a), on aliquots of the same homogenates. Each value is the mean ± SEM of 5–8 separate experiments.

obtained, so the HDC-containing cells could not be visualized by immunohistochemical methods. Nevertheless, the presence of HDC within specific sets of neurons in mammalian CNS was convincingly suggested by the distribution of HDC activity in a markedly heterogenous fashion between cerebral regions (Baudry et al., 1973; Schwartz et al., 1970; 1971; Taylor and Snyder, 1972b) (*see* Table 1). Heterogeneous distributions of the enzyme activity are also observed in analysis performed at a higher resolution level, i.e., among nuclei of regions like the hypothalamus (Pollard et al., 1976) or the upper brainstem (Pollard et al., 1978). In addition, subcellular fractionation studies indicated that a major portion of the HA-synthesizing enzyme is held in the cytoplasm of nerve terminals (Baudry et al., 1973; Snyder et al., 1974).

Lesion studies have allowed identification of several HDC-containing neuronal pathways. Lesions of the lateral hypothalamic area interrupting the medial forebrain bundle (MFB) lead to a substantial fall in specific HDC activity in various telencephalic areas (Garbarg et al., 1974 and 1976). Other lesions have confirmed that telencephalic areas do in fact contain histaminergic nerve-endings, as judged by the almost complete loss of HDC activity in the cerebral cortex of the cat or the hippocampal region of the rat (Barbin et al., 1975, 1976). More selective and combined lesions of the main afferent pathways to the hippocampus or the amygdaloid complex have permitted description of histaminergic afferents to these regions (Barbin et al., 1976; Ben Ari et al., 1977).

The localization of HA cell bodies has also been investigated, using several biochemical approaches. The first consisted of establishing a biochemical map of the hypothalamus (Brownstein et al., 1974; Pollard et al., 1976) and brainstem (Pollard et al., 1978), both regions likely to contain cell bodies. The mamillary bodies were found to contain by far the highest levels of HDC. The second approach consisted of selective destruction in the upper midbrain or hypothalamus, from which it was concluded that the cell bodies of the ascending histaminergic pathway were localized in the mesencephalic reticular formation and in the mamillary bodies.

These localizations have been confirmed by various approaches (Garbarg et al., 1980a): (1) iv administration of colchicine to inhibit axonal transport has resulted, 48 h later, in a significant decrease in telencephalic HDC; and (2) stereotaxic microinjection in the mamillary bodies of kainic acid, a neurotoxic analog of glutamate ablating selectively neuronal cell bodies, has led to significant decreases of HDC activity both locally and in telencephalon (Pollard et al., 1985; Schwartz et al., 1982). There is also evidence from lesion studies for a descending system innervating the brainstem (Pollard et al., 1978).

Nevertheless, it is clear that the complexity of CNS requires more powerful and precise methods than neurochemical approaches to localize chemically-identified neuronal pathways. In the last few years, several experimental approaches to this problem were reported, including immunohistochemical studies with polyclonal antibodies directed against either HA (Wilcox and Sybold, 1982; Leslie and Osborne, 1983; Panula et al., 1984; Steinbusch and Mulder, 1983, 1984) or the enzyme HDC (Tran and Snyder, 1981; Watanabe et al., 1983, 1984a) and attempts at localizing HDC autoradiographically with [^{3}H]-α-fluoromethylhistidine ([^{3}H]-α-FMH) (Schwartz et al., 1982), an irreversible inhibitor. Although the latter approach was unsuccessful, immunohistochemical studies have already provided some consistent information regarding HA-storing and synthesizing cells, particularly in the CNS. However, there have already been some discrepancies among data obtained with various antibodies. For instance, in the rat gastric mucosa, a tissue with high HDC activity, parietal cells were stained with one antibody (Tran and Snyder, 1981) and enterochromaffin-like cells with another (Watanabe et al., 1983), and results with antibodies against HA were not reported. Indeed, it is difficult to establish that antibodies raised against a small molecule like HA do not cross-react with other cerebral constituents. In the same way the specificity of polyclonal antibodies raised against a purified protein must always be questioned, particularly as long as the latter is not shown to be homogeneous in various sensitive tests, because antisera might contain antibodies directed against impurities in the antigen.

For these various reasons, the strategy used in this laboratory (Pollard et al., 1983) has been to develop monoclonal antibodies (Mab) against HDC because (1) the enzyme seems a highly specific marker of HA-synthesizing neurons in CNS, (2) as a result of its instability, HDC is difficult to purify to absolute homogeneity (Aures et al., 1970), and (3) with Mabs it is possible to avoid the development of antibodies against impurities and, therefore, to start from even a partially purified antigen, provided that the screening test allows identification of antibodies specifically directed against the enzyme molecule itself.

The Mab developed in this laboratory (HI 113-12) was used to localize HDC-immunoreactive structures in various mammalian tissue, including brain. Structures known to present a very high enzyme activity were clearly labeled with the antiHDC Mab HI 113-12. This was, for example, the case for rat fetal liver and cultured mouse mastocytoma cells (P 815 from BDA_2 mice), in which the HDC activity is clearly recognized by Mab HI 113-12 (*see* Table 2); the positive immunohistochemical response was located in the cytoplasm (Fig. 1).

In the rat stomach, another tissue with high enzyme activity, numerous HDC-immunoreactive mast cells were found in the submucosa (Fig. 2) and in the superficial portion of the mucosa. In some experiments staining was also observed at the level of cells located in the mid-portion of gastric glands (not shown), but this was inconsistent and requires confirmation, particularly in view of discrepant results with various polyclonal antibodies (Tran and Snyder, 1981; Watanabe et al., 1983).

In the rat brain, some HDC-immunoreactive structures could be visualized even without colchicine pretreatments. So far, HDC perikarya have been found only in the hypothalamus. A large group of HDC cell bodies were located in the posterior basal hypothalamus at the level of caudal magnocellular nucleus and around the arcuate nucleus at anteriorities comprised between A:2800 and A:3800 μm, according to Köning and Klippel (1970).

Table 2
Biochemical Specificity of the Monoclonal Antibody HI 113-12[a]

Preparation	Total decarboxylase activity, dpm/h × 10^3	Specifically immunoprecipitated decarboxylase activity	
		dpm/h × 10^3	% of total
Purified gastric HDC	6.2 ± 0.5	1.4 ± 0.1	24 ± 1
High speed supernatant from			
Rat gastric mucosa	3.9 ± 0.2	0.9 ± 0.1	24 ± 1
Rat hypothalamus	4.1 ± 0.3	1.0 ± 0.1	24 ± 1
Rat striatum	3.9 ± 0.4	1.0 ± 0.2	22 ± 4
Mouse mastocytoma	3.7 ± 0.1	0.6 ± 0.1	17 ± 1
Bacterial HDC (from *C. welchii*)	2.8 ± 0.2	Nondetectable	
Striatal glutamic acid decarboxylase	3.6 ± 0.4	Nondetectable	
Striatal DOPA decarboxylase	2.7 ± 0.2	Nondetectable	

[a]Decarboxylase activity specificially immunoprecipitated by Mab HI 113-12 (0.5 μg) was calculated by substracting the enzyme activity precipitated under similar conditions in the presence of a control ascitic fluid (Pollard et al., 1983).

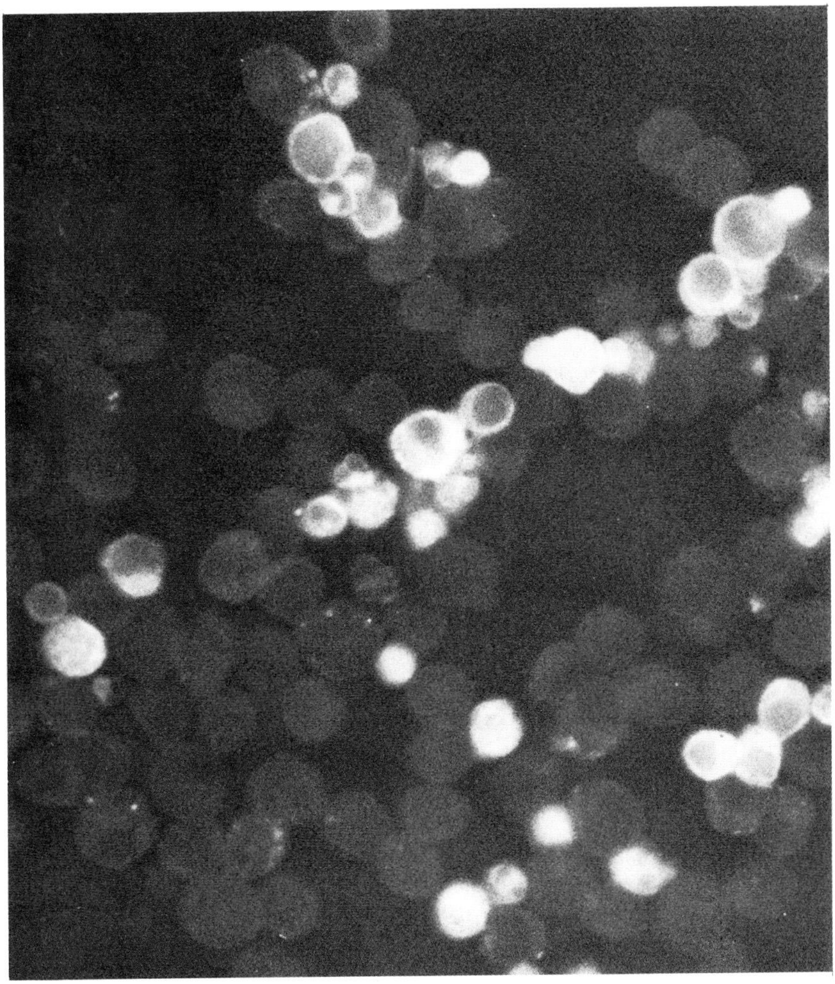

Fig. 1. Indirect immunofluorescence photomicrograph of mastocytoma cells (P 815, BDA_2 mice, kindly provided by Dr. Kinsky) after incubation with Mab HI 113-12 (×568).

The former nucleus was densely packed with large and fusiform cells that have large round nuclei and long processes (Fig. 3). More caudally, HDC cells appeared in the lateral mamillary nucleus. More rostrally, some HDC-immunoreactive neurons were present in the paraventricular nucleus and at the dorsal tip of the third ventricle. Immunoreactive varicose fibers restricted to the rat hypothalamus were detected entering into or emanating from the median eminence, and a dense network of HDC nerve terminals presenting a rather large aspect and surrounding unlabeled cells was widely distributed in the arcuate nucleus. This preliminary

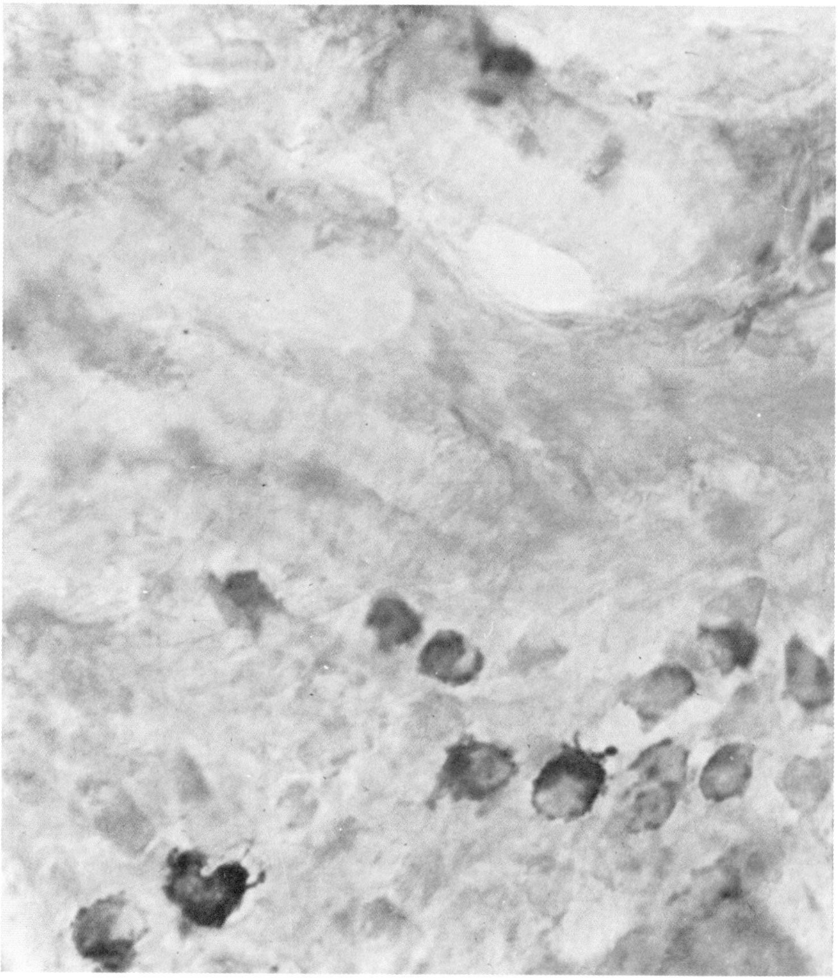

Fig. 2. Photomicrograph of transversal freeze microtome section (20 μm thickness) through the rat fundus after incubation with Mab HI 113-12 and immunostaining with the avidin biotin–peroxidase complex method. Note the presence of numerous HDC immunoreactive cells (apparently gastric mast cells) in the submucosa of the gland (×568).

distribution of HDC-containing neurons in rat brain, detected with our Mab HI 113-12, is in rather good agreement with findings obtained with polyclonal antibodies against either HDC (Watanabe et al., 1984a) or HA (Panula et al., 1984; Steinbusch and Mulder, 1983, 1984, 1985).

For example, Watanabe et al. (1983, 1984a) described a concentration of HDC cell bodies in the posterior hypothalamus, in posi-

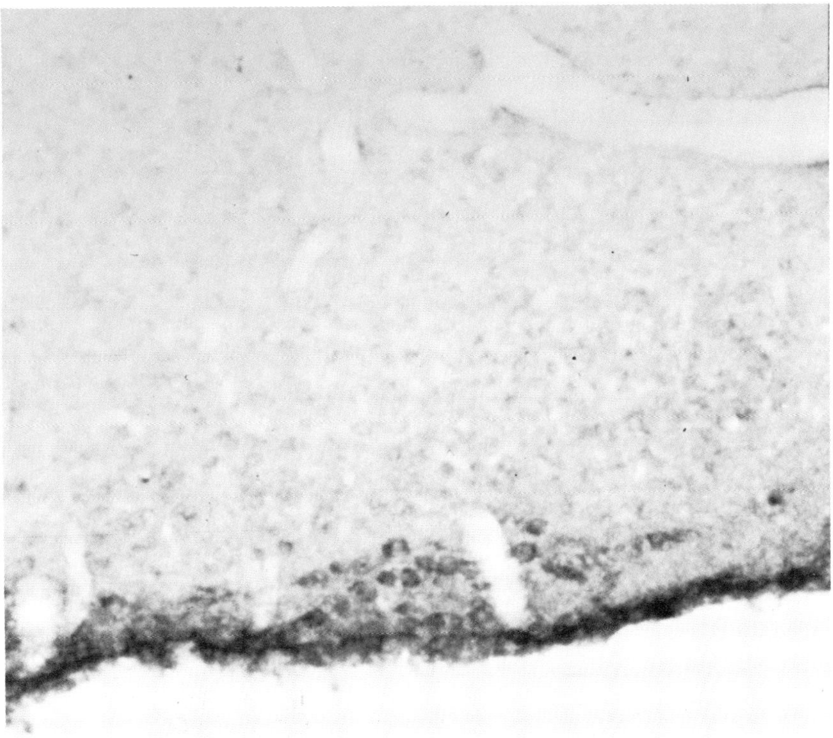

Fig. 3. Photomicrograph of a frontal freeze microtome section (20 μm thickness) through the posterior hypothalamus of the rat, after incubation with Mab HI 113-12 and immunostaining with the PAP-method. Note the presence of a group of HDC-immunoreactive neurons in the caudal magnocellular nucleus (×142).

tions corresponding to the tuberal, caudal, and postmamillary caudal magnocellular nuclei, and also indicated the presence of some scattered cells in the posterior and lateral hypothalamic nuclei. In addition, these authors suggested the coexistence of HDC and glutamate decarboxylase (GAD) immunoreactivities in the same neurons of the caudal magnocellular nucleus (Watanabe et al., 1984b).

Retrograde tracing experiments have recently shown the existence of an ascending pathway emanating from magnocellular neurons in the caudal mamillary region and directly innervating widespread regions of the neocortex (Vincent et al., 1983). Hence, the idea that ascending histaminergic neurons directly link the posterior hypothalamus to widespread telencephalic areas and provide them with limbic, emotional, and visceral information (Schwartz et al., 1980, 1984) now seems reinforced by a series of precise anatomical data.

4. Purification of HDC

The low content and extreme lability of the enzyme have so far hampered its purification to homogeneity. However, HDC has been highly purified from various tissues with high HDC activity, such as the hamster placenta (Håkanson, 1967a), rat whole fetus (Håkanson, 1963; Watanabe et al., 1979, 1982), rat fetal liver (Tran and Snyder, 1981; Taguchi et al., 1984), mouse mastocytoma (Hammar and Hjerten, 1980), or rat gastric mucosa (Savany and Cronenberger, 1978; Pollard et al., 1983, 1985).

Stabilizing conditions were particularly studied in several reports (Watanabe et al., 1979; Fukui et al., 1980; Hammar and Hjerten, 1980), and the essential role of dithiothreitol and polyethylene glycol were generally recognized. In addition, Hammar and Hjerten (1980) reported that phosphate ion stabilizes HDC, but sulfate destabilizes it. Tran and Snyder (1981) included EDTA in their buffers and Pollard et al. (1985), pyridoxal 5' phosphate (PLP) for the whole procedure.

After homogeneization of tissues and high-speed centrifugation, purification procedures generally include a double ammonium sulfate fractionation of the clear supernatant, similar to that initially used by Håkanson (1963), in which the 25–27% $(NH_4)_2SO_4$ supernatant and the 40–43% pellet are successively retained. Another common point of the various procedures is the ion-exchange chromatography (generally on DEAE cellulose) step—an important one allowing for separation of HDC from DOPA decarboxylase.

Affinity chromatography with the L-His-containing peptide carnosine as a ligand (Watanabe et al., 1979, 1982), hydroxylapatite column chromatography (Hammar and Hjerten, 1980), isoelectric focusing (Savany and Cronenberger, 1978, 1982), or chromatofocusing (Pollard et al., 1985) are more original purification steps that differentiate the procedures according to the source of the enzyme and the purpose of the preparation. Indeed, a high degree of purity (if not homogeneity of the preparation) is necessary to develop selective polyclonal antibodies against this antigen, whereas only a good recovery seems required to easily develop Mabs. A protocol for a rapid (partial) purification of HDC is proposed in section 8.

5. Properties of HDC

The various purified HDC preparations seem to display very similar catalytic properties (Table 3). As for other mammalian amino acid decarboxylases, the coenzyme of HDC from various origins seems to be PLP (Håkanson, 1967b; Aures et al., 1970; Palacios et al., 1976; Watanabe et al., 1979; Savany and Cronenberger, 1978; Hammar and Hjerten, 1980). However, no spectral study has as yet been carried out, and PLP has not been isolated from purified HDC. PLP seems bound to the apoenzyme in a relatively tight manner (Palacios et al., 1976) so that its addition to crude cerebral extracts does not result in marked increases in HDC catalytic activity (unpublished observation). The optimal pH for HDC from various origins is around 7, and the K_m value for L-His at this pH is in the millimolar range. However, the optimum pH depends on the substrate concentrations: The higher the substrate concentration, the lower the optimal pH (Håkanson, 1967a). And the lower the pH of the assay medium, the higher the K_m value. Ionic strength also influences HDC activity (Hammar and Hjerten, 1980; Watanabe et al., 1979): At high ionic strengths, the K_m and V_{max} values are increased. The HA-forming enzyme in brain homogenates from *Aplysia californica* (Weinreich and Yu, 1977), rat (Schwartz et al., 1970), mouse (Taylor and Snyder, 1972a), guinea pig (Schayer and Reilly, 1974), cat (White, 1960), and human (Barbin et al., 1980) displays catalytic properties similar to those found in purified mammalian preparation. In contrast, the physical properties of the enzyme appear to differ from one purified preparation to another. For instance, molecular weights of 90,000–110,000 (Hammar and Hjerten, 1980; Savany and Cronenberger, 1978, 1982; Watanabe et al., 1979; Pollard et al., 1985; Yamada et al., 1984) or 210,000 (Tran and Snyder, 1981) were reported, suggesting the possible existence of isoenzymes. This heterogeneity was further demonstrated by isoelectrofocusing: Savany and Cronenberger (1982) described the existence of three isoenzymes for rat gastric HDC with pH_i values of 5.9, 5.65, and 5.35, respectively, whereas Watanabe et al. (1982) proposed the existence of two types of isoenzymes, a fetal and a cerebral isoenzyme with pH_i values of 5.0 and 5.4, respectively, and suggested the presence of both types in the stomach. Interestingly, a Mab raised against the gastric enzyme

Table 3
Comparison of Some Properties of Purified L-Histidine Decarboxylases From Various Origins

Origin	Molecular weight, × 1000	Subunits, kdaltons	Cofactor	Affinity, μM	Optimum pH	pH Isoelectric	Purification factor, X-fold	References
Lactobacillus	208	26; 9	Pyruvic acid	900	4.7–5.2	—	200	Boeker et al., 1972; Snell, 1977
Rat fetal liver	210	145; 60	PLP	0, 1	—	—	1260	Tran and Snyder, 1981
	110	54 × 2	PLP	250	6.8	5.1	1490	Watanabe et al., 1984b
Rat whole fetus	90	Unknown	PLP	250	6.8	—	7700	Watanabe et al., 1979, 1982
Rat stomach	95	Unknown	PLP	154 ± 15; 181 ± 21; 213 ± 38	6.8	5.1; 5.6; 5.35	640	Savany and Cronenberger, 1978
	90–100	45–50	PLP	—	—	5.6–5.2	1080	Pollard et al., 1985
Rat brain	90	—	PLP	340 ± 10	—	5.0	160	Yamada et al., 1984
Mouse mastocytoma	110	56	PLP	—	—	5.0	560	Hammar and Hjerten, 1980

cross-reacts with HDC from rat brain or mouse mastocytoma, indicating that at least these enzymes share common antigenic determinants (Pollard et al., 1983). More work is probably needed to clarify the existence and properties of HDC isoenzymes.

6. Inhibitors of HDC

Because HA is synthetized in a single step and by a specific enzyme, HDC inhibitors are potentially useful tools with which to investigate the role of HA in CNS. Until recently, the only known HDC inhibitors were competitive inhibitors such as α-hydrazino-histidine or brocresine (Levine et al., 1969; Schwartz et al., 1970; Mole and Shepherd, 1972; Beaven, 1978; Kelly et al., 1981). However, the usefulness of in vivo experiments was limited by the low potency and relatively poor specificity of these inhibitors at efficient dosage. Recently, α-fluoromethylhistidine (α-FMH) was synthetized among a series of "suicide," i.e., K_{cat}, inhibitors of PLP-containing amino acid decarboxylases (Kollonitsch et al., 1978). This compound potently inhibits in a time-dependent, concentration-dependent, and nonreversible manner the activity from rat hypothalamus, whereas neither DOPA decarboxylase nor glutamate decarboxylase activities are significantly affected (Table 4) (Garbarg et al., 1980b). These data not only confirm the identity of cerebral HDC, but also indicate that, as previously assumed, the enzyme functions with PLP as a cofactor.

These various HDC inhibitors also interrupt, in a more or less selective manner, cerebral HA formation in vivo and elicit decreases in the endogenous amine level (although the decrease is a partial one restricted to the neuronal pool of cerebral HA) (Garbarg et al., 1980b; Maeyama et al., 1983). This was initially observed after administration of compounds like α-hydrazinohistidine or brocresine (Schwartz et al., 1972; Taylor and Snyder, 1971, 1972a), which display relatively low potency and selectivity. In addition, the irreversible inhibitor α-FMH, when administered systemically in low dosage, inactivates HDC almost completely in peripheral organs and brain regions in a long-lasting manner, restoration occurring progressively in the following days, most probably as a consequence of synthesis of new HDC molecules (Fig. 4). This inactivation process is accompanied, after a short delay, by substantial decreases in HA levels in various brain regions in which the amine synthesis is interrupted (Garbarg et al., 1974; Maeyama et al., 1982). Therefore, the data of Garbarg et al. (1980b) support the view that the time-course of the restoration of HDC activity in these

Table 4
Effect of a Treatment With (±)-α-Fluoromethylhistidine on Various Enzymatic Activities in Mouse Brain[a]

	L-Histidine decarboxylase, dpm/h/μg protein	Histamine *N*-methyltransferase, pmol/h/μg protein	L-DOPA decarboxylase, pmol/h/μg protein	L-Glutamate decarboxylase, pmol/h/μg protein
Cerebral cortex				
Control	10.2 ± 0.4	0.82 ± 0.04	14 ± 2	64 ± 3
Treated	1.1 ± 0.3[b]	0.98 ± 0.04[c]	12 ± 1[c]	79 ± 9[c]
Hypothalamus				
Control	50.9 ± 3.2	1.37 ± 0.06	31 ± 6	176 ± 16
Treated	5.8 ± 2.4[b] (89%)	1.35 ± 0.06[c]	29 ± 3[c]	153 ± 16[c]

[a]A group of five mice was injected ip with 100 mg/kg of (±)-α-fluoromethylhistidine and killed 4 h later. The various enzyme activities were measured in the same homogenate (Garbarg et al., 1980b).

[b]$p < 0.01$.

[c]Nonsignificant.

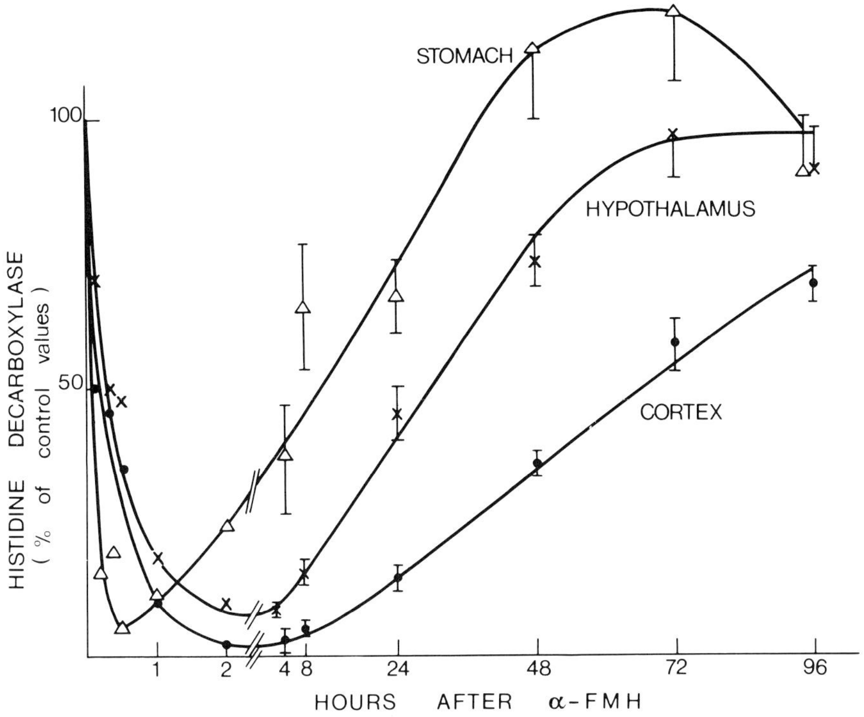

Fig. 4. Time-course of changes in L-histidine decarboxylase activity in mouse tissues after systemic administration of (+)-α-FMH. Groups of 5–15 mice received ip injections of (+)-α-FMH (10 mg/kg) and were killed at the indicated time intervals. Bars represent standard SEM values (from Garbarg et al., 1980b).

areas reflects the synthesis of new molecules of enzyme in the neuronal cell bodies and then their axonal transport to nerve terminals. Some 50% of initial HDC activity is recovered within 26 h in the hypothalamus (which contains HA cell bodies) and 58 h in the cerebral cortex (which contains nerve endings). Taking into account the distance of a few centimeters from hypothalamus to cortex, an approximative rate of 1 mm/h for axonal transport can be roughly calculated. This rate of transport is in rather good agreement with those reported for other soluble enzymes, such as tyrosine hydroxylase and DOPA decarboxylase (Oesch et al., 1973; Brimijoin and Wiermaa, 1977).

Taking into account the postulated mechanism of α-FMH, involving formation of a covalent link and its high selectivity, it seemed feasible to label the enzyme covalently with the radioactive inhibitor. [^{3}H]-α-FMH of high specific radioactivity (about 45 Ci/mmol) was synthesized (Schwartz et al., 1982). It was observed that [^{3}H]-α-FMH binds in vitro to brain slices, and its radioactivity is incorporated into the trichloroacetic acid (TCA)-insoluble material of a brain extract. However, there was no evidence for selective attachment to HDC as shown by: (1) the lack of protection by preincubation in the presence of the nonradioactive inhibitor in high concentration, and (2) the lack of correlation between binding and HDC activity in various regions (the activity in hypothalamus, the richest brain area, does not significantly differ from that in cerebral cortex or cerebellum).

The same remark holds true for in vivo studies: TCA-insoluble radioactivity was still detectable in mouse brain even several days after iv administration of [^{3}H]-α-FMH, but neither protection nor regional distribution paralleling that of HDC activity could be demonstrated for this binding (Schwartz et al., 1982).

Finally, autoradiography of the brain of mice injected with [^{3}H]-α-FMH did not show any evidence of selective labeling. These results suggest that significant nonselective binding of [^{3}H]-α-FMH occurs in all these experiments and predominates over the binding of the inhibitor to HDC molecules, which is presumably of low proportion. Hence, more classical approaches to visualize the enzyme are probably required.

7. HDC Activity Measurement

7.1. Existing Methods

Decarboxylation of L-His is the unique step in HA synthesis, and the same technical difficulties have been encountered in estimating HDC activity as for other rate-limiting enzymes, such as those involved in the formation of cerebral catecholamines or serotonin. HDC activity can be measured in various tissues by evaluating the reaction products, i.e., carbon dioxide or HA. Histidine in radioactive or nonradioactive form is used as a substrate. Although CO_2 formation can be estimated by the standard Warburg manometer technique, the utility of the method is strongly limited by its very low sensitivity. Alternatively, $^{14}CO_2$ production from His [^{14}C]-labeled in the carboxyl group is used in several procedures (Kobayashi, 1963; Aures et al., 1970; Henningson et al., 1974; Ritchie and Levy, 1975; Beaven, 1978). However, this method, commonly used for the measurement of high HDC activities, such as those found in rat stomach, does not appear suitable for estimating the low activities encountered in most brain areas because of the important—and sometimes erratic—nonenzymatic formation of $^{14}CO_2$ that results in high blank values and poor reproducibility (Rönnberg and Schwartz, unpublished observations).

Two types of reliable methods are currently available that depend on the use of nonisotopic and radiolabeled His, respectively, as substrate. Determination of the formed HA can be performed by either fluorometric or isotopic assays.

In the first type of method, the reaction product HA is isolated from the remaining L-His (saturating concentration) by extraction with organic solvent (Shore et al., 1959; Burkhalter, 1962; Werle and Lorenz, 1966; Schayer, 1971) or ion exchange chromatography (Schwartz et al., 1970; Yamatodani, 1978; Watanabe et al., 1979) and estimated fluorometrically after reaction with *o*-phthaldehyde.

In the single-isotope method of Taylor and Snyder (1972b), a saturating concentration of substrate is used, and the formed HA is radioenzymatically assayed by conversion into [^{3}H]-*tele*-methylhistamine by incubation with HA *N*-methyl transferase and [methyl-^{3}H]-*S*-adenosyl methionine as a methyl donor. In these methods, HDC activity is evaluated as the difference between HA levels at the end of the incubation and those already present in the tissue extract. This limits their reproducibility and sensitivity because in regions with low HDC activity, the values of assays are barely twice those of the corresponding "blanks."

This drawback is absent from methods using a radiolabeled amino acid as a substrate. Reilly and Schayer (1968) used [^{14}C]-His as a substrate, and the formed [^{14}C]-HA was isolated by derivatization and isotope dilution with successive recrystallizations until constant specific activity was attained. Although reliable, this method is extremely cumbersome. In order to avoid this limitation, a radiochromatographic procedure has been devised in this laboratory. The assay is based on the formation of [^{3}H]-HA from low concentrations of highly labeled [^{3}H]-L-His. In a second step, isolation of the synthesized [^{3}H]-HA is performed by ion-exchange chromatography (Baudry et al., 1973).

7.2. Protocol for the Radiochromatographic Assay

Tissues are homogenized by sonication in 5–20 vol of ice-cold 0.01*M* phosphate, pH 7.4. The incubations are run in triplicate and performed in 1-mL microtest tubes (Eppendorf). The incubation mixture (40 μL) consists of 25 μL of crude homogenate, 5 μL of 0.4*M* phosphate buffer, pH 7.4, containing 10 μ*M* PLP and 100 μ*M* nonradioactive HA (final concentrations), and 10 μL of [^{3}H]-His solution diluted in 0.01*M* phosphate buffer, pH 7.4, as 0.1 μ*M* final concentration. Blanks are obtained by substituting phosphate buffer for the tissue extracts in the incubation mixture.

The incubation is carried out for 60 min at 37°C and stopped by addition of 100 μL of ice-cold 0.68*N* perchloric acid containing 0.3*M* unlabeled His in order to minimize possible nonspecific decarboxylation of remaining [^{3}H]-His. The perchloric extracts are brought up to pH 8 by addition of 1 mL of a mixture containing 100 μL of 0.05*M* Tris-HCl buffer, pH 8, and potassium hydroxide. After centrifugation in order to eliminate the precipitated $KClO_4$ and proteins, the formed [^{3}H]-HA is separated from [^{3}H]-His by chromatography on Amberlite CG50 columns (Pasteur capillary pipets, d = 4 mm, h = 12 mm) freshly equilibrated in 0.05*M* Tris-HCl buffer, pH 8. The amine is absorbed onto the resin, the columns washed four times with 2 mL 0.02*N* HCl, and the [^{3}H]-HA eluted directly into counting vials with 2 mL *N* acetic acid.

Under these conditions, contamination by [^{3}H]-His in the eluates is less than 0.01% and recovery of [^{3}H]-HA from the columns more than 90%. Values are corrected for the "blanks" and recovery of HA and results expressed in dpm/h/μg protein. [^{3}H]-HA formation is linear with tissue concentration (at least up to 0.25 mg protein/25 μL) and with time (at least up to 2 h) (Garbarg et al., 1983). In all cases, less than 0.5% of the substrate is transformed. [^{3}H]-HA formation in a region known to present relatively low HDC activity, like the cerebral cortex, represents about three times the corresponding blank value. The substrate concentration in the radiochromatographic assay is kept well below the K_m value in order to maintain this adequate ratio.

The optimum pH for HDC has been shown to be dependent on the substrate concentrations. When saturating concentrations of His are used for HDC assay, the optimum pH is 6.8, whereas in the radiochromatographic method the optimum pH is 7.4.

7.3. Comments

Because of the different substrate concentrations used, it is difficult to directly compare HDC activities measured with the radiochromatographic assay and those found with assays using a nonradioactive substrate. However, the relative activities obtained with the various methods vary in parallel from one cerebral region to the other.

Provided that the substrate is previously purified on a small ion-exchange column, blank values are very low, and since the measure does not depend on the HA content of tissues, the radiochromatographic method is sensitive enough to measure HDC activity in restricted areas, such as hypothalamic nuclei (Pollard et al., 1976) or in small brainstem areas in which this activity is quite low (Pollard et al., 1978).

8. Protocol For a Rapid (Partial) Purification of HDC

We have deliberately aimed at a partial purification of the enzyme, starting from the rat gastric mucosa, a convenient source, and selecting a few rapid purification steps carried out at 4°C, which are described below:

8.1. Extraction

The scraped mucosa from normally fed rats is homogenized with a Polytron homogeneizer in 6 vol of 0.05*M* phosphate buffer, pH 7.0, in the presence of stabilizing agents (PLP, 10 μ*M*; dithiothreitol, 0.2 m*M*; EDTA, 10 m*M*; polyethylene glycol, 2%). The homogenate is centrifuged ($3.10^6 g \times$ min) and the pellet discarded.

8.2. Ammonium Sulfate Fractionation

The clear supernatant is treated with $(NH_4)_2SO_4$ (27% saturation) added gradually under gentle shaking. Twenty minutes later the mixture is centrifuged ($1.5 \cdot 10^6 g \times$ min), the supernatant collected and treated again with $(NH_4)_2SO_4$ (43% saturation). After centrifugation, as above, the pellet is dissolved in 10 mL 0.05*M* Tris-HCl buffer, pH 7.0 (containing PLP), and dialyzed overnight against 3 L of buffer with two changes.

8.3. Chromatofocusing

The dialysate is applied onto a polybuffer exchanger column (Pharmacia PBE 94), 1.5 × 15 cm, equilibrated with 0.05*M* Tris-HCl buffer, pH 7.0, containing PLP. After loading the column at a linear flow rate of 30 mL/h, the column is developed with 250 mL of the polybuffer 74 HCl, pH 5.0, containing PLP. Elution fractions (1 mL each) are collected in tubes containing 0.1*M* phosphate buffer, pH 7.4. The fractions, including the major peak of activity, are combined and the polybuffer separated from proteins using 60% ammonium sulfate precipitation.

8.4. Gel Filtration

For this final step, the pellet is dissolved in 3 mL of 0.05*M* phosphate buffer, pH 7.4, containing PLP. The sample is applied onto a Ultrogel AcA 44 (LKB) column (3 × 90 cm) equilibrated with the same buffer. Elution is developed with 100 mL of the same solution and the major peak of activity is concentrated to about 2 mL (Amicon membrane).

8.5. Comments

The entire procedure is summarized in Table 5. It results in a 1000-fold increase in the enzyme specific activity with a 25–30% recovery. Chromatofocusing is an important purification step in the procedure and is derived from the work of Savany and Cronenberger (1982), who carefully determined the isoelectric points for gastric HDC iosenzymes. Hence, although the enzyme is eluted in a rather broad pH range (between 5.6 and 5.2), there is a 25-fold factor increase in the specific activity at this step.

By calibration with standard proteins of the AcA 44 column used at the last step, the molecular weight of the native protein was estimated to correspond to 90,000–100,000. The final preparation submitted to SDS-PAGE under highly sensitive detection conditions (7 μg protein load/cm of gel and $AgNO_3$ staining) showed several bands, among which the major ones corresponded to apparent molecular weights of 95,000 and 45,000–50,000 (the latter presumably corresponding to the enzyme monomer) and represented about 5% of the total.

9. Protocol for Mab Development

The principle of the method developed by Köhler and Milstein (1975) is now well known. It involves obtaining from an immunized animal lymphoid cells that are immortalized by hybridization with an established, highly proliferating cell line and growing these hybrids in tissue culture so that antibodies produced by individual clones can be detected.

The problem of the mode of detection of antibodies in the cell cultures is a critical one since, in view of the large number of samples to be analyzed, a large-scale and not-too-tedious method is required. In addition, when starting from a partially purified antigen, the sensitivity and selectivity of detection may be crucial. An immunoprecipitation method has been chosen by which the mouse immunoglobulins directed toward the enzyme (from the semipurified gastric preparation) are precipitated with the latter by a (sheep) anti-mouse immunoglobulin and the HDC activity in the pellet detected by the radiochromatographic assay (Baudry et al., 1973). This screening method has the distinct advantage of detecting a large variety of antibodies, even nonprecipitant ones, directed against any part of the HDC molecule, except the active site.

Table 5
Rapid Purification of L-Histidine Decarboxylase From Rat Gastric Mucosa[a]

Steps	Specific activity, dpm/h/μg protein	Yield, %	Purification factor
Crude homogenate	188	100	1.0
High speed supernatant	434	57	2.3
0–27% $(NH_4)_2SO_4$ supernatant	806	48	4.3
27–43% $(NH_4)_2SO_4$ pellet	1766	33	6.5
Chromatofocusing	30,360	32	160
Gel filtration	205,454	30	1082

[a]All steps were carried out at 4°C in 3 d. Enzyme activity was measured by the radiochromatographic assay (Baudry and et al., 1973), using 0.1 μ*M* [^{3}H]-L-histidine (50 Ci/mmol) as substrate.

9.1. Immunization

Three-month old BALB/c mice are immunized by sc injections of purified gastric HDC (100 μg), coupled with a mitogenic derivative of bacterial lipopolysaccharide (Lange et al., 1983) and emulsified with an equal volume (0.2 mL) of Freund's complete adjuvant. Boosters are performed at 2-wk intervals, using the same amount of HDC preparation in Freund's incomplete adjuvant 7 d after each injection. Sera (1% dilution) are tested for the presence of polyclonal anti-HDC antibodies and, generally, positive responses are obtained after the second booster. Mice are bled 3 d following the last antigen administrations (50 μg ip and iv) (Buttin et al., 1978).

9.2. Cell Fusion and Cloning of Hybridoma

Sp_2/oAg myeloma cells and hybridoma are cultured in ER medium supplemented with 1 m*M* pyruvate and 2 m*M* glutamine and enriched with 10% horse serum.

Spleen cells from the immunized mouse are removed 3 d after

the final injection and fused with the nonsecreting mouse myeloma cell line Sp_2/oAg on membrane filters in the presence of polyethylene glycol (45%, w/v), as described by Buttin et al. (1978). Twenty-four hours after fusion, the cells are distributed in multiwell dishes and, 24 h later, the selective hypoxanthine (50 μ*M*) azaserine (10 μ*M*) medium is added.

Culture supernatants are taken for detection of anti-HDC antibodies, and cells from the positive wells are cloned by limit dilution (50 cells for 96 well-plates). Wells containing single clusters of hybridoma cells are tested again for the presence of anti-HDC antibodies.

9.3. Development of Ascites and Preparation of Immunoglobulin

Positive clones are injected ip into BALB/c mice previously treated with 2,6,10,14-tetramethylpentodecane (Pristane). Ascitic fluids are harvested and precipitated with sodium sulfate (18% saturation), the proteins dialyzed against 0.01*M* phosphate buffer, pH 7.4, and applied onto a DEAE A50 column (1 × 15 cm). Purified immunoglobulins are eluted without retention.

9.4. Detection of Anti-HDC Antibodies in Mouse Sera Cell Culture Supernatants or Ascitic Fluids

Screening for the secretion of anti-HDC antibodies is carried out using a coprecipitation antigen-binding assay. For this purpose, the purified enzyme (100 μL of a solution containing 1 μg protein) is incubated (24 h at 4°C) with the various samples (250 μL). Then, sheep anti-mouse immunoglobulins (100 μg in 10 μL, Cappel) are added alone or (in the case of hybridoma) with 10 μg of commercial mouse immunoglobulins as a carrier. Twenty-four hours later, the 50,000*g* × min pellets are washed with 2 × 1 mL phosphate buffer (0.05*M*, pH 7.4), resuspended in 50 μL buffer, and the immunoprecipitated HDC activity associated with the pellets assayed. This immunoprecipitation test was successfully applied to nonpurified mouse sera (1% dilution), undiluted hybridoma culture media, ascitic fluids, or purified monoclonal immunoglobulins. All assays included a positive control consisting of the donor mouse's immune serum. Negative controls consist of either preimmune serum or culture media of hybridoma cells generating antibodies against proteins completely different from the antigen, with obtained values subtracted to evaluate the specific immunoprecipitation.

9.5. Comments

Among more than 2000 hybridoma cell lines tested, only one clear anti-HDC antibody-producing clone was selected (Mab HI 113-12). Antibodies in undiluted culture media from positive hybridoma and clones were detected at the limit of sensitivity of the method (less than twice the corresponding negative controls) because of the limited antibody concentration in the culture medium and a nonspecific precipitation of HDC activity by culture medium constituents. In contrast, the immunoprecipitation by the much more concentrated ascitic fluid was unambiguous (Pollard et al., 1983).

This screening assay was chosen because it directly measures the most reliable marker of HDC, i.e., its catalytic activity. Alternatively, it might have been feasible to use an assay measuring inhibition of enzyme activity by antibodies. However, some of the previous assays with immune rabbit or mouse antisera failed to detect the presence of antibody populations inactivating HDC (Tran and Snyder, 1981; Watanabe et al., 1982; unpublished results from our laboratory).

The Mab HI 113-12 recognized not only the purified enzyme, but also the HDC activities of crude extracts from rat gastric mucosa, striatum, or hypothalamus, as well as from mouse mastocytoma (*see* Table 2). In all cases, about 20% of the initial enzyme activity was specifically immunoprecipitated under parallel experimental conditions, indicating the presence of common antigenic constituents on the HDC from various sources, which seems important for immunohistochemical studies in various tissues because the occurrence of several HDC isoenzymes has been suggested (Savany and Cronenberger, 1982; Yamada et al., 1984) (*see* section 5).

Apparently, the Mab was not directed against the enzyme catalytic site since it did not inhibit the activity as was, in fact, anticipated from the immunoprecipitation test used to screen the clones. In addition, when the anti-mouse immunoglobulin was omitted from the assay, no HDC activity was specifically precipitated, indicating that the Mab was nonprecipitant. Mab HI 113-12 did not recognize either the HDC from *Clostridium welchii*, an enzyme largely different from mammalian HDC, or the glutamate and DOPA decarboxylases from rat striatum (Table 2).

References

Aures D., Håkanson R., and Clark W. G. (1970) Histidine Decarboxylase and DOPA Decarboxylase, in *Handbook of Neurochemistry*. Lajtha A., (ed.), vol. 4, Plenum, New York.

Barbin G., Hirsch J. C., Garbarg M., and Schwartz J. C. (1975) Decrease in histamine content and decarboxylase activities in an isolated area of the cerebral cortex of the cat. *Brain Res.* **92,** 170–174.

Barbin G., Garbarg M., Schwartz J. C., and Storm-Mathisen J. (1976) Histamine-synthetizing afferents to the hippocampal region. *J. Neurochem.* **26,** 259–263.

Barbin G., Palacios J. M., Garbarg M., Schwartz J. C., Gaspar P., Javoy-Agid F., and Agid Y. (1980) L-Histidine decarboxylase in the human brain: Properties and localization. *J. Neurochem.* **35,** 400–406.

Baudry M., Martres M. P., and Schwartz J. C. (1973) The subcellular localization of histidine decarboxylase in various regions of rat brain. *J. Neurochem.* **21,** 1301–1309.

Beaven M. A. (1978) Histamine: Its Role in Physiological and Pathological Processes, in *Monographs in Allergy*. (Dukor P., Kallos P., Trnka Z., Waksman B. H., and De Weck A. L. eds.), S. Karger, Basel.

Ben Ari Y., Le Gal La Salle G., Barbin G., Schwartz J. C., and Garbarg M. (1977) Histamine synthesizing afferents within the amygdaloid complex and bed locus of the stria terminalis of the rat. *Brain Res.* **138,** 285–294.

Boeker E. A. and Snell E. E. (1972) Amino acid decarboxylase, in *The enzymes*. (Boyer P. D., ed.), vol. 6, Academic, New York.

Brimijoin S. and Wiermaa M. J. (1977) Rapid axonal transport of tyrosine hydroxylase in rabbit sciatic nerves. *Brain Res.* **120,** 77–96.

Brownstein M. J., Saavedra J. M., Palkovits M., and Axelrod J. (1974) Histamine content of hypothalamic nuclei of the rat. *Brain Res.* **77,** 151–156.

Burkhalter A. (1962) The formation of histamine by fetal rat liver. *Biochem. Pharmacol.* **11,** 315–322.

Buttin G., Leguern C., Phalente L., Lin E. C. C., Medrano L., and Cazenave P. A. (1978) Production of Hybrids Secreting Monoclonal Antiidiotypic Antibodies by Cell Fusion on Membrane Filter, in *Current Topics in Microbiology and Immunology*. (Potter F. M. and Warner N., eds.), vol. 81, Springer Verlag, New York.

Fukui H., Watanabe T., and Wada H. (1980) Immunochemical cross-reactivity of the antibody elicited against L-histidine decarboxylase purified from the whole bodies of fetal rats with the enzyme from rat brain. *Biochem. Biophys. Res. Commun.* **93,** 333–339.

Garbarg M., Barbin G., Feger J., and Schwartz J. C. (1974) Histaminergic pathways in rat brain evidenced by lesions of the MFB. *Science* **186,** 833–835.

Garbarg M., Barbin G., Bischoff S., Pollard H., and Schwartz J. C. (1976) Dual localization of histamine in an ascending neuronal pathway and in non-neuronal cells evidenced by lesions in the lateral hypothalamic area. *Brain Res.* **106,** 333–348.

Garbarg M., Barbin G., Llorens C., Palacios J. M., Pollard H., and Schwartz J. C. (1980a) Recent Developments in Brain Histamine Research: Pathways and Receptors, in *Neurotransmitters, Receptors, and Drug Action.* (Essman B., ed.), Spectrum, New York.

Garbarg M., Barbin G., Rodergas E., and Schwartz J. C. (1980b) Inhibition of histamine synthesis in brain by α-fluoromethylhistidine, a new irreversible inhibitor: In vitro and in vivo studies. *J. Neurochem.* **35,** 1045–1052.

Garbarg M., Pollard H., Quach T. T., and Schwartz J. C. (1983) Methods in Brain Histamine Research, in *Methods in Biogenic Amine Research.* (Parvez S., Nagatsu T., Nagatsu I., and Parvez H., eds.), Elsevier Science, Amsterdam.

Green J. P. (1970) Histamine, in *Handbook of Neurochemistry.* (Lajtha A., ed.), 2nd Ed., vol. 4, Plenum, New York.

Håkanson R. (1963) Histidine decarboxylase in the fetal rat. *Biochem. Pharmacol.* **12,** 1289–1292.

Håkanson R. (1967a) Kinetic properties of mammalian histidine decarboxylase. *Eur. J. Pharmacol.* **1,** 42–46.

Håkanson R. (1967b) Mammalian histidine decarboxylase: Interaction between apoenzyme and pyridoxal-5'-phosphate. *Eur. J. Pharmacol.* **1,** 383–390.

Håkanson R. and Owman C. (1966) Distribution and properties of amino acid decarboxylase in gastric mucosa. *Biochem. Pharmacol.* **15,** 489–499.

Hammar L. and Hjerten S. (1980) Purification and immunochemical analysis of histidine decarboxylase from murine mastocytoma. *Agents Actions* **10,** 92–98.

Henningson S., Lindel L., and Rosengren E. (1974) Comparative study of two isotopic methods determining histamine formation in vitro. *Biochem. Pharmacol.* **23,** 2671–2678.

Hough L. B. and Green J. P. (1984) Histamine and Its Receptors in the Nervous System, in *Handbook of Neurochemistry.* (Lajtha A., ed.), vol. 6, Plenum, New York.

Kelly J. L., McLean E. W., and Miller C. A. (1981) Inhibition of histidine decarboxylase by imidazole derivatives. *J. Med. Chem.* **70,** 341–343.

Kobayashi Y. (1963) Determination of histidine decarboxylase activity by liquid scintillation counting of ^{14}C-CO_2. *Anal. Biochem.* **5,** 284.

Köhler G. and Milstein C. (1975) Continuous cultures of fused cells secreting antibody of predefined specificity. *Nature* **256,** 495–497.

Kollonitsch J., Patchett A. A., Marburg S., Maycock A. L., Perkins L. M., Doldouras G. A., Duggan D. E., and Aster S. D. (1978) Selective inhibitors of biosynthesis of aminergic neurotransmitters. *Nature* **274,** 906–908.

König J. F. R. and Klippel R. A. (1970) The rat brain, a stereotaxic atlas of the forebrain and lower parts of the brainstem. R. E. Krieger, Huntington, New York.

Lange M., Le Guern C., and Cazenave P. A. (1983) Covalent coupling of antigens to chemically activated lipopolysaccharide: A tool for in vivo and in vitro specific B cell stimulation. *J. Immunol. Methods* **63,** 123–131.

Leslie R. and Osborne N. (1983) An immunohistochemical study to localise histaminergic neurones in invertebrate and vertebrate nervous systems. Proceedings of the British Pharmacological Society, C80.

Levine R. J. and Noll W. W. (1969) Histidine decarboxylase and its inhibition. *Ann. NY Acad. Sci.* **166,** 246–256.

Lovenberg W., Barchas J., Weissbach Y., and Udenfriend S. (1962) Aromatic L-amino acid decarboxylase. *J. Biol. Chem.* **237,** 83–93.

Maeyama K., Watanabe T., Taguchi A., Yamatodani A., and Wada H. (1982) The effect of α-fluoromethylhistidine, a suicide inhibitor of histidine decarboxylase, on histamine levels in mouse tissues. *Biochem. Pharmacol.* **31,** 2367–2370.

Maeyama K., Watanabe T., Yamatodani A., Taguchi Y., Kambe H., and Wada H. (1983) Effect of α-fluoromethylhistidine on the histamine content of the brain of w/w^v mice devoid of mast-cells: Turnover of brain histamine. *J. Neurochem.* **41,** 128–134.

Mole K. H. and Shepherd D. M. (1972) Inhibition of histidine decarboxylase in vitro. *Arch. Int. Pharmacodyn.* **195,** 109–147.

Oesch F., Otten U., and Thoenen H. (1973) Relationship between the rate of axoplasmic transport and subcellular distribution of enzymes involved in the synthesis of norepinephrine. *J. Neurochem.* **20,** 1691–1706.

Palacios J. M., Mengod G., Picatoste F., Grau M., and Blanco I. (1976) Properties of rat brain histidine decarboxylase. *J. Neurochem.* **27,** 1455–1460.

Panula P., Yang H. Y. T., and Costa E. (1984) Histamine-containing neurons in the rat hypothalamus. *Proc. Natl. Acad. Sci. USA* **81,** 2572–2576.

Pollard H., Bischoff S., Llorens-Cortes C., and Schwartz J. C. (1976) Histamine and histidine decarboxylase in discrete nuclei of rat hypothalamus and the evidence for mast cells in the median eminence. *Brain Res.* **118,** 509–513.

Pollard H., Llorens C., Barbin G., Garbarg M., and Schwartz J. C. (1978) Histamine and histidine decarboxylase in brainstem nuclei: Distribution and decrease after lesions. *Brain Res.* **157,** 178–181.

Pollard H., Pachot I., Schwartz J. C., Lange M., Legrain P., and Buttin G. (1983) Production d'anticorps monoclonaux anti L-Histidine decarboxylase. *C.R. Acad. Sci., Paris* **297,** 553–556.

Pollard H., Pachot I., Legrain P., Buttin G., and Schwartz J. C. (1985) Development of a Monoclonal Antibody Against L-Histidine Decarboxylase as a Selective Tool for the Localisation of Histamine Synthesising Cells, in *Frontiers in Histamine Research.* (Ganellin C. R. and Schwartz J. C., eds.), Pergamon, New York.

Reilly M. A. and Schayer R. W. (1968) Further studies on the histidine–histamine relationship in vivo: Effects of endotoxin and of histidine decarboxylase inhibitors. *Br. J. Pharmacol.* **34,** 551–563.

Ritchie D. G. and Levy D. A. (1975) A microassay for mammalian histidine decarboxylase. *Anal. Biochem.* **66,** 194–205.

Savany A. and Cronenberger L. (1978) Purification et detection de plusieurs formes moléculaires de l'histidine decarboxylase de la muqueuse gastrique de rat. *Biochem. Biophys. Acta* **526,** 247–258.

Savany A. and Cronenberger L. (1982) Isolation and properties of multiple forms of histidine decarboxylase from rat gastric mucosa. *Biochem. J.* **205,** 405–412.

Schayer R. W. (1971) Determination of Histidine Decarboxylase Activity, in *Methods of Biochemical Analysis,* (Glick D., ed.), (suppl.). Interscience, New York.

Schayer R. W. and Reilly M. A. (1974) Histamine formation in guinea pig brain and other tissues: Effect of α-methyldopa. *Agents Actions* **4,** 193–242.

Schwartz J. C. (1975) Histamine as a transmitter in brain. *Life Sci.* **17,** 503–518.

Schwartz J. C., Garbarg M., Lebrecht V., Nowak J., Pollard H., Rodergas E., Rose C., Quach T. T., Morgat J. L., and Roy J. (1982) Histaminergic Systems in Brain: Studies on Localisation and Actions, in *Advances in the Biosciences.* (Uvnas B. and Tasaka K. eds.), vol. 33, Pergamon, Oxford.

Schwartz J. C., Garbarg M., and Pollard H. (1984) Histaminergic Transmission in Brain, in *The Handbook of Physiology.* (Bloom F. E., ed.), vol. The Intrinsic Regulatory System of the Brain. American Physiological Society, (in press).

Schwartz J. C., Lampart C., and Rose C. (1970) Properties and regional distribution of histidine decarboxylase in rat brain. *J. Neurochem.* **17,** 1527–1534.

Schwartz J. C., Lampart C., Rose C., Rehault M. C., Bischoff S., and Pollard H. (1971) Histamine formation in rat brain during development. *J. Neurochem.* **18,** 1787–1789.

Schwartz J. C., Lampart C., and Rose C. (1972) Histamine formation in rat brain in vivo: Effects of histidine loads. *J. Neurochem.* **19,** 801–810.

Schwartz J. C., Pollard H., and Quach T. T. (1980) Histamine as a neurotransmitter in mammalian brain: Neurochemical evidence. *J. Neurochem.* **35,** 26–33.

Schwartz J. C., Rose C., and Caillens H. (1973) Metabolism of methylhistamine formed through a new pathway: Decarboxylation of L-3-methylhistidine. *J. Pharmacol. Exp. Ther.* **184,** 766–779.

Shore P. A., Burkhalter A., and Cohn V. H. (1959) A method for the fluorometric assay of histamine in tissues. *J. Pharmacol. Exp. Ther.* **127,** 182–186.

Snell E. E. (1977) Pyruvate-containing enzymes. *Trends Biochem. Sci.* **2,** 131–135.

Snyder S. H., Brown B., and Kuhar M. J. (1974) The subsynaptosomal localization of histamine, histidine decarboxylase, and histamine *N*-methyltransferase in rat hypothalamus. *J. Neurochem.* **23,** 37–45.

Steinbusch H. W. M. and Mulder A. H. (1983) Localization of histaminergic neurons in the rat brain: An immunohistochemical study. *Neurosci. Lett.* Suppl. **14,** S356.

Steinbusch H. W. M. and Mulder A. H. (1984) Immunohistochemical Localization of Histamine in Neurons and Mast Cells in the Rat Brain, in *The Handbook of Chemical Neuroanatomy.* (Bjorklund A. and Hokfelt T., eds.), vol. 2, Elsevier, Amsterdam 726–740.

Steinbusch H. W. and Mulder A. H. (1985) Localization and projections of histamine-immunoreactive neurons in the central nervous system of the rat, in *Frontiers in Histamine Research* (Ganellin C. R. and Schwartz J. C., eds.), Pergamon, New York.

Taguchi Y., Watanabe T., Kubota H., Hayayuki H., and Wada H. (1984) Purification of histidine decarboxylase from the liver of fetal rats and its immunochemical and immunohistochemical characterizations. *J. Biol. Chem.* **259,** 5214–5221.

Taylor K. M. and Snyder S. H. (1971) Histamine in rat brain: Sensitive assay of endogenous levels formation in vivo and lowering by inhibitors of histidine decarboxylase. *J. Pharmacol. Exp. Ther.* **179,** 619–633.

Taylor K. M. and Snyder S. H. (1972a) Dynamics of the regulation of histamine levels in mouse brain. *J. Neurochem.* **19,** 341–354.

Taylor K. M. and Snyder S. H. (1972b) Isotopic microassay of histamine, histidine decarboxylase and histamine transferase in brain tissue. *J. Neurochem.* **19,** 1343–1358.

Tran V. T. and Snyder S. H. (1981) Histidine decarboxylase, purification from fetal rat liver, immunologic properties, and histochemical localization in brain and stomach. *J. Biol. Chem.* **256,** 680–686.

Vincent S. R., Hokfelt T., Skirboll L. A., and Yenwu J. (1983) Hypothalamic γ-aminobutyric acid neurons project to the neocortex. *Science* **220,** 1309–1310.

Watanabe T., Nakamura H., Liang L. H., Yamatodani A., and Wada H. (1979) Partial purification and characterization of L-histidine decarboxylase from fetal rats. *Biochem. Pharmacol.* **28,** 1149–1155.

Watanabe T., Maeyama K., Yamatodani A., Yamada M., Kitamura Y., and Wada H. (1980) Histidine decarboxylase activities in mutant mice deficient in mast cells. *Life Sci.* **16,** 1569–1574.

Watanabe T., Yamada M., Taguchi Y., Kubota H., Maeyama K., Yamatodani A., Fukui H., Shiosaka S., Tohyama M., and Wada H. (1982) Purification and Properties of Histidine Decarboxylase Isozymes and Their Pharmacological Significance, in *Advances in the Biosciences.* (Uvnas B. and Tasaka K., eds.), vol. 33, Pergamon, Oxford.

Watanabe T., Taguchi Y., Hayashi H., Wada H., Kubota H., Terano Y., Tanaka J., Shiosaka S., and Tohyama M. (1983) Evidence for the presence of histaminergic nervous system in the brain: An immunochemical analysis. *Neurosci. Lett.* **39,** 249–254.

Watanabe T., Taguchi Y., Shiosaka S., Tanaka J., Kubota H., Terano Y., Tohyama M., and Wada H. (1984a) Distribution of the histaminergic neuron system in the central neurons system of rats: A fluorescent immunohistochemical analysis with histidine decarboxylase as a marker. *Brain Res.* **29,** 13–25.

Watanabe T., Taguchi Y., Takeda N., Shiosaka S., Tohyama M., and Wada H. (1984b) Purification of and Antibodies Against L-Histidine Decarboxylase, in *Frontiers in Histamine Research.* (Ganellin C. R. and Schwartz J. C., eds.), Pergamon, New York (in press).

Werle E., and Lorenz W. (1966) Histamin and histidindecarboxylasen in Schilddrüse and Thymus. *Biochem. Pharmacol.* **15,** 1059–1070.

Weinreich W. and Yu Y. T. (1977) The characterization of histidine decarboxylase and its distribution in nerves, ganglia, and in single neuronal cell bodies from the CNS of *Aplysia californica. J. Neurochem.* **28,** 361–369.

White T. (1960) Formation and catabolism of histamine in cat brain in vivo. *J. Physiol.* **152,** 299–308.

Wilcox B. J. and Seybold V. S. (1982) Localization of neuronal histamine in rat brain. *Neurosci. Lett.* **29,** 105–110.

Yamada M., Watanabe T., Fukui H., Taguchi Y., and Wada H. (1984) Comparison of histidine decarboxylases from rat stomach and brain with that from whole bodies of rat fetus. *Agents Actions* **14,** 143–152.

Yamatodani A. (1978) A systematic determination method of biogenic amines and its application to pharmacological studies. *Osaka Daigaku Igaka Iasshi* (in Japanese) **30,** 155–176.

Chapter 7
Monoamine Oxidase

Peter H. Yu

1. Introduction

Monoamine oxidase (MAO; EC 1.4:3.4; amine:oxygen oxidoreductase; deaminating; flavine-containing) is an enzyme that oxidizes a variety of monoamine neurotransmitters and neuromodulators, as well as exogenous bioactive monoamines. It was first characterized by Hare (1928) as tyramine (TA) oxidase and was later called monoamine oxidase by Zeller (1938). A considerable interest in the enzyme developed after the discovery that MAO inhibitors alleviated clinical symptoms of depression (Loomer et al., 1957). Unfortunately, MAO inhibitors are not widely prescribed as therapeutic agents for the treatment of depression because a number of adverse reactions occur, particularly the severe hypertensive effect that may result from ingestion of TA-rich food (Blackwell and Marley, 1966). The search continues, however, for new MAO inhibitor that have antidepressant properties without adverse side effects, and, in fact, a large number of highly active, specific, reversible or irreversible, prodrug MAO inhibitors have been found.

Human platelet MAO activity has been found to correlate with a variety of mental disorders and personality traits (Sandler et al., 1981; Fowler et al., 1982). The enzyme has been assumed to exist in multiple forms (Collins et al., 1970; Yang and Neff, 1973), based on the properties of substrate preference and sensitivity toward specific MAO inhibitors. Monoamine oxidase-A preferentially oxidizes 5-hydroxytryptamine (5-HT) and is very sensitive to the MAO-A inhibitor clorgyline, whereas MAO-B preferentially oxidizes β-phenylethylamine, and is more sensitive to the MAO-B inhibitor deprenyl. The molecular basis of the multiplicity of the enzyme has been the focus of recent research and controversy. The enzyme is bound to the outer mitochondrial membrane (Greenawalt, 1972), and is widely distributed in different tissues of vertebrates.

A wide variety of methods for the estimation of MAO activity based on different principles and techniques has been developed in the last two decades (*see* review detailed by Tipton and Youdim, 1983). The radioenzymatic method for MAO assay is simple, specific, sensitive, and commonly used; a detailed procedure is, therefore, described. The solubilization and purification procedures of MAO and the problem of the multiplicities at a molecular level are also discussed in this chapter.

2. Oxidative Deamination Reaction

MAO is a flavine-dependent enzyme. The flavine moiety is reduced as the amine is oxidized to iminium ion (Yasunobu and Gomes, 1971). The iminium ion is then hydrolyzed to aldehyde (Smith et al., 1962), and the reduced flavine is reoxidized enzymatically with molecular oxygen (Houslay and Tipton, 1973a). It has been suggested that oxidation of the amine occurs when it is in the free form (McEwen et al., 1969). Recently it has been proposed that MAO catalyzes the oxidation of monoamines by two one-electron transfers from the substrates to flavine (Silverman et al., 1980). As indicated below, in the slow step a nonbonded electron of the amine nitrogen is transferred to the flavine to give the amine radical cation and the flavine semiquinone radical. This renders the α protons of the amine much more acidic, and, therefore, proton loss is facilitated with the subsequent generation of the iminium ion and the reduction of flavine.

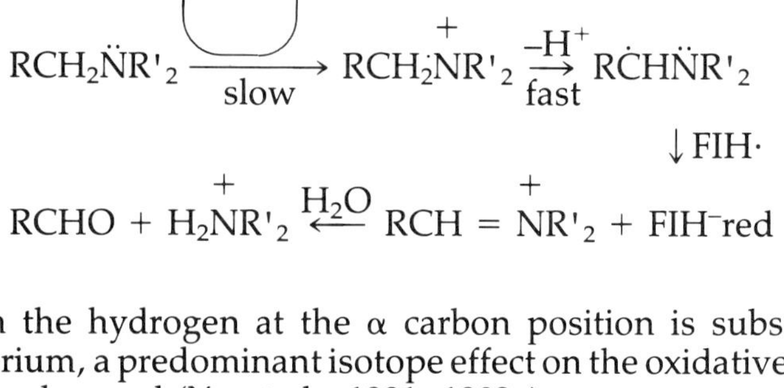

When the hydrogen at the α carbon position is substituted by deuterium, a predominant isotope effect on the oxidative deamination is observed (Yu et al., 1981, 1982a).

3. Occurrence and Distribution

MAO is widely distributed in mammalian tissues, but the distribution of the two forms of the enzyme in these tissues is rather confusing (Fowler et al., 1978). The classification of the two forms of MAO on the basis of substrate preference and inhibitor sensitivity may be an oversimplification. Uncritical use of this classification to extend conclusions obtained from one species to another can be quite misleading (Fowler et al., 1978). In the rat, for example, the ratio of MAO-A to MAO-B is similar in both the brain and liver (Squires, 1972), whereas the spleen and heart possess mostly MAO-A (Edwards et al., 1979), and the pineal gland contains predominantly MAO-B (Goridis and Neff, 1971). In humans, the placenta and platelet contain predominantly MAO-A (Bathina et al., 1975) and MAO-B (Donnelly and Murphy, 1977), respectively, and the liver, brain, and heart contain both MAO-A and MAO-B (Hall et al., 1969; Yu, unpublished). The enzyme is distributed heterogeneously throughout the rat brain, with slightly higher activity in the thalamic and hypothalamic regions and lower activity in the cerebellum (La Motte et al., 1969). It has generally been assumed that MAO-A is intraneuronal (Jarrott and Iversen, 1971). This conclusion is based on the following observations: (a) the neurotransmitters 5-HT and noradrenaline (NA) are typical MAO-A substrates; (b) MAO-A activity can be enriched in the isolated synaptosomal fraction from sucrose density gradient centrifugation (Student and Edwards, 1977); (c) lesions of the nigrostriatal pathway cause a selective reduction of MAO-A activity (Demarest and Azzaro, 1979); and (d) the neuroblastoma culture cells (Donnelly et al., 1976) and primary culture of neuronal cells (Yu and Hertz, unpublished) exhibit MAO-A activity exclusively. Although MAO-B was supposed to be most abundant in the glial cells, MAO-A was exclusively observed in the primary glial culture cells (Yu and Hertz, 1982) and C_6 glioma cells (Murphy et al., 1976). MAO-B appears only in long-term cultures or dibutyryl cyclic AMP-induced differentiated glial cultures (Yu and Hertz, 1983). The culture condition may provide a different mitochondrial membrane microenvironment for MAO, which may vary with the stage of development. Predominantly MAO-A has also been found in capillaries of young and adult rats, obtained from bulb isolation (Yu, 1984). No general conclusion can be drawn from among different species with respect to MAO distribution.

4. Multiplicity of Monoamine Oxidase

4.1. Classification of MAO

Since MAO is tightly bound to the mitochondrial outer membrane (Schnaitman et al., 1967), vigorous treatment, such as prolonged sonication or application of detergents, is necessary in order to solubilize the enzyme. Solubilized rat liver mitochondrial proteins can be separated by polyacrylamide gel electrophoresis, and a number of discrete bands of MAO activity that differ in their activities toward different substrates can be differentiated (Youdim et al., 1969; Kim and D'Iorio, 1968). These separated forms of MAO differ considerably in their phospholipid content, and it has been suggested that they could be artifacts of the purification procedure (Houslay and Tipton, 1973b).

Another aspect of the multiplicity of MAO is the substrate preference of the enzyme and the inhibition of MAO activity by specific inhibitors. Johnston (1968) demonstrated that clorgyline is an irreversible inhibitor of MAO. Preincubation of MAO with clorgyline resulted in inhibition of the enzyme activity toward 5-HT at much lower concentrations than that required to inhibit the activity toward benzylamine. Deprenyl, on the other hand, exhibited much higher inhibitory activity toward phenylethylamine and benzylamine than toward 5-HT (Knoll and Magyar, 1972). In addition, oxidation of TA (a mixed type of substrate) by rat liver showed a biphasic inhibition response curve to either clorgyline or deprenyl. Eventually, different amine substrates were grouped into A substrates (5-HT, NA, adrenaline, octopamine), B substrates (phenylethylamine, benzylamine), and mixed-type substrates (TA, dopamine). This classification is occasionally ambiguous with respect to different species of animals (for review, *see* Fowler et al., 1978). It is also interesting that phenylethylamine was shown to be a MAO-B substrate only at relatively low concentrations (<0.01 m*M*). At higher concentrations, phenylethylamine exhibited a biphasic inhibition response curve, indicating a mixed-type substrate profile (Kinemuchi et al., 1980). MAO inhibitors can be categorized into A and B types. The classification of different types of MAO inhibitors will be described later. It has been shown that specific MAO inhibitors can differentially affect MAO activity in vivo. The physiological function of MAO-A and MAO-B, however, is still not fully understood.

4.2. Nature of the Multiple Forms

A considerable amount of new knowledge regarding the nature of MAO-A and MAO-B has been accumulated in recent years. Although a method for the complete separation of functionally active MAO-A and MAO-B from a tissue containing both subtypes is yet to be developed, MAO-A and MAO-B subunits can be readily separated after irreversible linkage with tritium-labeled pargyline (Callingham and Parkinson, 1979). Antisera or monoclonal antibodies against MAO-B can selectively precipitate MAO-B, but not MAO-A (Denney et al., 1982). Furthermore, only those tissues rich in MAO-B appear to be stained by the immunochemical method with MAO-B antisera (Pintar et al., 1983). The amino acid sequence of the structures adjacent to the flavine moiety of both MAO-A and MAO-B, however, appear to be identical (Yu, 1981a). Peptide mapping of fragments released from [^{3}H]-pargyline-labeled subunits by limited proteolysis, however, reveals some differences between the two (Cawthon and Breakfield, 1979). Although MAO-A and MAO-B are distinct enzyme entities, a conclusion cannot be drawn as to whether MAO-A and MAO-B are derived from the same gene loci or are products of posttranslational modification. The entire primary structure of the two isoenzymes is yet to be established. There is evidence indicating that MAO is associated with other components, such as phospholipids, which could not only modify catalytic properties, but may also affect the antigenic conformation and susceptibility, subject to limited proteolysis.

The involvement of lipids in the expression of MAO activity has been investigated by examining the effects of detergents and chaotrophic agents (Houslay and Tipton, 1973b), different phospholipases (White and Stine, 1984; Navarro-Welch and McCauley, 1982; Huang and Faulkner, 1980), lipid depletion (Yu, 1979), and the replacement of different phospholipids (Buckman et al., 1983; Houslay, 1980). Despite the fact that MAO-A and MAO-B are not interconvertible under any of these conditions, the MAO activities are regulated by modification of the mitochondrial microenvironment. Similar kinds of regulation may occur in vivo because both endogenous inhibitors (Glover et al., 1980) and activators (Yu and Boulton, 1979) have been observed.

4.3. Some Properties of MAO

4.3.1. Molecular Weight

Because of the difficulty of solubilizing MAO from the mitochondria, as well as aggregation and possible interspecies differences, it is difficult to generalize the molecular weight of MAO. It is generally agreed, on the basis of the flavine content, that bovine liver MAO has a molecular weight of about 100,000 (Tipton, 1968; Gomes et al., 1969; White and Stine, 1982). The enzyme consists of two subunits of similar molecular weight (only one subunit linked to flavine) when determined by sodium dodecyl polyacrylamide gel electrophoresis (SDS-PAGE). The molecular weight of the flavine-containing subunit has also been determined by incorporating [^{3}H]-pargyline into the enzyme and then separating the labeled subunit by SDS-PAGE. The molecular weight of the MAO-A subunit from human placenta has been reported to be 67,000 (Brown et al., 1980) or 63,000 (Cawthon et al., 1981), whereas that from rat liver is 60,000 (Callingham and Parkinson, 1979). The molecular weight of the MAO-B subunit from human platelet has been variously reported to be 63,000 (Brown et al., 1980), 60,000 (Cawthon et al., 1981), or 55,000 (Callingham and Parkinson, 1979). The slight discrepancies may be a result of some variation in the electrophoretic technique applied, the association of other components, or the distribution and abundance of charged or hydrophobic amino acids.

4.3.2. Cofactors

Flavine has been found to be covalently linked to MAO through the 8α-CH_3 group of riboflavine. Cysteine is substituted on the 8α-CH_3 group of flavine through a thioether linkage, and the sequence of the peptide has been established as (Walker et al., 1971):

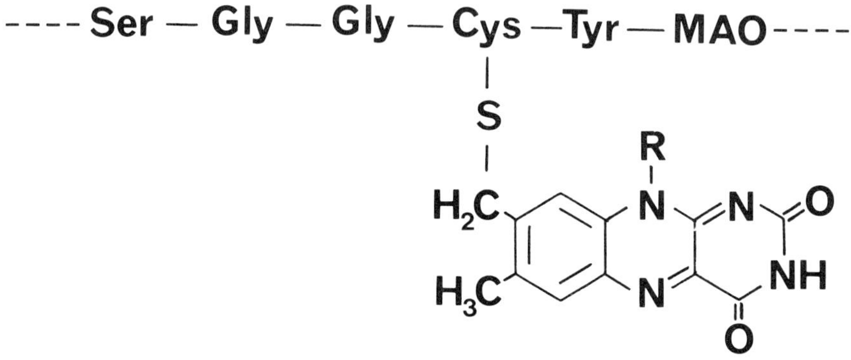

Copper and iron are apparently not components of the enzyme in highly purified bovine liver MAO (Weyler and Salah, 1981), although metal chelators have been reported to inhibit MAO activity (Youdim and Sourkes, 1966), and in iron-deficient animals the level of MAO activity is reduced (Symes et al., 1969).

Table 1
Kinetic Parameters of Rat Liver MAO-A and MAO-B Toward Different Substrates

Substrate	MAO forms[a]	K_m, μM	V_{max}	Reference
5-Hydroxytryptamine	A + B	173	2.42	Tipton et al., 1982
	A	161	1.96	
	B	2032	0.22	
Dopamine	A + B	220	8.0	Fowler and Tipton, 1984
	A	140	3.6	
	B	430	4.9	
p-Tyramine	A + B	313	3.01	Tipton et al., 1982
	A	107	1.06	
	B	579	2.01	
Tryptamine	A + B	—	—	Suzuki et al., 1981
	A	3	4.95	
	B	7	3.31	
β-Phenylethylamine	A + B	20	2.65	Tipton et al., 1982
	A	280	0.35	
	B	20	2.30	

[a]MAO-A and MAO-B obtained from rat liver after irreversible inhibition of MAO-B with deprenyl and MAO-A with clorgyline.

4.3.3. Kinetic Parameters

The K_m and V_{max} of the two types of MAO can be determined by use of selective irreversible MAO inhibitors. There are numerous reports of these values in different tissues and species. The K_m and V_{max} values for rat liver MAO are listed in Table 1.

4.3.4. Effect of pH and Oxygen on MAO Activity

The pH optima values vary with different amine substrates, and substrate concentrations also depend upon the buffer or tissue used. Bovine liver MAO, for example, catalyzes phenylethylamines (TA, dopamine) most actively at lower pH ranges, 7.4–7.5, indoleamines (5-HT, tryptamine) at 8.2–8.5, and kynuramine at 9.1 (Achee et al., 1974).

The activity of rat liver MAO is increased uncompetitively with increasing concentrations of oxygen. The Michaelis constant for oxygen is dependent upon the amine substrates used, and is related to K_{cat} values of the enzyme toward the different amine substrates (Fowler and Oreland, 1980). The two forms of enzyme are interactive and may display halfsites reactivity.

5. MAO Inhibitors

MAO is perhaps the enzyme with the largest number and widest variety of inhibitors (*see* review by Ho, 1972). A few of the important MAO inhibitors are listed in Table 2. Iproniazid, tranylcypromine, and phenelzine are MAO inhibitors that are used clinically as antidepressants (*see* review by Quitkin et al., 1979). They have no selective inhibitory activity toward MAO-A and MAO-B and represent "first generation" MAO-inhibitors. In the late 1960s and early 1970s, selective MAO inhibitors for MAO-A, such as clorgyline [*N*-methyl-*N*-propargyl-3-(2,4-dichlorophenoxy)-propylamine] (Johnson, 1968), Lilly 51641 [*N*-(2-[*O*-chlorphenoxy]-ethyl)-cyclopropylamine] (Fuller, 1968), PCO [5-phenyl-3-(*N*-cyclopropyl)-ethylamine-1,2,4-oxadiazole] (Mantle et al., 1975), and selective MAO-B inhibitors, such as deprenyl [phenylisopropyl-methyl-propinylamine] (Knoll and Magyar, 1972), pargyline [*N*-methyl-*N*-benzyl-propynylamine] (Fuller et al., 1970), and so on, were discovered. Clorgyline and deprenyl are widely applied for the characterization of different types of MAO. These MAO inhibitors all form irreversible links with the enzymes. The reversible MAO inhibitors, which form another category of inhibitors, have recently attracted attention because of their potentially improved clinical properties. Included in this category are the reversible MAO-A inhibitors harmine (Horita and McGrath, 1960), harmaline (Fuller, 1972), cimoxatone(3-[4-[5-(methoxymethyl)-2-oxo-3-oxazolidinyl]-phenoxy]-methyl]-benzonitrile) (Kan and Strolin-Benedetti, 1981), CGP11305A (4-[5-methoxy-7-bromobenzofuranyl]-2-piperidine HCl) (Waldmeier et al., 1983), amiflamine (*S*-(+)-4-dimethylamine-2,α-dimethyl-phenylethylamine (+)-hydrogen tartrate) (Ask et al., 1982), moclobemide (*p*-chloro-*N*-[2-morpholinoethyl]-benzamide) (Da Prada et al., 1982), and the reversible MAO-B inhibitors, such as MD 780236 (3-[4-([3-chlorophenyl]methoxy)phenyl]-5-[(methylamino) methyl]-2-oxazolidinone methanesulfonate) (Strolin-Benedetti et al., 1982). Interestingly, the tricyclic antidepressants imipramine and amitriptyline (Roth and Gillis, 1974; Edwards and Burns, 1974) and the stimulant amphetamine (Mantle et al., 1976) also possess reversible inhibitory activity toward MAO. Recently, an MAO inhibitory prodrug has been discovered. MDL-72394 [(E)-β-fluoromethylene-*m*-tyrosine] can be decarboxylated by aromatic L-amino acid decarboxylase and forms a potent irreversible MAO inhibitor (E)-β-fluoromethylene-*m*-tyramine (McDonald et al., 1984). This inhibitor has been shown to be neuronal selective.

6. Clinical Significance

6.1. MAO Inhibitors as Antidepressants

In 1952, Zeller and Barsky found that MAO activity was inhibited by iproniazid, and Selikoff et al. (1952) noticed an elevation in mood in tuberculosis patients treated with this drug. Later, iproniazid was used to treat depressed patients, with favorable results (Crane, 1957; Kline, 1958). A great deal of controversy, however, has revolved around the use of MAO inhibitor antidepressants. The hepatic toxicity and occasional hypertensive crises associated with the use of MAO inhibitor antidepressants eventually resulted in the withdrawal of iproniazid from the market (*see* review by Blackwell et al., 1967). Other MAO inhibitors, such as phenelzine (Nardil) and tranylcypromine (Parnate), although still available for prescription as antidepressants, are not prescribed with anything like the frequency of the tricylics. These drugs are, however, particularly effective in neurotic and atypical depressions (*see* review by Quitkin et al., 1979); to obtain maximum benefit, however, a more rational approach in the use of MAO inhibitors will be required. Patients suitable for treatment cannot be adequately classified using conventional diagnostic labels. Patients with primary symptoms such as hypochondriasis, agoraphobia, social phobia, irritability, somatic anxiety, and anergia respond well, whereas those with primarily depressed mood, guilt, and ideas of reference and personality disorders seldom respond (Tyrer, 1976). Recent developments of specific, reversible MAO inhibitors or prodrugs (*see* section 5) and their use as antidepressants seem encouraging. For example, deprenyl, a type B MAO inhibitor lacking the TA pressor reaction (cheese effect) has been shown to be an effective antidepressant (Mendelwicz and Youdim, 1983), and beneficial effects have been observed in over 90% of patients treated with deprenyl in combination with the essential amino acid phenylalanine (Birkmayer et al., 1984).

6.2. Platelet MAO as a Marker for Psychiatric Disorders

Since the original report by Murphy and Wyatt in 1972, there have been numerous reports confirming low platelet MAO activity in schizophrenia and other mental disorders (for reviews, *see* Wyatt et al., 1979; Sandler et al., 1981; Fowler et al., 1982). Human platelets are easily isolated and have been proposed as a model for central serotonergic neurons (*see* review by Stahl, 1977). Human

Table 2
Some Important MAO Inhibitors

Inhibitor	Chemical nomenclature	Type	Reversibility	Structure
Iproniazid	"Marsilid"; N^1-Isonicotinosyl-N^2-isopropylhydrazine	A + B	Irr	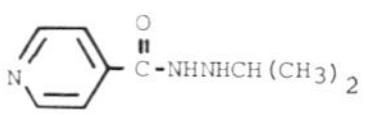
Phenelzine	"Nardil"; Phenethylhydrazine	A + B	Irr	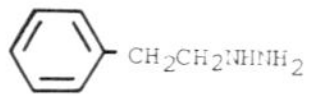
Tranylcypromine	"Parnate"; 2-Phenylcyclopropylamine	A + B	Irr	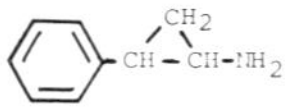
Clorgyline	*N*-(2,4-Dichloro-phenoxy-*n*-propyl)-*N*-methylpropargylamine	A	Irr	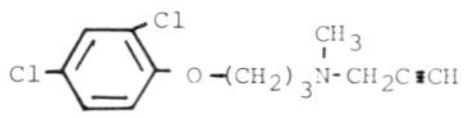
Lilly 51641	*N*(2-(*O*-Chlorophenoxylethyl)cyclopropylamine	A	Irr	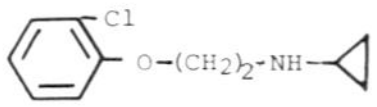
PCO	5-Phenyl-3-(*N*-cyclopropyl)ethylamine-1,2,4-oxaldiazole	A	Irr	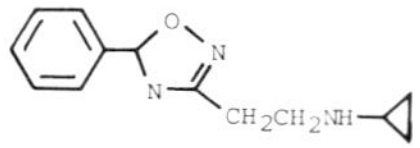
Deprenyl	Phenylisopropylmethylpropinylamine	B	Irr	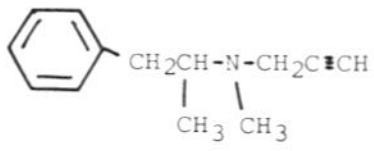

Pargyline	*N*-Methyl-*N*-propynylbenzylamine	B	Irr	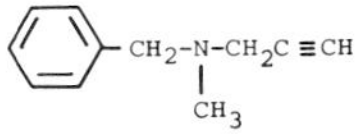
Harmaline	4,9-Dihydro-7-methoxy-1-methyl-3H-pyrido (3,4-b) indole	A	Rev	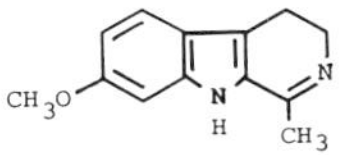
Harmine	7-Methoxy-1-methyl-9-pyrid (3,4-b) indole	A	Rev	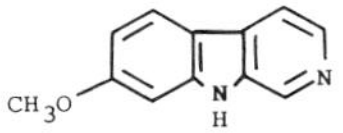
Cimoxatone	(3-(4-(5-(Methoxylmethyl)-2-oxo-oxazolidinyl)phenoxy) methyl)-benzonitrile	A	Rev	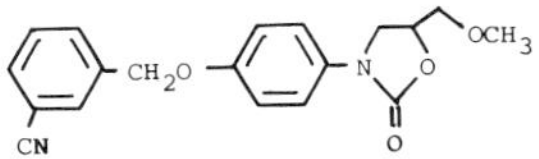
CGP 11305A	4-(5-Methoxy-7-bromo-benzofuranyl)-2-piperidine HCl	A	Rev	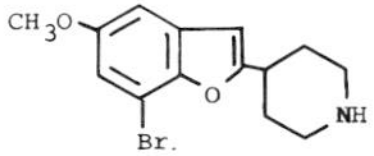
Amiflamine	(+)-4-Dimethylamine-2,α-dimethylphenylethylamine	A	Rev	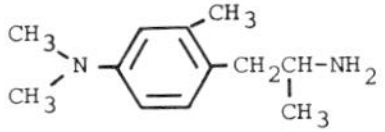
Moclobemide	*p*-Chloro-*N*-(2-morphoilinoethyl) benzamide	A	Rev	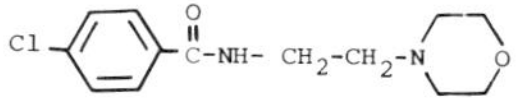
MD780236	3-(4-[(3-Chlorophenyl) methoxy]-phenyl)-5-([methyl-amino]-methyl)-2-oxazolidinonemethane sulfonate	B	Rev	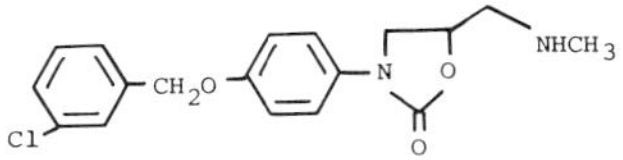
MDL-72394	(E)-β-Fluoromethylene-*m*-tyrosine	A	Irr	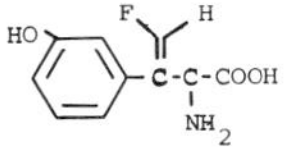

platelet MAO is genetically linked, as confirmed by twin (Hussein et al., 1980) and family studies (Pandey et al., 1979). Furthermore, it has been suggested that platelet MAO is controlled by a major locus (Goldin et. al., 1982; Rice et al., 1984). Low platelet MAO activity has also been observed in alcoholics and their first degree relatives (Alexopoulos et al., 1983), migraine patients (Littlewood et al., 1984), and prison inmates labeled aggressive (Boulton et al., 1983), whereas high platelet MAO has been described in unipolar depressive patients (Edwards et al., 1978) and agoraphobic patients (Yu et al., 1983). There is no indication that platelet MAOs obtained from high and low activity subgroups are biochemically different (Yu et al., 1982a), although it must be admitted that a comparison of platelet MAO in different mental disorders is complicated because of diagnosis difficulties. Furthermore, it is essential to choose meaningful control subjects, so that for a study of depressives, for example, patient variables, such as age, gender, severity of illness, diagnostic category, number of psychotic episodes, length of current clinical episode, degree of chronicity, medication status, previous treatment and response, ovarian status, and inpatient or outpatient status should be carefully assessed, recorded, and grouped (Kupfer and Rush, 1984).

7. Assay of MAO Activity

MAO activity can be measured by a large number of methods, most of which have recently been extensively reviewed by Tipton and Youdim (1983). The principle of the measurement of MAO is based on the following reaction:

$$RCH_2NH_2 + O_2 + H_2O \xrightarrow{MAO} RCHO + NH_3 + H_2O_2$$

$$RCOH \xleftarrow{\text{alcohol dehydrogenase}} RCHO \xrightarrow{\text{aldehyde dehydrogenase}} RCOOH$$

The consumption of oxygen, the disappearance of amine substrates, the formation of aldehyde, ammonia, and hydrogen peroxide (H_2O_2), and the acidic or alcoholic metabolite have all been applied in the detection of MAO activity. A survey of the different methods is listed in Table 3. A good method should be simple, specific, sensitive, suitable for different substrates, rapid, and highly reproducible. Some of the spectrophotometric and fluorometric methods can only be used with particular substrates or substrate derivatives; in addition, they lack sensitivity, and

Table 3
Assay of Monoamine Oxidase

Principle	Method	Reference
Consumption of oxygen	Manometry	Creasey, 1956
	Oxygen electrode	Tipton and Dawson, 1968
Disappearance of amine substrates	Direct UV measurement (Kynuramine)	Weissbach et al., 1960
	Ninhydrin reagent (spectrophotometry)	Udenfriend et al., 1955
	2,4,6-Trinitrobenzene-1-sulphonic acid (TNBS) reagent (spectrophotometry)	Obata et al., 1971
	HPLC (electrochemical)	Yu et al., 1982b
Formation of ammonium	Nessler's color reaction (spectrophotometry)	Cotzias and Greenough, 1959; Guha and Krishamurti, 1965
	Indolephenol color reaction (spectrophotometry)	Nagatzu and Yagi, 1966
	Coupled with glutamate dehydrogenase system (spectrophotometry)	Yamada et al., 1972
	Fluorometry	Harada and Nagatzu, 1973
	Ammonium sensitive	Meyerson et al., 1978; Nimmo and Tipton, 1979
Formation of aldehydes	Direct UV measurement with various amine derivatives as substrates	Tabor et al., 1954; Deitrich et al., 1962; McEwen and Cohen, 1963; Zeller et al., 1965; Houslay and Tipton, 1974

(continued on next page)

Table 3 (*continued*)

Principle	Method	Reference
	p-Dimethylaminobenzalamine as substrate (colorimetry)	Deitrich and Erwin, 1969
	2,4-Dinitrophenylhydrazine reagent (colorimetry)	Green and Haughton, 1961
	Formation of 4-hydroxyquinoline from kynuramine (fluorometry)	Kraml, 1965
	Formation of 4,5-dihydroxyquinoline from 5-hydroxykynuramine (fluorometry)	Takahashi and Takahara, 1968
	Coupled with aldehyde dehydrogenase system (spectrophotometry)	Houslay and Tipton, 1973b; Anderson et al., 1976
	Fluorometry	Udenfriend, 1962
	Bioluminesence	Tenne et al., 1985
	HPLC (Spectrophotometry)	Nissinen, 1984
	HPLC (Fluorophotometry)	Nohta et al., 1983
	HPLC (Electrochemical detection)	Yu et al., 1986
Formation of H_2O_2	Amperometry	Mason and Olson, 1970
	Homovanillic acid as fluorescent acceptor (fluorometry)	Snyder and Hendley, 1968; Guibaut et al., 1968; Tipton, 1969

	Seueo-2',7'-dichlorofluorescein as fluorescent acceptor (fluorometry)	Köhli and Cooper, 1978
	Seopolin as fluorescent acceptor (fluorometry)	Harris and Cooper, 1982
Formation of acid or alcohol metabolites	Direct fluorometry of indoleacetic acid or 5-hydroxyindoleacetic acid	Lovenberg et al., 1964; Hidaka et al., 1967
	Radioenzymatic (solvent extraction)	Wurtman and Axelrod, 1963; Otsuka and Kobayashi, 1964; Tipton and Youdim, 1976; Fowler and Oreland, 1980
	Radioenzymatic (ion exchange chromatography)	Robinson et al., 1968; Southgate and Collins, 1969; Goridis and Neff, 1971
	Radioenzymatic (liquid chromatography)	Jain et al., 1973; Wu and Dyck, 1975
	Gas chromatography	Uji et al., 1980
	HPLC	Kobes et al., 1980; Kawai and Nagatzu, 1982; Yu et al., 1986

crude enzyme preparations often cause interference. It is also very important to measure activity using type A, type B, and mixed-type substrates. High-pressure liquid chromatography (HPLC) in combination with electrochemical detection now seems to be the method of choice, particularly when identifying catecholic, phenolic, and indolic substrates (Kawai and Nagatzu, 1982). The typical MAO-B substrate phenylethylamine, along with its acidic metabolite phenylacetic acid, however, cannot be detected with this method.

During the past decade, the radioenzymatic method developed by Wurtman and Axelrod (1963) has been most commonly used. The principle of this procedure involves incubating substrates labeled with ^{14}C with MAO, after which the radioactively labeled products are separated, usually by extraction, and counted. This procedure is quite reliable in many cases. A modification is outlined below.

7.1. Radioenzymatic Assay Procedure

7.1.1. Enzyme Preparation

Crude tissue homogenates (tissues homogenized in 0.32*M* sucrose or 0.11*M* KCl), partially purified mitochondrial membrane fragments (see below), fractions prepared by solubilization in detergents, as well as enzymes that have been purified to different degrees of purity can all be used.

7.1.2. Reagents

1. Phosphate buffer (0.2 mol/L)
2. Substrate solution: This depends on which substrate is being utilized and the purpose of the study (often the final concentration of the substrate can be set at twice its K_m value)
3. ^{14}C-Labeled substrates: specific activity approximately 50 mCi/mmol, concentration 100 μCi/mL
4. Citric acid, 2*N*
5. Extraction solvent, i.e., toluene:ethylacetate (1:1)
6. Scintillation cocktail

7.1.3. Assay Procedure

The following incubation mixture (200 μL) is prepared in conical glass centrifuge tubes in an icebath.

	Volume, μL	Final concentration
Enzyme (suitably diluted)	50	—
Phosphate buffer (0.2*M*, pH 7.4)	50	$5 \times 10^{-2}M$
Substrate ($8–20 \times 10^{-4}M$)	49	$2–5 \times 10^{-4}M$
^{14}C-substrate (premixed with the nonlabeled substrate)	1	0.1 μCi
H_2O	50	—

In the incubation mixture used to obtain blank values, the MAO inhibitor pargyline (0.1 m*M*) is included. The incubation is carried out at 37°C for 30 min and terminated by adding 200 μL 2*N* citric acid and 1 mL extraction solvent. After vigorous mixing on a Vortex mixer, the mixtures are centrifuged at 3000*g* for 5 min in a clinical centrifuge. An aliquot (0.5 mL) of the organic phase is then transferred to a counting vial, toluene-based scintillation fluid (10 mL) is added, and the mixture is subsequently counted in a liquid scintillation counter.

7.1.4. Calculation

If the scintillation counter is not equipped with automatic quenching compensation, the counting efficiency (%) (or quenching correction) must be determined as described by Peng (1970).

$$\text{Specific MAO activity (nmoles/min/mg)} = \text{(Exp–blank) dpm} \times \frac{1}{\text{Specific radioactivity}} \times \frac{1}{\text{Time of incubation (min)}} \times$$

$$\text{Extraction coefficient} \times \frac{1.2}{0.5} \times \frac{1}{\text{protein (mg)}}$$

7.1.5. Extraction of Product

It has generally been considered (although it may not be true for some substrates) that the aldehydes formed in the above incubation conditions are unstable and are rapidly oxidized in air to their corresponding acids or alcohols in the presence of aldehyde reductase and NADH or NADPH. The extraction conditions are such that the amine substrates remain associated with the aqueous phase and the acidic or neutral metabolites (i.e., acids, alcohols, or aldehydes) extract into the organic solvent phase. Ion-exchange resins and liquid ion exchangers have also been used to separate the reaction products from the substrates. Comparisons of these various methods indicate that they are all rather similar in their efficiency of extraction (Anderson, 1983), and it is a fact that they are relatively easy to perform and apparently suitable for measuring both low and high MAO activity.

Unfortunately, extraction with solvents is never 100% effective, and, indeed, the extraction efficiency depends on the substrates and solvents used (Tipton and Youdim, 1983). A number of different solvents have been employed (Wurtman and Axelrod, 1963; Otsuka and Kobayashi, 1964; McCaman et al., 1965; Southgate and Collins, 1969; Jain et al., 1973; Callingham and Laverty, 1973; Tipton and Youdim, 1976; Fowler and Oreland, 1980). To illustrate the point, approximate recoveries with different solvents for 5-HT, β-phenylethylamine, and *p*-TA are listed in Table 4. It is clear, therefore, that the extraction coefficients must be established for each substrate and solvent before proceeding to finalize any MAO activities.

7.1.6. Validation of Method

This radioenzymatic method is highly sensitive, specific, reproducible, relatively simple, and convenient because a wide variety of radioactively labeled amine substrates are commercially available. There are, however, several sources of error that may produce misleading results.

Although tritium-labeled amine substrates are commercially available and offer much higher specific radioactivities, their use in the radioenzymatic assay for MAO is not recommended. First, the

Table 4
Extraction of Different Amine Metabolites by Different Solvents[a]

Substrate used	Recovery of metabolite (%)	
	Toluene : ethyl acetate (1 : 1)	Benzene : ethyl acetate (1 : 1)
5-Hydroxytryptamine	74	91
β-Phenylethylamine	92	—
p-Tyramine	96	65
Dopamine	54	18

[a]Data from Otsuka and Kobayashi (1964), Tipton and Youdim (1976), and Fowler and Oreland (1980).

dpm values obtained after extraction are unreliable because of exchange of the isotope in water, even in the absence of enzyme (May, 1980). Second, it has recently been observed that when the label is on the α-carbon position, a significant isotope effect occurs during the enzyme reaction (Yu et al., 1981, 1982b). It has been shown, for example, that the rates of enzymatic deamination of those amines in which the α-side chain hydrogens had been replaced by deuterium were much slower (approximately one-third) than those exhibited with unlabeled amines. Compounds labeled with tritium would probably exhibit an even more pronounced isotope effect since the carbon-tritium bond is even stronger.

A further disadvantage of this method is the fact that it is a discontinuous procedure, i.e., incubation is carried out for 20–30 min and then the reaction is stopped. It is also necessary to establish the time course in order to ensure that the reaction is linear. Often the linearity disappears when a proportion of substrate has been consumed. It is also dependent on the amount of substrate added and the amount of enzyme present in the incubation mixture.

Finally, an additional criterion relates to substrate purity for both unlabeled and labeled substrates. Some contaminants are potent MAO inhibitors and frequently appear as substrate inhibitors in kinetic studies. Impurities in the radioactive amines often cause high background errors in the calculation of enzyme activity. In such cases the substrate should be purified by recrystallization, extraction with solvents, or appropriate chromatography.

7.2. HPLC: Electrochemical Detection Method

A direct measurement of the aldehyde or acid products from the deamination of the amine substrates has been developed (Yu et al., in press). In this method the enzyme reactions are carried out as described in section 7.1.3. The enzyme reaction is terminated by addition of 800 μL 0.1*N* perchloric acid containing 0.25 m*M* EDTA and 0.1 m*M* sodium metabisulfite ($Na_2O_5S_2$). After centrifugation, 20 μL of the supernatant is subjected to HPLC analysis.

In the case of measurement of acid metabolites, yeast aldehyde dehydrogenase (2.5 units) and 1.5 μmol β-nicotinamide adenine dinucleotide (β-NAD) should be added to the enzyme reaction mixture.

7.2.1. HPLC Analysis

The chromatographic separations are performed at ambient temperatures on a 250 × 4.6 mm id Ultrasphere I.P. analytical column packed with C_{18} spherical 5-μ*M* particles (Beckman, Toronto, Ont.). An SSI precolumn filter (0.5 μ*M* filter elements) (Terochem, Rexdale, Ont.) and a 30 × 4.6 mm id Brownlee MPLC RP-18 SPHERI-5 guard column (TMA, Calgary, Alberta) are installed between the WISP 710B automated sample injector and analytical column (a 20-μL aliquot of each prepared sample is injected onto the column). The mobile phase composed of 75 m*M* monobasic sodium phosphate, 1 m*M* sodium octyl sulfate, 500 μ*M* EDTA, and 12.5% acetonitrile, and adjusted to a final pH value of 2.75 with phosphoric acid is pumped through the column at 1.0 mL/min, using a Waters M45 solvent delivery system (Millipore, Mississauga, Ont.). Prior to the addition of acetonitrile, the buffer is filtered through a 0.22 μ*M* millipore filter. Degassing of the mobile phase is accomplished by either vacuum pressure or helium sparging. The detector consists of a BAS model TL-5A thin-layer amperometric electrode (Mandel, Rockwood, Ont.) that is controlled at 0.7 V with dopamine and 5-HT and at 0.9 V with TAs, synephrines, and so on, as substrates vs a Ag/AgCl reference electrode. Signals for the detector are integrated by peak height, using a Shimadzu C-R3A recording data processor (Tekscience, Oakville, Ont.).

7.2.2. Calculation

$$\text{Specific MAO activity (nmoles/min/mg)} = \text{Peak height of product} \times \frac{\text{nmol}}{\text{Peak height standard}} \times \frac{1}{\text{Time of incubation (min)}} \times \text{Dilution factor} \times \frac{1}{\text{Protein (mg)}}$$

7.2.3. Validation of Method

The HPLC-electrochemical detection method is very simple, rapid, sensitive, and selective. The enzyme reactions are stopped by perchloric acid, and the supernatant can be directly injected into HPLC column for analysis. The unreacted amine substrate does not interfere with the separation and quantitation of the corresponding aldehyde or acid products; therefore, a preseparation of the products from their substrates is unnecessary. One HPLC analysis requires 15 min. If the HPLC is equipped with automated injector, over 90 samples can be analyzed in one day.

The detection with electrochemical method is extremely sensitive, i.e., in the picogram range. It is at least comparable to the radiometric method for MAO assay using ^{14}C-labeled substrate.

When the crude tissue homogenates are used for the determination of MAO activity, it is recommended that the acid products be measured. Because the conversion of aldehyde to acid is not always complete, aldehyde dehydrogenase and β-NAD should be included. When washed mitochondrial membranes or purified, or partially purified, MAO are used, a direct measurement of aldehyde product can be adopted. The aldehyde is quite stable in perchloric acid containing an antioxidant and EDTA.

This method is very useful in determination of MAO activity toward catecholamines and 5-HT because of the active electrochemical properties of these substances. Phenolic amines and their deaminated products are also electrochemically active, but a higher voltage (0.9 V) is required. The limitation of this method is that it cannot measure β-phenylethylamine, a typical MAO-B substrate, and its metabolites because they are not detectable electrochemically. Therefore, relatively less sensitive HPLC methods, i.e., spectrophotometric (Nissinen, 1984) or fluorometric (Nohta et al., 1983) procedures, can be applied.

8. Histochemical and Cytochemical Localization of MAO

In order to understand the physiological function(s) of MAO, it is necessary to determine the cellular and subcellular distribution of the enzyme or subtypes of the enzyme. Most of the existing methods are based on the measurement of the aldehyde and H_2O_2 formed during the enzyme reaction. Many different techniques have been used, as summarized in Table 5. Tetrazolium salts (Glenner et al., 1957; Shannon et al., 1974; Lowe et al., 1975; Williams et al., 1975) and ferricyanide (Hanker et al., 1973) have been commonly used as electroacceptors for the structural localization of MAO. Some of these procedures, however, gave inconsistent results, and in some of the others the electron density of the reaction products was not sufficiently high to permit reaction and identification. An alternative is the cytochemical procedure described by Fujimoto et al. (1982), which is based on the formation of a complex between cerous ions and H_2O_2, which is produced in the reaction. Another alternative is to couple H_2O_2 with horseradish peroxidase-catalyzed peroxidation of 3,3'-diaminobenzidine tetrahydrochloride (Kishimoto et al., 1983). The tetrazolium salts method, utilizing the aldehyde group produced in the reaction, can produce different results when various substrates are used. Indoleamines, such as 5-HT and tryptamine, produce aldehydes with strong reduction potential, so they can be used. *p*-TA, although a common substrate for both MAO-A and MAO-B, can-

Table 5
Histochemical and Cytochemical Localization of MAO

Principle	Method	Reference
Measurement of aldehyde	Tetrazolium salts	Glenner et al., 1957; Shannon et al., 1974; Lowe et al., 1975; Williams et al., 1975
	Ferricyanide	Hanker et al., 1973
Measurement of hydrogen peroxide	Cerium	Fujimoto et al., 1982
	Coupled peroxidation (peroxidase)	Graham and Karnovsky, 1965; Williams et al., 1979; Kishimoto et al., 1983
Immunochemical	MAO-B antisera against bovine liver MAO	Pintar et al., 1983

not be used because the aldehyde is too weak to reduce tetrazolium salts to formazan precipitates. The measurement of H_2O_2 is, therefore, preferable in the histochemical or cytochemical study of MAO. One problem is that the precise localization of the original sites of the enzyme is difficult to determine if the tissues are not fixed, but the common fixation procedures employed would likely cause certain degrees of inactivation of MAO. A careful choice of *p*-formaldehyde and glutaraldehyde is necessary (Yoo et al., 1974).

More recently, immunohistochemical and immunocytochemical methods have been successfully applied to the analysis of MAO in tissues (Pintar et al., 1983). Antisera have been prepared against purified bovine MAO-B and appear to react selectively with MAO-B and not with MAO-A. In this way the MAO-B enzyme can be selectively localized.

8.1. Histochemical Identification of MAO by a Coupled Peroxidation Method

8.1.1. Fixation

According to the method of Kishimoto et al. (1983), guinea pigs are anesthetized with ip injections of sodium pentabarbitone (30 mg/kg). This is followed by transcardial perfusion for fixation purposes with cold (2°C) 1% *p*-formaldehyde and 1% glutaldehyde, maintaining a constant flow rate (50 mL/min for 8 min). After dissection of selected brain regions, they are cut into small blocks (5–7 mm) that are immersed overnight in 0.2% *p*-formaldehyde and 15% sucrose in 0.1*M* phosphate buffer (pH 7.4) at 3–5°C. Sections are then cut from the blocks in a cryostat, and the fixatives removed by soaking in ice-cold 15% sucrose in 0.1*M* phosphate buffer (pH 7.4) for several hours.

8.1.2. Incubation Reaction

A typical incubation mixture for determination of MAO-A and MAO-B is: *p*-TA HCl (5 mg), horseradish peroxidase (10 mg), and 3,3'-diaminobenzidine tetrahydrochloride (0.5 mg) dissolved in 0.05*M* Tris-HCl buffer (pH 7.6) (10 mL). Sections incubated for 90 min at 37°C with this mixture show satisfactorily intense staining with negligible nonspecific staining. If it is desired to stain either MAO-A or MAO-B, one of the specific MAO inhibitors (clorgyline for MAO-A or deprenyl for MAO-B) is injected ip 1 h before sacrifice of the animal.

9. Purification of Enzyme

9.1. Preparation of Mitochondria

Since MAO is bound to the outer membrane of mitochondria, all purification procedures should begin with the preparation of mitochondria or outer mitochondrial membrane fragments. Rat liver mitochondria, for example, may be prepared by differential centrifugation (Kropp and Wilson, 1970). Briefly, rat liver homogenized in 0.25*M* sucrose is centrifuged at 600*g* for 15 min; after removing the nuclear fraction sediment, the crude mitochondria in the supernatant are isolated by centrifuging at 6500*g* for 20 min. The mitochondrial pellet can then be washed with isotonic sucrose solution and the resultant pellet suspended in chilled, distilled water. After lysis, the mitochondrial membrane fragments are then obtained by centrifugation at 30,000*g* for 30 min.

Sucrose density gradient centrifugation is then used to isolate the mitochondrial outer membrane fragments (Felgner et al., 1979). The crude mitochondrial membrane fragment fraction, as isolated above, is applied to the top of three layers of sucrose solution (0.76, 1, and 1.32*M*). After centrifuging at 63,500*g* for 120 min, the outer mitochondrial membrane fraction migrates to the interface between 0.76 and 1*M* sucrose layers.

9.2. Preparation of Human Platelets

Human platelets have been used to study MAO activity as an index of mental abnormality. Variability in the results obtained can often be related to the method used in the preparation of the platelets. A convenient, quick, and reproducible procedure for isolating and disrupting human platelets (see below) has been described (Yu and Boulton, 1979). Blood samples (20 mL) from fasting or nonfasting subjects are collected in vacuum blood collection tubes (Venoject) containing sodium citrate as anticoagulant. Within an hour the blood is transferred to plastic centrifuge tubes and centrifuged at 200*g* for 10 min; the supernatant (platelet-rich plasma) is carefully poured into clean plastic centrifuge tubes and centrifuged again at 2000*g* for 10 min. The supernatant is discarded and the platelet pellets washed twice with 0.32 sucrose (10 mL) by suspension, followed by centrifugation at 2000*g* for 10 min. The pellets are finally resuspended in 0.32*M* sucrose (2 mL) and homogenized by sonication at 75 W (P.E.P.) for 20 s, using a needle probe tip (Braun-sonic 1510). This preparation may be kept at –70°C for up to several months before analysis.

9.3. Solubilization of MAO

Solubilization of MAO from the mitochondrial membrane can be achieved using nonionic detergents and prolonged sonication. Among the detergents used are Cutscum, desoxycholate, cholate, OP-10, Non-ion NS 210, digitonin, and Triton X-100. Triton X-100 is the most commonly used. The disadvantage of using these nonionic detergents is that they tend to bind strongly to proteins and are very difficult to remove (Sourkes, 1968). Triton X-100 can also cause selective inactivation of MAO-A (Yu, 1981b). Repeated sonication, freezing, and thawing has been used to solubilize bovine liver MAO (Sourkes, 1968); this procedure, however, is not as successful in solubilizing brain mitochondrial MAO (Nagatzu, 1966). Recently, octyl-β-glucoside has been introduced as a solubilizer (Fritz and Abell, 1982). This glucoside seems to be equivalent to purified Triton X-100; it is probably the best one to use for enzyme characterization studies because of its well-defined chemical composition and ease of removal by dialysis.

9.4. Purification of MAO-B From Bovine Liver

Minamiura and Yasunobu (1978) have improved their previous procedure (Gomes et al., 1969) for the purification of MAO from bovine liver mitochondria. The modified procedure is as follows. The purified mitochondrial pellets are first washed with a 0.1*M* phosphate buffer solution containing 0.5% Triton X-100 at pH 8.2 to remove lipids. These washed mitochondria are then suspended in 0.1*M* buffer containing 1% Triton X-100 and 0.5% potassium cholate. After stirring for 4 h, the suspension is frozen overnight, thawed the next morning, and precipitated, using solid ammonium sulfate between 25–40% saturation at pH 7.5. The enzyme that floats to the surface is collected and dialyzed against 0.02*M* phosphate buffer at pH 7.6. Prewashed calcium-phosphate gel is then added to the dialyzed enzyme and the mixture is gently stirred, without foaming, for 15 min; this is followed by centrifugation at 5000*g* for 5 min. The supernatant is saved and the calcium-phosphate treatment repeated. The resulting supernatant is applied to a DEAE-cellulose column that has been equilibrated with 0.02*M* buffer at pH 7.5. After the column has been washed with the equilibrating buffer, a linear gradient of sodium chloride (0–0.5*M*) in the same buffer containing 1% Triton X-100 is applied to elute the enzyme. The MAO fractions are collected, precipitated with ammonium sulfate, and dialyzed against 0.01*M* phosphate buffer at pH 7.4. The enzyme is then further subjected to hydroxyapatite chromatography. This column is preequilibrated with 0.01*M* phosphate buffer at pH 8.2; after application of the samples, it is washed successively with 0.1 and 0.2*M* buffer and 0.2*M* buffer containing 0.2% potassium cholate. The highest specific activity MAO is eluted in 0.2*M* buffer. An 82-fold purification has been recorded. The enzyme purified at this step is homogeneous, as indicated by SDS-PAGE.

9.5. Purification of MAO-B From Human Platelet

Human platelets contain exclusively type B MAO. A procedure has been developed to isolate homogenous MAO-B from them (Ansari et al., 1983).

The washed platelet suspension is frozen at –20°C overnight, thawed, and centrifuged at 35,000*g* for 60 min. The pellets are then suspended in 0.05*M* phosphate buffer at pH 8.0 containing 0.1% Triton X-100 at 4°C for 60 min. They are then recentrifuged at 35,000*g* for 30 min. The pellets, after suspension in the same buffer containing 0.5% Triton, are stirred for 60 min at 4°C and centrifuged at 150,000*g* for 60 min. The supernatant containing solubilized MAO is dialyzed against 0.01*M* phosphate buffer at pH 8.0. The MAO-B in the dialysate is then fractionated on a DEAE-Sephacel (Pharmacia) column that has been previously equilibrated with 0.01*M* phosphate buffer at pH 8.0. The column is developed by stepwise elution with 0.01*M*, 0.1*M* phosphate buffer (pH 8.0), and 0.1*M* phosphate buffer (pH 8.0) containing 0.25% Triton X-100. The collected enzyme fractions are precipitated by addition of ammonium sulfate to 50% saturation. The precipitate that floats to the surface is dissolved in 0.05*M* phosphate buffer (pH 8.0) containing 1% octylglucoside and dialyzed against 0.025*M* Tris-acetate buffer at pH 7.4. Solid octylglucoside is then added to the dialyzed solution to a final concentration of 1%. This preparation is then further fractionated on a Polybuffer Exchanger 94 (Pharmacia) chromato-focusing column according to the instructions provided by Pharmacia Fine Chemicals. Diluted (1:8) Polybuffer 7.4 (Pharmacia) pH 4 (adjusted with 1*N* HCl) containing 1% octylglucoside is used to elute the enzyme. Those fractions exhibiting high MAO-B activity (i.e., near pH 5.3) are pooled and concentrated by ammonium sulfate precipitation. The floating precipitate is collected, dissolved in 0.05*M* phosphate buffer at pH 7.4 containing 1% octylglucoside and dialyzed against pH 7.4 buffer, and then subjected to HPLC separation performed on a Synchropah AX-300 column in 0.1*M* phosphate buffer (pH 7.4), followed by a gradient (0–1%) of octylglucoside in the same buffer. The MAO-B active fractions are pooled, dialyzed, and lyophilized. This method yields a 43-fold purification, but only a 0.25% recovery.

9.6. Other Methods

Catalytically active isozymes from rat liver MAO have been purified from the outer mitochondrial membrane by repetitive solubilization with octyl-β-glucoside, followed by reconstitution into lipid vesicles (Stadt et al., 1982). This method appears to be simple, but has not yet been widely adopted. Salach (1979) has developed a method for purifying bovine liver MAO by utilizing polymer partition and treatment with phospholipases. Dextran, Ficoll, or polyethylene glycol layers are then used to separate the MAO by centrifugation. The purified enzyme appears as a single band in SDS-PAGE. A 157-fold purification and 55% yield are achieved.

A [^{3}H]-pargyline-labeled MAO subunit (catalytically inactive) can also be purified by DEAE-cellulose chromatography and gel filtration (Yu, 1981a).

The purification of MAO-B from bovine liver and human platelets has been well described. A purification of MAO-A, however, has yet to be established. Recent reports suggest that MAO-B can be separated from mixed-type MAO, such as human brain, by DEAE-Sepharose CL-6B column chromatography (Pearce and Roth, 1984) and from monkey brain by preparative isoelectric focusing (Egashira et al., 1984).

References

Achee F. M., Togulga G., and Gabay S. (1974) Studies of monoamine oxidases: Properties of the enzyme in bovine and rabbit brain mitochondria. *J. Neurochem.* **22,** 651–661.

Alexopoulos G. S., Lieberman K. W., and Frances R. J. (1983) Platelet MAO activity in alcoholic patients and their first-degree relatives. *Am. J. Psychiatry* **140,** 1501–1504.

Anderson S. M. (1983) The measurement of monoamine oxidase activity: A comparison of the ion exchange resin column method and the solvent extraction method. *J. Pharmacol. Methods* **10,** 271–275.

Anderson R. A., Meyerson L. R., and Tabakoff B. (1976) Characteristics of enzyme forming 3-methoxy-4-hydroxylphenylethylethyleneglycol (MOPEG) in brain. *Neurochem. Res.* **1,** 525–540.

Ansari G. A. S., Patel N. T., Fritz R. R., and Abell C. W. (1983) Purification of human platelet monoamine oxidase B by high performance liquid chromatography. *J. Liq. Chromatogr.* **6,** 1407–1419.

Ask A. L., Hogberg K., Schmidt L., Kiessling H., and Ross S. B. (1982) (+)-4-Dimethylamino-2,α-dimethyl phenylethylamine (FLA 336H), a selective inhibitor of the A form of monoamine oxidase in the rat brain. *Biochem. Pharmacol.* **31,** 1401–1406.

Bathina H. B., Huprikar S. V., and Zeller E. A. (1975) New approaches to the characterization of mitochondrial monoamine oxidase (MAO) type A and B. *Fed. Proc.* **34,** 293.

Birkmayer W., Riederer P., Linauer W., and Knoll J. (1984) l-Deprenyl plus L-phenylanine in the treatment of depression. *J. Neural Transm.* **59,** 81–87.

Blackwell B. and Marley E. (1966) Interaction of cheese and its constituents with monoamine oxidase inhibitors. *Br. J. Pharmacol. Chemother.* **26,** 120–141.

Blackwell B., Marley E., and Price J. (1967) Hypertensive interactions between monoamine oxidase inhibitors and food stuffs. *Br. J. Psychiatry* **113,** 349–365.

Boulton A. A., Davis B. A., Yu P. H., Wormith J. S., and Addington D. (1983) Trace acid levels in the plasma and MAO activity in the platelet of violent offenders. *Psychiatry Res.* **8,** 19–23.

Brown G. K., Powell J. F., and Craig I. W. (1980) Molecular weight differences between human platelet and placental monoamine oxidase. *Biochem. Pharmacol.* **29,** 2595–2603.

Buckman T. D., Eiduson S., Sutplein M. S., and Chang R. (1983) Selective effects of catalysis by multiple forms of monoamine oxidase produced by interactions of acidic phospholipids with mitochondrial membranes. *J. Biol. Chem.* **258,** 8670–8676.

Callingham B. A. and Laverty R. (1973) Studies on the nature of the increased monoamine oxidase activity in rat heart after adrenalactory. *J. Pharm. Pharmacol.* **25,** 940–947.

Callingham B. A. and Parkinson D. (1979) Tritiated Pargyline Binding to Rat Liver Mitochondrial MAO, in *Monoamine Oxidase: Structure, Function, and Altered Function.* (Singer T. P., Von Korff R. W., and Murphy D. L., eds.), Academic, New York.

Cawthon R. M. and Breakfield X. O. (1979) Differences in A and B forms of monoamine oxidase revealed by limited proteolysis and peptide mapping. *Nature* **281,** 692–694.

Cawthon R. M., Pintar J. E., Haseltine F. P., and Breakfield X. O. (1981) Differences in the structure of A and B forms of human monoamine oxidase. *J. Neurochem.* **37,** 363–372.

Collins G. G. S., Sandler M., Williams E. D., and Goudim M. G. H. (1970) Multiple forms of human brain mitochondrial monoamine oxidase. *Nature* **225,** 817–820.

Cotzias G. C. and Greenough J. J. (1959) Quantitative estimation of amine oxidase. *Nature* **183,** 1732–1733.

Crane G. E. (1957) Iproniazid (Marilid) phosphate, a therapeutic agent for mental disorders and debilitating disease. *Psychiat. Res. Report* **8,** 142–152.

Creasey N. W. (1956) Factors which interfere in the monometric assay of monoamine oxidase. *Biochem. J.* **64,** 178–183.

Da Prada M., Keller M. H., Kettler R., Schaffner R., Pieri M., Burkard W. P., Korn A., and Haefely W. E. (1982) Ro 11-1163, a Specific and Short-Acting MAO Inhibitor With Antidepressant Properties, in *Monoamine Oxidase: Basic and Clinical Frontiers.* (Kamijo K., Usdin E., and Nagatzu T., eds.), Excerpta Medica, Amsterdam.

Deitrich R. A. and Erwin V. G. (1969) A convenient spectrophtometric assay for monoamine oxidase. *Anal. Biochem.* **30,** 394–402.

Deitrich R. A., Hellerman L., and Wein J. (1962) Diphosphopyridine nucleotide-linked aldehyde dehydrogenase. I. specificity and sigma-rho function. *J. Biol. Chem.* **237,** 560–564.

Demarest K. T. and Azzaro A. J. (1979) The Association of Type A Monoamine Oxidase With the Nigrostriatal Dopamine Neuron, in *Monoamine Oxidase: Structure, Function, and Altered Functions.* (Von Korff R. W., Singer T. P., and Murphy D. L., eds.), Academic, New York.

Denney R. M., Fritz R. R., Patel N. T., and Abell G. W. (1982) Human liver MAO-A and MAO-B separated by immunoaffinity chromatography with MAO-B specific monoclonal antibody. *Science* **215,** 1400–1403.

Donnelly C. H. and Murphy D. L. (1977) Substrate- and inhibitor-related characteristics of human platelet monoamine oxidase. *Biochem. Pharmacol.* **26,** 853–858.

Donnelly C. H., Richelson E., and Murphy D. L. (1976) Properties of monoamine oxidase in mouse neuroblastoma NIE-115 cells. *Biochem. Pharmacol.* **25,** 1639–1643.

Edwards D. J. and Burns M. O. (1974) Effects of tricyclic antidepressants upon human platelet monoamine oxidase. *Life Sci.* **15,** 2045–2058.

Edwards D. J., Pak K. Y., and Venetti M. C. (1979) Developmental aspects of rat heart monoamine oxidase. *Biochem. Pharmacol.* **28,** 2337–2343.

Edwards D. J., Spiker D. G., Kupfer D. J., Foster G., Neil J. F., and Abram L. (1978) Platelet monoamine oxidase in affective disorders. *Arch. Gen. Psychiatry* **35,** 1443–1446.

Egashira T., Yamamoto T., and Yamanake Y. (1984) Isoelectric focussing of isoenzymes of monkey brain monoamine oxidase. *Life Sci.* **34,** 915–921.

Felgner P. L., Messer J. L., and Wilson J. E. (1979) Purification of hexokinase binding protein from outer mitochondrial membrane. *J. Biol. Chem.* **254,** 4946–4949.

Fowler C. J. and Oreland L. (1980) The nature of the substrate selective interaction between rat liver mitochondrial monoamine oxidase and oxygen. *Biochem. Pharmacol.* **29,** 2225–2233.

Fowler C. J. and Tipton K. F. (1984) On the substrate specificities of the two forms. *J. Pharm. Pharmacol.* **36,** 111–115.

Fowler C. J., Callingham B. A., Mantle T. J., and Tipton K. F. (1978) Monoamine oxidase A and B: A useful concept? *Biochem. Pharmacol.* **27,** 97–101.

Fowler C. J., Tipton K. F., Mackay V. P., and Youdim B. H. (1982) Human platelet monoamine oxidase—a useful enzyme in the study of psychiatric disorders? *Neuroscience* **7,** 1577–1594.

Fritz R. R. and Abell C. W. (1982) Solubilization of monoamine oxidase by octylglucoside and Triton X100. *Biochem. Med.* **28,** 94–108.

Fujimoto T., Inomata K., and Ogawa K. (1982) A cerium method for the ultracytochemical localization of monoamine oxidase activity. *Histochem. J.* **14,** 87–98.

Fuller R. W. (1968) Kinetic studies and effects in vivo of a new MAO inhibitor *N*-(2-(*O*-chlorophenoxy)-ethyl)-cyclopropylamine. *Biochem. Pharmacol.* **17,** 2097–2106.

Fuller R. W. (1972) Selective Inhibition of Monoamine Oxidase, in *Advances in Biochemistry and Psychopharmacology,* **5,** (Costa E. and Greengard P., eds.), Raven, New York.

Fuller R. W., Warren B. J., and Molloy B. B. (1970) Selective inhibition of MAO in rat brain mitochondria. *Biochem. Pharmacol.* **19,** 2934–2936.

Glenner G. G., Burtner H. J., and Brown G. W. (1957) The histochemical demonstration of monoamine oxidase activity by tetrazolium salts. *J. Histochem. Cytochem.* **5,** 591–600.

Glover V., Reveley M. A., and Sandler M. (1980) A monoamine oxidase inhibitor in human urine. *Biochem. Pharmacol.* **29,** 467–470.

Goldin L. R., Gershon E. S., Lake C. R., Murphy D. L., McGinniss M., and Sparkes R. S. (1982) Segregation and linkage studies on plasma dopamine-β-hydroxylase (DBH), erythrocyte catechol-*O*-methyltransferase (COMT), and platelet monoamine oxidase (MAO): Possible linkage between ABO locus and a gene controlling DBH activity. *Am. J. Hum. Genet.* **34,** 250–262.

Gomes B., Zgane I., Kloepper H. G., and Yasunobu K. T. (1969) Amine oxidase XIV. Isolation and characterization of the multiple beef liver amine oxidase components. *Arch. Biochem. Biophys.* **132,** 16–27.

Goridis C. and Neff N. H. (1971) Evidence for a specific monoamine oxidase associated with sympathetic nerves. *Neuropharmacology* **10,** 557–564.

Graham R. C., Jr. and Karnovsky M. J. (1965) The histochemical demonstration of monoamine oxidase activity by coupled peroxidatic oxidation. *J. Histochem. Cytochem.* **13,** 604–605.

Green A. L. and Haughton T. M. (1961) A colorimetric method for the estimation of monoamine oxidase. *Biochem. J.* **78,** 172–175.

Greenawalt J. W. (1972) Localization of Monoamine Oxidase in Rat Liver Mitochondria, in *Advances in Biochemistry and Psychopharmacology,* **5,** (Costa E. and Greengard P., eds.), Raven, New York.

Guha S. R. and Krishamurti C. R. (1965) Purification and solubilization of monoamine oxidase of rat liver mitochondria. *Biochem. Biophys. Res. Commun.* **18,** 350–354.

Guibault G. G., Brignac P. J., Jr., and Juneau M. (1968) New substrates for the fluorometric determination of oxidative enzymes. *Anal. Chem.* **40,** 1256–1265.

Hall D. W. R., Logan B. W., and Parsons G. H. (1969) Further studies on the inhibition of MAO by M and B 9302 (clorgyline) 1. Substrate specificity in various mammalian species. *Biochem. Pharmacol.* **18,** 1447–1454.

Hanker J. S., Kusuk C. J., Bloom F. E., and Pearse A. G. E. (1973) The demonstration of dehydrogenases and monoamine oxidase by the formation of osmium blacks at the sites of Hatchete's brown. *Histochemie* **33,** 205–230.

Harada M. and Nagatzu T. (1973) A sensitive fluorometric assay for monoamine oxidase activity. *Anal. Biochem.* **56,** 283–288.

Hare M. L. C. (1928) Tyramine oxidase: New enzyme in liver. *Biochem. J.* **22,** 968–979.

Harris E. J. and Cooper M. B. A. (1982) Monoamine oxidase activity of mitochondria prepared from rat liver and rat heart. *J. Neurochem.* **38,** 1068–1071.

Hidaka H., Nagatzu T., and Yagi K. (1967) Micro-determination of monoamine oxidase using serotonin as substrate. *J. Biochem.* (Tokyo), **62,** 621–623.

Ho B. T. (1972) Monoamine oxidase inhibitors. *J. Pharm. Sci.* **61,** 821–837.

Horita A. and McGrath W. R. (1960) The interaction between reversible and irreversible monoamine oxidase inhibitors. *Biochem. Pharmacol.* **3,** 206–211.

Houslay M. D. (1980) Lipid substitution of mitochondrial monoamine oxidase can lead to the abolition of clorgyline selective inhibition in the A/B ratio assessed by substrate utilization. *Biochem. Pharmacol.* **29,** 3211–3213.

Houslay M. D. and Tipton K. F. (1974) A kinetic evaluation of monoamine oxidase activity in rat liver mitochondrial outer membranes. *Biochem. J.* **139,** 645–652.

Houslay M. D. and Tipton K. F. (1973a) The reaction pathway of membrane-bound rat liver mitochondrial monoamine oxidase. *Biochem. J.* **135,** 735–750.

Houslay M. D. and Tipton K. F. (1973b) The nature of the electrophoretically separable multiple forms of rat liver monoamine oxidase. *Biochem. J.* **135,** 173–186.

Huang R. H. and Faulkner R. (1980) Lipid-protein interactions in the multiple form of monoamine oxidases: Lipases as probes using purified rat brain mitochondria. *Mol. Pharmacol.* **18,** 267–273.

Hussein L., Sindarto E., and Goedde H. W. (1980) Twin studies and substrate differences in platelet monoamine oxidase activity. *Hum. Hered.* **30,** 65–70.

Jain M., Sands F., and von Korff R. W. (1973) Monoamine oxidase activity measurements using radioactive substrates. *Anal. Biochem.* **52,** 542–554.

Jarrott B. and Iversen L. L. (1971) Noradrenaline metabolising enzymes in normal and sympathetically denervated vas deferens. *J. Neurochem.* **18,** 1–6.

Johnston J. P. (1968) Some observations upon a new inhibitor of monoamine oxidase in brain tissue. *Biochem. Pharmacol.* **17,** 1285–1297.

Kan J. P. and Strolin-Benedetti M. (1981) Characteristics of the inhibition of rat brain monoamine oxidase in vitro by MD 780515. *J. Neurochem.* **36,** 1561–1571.

Kawai S. and Nagatzu T. (1982) Gas Chromatographic and High-Performance Liquid Chromatographic Determination of Monoamine Oxidase Activity Using Mixed Substrates, in *Monoamine Oxidase: Basic and Clinical Frontiers.* (Kamijo K., Usdin E., and Nagatzu T., eds.), Excerpta Medica, Amsterdam.

Kim H. C. and D'Iorio A. (1968) Possible isoenzymes of monoamine oxidase in rat tissue. *Can. J. Biochem.* **46,** 295–297.

Kinemuchi H., Wakui W., and Kamijo K. (1980) Substrate selectivity of type A and type B monoamine oxidase in rat brain. *J. Neurochem.* **35,** 109–115.

Kishimoto S., Kimura H., and Maeda T. (1983) Histochemical demonstration for monoamine oxidase (MAO) by a new coupled peroxidation method. *Cell. Mol. Biol.* **29,** 61–69.

Kline N. S. (1958) Clinical experience with iponiazid (Marsilid). *J. Clin. Exp. Psychopathol.* **19,** (Suppl.), 72–78.

Knoll J. and Magyar K. (1972) Some Puzzling Pharmacological Effects of Monoamine Oxidase Inhibitors, in *Monoamine Oxidases—New Vistas. Advances in Biochemical Psychopharmacology,* (Costa E. and Sandler M., eds.), Raven, New York.

Kobes R. D., Wyatt R. J., and Neckers L. M. (1980) A sensitive rapid fluorometric assay for monoamine oxidase utilizing high-pressure liquid chromatography. *J. Pharmacol. Methods* **3,** 305–310.

Köchli H. and von Wartburg J. P. (1978) A sensitive photometric assay for monoamine oxidase. *Anal. Biochem.* **84,** 127–135.

Kraml M. (1965) A rapid microfluorometric determination of monoamine oxidase. *Biochem. Pharmacol.* **14,** 1683–1685.

Kropp E. S. and Wilson J. E. (1970) Hexokinases binding sites on mitochondrial membrane. *Biochem. Biophys. Res. Commun.* **38,** 74–79.

Kupfer D. S. and Rush A. J. (1984) Recommendations for scientific reports on depression. *Am. J. Psychiatry* **140,** 1327–1328.

La Motte R. H., Schmidt D. E., and Ruliffson W. S. (1969) Multiple substrate determination of MAO distribution and iproniazid inhibition in rat brain. *J. Neurochem.* **16,** 725–730.

Littlewood J., Glover V., Sandler M., Peatfield R., Petty R., and Rose F. C. (1984) Low platelet monoamine oxidase activity in headache: No correlation with phenolsulfotransferase, succinate dehydrogenase, platelet preparation method, or smoking. *J. Neurol. Neurosurg. Psychiatry* **47,** 338–343.

Loomer H. P., Saunders J. C., and Kline N. S. (1957) A clinical and pharmacodynamic evaluation of iproniazid as a psychic energizer. *Am. Psychiatry Assn. Psychiatric Res. Rep.* **8,** 129.

Lovenberg W., Levine R. J., and Sjoerdsma A. (1964) A sensitive assay of monoamine oxidase activity: In vitro application to heart and sympathetic ganglia. *J. Pharmacol. Exp. Ther.* **135,** 7–10.

Lowe M. C., Reichenback D. D., and Horita A. (1975) Extraneuronal monoamine oxidase in rat heart: Biochemical characterization and electron microscopic localization. *J. Pharmacol. Exp. Ther.* **194,** 522–526.

Mason W. D. and Olson C. L. (1970) Differential amperometric measurement of monoamine oxidase activity at tubular carbon electrode. *Anal. Chem.* **42,** 488–494.

Mantle T. J., Tipton K. F., and Garrett N. J. (1976) Inhibition of monoamine oxidase by amphetamine and related compounds. *Biochem. Pharmacol.* **25,** 1073–2077.

Mantle T., Wilson K., and Long R. (1975) Studies on the selective inhibition of membrane-bound rat liver monoamine oxidase. *Biochem. Pharmacol.* **24,** 2031–2038.

May M. E. (1980) A caution in the use of tritiated substrates for monoamine oxidase assays. *J. Neurochem.* **35,** 1453–1454.

McCaman R. E., McCaman M. W., Hunt J. M., and Smith M. S. (1965) Microdetermination of monoamine oxidase and 5-hydroxytryptophan decarboxylase activities in nervous tissue. *J. Neurochem.* **12,** 15–23.

McDonald I. A., Lacoste J. M., Bey P., Wagner J., Zreika M., and Palfreyman M. G. (1984) (E)-β-(Fluoromethylene)-*m*-tyrosine: A substrate for aromatic L-amino acid decarboxylase liberating an enzyme-activated irreversible inhibitor of monoamine oxidase. *J. Am. Chem. Soc.* **106,** 3354–3356.

McEwen C. M. and Cohen J. D. (1963) An amine oxidase in normal human serum. *J. Lab. Clin. Med.* **62,** 766–776.

McEwen C. M., Jr., Sasaki G., and Jones D. C. (1969) Human liver mitochondrial monoamine oxidase. II. Determinants of substrate and inhibitor specificities. *Biochemistry* **8,** 3952–3962.

Mendlewicz J. and Youdim M. B. H. (1983) L-Deprenil, a selective monoamine oxidase type B inhibitor in the treatment of depression: A double blind evaluation. *Br. J. Psychiatry* **142,** 508–511.

Meyerson L. R., McMurtrey K. D., and Davis V. E. (1978) A rapid and sensitive potentiometric assay for monoamine oxidase using an ammonium-sensitive electrode. *Anal. Biochem.* **86,** 287–297.

Minamiura N. and Yasunobu K. T. (1978) Bovine liver monoamine oxidase: A modified purification procedure and preliminary evidence for two subunits and one FAD. *Arch. Biochem. Biophys.* **189,** 481–489.

Murphy D. L. and Wyatt R. J. (1972) Reduced platelet monoamine oxidase activity in chronic schizophrenia. *Nature* **238,** 225–226.

Murphy D. L., Donnelly C. H., and Richelson E. (1976) Substrate- and inhibitor-related characteristics of monoamine oxidase in C6 rat glial cells. *J. Neurochem.* **26,** 1231–1235.

Nagatzu T. (1966) Partial separation and properties of mitochondrial monoamine oxidase in brain. *J. Biochem.* (Tokyo) **59,** 606–612.

Nagatzu T. and Yagi K. (1966) A simple assay of monoamine oxidase and D-amino acid oxidase by measuring ammonia. *J. Biochem.* (Tokyo) **60,** 219–221.

Navarro-Welch C. and McCauley R. B. (1982) An evaluation of phospholipids as regulators of monoamine oxidase A and monoamine oxidase B activities. *J. Biol. Chem.* **257,** 13645–13649.

Nimmo G. A. and Tipton K. F. (1979) The distribution of soluble and membrane bound forms of glutaminase in pig brain. *J. Neurochem.* **33,** 1083–1094.

Nissinen E. (1984) Determination of monoamine oxidase B activity by high-performance liquid chromatography. *J. Chromatogr.* **309,** 156–159.

Nohta H., Zaitzu K., Tsuruta Y., and Okura Y. (1983) Assay for monoamine oxidases A and B by high-performance liquid chromatography with fluorescence detection. *J. Chromatogr.* **280,** 343–349.

Obata F., Ushiwata A., and Nakamura Y. (1971) Spectrophotometric assay of monoamine oxidase using 2,4,6-trinitrobenzene-1-sulfonic acid. *J. Biochem.* (Tokyo) **69,** 349–354.

Otsuka S. and Kobayashi Y. (1964) A radioisotopic assay for monoamine oxidase determinations in human plasma. *Biochem. Pharmacol.* **13,** 995–1006.

Pandey E. N., Dorus E., Shaughnessy R., and Davis J. M. (1979) Genetic control of platelet monoamine oxidase activity: Studies on normal families. *Life Sci.* **25,** 1173–1178.

Pearce L. B. and Roth J. A. (1984) Monoamine oxidase: Separation of type A and B activities. *Biochem. Pharmacol.* **33,** 1809–1811.

Peng C. T. (1970) A Review of Methods of Quenching Correction in Liquid Scintillation Counting, in *The Current Status of Liquid Scintillation Counting.* (Bransome E. P., ed.), Crune and Stratton, New York.

Pintar J. E., Levitt P., Salach U. I., Weyer W., Rosenberg M. B., and Breakfield X. O. (1983) Specificity of antisera prepared against pure bovine MAO-B. *Brain Res.* **276,** 127–139.

Quitkin F., Rifkin A., and Klein D. G. (1979) Monoamine oxidase inhibitors: A review of antidepressant effectiveness. *Arch. Gen. Psychiatry* **36,** 749–760.

Rice J., McGuffin P., Goldin L. R., Shaskan E. G., and Gershon E. S. (1984) Platelet monoamine oxidase (MAO) activity: Evidence for a single major locus. *Am. J. Hum. Genet.* **36,** 36–43.

Roth J. A. and Gillis C. N. (1974) Inhibition of rabbit mitochondrial MAO by tricyclic antidepressant drugs. *Fed. Proc.* **23,** 498.

Salach J. I. (1979) Monoamine oxidase from beef liver mitochondria: Simplified isolation procedure, properties, and determination of its cysteinyl flavine content. *Arch. Biochem. Biophys.* **912,** 128–137.

Sandler M., Reveley M. A., and Glover V. (1981) Human platelet monoamine oxidase activity in health and disease: A review. *J. Clin. Pathol.* **34,** 292–302.

Schnaitman C., Erwin V. G., and Greenawalt J. W. (1967) The submitochondrial localization of MAO. An enzymatic marker for the outer membrane of rat liver mitochondria. *J. Cell Biol.* **32,** 719–735.

Selikoff I. J., Robitzen E. H., and Ornstein G. G. (1952) Toxicity of hydrazine derivatives of isonicotinic acid in chemotherapy of human tuberculosis. *Q. Bull. Sea View Hosp.* **13,** 17.

Shannon W. A., Jr., Wasserkrug H. L., and Seligman A. M. (1974) The ultrastructural localization of monoamine oxidase (MAO) with tryptamine and a new tetrazolium salt, 2-(2'-benzothiazolyl)-5-styryl-3-(4'-phthalhydrzidyl) tetrazolium chloride (BSPT). *J. Histochem. Cytochem.* **22,** 170–182.

Silverman R. B., Hoffman S., and Catus W. B., III. (1980) A mechanism for mitochondrial monoamine oxidase-catalyzed amine oxidation. *J. Am. Chem. Soc.* **102,** 7126–7128.

Smith T. E., Weissbach H., and Udenfriend S. (1962) Studies on the mechanism of action of monoamine oxidase: Metabolism of *N,N*-dimethyltryptamine and *N,N*-dimethyltryptamine-*N*-oxide. *Biochemistry* **1,** 137–143.

Snyder S. H. and Hendley E. D. (1968) A simple and sensitive fluorescence assay for monoamine oxidase and diamine oxidase. *J. Pharmacol. Exp. Ther.* **163,** 386–392.

Southgate J. and Collins G. G. S. (1969) The estimation of monoamine oxidase using ^{14}C-labelled substrates. *Biochem. Pharmacol.* **17,** 109–119.

Sourkes T. L. (1968) Properties of the monoamine oxidase of rat liver mitochondria. *Adv. Pharmacol.* **6A,** 61–69.

Squires R. F. (1972) Multiple Forms of Monoamine Oxidase in Intact Mitochondrial as Characterized by Selective Inhibitors and Thermal Stability: A Comparison of Eight Mammalian Species, in *Advances in Biochemistry and Psychopharmacology* **5,** (Costa E. and Greengard, P., eds.), Raven, New York.

Stadt M. A., Banks P. A., and Kobes R. D. (1982) Purification of rat liver monoamine by octyl glucoside extraction and reconstruction. *Arch. Biochem. Biophys.* **214,** 223–230.

Stahl S. M. (1977) The human platelet. A diagnostic and research tool for the study of biogenic amines in psychiatric and neurologic disorders. *Arch. Gen. Psychiatry* **34,** 509–516.

Strolin-Benedetti M., Dostert P., Boucher T., and Guffroy C. (1982) A New Reversible Selective Type B Monoamine Oxidase Inhibitor MD 780236, in *Monoamine Oxidase: Basic and Clinical Frontiers.* (Kamijo K., Usdin E., and Nagatzu T., eds.), Excerpta Medica, Amsterdam.

Student A. K. and Edwards D. J. (1977) Subcellular localization of type A and type B MAO in rat brain. *Biochem. Pharmacol.* **26,** 2337–2342.

Suzuki O., Katzumata Y., and Oya M. (1981) Characterization of eight biogenic indolamines as substrates for type A and type B monoamine oxidase. *Biochem. Pharmacol.* **30,** 1353–1358.

Symes A. L., Sourkes T. L., Youdim M. B. H., Gregoriadis G., and Birnbaum H. (1969) Decreased monoamine oxidase activity in liver of iron-deficient rat. *Can. J. Biochem.* **47,** 999–1002.

Tabor C. W., Tabor H., and Rosenthal S. M. (1954) Purification of amine oxidase from beef plasma. *J. Biol. Chem.* **208,** 645–661.

Takahashi H. and Takahara S. (1968) A sensitive fluorometric assay for monoamine oxidase based on the formation of 4,6-quinolinediol from 5-hydroxykynurenamine. *J. Biochem.* (Tokyo) **64,** 7–11.

Tenne M., Finberg J. P. M., Youdim M. B. H., and Ulitzur S. (1985) A new rapid and sensitive bioluminescene assay for monoamine oxidase activity. *J. Neurochem.* **44,** 1378–1384.

Tipton K. F. (1968) The purification of pig brain mitochondrial monoamine oxidase. *Eur. J. Biochem.* **4,** 103–107.

Tipton K. F. (1969) A sensitive fluorometric assay for monoamine oxidase. *Anal. Biochem.* **28,** 318–325.

Tipton K. F. and Dawson A. P. (1968) The distribution of monoamine oxidase and α-glycerophosphate dehydrogenase in pig brain. *Biochem. J.* **108,** 95–99.

Tipton K. F. and Youdim M. B. H. (1983) The Assay of Monoamine Oxidase Activity, in *Methods in Biogenic Amine Research.* (Parvez S., Nagatzu T., Nagatzu I., and Parvez H., eds.), Elsevier Science, Amsterdam.

Tipton K. F. and Youdim M. B. H. (1976) Assay of monoamine oxidase. *Ciba Found. Symp.* (New Ser.) **39,** 393–403.

Tipton K. F., Fowler C. J., and Houslay M. D. (1982) Specificities of the Two Forms of Monoamine Oxidase, in *Monoamine Oxidase: Basic and Clinical Frontiers.* (Kamijo K., Usdin E., and Nagatsu T., eds.), Excerpta Medica, Amsterdam.

Tyrer P. (1976) Towards rational therapy with monoamine oxidase inhibitors. *Br. J. Psychiatry* **128,** 354–360.

Udenfriend S. (1962) *Fluorescence Assay in Biology and Medicine.* Academic, New York.

Udenfriend S., Weissbach H., and Clark C. T. (1955) The estimation of 5-hydroxytryptamine (serotonin) in biological tissue. *J. Biol. Chem.* **215,** 337–344.

Uji A., Kawai S., and Nagatzu T. (1980) Gas chromatographic determination of monoamine oxidase activity using mixed substrates. *J. Chromatogr.* **221,** 155–160.

Waldmeier P. C., Baumann P. A., Delini-Stula A., Bernasconi R., Sigg K., Buech O., and Felner A. E. (1983) Characterization of New, Short-Acting and Specific Inhibitor of Type A Monoamine Oxidase, in *Monoamine Oxidase and its Selective Inhibitors.* (Beckmann H. and Riederer P., eds.), Karger Verlag, Basel.

Walker W. H., Dearney E. B., Seng R. L., and Singer T. P. (1971) Covalently-bound flavin of hepatic monoamine oxidase. 2. Identification and properties of riboflavin cysteinyl. *Eur. J. Biochem.* **24,** 328–331.

Weissbach H., Smith T. E., Daly J. W., Witkop B., and Udenfriend S. (1960) A rapid spectrophotometric assay of monoamine oxidase based on the disappearance of kyneuramine. *J. Biol. Chem.* **235,** 1160–1163.

Weyler W. and Salach J. I. (1981) Ion content and spectral properties of highly purified bovine liver monoamine oxidase. *Arch. Biochem. Biophys.* **212,** 147–153.

White H. L. and Stine D. K. (1982) Monoamine oxidase A and B as components of a membrane complex. *J. Neurochem.* **38,** 1429–1436.

White H. L. and Stine D. K. (1984) Selective effects of proteases and phosphalipase A_2 on monoamine oxidases A and B of human brain and liver. *J. Neurochem.* **42,** 1743–1751.

Williams D., Gascoigne J. E., Street M., and William E. D. (1979) Histochemical characterization of monoamine oxidase in ependyma of rat hypothalamus. *Histochem. J.* **11,** 83–95.

Williams D., Gascoigne J. E., and Williams E. D. (1975) A tetrazolium technique for the histochemical demonstration of multiple forms of rat brain monoamine oxidase. *Histochem. J.* **7,** 585–597.

Wu P. H. and Dyck L. E. (1975) Microassay for the estimation of monoamine oxidase activity. *Anal. Biochem.* **72,** 637–642.

Wurtman R. J. and Axelrod J. (1963) A sensitive and specific assay for the estimation of monoamine oxidase. *Biochem. Pharmacol.* **12,** 1439–1441.

Wyatt R. J., Potkin S. G., and Murphy D. L. (1979) Platelet MAO activity in schizophrenia: A review of the data. *Am. J. Psychiatry* **136,** 377–385.

Yamada H., Suzuki H., and Ogura Y. (1972) Amine Oxidases From Microorganism: Stoichiometry of the Reaction Catalyzed by Amines Oxidase of *Aspergillus niger,* in *Advances in Biochemistry and Psychopharmacology,* **5,** (Costa E. and Greengard, P., eds.), Raven, New York.

Yang H. Y. and Neff N. H. (1973) β-Phenylethylamine: A specific substrate for type B monoamine oxidase of brain. *J. Pharmacol. Exp. Ther.* **187,** 365–371.

Yasunobu K. T. and Gaomes B. (1971) Mitochondrial amine oxidase (monoamine oxidase) (beef liver). *Methods of Enzymol.* **17B,** 709–717.

Yoo B. Y., Oreland L., and Persson A. (1974) Effects of formaldehyde and glutaldehyde on the monoamine oxidase activity in isolated rata liver mitochondria. *J. Histochem. Cytochem.* **22,** 445–446.

Youdim M. B. H. and Sourkes T. L. (1966) Properties of purified, soluble monoamine oxidase. *Can. J. Biochem.* **44,** 1399–1400.

Youdim M. B. H., Collins G. G. S., and Sandler M. (1969) Multiple forms of rat brain monoamine oxidase. *Nature* **223,** 626–628.

Yu P. H. (1979) Effects of Lipid Depletion on Type A and Type B

Monoamine Oxidase of Rat Heart and Bovine Liver Mitochondria, in *Monoamine Oxidase: Structure, Function, and Altered Functions.* (Singer T. P., vonKorff R. W., and Murphy D. L., eds.), Academic, New York.

Yu P. H. (1981a) Studies on the pargyline-binding site of different types of monoamine oxidase. *Can. J. Biochem.* **59,** 30–37.

Yu P. H. (1981b) Some effects of non-ionic detergent Triton X-100 on rat liver monoamine oxidase. *Biochem. Pharmacol.* **30,** 3089–3094.

Yu P. H. (1984) Monoamine oxidase in young and adult rat brain capillaries. *J. Neural Transm.* **60,** 239–245.

Yu P. H. and Boulton A. A. (1979) Activation of platelet monoamine oxidase by plasma in the human. *Life Sci.* **25,** 31–36.

Yu P. H. and Hertz L. (1982) Differential expression of type A and type B monoamine oxidase of mouse astrocytes in primary cultures. *J. Neurochem.* **37,** 1492–1495.

Yu P. H. and Hertz L. (1983) Type A and B monoamine oxidase in glial cells in long term culture. *Prog. Neuro.-Psychopharmacol. Biol. Psychiatry* **7,** 687–690.

Yu P. H., Bailey B. A., and Durden D. A. (1986) High performance liquid chromatography of aldehydes and acids in monoamine oxidase catalyzed reactions. *Anal. Biochem.* (in press).

Yu P. H., Barclay S., Davis B. A., and Boulton A. A. (1981) Deuterium isotope effects on the enzymatic oxidation of trace amines. *Biochem. Pharmacol.* **30,** 3089–3094.

Yu P. H., Bowen R., Carlson K., O'Sullivan K., and Boulton A. A. (1982a) A comparison of some biochemical properties of platelet monoamine oxidase in some mentally disordered and healthy individuals. *Psychiatry Res.* **6,** 107–121.

Yu P. H., Kazakoff C. W., Davis B. A., and Boulton A. A. (1982b) Deuterium isotope effect enzymatic oxidation of dopamine and serotonin. *Biochem. Pharmacol.* **31,** 3697–3698.

Yu P. H., Bowen R., Davis B. A., and Boulton A. A. (1983) Platelet monoamine oxidase activity and trace acid levels in plasma of agoraphobic patients. *Acta Psychiatr. Scand.* **67,** 188–194.

Zeller E. A. (1938) Uber den enzymatischen abbau von Histamin und Diamine. *Helv. Chim. Acta* **21,** 880–890.

Zeller E. A. and Barsky J. (1952) In vivo inhibition of liver and brain monoamine oxidase by 1-isonicotinyl-2-isopropylhydrazine. *Proc. Soc. Exp. Biol. Med.* **81,** 459–461.

Zeller V., Ramachander G., and Zeller E. A. (1965) Amine oxidases. XXI. A rapid method for the determination of the activity of monoamine oxidase and monoamine oxidase inhibitors. *J. Med. Chem.* **8,** 440–450.

Chapter 8
Choline Acetyltransferase

Henry G. Mautner

1. History

The central role of actylcholine (Ach) in the transmission of the nerve impulse has been accepted ever since the pioneering experiments of Loewi (1921). In 1953, Nachmansohn postulated that in electrically excitable membranes, the attachment of Ach to a protein, the Ach receptor (AchR), induced a change in membrane cation permeability. He also recognized that the functions of the AchR, as well as of the enzyme responsible for the synthesis of Ach, choline acetyltransferase (ChA, EC 2.3.1.6), and the enzyme responsible for the hydrolysis of Ach (acetylcholinesterase), form a coordinated, cyclic process. The flow of cations triggered by the attachment of Ach to the AchR is passive, whereas Ach is a high–energy compound. Metabolic energy is required to restore ion gradients to the state existing before the transient Ach–induced permeability change, and metabolic energy is required for the synthesis of Ach. Thus, ChA is a key link between the functions of the Ach cycle and the metabolism of the cell.

ChA was discovered by Nachmansohn and Machado (1943). Their report that ATP catalyzed the acetylation of choline in a cell–free system seemed so surprising at the time that *Science*, the *Journal of Biological Chemistry*, and *Proceedings of the Society of Experimental Biology and Medicine* refused to publish it (Nachmansohn, 1972). They also noted that, upon dialysis, ChA activity was lost rapidly, suggesting the requirement for a coenzyme in this system (Nachmansohn et al., 1943; Nachmansohn and Machado, 1943). This cofactor proved to be identical with the factor, which he named coenzyme A (CoA), found by Lipmann to be essential for the acetylation of sulfanilamide (Lipmann, 1945).

Lynen and his collaborators showed that CoA, a thiol, forms thiolesters capable of transferring acyl groups to suitable acceptors and capable of undergoing Aldol condensations (Jaenicke and Lynen, 1960). The acetylation of choline by acetyl CoA:

$$(CH_3)_3\overset{+}{N}CH_2CH_2OH + CoAS\overset{\overset{\displaystyle O}{\|}}{C}CH_3 \longleftrightarrow (CH_3)_3\overset{+}{N}CH_2CH_2O\overset{\overset{\displaystyle O}{\|}}{C}CH_3 + CoASH$$

represents the alcoholysis of a thiolester to form an ester and a thiol (CoA).

A great deal of research has been centered on ChA in recent years, not only because of the central role of Ach in the propagation of the nerve impulse, but also because of the great usefulness of antibodies to ChA in mapping cholinergic neurons (Wainer et al., 1984). Claims that the development of Alzheimer's disease (senile dementia) can be related to a dramatic drop in ChA in the cortical regions (Davies and Maloney, 1976; Bowen et al., 1976; Bartus et al., 1982; Terry and Katzman, 1983) have added further interest to studies of ChA.

In addition, it should be noted that the presence of ChA in a variety of nonneuronal tissues makes it apparent that many of the functions of ChA still remain to be defined.

In spite of ChA having been known for more than 40 yr, the enzyme has not been crystallized, and many questions about its structure and mechanism of action remain to be answered.

2. Sources of Choline Acetyltransferase

Since the initial discovery of ChA in rabbit brain (Nachmansohn and Machado, 1943), ChA has been found in varying amounts in all electrically excitable tissues, including motor and sensory nerves (Nachmansohn, 1959). The ubiquity of ChA and acetyltransferase in nervous tissue led to the proposal that Ach is essential for linking the permeability changes of axonal membranes to the propagation of the nerve impulse (Nachmansohn, 1954). However, the relatively low levels of ChA in sensory fibers were cited as evidence for the noninvolvement of Ach in axonal conduction (Hebb, 1972). Although axonal membranes contain not only ChA, but also acetylcholinesterase (De Lorenzo et al., 1969) and biopolymers capable of binding cholinergic ligands (Denburg, 1972; Jumblatt et al., 1981), no convincing evidence linking the ChA cycle to the transmission of the action potential in nerves has yet appeared.

That the functions of Ach are not limited to the induction of cation permeability changes in synaptic membranes is emphasized by the presence of ChA in a wide variety of nonnervous tissue. Thus, ChA has been found in microorganisms (Stephenson and Rowatt, 1947; Alpert et al., 1966; White and Cavallito, 1970a), nettles (Barlow and Dixon, 1973), mung beans (Jaffe, 1970), moss callus (Hartmann and Kilbinger 1974a,b), peas, spinach, sunflowers, blue-green algae (Smallman and Maneckjee, 1981), and, with acetylcholinesterase, in numerous other plants (Miura and Shih, 1984). An interesting observation reports that light regulates Ach levels in plants (Hartmann and Kilbinger, 1974a,b). The presence of ChA has also been reported in *Schistosomes* (Bueding, 1952) and the protozoa *Paramecium* and *Trypanosoma rhodesiense* (Bulbring et al., 1949). The isolation and partial purification of ChA from the nematode *Caenorhabditis elegans* was reported recently (Rand and Russell, 1985). A particularly rich, nonneuronal source of ChA is provided by placentae (Morris, 1966; Hersh et al., 1978). ChA is also found in spleen (Stephens-Newsham et al., 1979).

The recent report of the presence of high levels of ChA in spermatozoa from a variety of sources (Bishop et al., 1976; Harbison et al., 1976; Rama-Sastry et al., 1981) provides a good example for the caution with which such reports must be treated. It is easy to confuse ChA with carnitine acetyltransferase (Goodman et al., 1984) and, possibly, with other acetyltransferases as well. Similarly, reports of high levels of ChA in heart tissue ignored the fact that this tissue contains much more carnitine acetyltransferase than ChA. The problem of possibly mistaking other acetyltransferases for ChA will be discussed elsewhere.

Even though some of the older literature reporting ChA in various cell types bears reevaluation, the presence of ChA in nonneuronal tissue is well established. The presence of enzymes catalyzing the synthesis and the breakdown of Ach in a wide variety of cells suggests that Ach-mediated control of cell permeability to cations is a general phenomenon of cells, whether innervated or not. A possible link of the Ach cycle to chemotaxis has also been proposed (Mautner, 1977) because changes in membrane potential are claimed to be responsible for coupling sensory receptors to the rotation of flagellae in chemotactic cells (Doetsch, 1972).

A recent review discusses the presence of a whole series of extracellular messengers, including peptide hormones, Ach, catecholamines, and steroids, in unicellular organisms (Roth et al., 1982). It seems likely that many of the functions we associate with intercellular communication must have been preceded by intracellular functions. In the case of Ach, control of cell membrane permeability and chemotaxis seems a likely candidate for such functions. A summary of work on the presence of cholinergic systems in non-nervous tissue has appeared (Rama-Sastry and Sadavongvivad, 1979).

3. Methods of Determining Choline Acetyltransferase

A few years after the discovery of ChA, the Hestrin method for the chemical determination of Ach was introduced (Hestrin, 1949). In this procedure, the Ach produced by ChA is permitted to react with hydroxylamine to form hydroxamic acid that produces a readily quantitated red color in the presence of ferric iron. This procedure, although relatively insensitive, has the advantage of permitting the study of fast initial kinetics.

More sensitive procedures were introduced when [^{14}C]-acetyl-CoA became available. [^{14}C]-Acetyl-CoA was used to prepare [^{14}C]-Ach that was precipitated as the reineckate salt (McCaman and Hunt, 1965). This procedure was soon replaced by the use of anion-exchange resins for the selective retention of [^{14}C]-Ach formed by the ChA-catalyzed acetylation of choline (Schrier and Shuster, 1967).

A widely used assay is based on the selective ability of Ach to form a lipid-soluble complex with tetraphenyl boron. Labeled Ach, formed by the action of ChA, interacts with sodium tetraphenyl boron in acetonitrile and is extracted into a toluene scintillation mixture while acetyl CoA remains in the aqueous phase (Fonnum, 1969, 1975). Liquid scintillation counting permits the use of either [^{3}H]-acetyl-CoA or [^{14}C]-acetyl-CoA in this assay. Because the reaction takes place in a closed vessel, exposure of personnel to radioactivity and organic solvents is minimized. The primary advantage of the Fonnum procedure resides in the fact that if extraction of the tetraphenyl boron–Ach complex is carried out at neutral pH, the presence of a carboxyl group in acetylcarnitine prevents complex formation at that pH (a pH of 2.8 is needed). Thus, the Fonnum assay permits the differential determination of Ach and acetylcarnitine.

All the radiochemical determinations of ChA activity share the disadvantage of not being suitable for rapid rate determinations used to probe intermediate steps in the catalytic process. To circumvent this problem, an assay was developed in which Ach synthesis was coupled to the ATP citrate lyase and malic dehydrogenase reactions. Ach formation could be followed by tracing the disappearance of NADH at 340 nm (Hersh and Peet, 1977):

AcCoA + Choline → Ach + CoA

CoA + Citrate + ATP → Oxaloacetate + AcCoA + ADP + P_i

Oxaloacetate + NADH → Malate + NAD

This procedure proved very useful in Hersh's laboratory for studying the kinetic mechanisms and, in particular, the ordering of the ChA reaction. However, the use of coupled reactions may complicate the interpretation of the action of enzyme inhibitors.

Alternate procedures include a single-vial extraction assay using [^{3}H]-choline as the labeled substrate (Rand and Johnson, 1981). After incubation with an excess of acetyl-CoA for a suitable period, unreacted [^{3}H]-choline is quickly and quantitatively converted to phosphoryl-[^{3}H]-choline by the addition of an excess of choline kinase. Addition of scintillation fluid containing sodium tetraphenyl boron permits the determination of [^{3}H]-Ach that, in contrast to phosphoryl choline, forms a lipid-soluble complex.

Another assay tries to minimize the problems introduced by the possible presence of carnitine acetyltransferase by relying on the ability of *Electrophorus* acetylcholinesterase to hydrolyze Ach, but not acetylcarnitine. Assay of ChA by the Schrier and Shuster method is carried out in (a) the presence of neostigmine (an acetylcholinesterase inhibitor), and (b) the presence of electric organ acetylcholinesterase. After passing the reaction mixtures over anion exchange columns, the difference in the radioactivities of the effluents is proportional to ChA activity (Hamprecht and Amano, 1974; Tsuchida et al., 1982).

As already noted, contamination with carnitine acetyltransferase can lead to considerable errors in the estimation of ChA in various tissues (Goodman et al., 1984). Two types of error are encountered: (1) If labeled acetyl-CoA is used, the presence of carnitine acetyltransferase may lead to the formation of labeled acetylcarnitine, which can be mistaken for labeled Ach, and (2) Carnitine acetyltransferase can acetylate choline as well as carnitine, although acetylation of the former is relatively inefficient. Thus, Ach can be formed in the absence of ChA. Although saturation of brain ChA is seen with 2–5 m*M* choline, saturation of carnitine acetyltransferase requires 800–1000 m*M* choline. Therefore, in the presence of carnitine acetyltransferase, care must be taken to see that, at the choline concentration used, acetylation is independent of choline concentration.

As already discussed, errors of this type are avoided more

readily with the tetraphenyl boron extraction method (Fonnum, 1975) than with assays using anion exchange columns (Schrier and Shuster, 1967). In any case, it is essential to establish that the product of the reaction is indeed Ach and that the enzyme responsible for its formation is indeed ChA. A chemiluminescent procedure for the determination of Ach was introduced by Israël and Lesbats (1980, 1981, 1982; Birman, 1985). This procedure is based on the conversion of choline to betaine and H_2O_2 by choline oxidase (EC 1.1.3.17). H_2O_2 is then determined by one of several available luminescence assays, of which the luminol-peroxidase reaction is reported to be the most convenient.

There remains a need for a sensitive and highly specific assay for ChA in which the disappearance of reactants and the formation of products can be followed directly without either a separation step or coupling to other enzyme reactions.

Numerous publications have discussed the effect of salt concentration on the activity of ChA, with different groups reporting activation, inhibition, no effect, ion specificity, or lack of ion specificity (for review, *see* Mautner, 1977). The origin of this problem will be discussed in the section dealing with the ordering of the ChA reaction.

It is essential that the salt concentration is rigidly controlled in all studies with ChA. It should also be remembered that salt effects alter with the concentration of cofactor in this reaction. A further complication is introduced by the observation that ChA may be bound to membranes in a process dependent on ionic strength or pH (Fonnum, 1968; Benishin and Carroll, 1983). This problem is particularly important for investigators studying Ach synthesis in membrane preparations such as synaptosomes. Again, possible confusion of ChA with other acetyltransferases must be guarded against.

4. Methods of Purifying Choline Acetyltransferase

In spite of ChA having been discovered more than forty years ago, relatively pure enzyme became available only during the last decade. The slowness of this progress is caused primarily by the very low levels of ChA in most tissues and the tendency of the enzyme to become inactivated during most purification procedures.

The question of deciding when a "pure" enzyme is indeed pure confused some of the earlier work with ChA. Claims appeared that homogenous ChA could be obtained by using sim-

Table 1
Purification of Choline Acetyltransferase

Source of enzyme[a]	Final activity, μ*M*/min/mg protein	Reference
Placenta (human)	92.7	Hersh et al., 1978b
Brain (bovine)	54.0	Ryan and McClure, 1979
	142.0	Cozzari and Hartman, 1983a
Brain (rat)	40.2	Ryan and McClure, 1979
	70–80	Dietz and Salvaterra, 1980
Brain (pig)	135	Eckenstein et al., 1981
Brain (human)	24	Peng et al., 1983
Head *Drosophila*	500	Slemmon et al., 1982
Optic ganglia (squid)	66.7	Husain and Mautner, 1973
Nematode (*C. elegans*)	3.7	Rand and Russell, 1985

[a]This list is incomplete and includes only references reporting unusually high enzyme activity for the tissue in question.

ple purification methods (Chao and Wolfgram, 1973; Roskoski et al., 1975) even though the enzyme obtained had only minimal activity. It later became apparent that some of these preparations were either proteins containing ChA as an impurity or ChA that had been extensively decomposed during purification. For this reason, skepticism is needed in considering some of the early work dealing with antibodies to ChA and their use for the immunohistochemical localization of cholinergic neurons (Rossier, 1975; Chao et al., 1977).

Since reviews of the earlier literature dealing with the purification of ChA have appeared (Mautner, 1977; Rossier, 1977a), this summary will concentrate on recent purification procedures (*see* Table 1).

Because the lack of stability of ChA is a major problem during purification, several attempts have been made to stabilize the enzyme. Some groups have added extraneous protein, whereas others added rather high concentrations of sucrose (Husain and Mautner, 1973; Shuster and O'Toole, 1974), glycerol (Hersh et al., 1978b), or ethylene glycol (Dietz and Salvaterra, 1980). Chromatography of ChA has also been carried out in the presence of dilute ethanol, which was claimed to minimize the interaction of the enzyme with other proteins (Cozzari and Hartman, 1983a). The ability of dilute organic solvents, such as ethanol, *n*-butanol, acetone, or ether, to stimulate ChA activity was recognized some time ago (Berman-Reisberg, 1974).

Dithiothreitol and other thiols have also been used to stabilize ChA. However, thiols do not stabilize all forms of the enzyme. Furthermore, rapid thiolysis of acetyl-CoA may complicate assays of ChA activity. Metal chelating agents such as EDTA or EGTA have also been used widely during ChA purification procedures to minimize the effect of calcium-requiring proteases.

The literature dealing with the purification of ChA variously holds this enzyme to be unstable during early stages of purification, to be unstable during late stages of purification, or to be fairly stable. Recently, it was claimed that enzyme stability depends on the method of purification, rather than on the degree of purification achieved (Cozzari and Hartman, 1983a,b), possibly because the presence of contaminating proteins and the order in which they are removed affects enzyme activity. It was reported that some of the loss of activity seen after dye-ligand chromatography or other affinity-chromatographic procedures may be because of copurification of inhibitory compounds. The presence of endogenous ChA inhibitors in brain has been claimed (Cozzari and Hartman, 1983c).

Unfortunately, no source of ChA has been found in which this enzyme is present in amounts equivalent to the plentitude of acetylcholinesterase and AchR in the electric organs of electric fish. Because tissue Ach is stored until used, there is no need for ChA to be present in the amounts in which acetylcholinesterase and AchR must be localized in their effector region.

The richest sources of ChA known at this time are the head ganglia of squid (Nachmansohn and Weiss, 1948; Husain and Mautner, 1973) and *Drosophila* heads (Driskell et al., 1978; Slemmon et al., 1982). In view of the difficulties encountered in obtaining ChA in amounts adequate for crystallization and for physicochemical studies of the substrate interactions of this enzyme, continued screening for ChA in various organisms and tissues is warranted.

4.1. Initial Stages of Choline Acetyltransferase Purification

Acetone powders provided the earliest method of concentrating ChA activity (Nachmansohn and Weiss, 1948). At least one recent purification (Cozzari and Hartman, 1983a) uses this procedure. The high stability of ChA in the presence of organic solvents has already been noted.

A combination of ammonium sulfate precipitation and extraction, following homogenization and centrifugation of the tissue from which ChA is isolated, was used during most of the recent purifications of the enzyme. Precipitation with polyethylene glycol proved a useful step during the purification of ChA from rat brain (Malthe-Sørenssen, et al., 1978), *Drosophila* heads (Slemmon et al., 1982), squid head ganglia, and placentae (Mautner and Shachter, unpublished data).

4.2. Affinity Chromatography

In addition to the "usual" columns, such as CM-cellulose, CM-Sephadex, Sephadex G-150, DEAE cellulose, cellulose phosphate, and hydroxylapatite, several purification procedures describe the use of columns carrying inhibitors of ChA. For instance, the purification of ChA from squid head ganglia used a Sepharose column to which the reversible ChA inhibitor *m*-bromostyryl pyridinium had been attached (Husain and Mautner, 1973). More recently, a naphthylvinyl pyridinium column was used for the purification of rat brain ChA (Cozzari and Hartman, 1983a,b). In the former purification, the affinity-ligand was attached through a six-carbon chain, and in the latter through a ten-carbon chain. Because ChA binds strongly to columns carrying long-chain alkyl groups, it is not clear whether in these procedures true affinity chromatography or hydrophobic chromatography was taking place. It seems that, for effective affinity-chromatography of ChA, the ligand should be attached through a relatively short alkyl chain. It should also be noted that some "specific" ChA inhibitors will not inhibit all forms of this enzyme.

Since CoA is bound relatively tightly to ChA, Ryan and McClure (1979) and Driskell et al. (1978) successfully used CoA-Sepharose columns for the purification of ChA. The column was prepared by the reaction of cyanogen bromide-activated Sepharose 4B with CoA (Chibata et al., 1974). Although it was originally assumed that the cofactor is attached through the 6-amino group, our inability to link the inhibitor CoA methyldisulfide by this procedure suggested that sulfur is the point of attachment of CoA to these columns (Mautner, 1977; Mautner and Currier, unpublished data). Purification of ChA on CoA columns is facilitated by the presence of choline in the elution buffer (Ryan and McClure, 1979).

Since organic mercurials are inhibitors of ChA, with inhibition being reversible by thiols, the usefulness of organomercurial columns for the purification of ChA was explored. Such columns were introduced for the separation of thiol proteins from nonthiol proteins (Sluyterman and Wijdenes, 1970). Organomercurial columns were used for the purification of ChA from squid head ganglia (Husain and Mautner, 1973), cow brains (Chao and Wolfgram, 1973; Malthe-Sørenssen et al., 1978), and *Drosophila* heads (Driskell et al., 1978), with elution being achieved in variable yield with 2-mercaptoethanol or cysteine. Since the mercurial columns used were generally prepared by the procedure of Cuatrecasas (1970) involving "long-armed" attachment, the preparation and testing of "short-armed" mercurial columns would be worthwhile.

4.3. Hydrophobic Chromatography

As already noted, ChA binds strongly to columns carrying alkyl groups, with binding becoming tighter as the number of carbons in the chains is increased (Mautner and Currier, 1976; Currier and Mautner, 1977; Rossier, 1977b). For this reason, some of the reported procedures using affinity chromatography may have used "hydrophobic chromatography" (Shaltiel, 1974) as a purification step.

Once the hydrophobic nature of ChA was recognized (Mautner and Currier, 1976; Rossier, 1977b), hydrophobic chromatography began to be used for the purification of this enzyme. Thus, phenyl Sepharose was used for the purification of rat brain ChA (Dietz and Salvaterra, 1980), hexyl Sepharose for the purification of pig brain ChA (Eckenstein et al., 1981), and octyl Sepharose for the purification of *Drosophila* ChA (Slemmon et al., 1982). We have also found phenyl Sepharose columns useful for the purification of ChA from placentae and squid head ganglia (Mautner and Shachter, unpublished data).

4.4. Dye Ligand Chromatography

Glazer (1970a) noted that some aromatic dyes, with no obvious relationship to normally bound ligands, were capable of tight attachment to globular proteins, predominantly to areas overlapping cofactor or substrate binding sites. At about the same time, it was noted fortuitously that Cibacron blue F3GA, a textile dye, when attached to column matrices, could be used for the purification of numerous enzymes (for review, *see* Stellwagen, 1977; Dean and Watson, 1979; Subramanian, 1984). Roskoski et al. (1975) noted that blue dextran columns could be used for the purification of placental ChA that was eluted from the column with a linear 0–0.5M KCl gradient. This procedure was improved by Hersh et al. (1978b), who eluted the enzyme with a 0–0.20 μM linear gradient of acetyl-CoA. Since then, several groups have found columns carrying Cibacron blue F3GA useful for the purification of ChA (Dietz and Salvaterra, 1980; Eckenstein et al., 1981; Slemmon et al., 1982), using acetyl-CoA or CoA as the eluant.

After it was noted that the simple, fluorescent dye 1,8-anilino naphthalene sulfonate was a competitive antagonist of the binding of acetyl-CoA to ChA (Currier and Mautner, 1977), a systematic study of the specificity of the binding of aromatic dyes to this enzyme was carried out. It was found that Procion red and Congo red bound even more tightly to ChA than Cibacron blue F3GA (Mautner et al., 1981a), and bis-ANS and tetraiodofluorescein were also good inhibitors. ChA-dye interactions will be discussed in more detail elsewhere.

Green A-agarose columns were used for the purification of *Drosophila* ChA (Slemmon et al., 1982). Orange A-agarose binds carnitine acetyltransferase, but few other enzymes (Amicon, 1980). It can be used not only as a step during the purification of ChA, but also for the separation of ChA from carnitine acetyltransferase (Mautner and Shachter, unpublished data).

4.5. Immunoaffinity Chromatography

Such columns were used for the purification of ChA some time ago. Rossier et al. (1973) obtained antibodies to contaminating proteins using a relatively impure preparation of ChA; these were attached to chromatographic columns. ChA passed through this column, which retained only the contaminants. This procedure provided a good purification step for rat brain ChA and emphasizes the caution with which "specific" antibodies to ChA must be viewed, as noted by Rossier (1975, 1981).

Two major problems encountered in using ChA-antibody columns for the purification of this enzyme are: (1) Some of the antibodies to ChA reported in the earlier literature are probably antibodies to contaminating proteins, and (2) When authentic antibodies to ChA are obtained, the enzyme–antibody complex is so tight that harsh means may have to be used for elution to be achieved.

The first successful preparation of monoclonal antibodies (Mab) to ChA was reported by Levey et al. (1982) and Levey and Wainer (1982). This group tried to use Mab IgG Sepharose for the purification of this enzyme (Levey et al., 1981) and found that ChA could not be eluted from these columns at low pH or at high salt concentration. Active enzyme could be removed from Mab immunoabsorbants only by competition with other free Mabs to ChA. Although the Mabs produced initially were reactive primarily with bovine brain ChA, Levey et al. (1983) later produced Mabs showing reactivity with all mammalian ChA species tested. These Mabs proved to be reliable reagents for the immunohistochemical localization of cholinergic neurons (Wainer et al., 1984).

Monoclonal antibodies to ChA were also prepared by Eckenstein and Thoenen (1982), Crawford et al. (1982), and Ichikawa et al. (1983). Eckenstein and Thoenen adsorbed anti-pig brain ChA IgG to a protein A Sepharose column and found that the enzyme–antibody complex could be eluted at a pH of 6 because the association of the complex with protein A is relatively loose.

To facilitate elution from immunoaffinity columns, antigen binding fragments (Fab) were prepared by mercuripapain digestion of antiChA IgG. The fragments were coupled to cyanogen bromide-activated Sepharose, yielding Fab columns capable of binding ChA from human brain or placentae. Elution of the enzyme could be achieved with 50% recovery using Tris-buffer (pH 9) or phosphate citrate buffer (pH 5) (Peng et al., 1983).

This review will not cover the use of Mabs and polyclonal Abs of ChA for the immunohistochemical localization of cholinergic neurons (Rossier, 1981; Wainer et al., 1984; McGeer et al., 1984).

4.6. Other Purification Procedures

High-pressure liquid chromatography using a Waters I-125 exclusion column, was used as the final step of the purification of pig brain ChA (Eckenstein and Thoenen, 1982). Isoelectric focusing of the *Drosophila* enzyme yielded two peaks of activity (Slemmon et al., 1982). Chromatofocusing of bovine brain and human placental ChA yielded multiple isoelectric and molecular weight variants of these enzymes (Hersh et al., 1984). This latter paper also emphasizes the care that should be taken to minimize proteolysis during ChA purification.

5. Conformational Properties of Choline and Acetylcholine

Under physiological conditions, the formation of Ach by the acetylation of choline by acetyl-CoA versus the reverse reaction is favored only slightly. The equilibrium constant is only 12 under physiological conditions (Pieklik and Guynn, 1975). The reason for this is that Ach, almost to the same extent as acetyl-CoA, is a high-energy compound. The free energy of hydrolysis of Ach, under physiological conditions, is –6.99 kcal, as compared to –7.4 kcal for ATP and –7.6 kcal for acetyl-CoA. For this reason, the acetylation of choline is freely reversible under physiological conditions.

X-ray diffraction studies of crystals (Canepa et al., 1966) and NMR studies of solutions (Culvenor and Ham, 1966) show that the -OCCN- grouping of Ach is in the *gauche* conformation, whereas the thiolester and selenolester analogs of Ach are in the fully extended *trans* conformation (Cushley and Mautner, 1970). The folded *gauche* conformation is seen only when the nitrogen of 2-dimethylaminoethyl esters is protonated. In the absence of protonation there is free rotation around the central ethylene group (Makriyannis and Mautner, 1972) (*see* Fig. 1).

These results suggest that the proximity of the acyloxy oxygen and the trimethylammonium group of Ach involves a "hard acid–hard base" interaction, and that Ach, to some degree, can be thought of as an inner salt (Mautner et al., 1972; Mautner, 1980). This is the reason Ach is slightly soluble in nonpolar solvents. Presumably, the attachment of the trimethylammonium group of Ach to a negatively charged binding site of a biopolymer (such as ChA or the AchR) will result in a conformational change of this ligand because the charge of the trimethylammonium group would now be neutralized, minimizing interaction with the ester group.

It also seems likely that the coulombic interaction between the trimethylammonium group and the acyloxy oxygen of Ach plays a part in activating the carbonyl carbon of Ach to nucleophilic attack, and in making this a high–energy ester. The rotational barriers around the central ethylene group have been studied using NMR spectrometry. The lower limits are 3.5–4.0 kcal (Behr and Lehn, 1972; Makriyannis, 1974), whereas the upper limit is approximately 8 kcal (Makriyannis et al., 1979). It should be noted that one of the

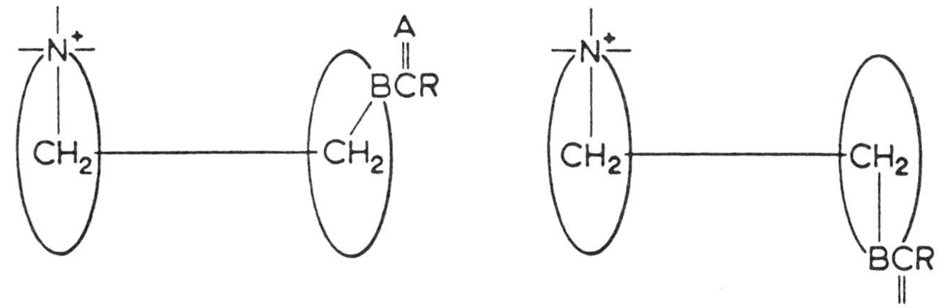

B = O
A = O, S, Se
R = $-CH_3$, $-C_6H_5$

B = S, Se
A = O, S, Se
R = CH_3, C_6H_5

Fig. 1. Conformation in the crystal or in solution of acetylcholine and a series of closely related analogs (from Mautner, 1980).

key factors in the operation of the "Ach cycle" resides in the enormous difference in the depolarizing abilities of choline and Ach. Choline is one of the few trimethylammonium compounds known that is an extremely poor depolarizing agent. This selective inability of choline to alter the cation permeability of synaptic membranes is not shared even by molecules as closely related as cholinethiol, methoxycholine, or homocholine (Webb and Mautner, 1966; Mautner, 1974, 1980).

The fact that acyl derivatives of CoA are high–energy compounds, with respect to analogous esters, is generally attributed to the lack of resonance stabilization of thiolesters as compared to esters (Jaenicke and Lynen, 1960). Since sulfur, in contrast to oxygen, can accommodate electrons in its *d* orbitals, charge-separated forms of the thiolester group, with a partial negative charge on the sulfur, are favored.

It should be remembered that thermodynamic information cannot be used to predict reaction rates. Although "high-energy thiolesters" react much more rapidly than esters with amines (Hawkins and Tarbell, 1953) and carbanions (Lienhard and Jencks, 1965), they react more slowly than analogous esters with hydroxide ions (Connors and Bender, 1961; Chu and Mautner, 1966) or alcohols.

The primary reason for the central metabolic importance of thiolesters derived from CoA is that such compounds are stable to hydrolysis, whereas, at the same time, they are very reactive with amines, carbanions, and other biologically relevant nucleophiles.

6. Kinetics of the Choline Acetyltransferase-Catalyzed Formation of Acetylcholine

Since the ChA-catalyzed formation of Ach involves two substrates (choline and acetyl-CoA) and two products (Ach and CoA), much effort was expended on studies of the ordering of the reaction. Potter et al. (1968) suggested that both substrates combine with the enzyme before any product is released. Later it was shown that Ach is a competitive antagonist of choline binding and CoA a competitive antagonist of acetyl-CoA binding. On the other hand, Ach is a noncompetitive antagonist of acetyl-CoA, and CoA is a noncompetitive antagonist of choline (Morris et al., 1971; Rama-Sastry and Henderson, 1972). Morris and Grewaal (1971) found that ChA catalyzed the isotopic exchange between [^{14}C]-choline and Ach only in the presence of CoA. These findings suggested a

Theorell-Chance bi–bi mechanism in which the steady-state concentration of the ternary complex is low.

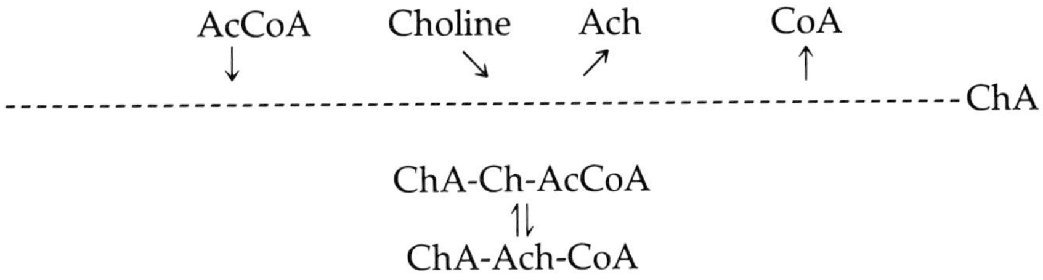

This mechanism was reexamined in Hersh's laboratory (Hersh and Peet, 1977; Hersh et al., 1978a; Hersh, 1982). They found that although the choline–Ach and acetyl CoA–CoA exchange patterns were consistent with a Theorell-Chance mechanism, it is a random Theorell-Chance mechanism. The predominant path, in which acetyl-CoA is the first substrate to bind and CoA is the last product to leave, is indicated with a heavy line (Hersh, 1982) (*see* Fig. 2).

ChA catalysis is complicated by salt effects that alter the binding of substrates and products. Addition of salt increases the K_m, both for the substrates and the products of this reaction, and increases the maximal velocity of the reaction (Hersh and Peet, 1978; Hersh, 1979; Hersh et al., 1978a). Evidence was presented that, at low salt concentration, the rate-limiting step is CoA release. On the other hand, at higher salt concentration, the rate-limiting step becomes the interconversion of the ternary complex. As will be discussed later, the effect of salt on the binding of CoA involves competition by salt for the attachment of the 3'-phosphate group of CoA with an active-site arginine residue (Mautner et al., 1981b).

Considerable controversy surrounded the ion specificity of the effect of salt on ChA catalysis (Mautner, 1977; Rossier, 1977a). Rossier made the tempting suggestion that ChA was regulated by chloride ion. It was proposed that, in a hypothetical presynaptic–enzyme complex in which ChA is bound to the presynaptic mem-

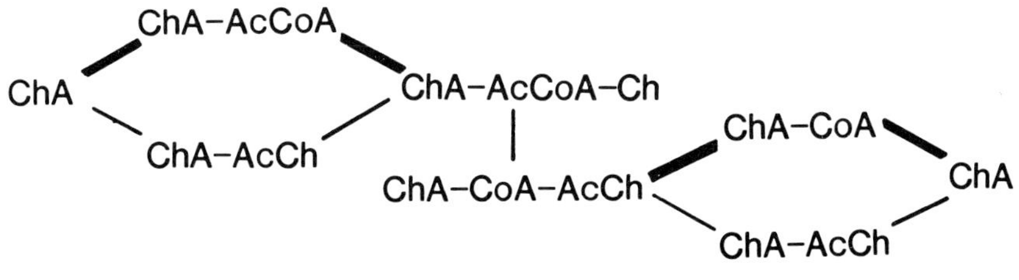

Fig. 2. Kinetic mechanism of choline acetyltransferase reaction (as redrawn from Hersh, 1982).

brane, the increase in chloride induced by depolarization activates ChA activity (Rossier et al., 1977; Rossier, 1977a). It was claimed that chloride was more effective than other anions in activating the enzyme; however, this effect could be seen only with "fresh and unpurified enzyme."

No ion specificity was seen when Hersh and Peet (1978) examined the effects of salt on ChA activity. Nor could any ion specificity be detected in studies of the protective effect of salts on the inactivation of ChA by arginine reagents (Mautner et al., 1981b).

In view of the possible involvement of ChA in the development of senile dementia of the Alzheimer type, a recent study compared the kinetic parameters of ChA derived from the frontal cortex and caudate nucleus of patients with this disease with ChA from control samples. No significant changes were found in the K_m of choline or acetyl-CoA or in the maximal velocity. There was also no evidence for the presence of inhibitors of ChA in Alzheimer's disease brain samples (Hurd and McClure, personal communication).

7. Active Sites Involved in Choline Acetyltransferase Function

7.1. Choline Binding Site

Berman et al. (1953) found that, in the presence of ChA, choline is acetylated much more rapidly than its dimethylamino and monomethylamino analogs and proposed the presence of an "anionic binding site" in this enzyme. The maximal ChA-catalyzed acetylation rate is as high for the dimethylsulfonium analog of choline as it is for choline itself (Frankenberg et al., 1973), even though the binding of the dimethylsulfonium compound is looser than that of choline. Thus, the presence of the third methyl group in choline, although favoring tight binding to the enzyme, does not affect the rate of acetylation. Deuteration of the methyl groups of choline does not change acetylation rate (Garrison-Gund and Nef, 1976).

That the precision of the alignment of the onium group of the substrate on the anionic binding site of ChA is relatively unimportant is emphasized by the relatively low stereospecificity of α-methylcholine (Hemsworth and Smith, 1970). The absolute configurations of the α- and β-methyl derivatives of choline are known (Becket et al., 1963).

On the other hand, alignment between the hydroxy group of choline and the thiolester portion of acetyl-CoA is extremely important, as emphasized by the great stereospecificity seen when the β-carbon of choline is methylated. L-β-methylcholine, in contrast to the D-α-methyl enantiomer, is not acetylated by ChA (Hemsworth and Smith, 1970). Similarly, the distance between the onium group and the hydroxy group of choline is critical. Homocholine (in which three rather than two carbon atoms separate these groups) shows no ChA-catalyzed acetylation (Currier and Mautner, 1974).

$$(CH_3)_3\overset{+\,\alpha}{N}CH_2\overset{\beta}{C}H_2OH \qquad (CH_3)_2\overset{+}{S}CH_2CH_2OH$$

Choline — Dimethylsulfonium analog of choline

$$(CH_3)_3\overset{+}{N}CH_2CH_2CH_2OH$$

Homocholine

7.2. Acetyl Coenzyme A Binding Site

CoA and acetyl-CoA are bound much more tightly to ChA than either choline or Ach. The effects of salt concentration on substrate and product binding have already been discussed. Studies of the specificity of the interactions of CoA with ChA have been carried out. As would be expected, dethiation of CoA did not affect binding to ChA. The similarity in the binding interactions with ChA of CoA and its 1,N^6-etheno derivative shows that the 6-amino group on the purine ring is relatively unimportant. Very important for efficient binding of CoA to ChA is the contribution of the 3'-phosphate on the ribose ring. CoA binds almost 100 times more tightly than its dephospho analog (Currier and Mautner, 1977; Mautner, 1977; Rossier, 1977b).

Although salt affects the affinity of CoA for ChA, it does not affect the binding of 3'-dephospho CoA (Hersh, 1979). CoA is a better protecting agent than dephospho CoA against enzyme inactivation by arginine reagents (Mautner et al., 1981b). It is clear that the 3'-phosphate of CoA is of great importance in the binding of this cofactor to ChA and probably in the regulation of this enzyme.

7.3. Thiol Groups

The inhibition of ChA activity by thiol reagents was noted a long time ago (Berman-Reisberg, 1957). Mannervik and Sörbo reported (1970) that choline and Ach did not protect ChA against inhibition by thiol reagents, but noted some protection by acetyl-CoA. On the other hand, Roskoski (1974a) claimed protection by acetyl-CoA, Ach and acetylthiolcholine. He also claimed (1973) isolation of an acetythio derivative of ChA, the deacylation of which was the rate-limiting step in the ChA reaction.

Unfortunately, these claims could not be substantiated when the interaction of thiol reagents with ChA of substantial purity was investigated (Currier and Mautner, 1974; Malthe-Sørenssen, 1976a). Roskoski et al. (1975) later raised the possibility that "the acetyl thioprotein intermediate is adventitious."

We investigated the essentiality of the thiol groups of ChA by the use of alkyl alkanethiolsulfonates. These reagents are easy to synthesize, label, and remove. They can be used to attach alkylthio groups of varying size to thiol groups of biopolymers to form mixed disulfides. If thiol groups are found within active sites, methyldisulfide formation should provide efficient inhibition. If thiol groups are outside of active sites, inhibition should not be seen until the size of the alkyl groups is increased (Smith et al., 1975),

$$\text{P—SH} + \text{RSSO}_2\text{R}^1 \rightarrow \text{P—SSR} + \text{R}^1\text{SO}_2^- + \text{H}^+$$

Methyl methanethiolsulfonate proved to be an excellent inhibitor of ChA from squid ganglia (Currier and Mautner, 1976). However, inhibition showed a lag period, raising the possibility that this reagent might not be reacting with a thiol group of the enzyme, but rather with the thiol group of CoA, one of the reaction products. That this was the case could be proved by the synthesis of a series of alkyl disulfides of CoA that proved to be potent ChA-inhibitors, showing no lag period (Currier and Mautner, 1976, 1977). This work was later extended by Hersh et al. (1979), who investigated the interactions of ChA with methoxycarbonyl alkyl disulfides, methoxycarbonyl CoA disulfide, and DTNB.

CoASH	CoASSR	$\text{CoASSC}(=\text{O})\text{R}$
Coenzyme A	CoA alkyl (aryl) disulfide	Alkoxycarbonyl CoA disulfide

They concluded that ChA contained two classes of reactive thiol groups, neither of which is attached to the active site.

7.4. Imidazole Groups

It was noted that the ChA-catalyzed acetylation of choline and aminocholine showed a considerable D_2O effect, greatest at the pH (pD) at which enzyme activity was maximal (Currier and Mautner, 1974). Although interpretations of isotope effects must be treated with caution (Jencks, 1969), D_2O isotope effects normally indicate that general-base catalysis, with rate-limiting proton-transfer, is taking place (Bender, 1971). Since the pH (pD) rate maxima occurred near neutrality, it seemed likely that here, as in many hydrolytic enzymes (Blow, 1976), the imidazole group of a histidine residue was involved in proton transfer. In such enzymes, imidazole, through general-base removal of the proton of the hydroxy group of serine, enhances the nucleophilicity of this residue. What is unusual in the case of ChA is that the imidazole interacts not with the hydroxy group of serine but with the hydroxy group of the substrate choline, attached to its binding site.

That imidazole is indeed essential to the catalytic functions of ChA, could be shown by the observation that rose bengal and, to a lesser extent, methylene blue inactivated ChA in the presence, but not in the absence, of light (Currier and Mautner, 1974). It is known that such dyes are relatively specific reagents for the inactivation of histidine residues (Westhead, 1965; Glazer, 1970b), presumably through the generation of singlet oxygen (Schaap et al., 1975). The essentiality of imidazole groups to ChA function was confirmed by Roskoski (1974b), who used ethoxyformic anhydride as a histidine-directed reagent, and by Malthe-Sørenssen (1976a). We suggested the interaction between enzyme-bound choline and acetyl-CoA and an active site histidine (Currier and Mautner, 1974).

Alternatively, Malthe-Sørenssen (1976a) suggested rate-limiting nucleophilic attack of histidine on enzyme-bound acetyl-CoA, with formation of an acetyl-imidazolyl intermediate. However, the D_2O isotope effect observed makes a mechanism in which the rate-determining step would be the nucleophilic attack of histidine unlikely. In nucleophilic attack of this type, no D_2O isotope effect would be expected (Bender et al., 1962).

7.5. Arginine Residues

An X-ray diffraction study of liver alcohol dehydrogenase, which had been cocrystallized with Cibacron blue F3GA, showed a sulfonate group attached to ring A interacting with an arginine residue of the enzyme (Biellman et al., 1979). Since we had found this dye to be an excellent inhibitor of ChA, capable of competing with CoA binding (Currier and Mautner, 1977; Mautner et al., 1981a), it seemed of interest to study the inhibition of ChA by arginine-specific reagents.

A series of dicarbonyl compounds have been reported to be "specific" arginine-inactivating reagents (Takahashi, 1968). We found 2,3-butanedione, phenylglyoxal, and camphorquinone-10-sulfonic acid to be irreversible inhibitors of ChA (Mautner, 1980; Mautner et al., 1981b). The enantiomers of the camphorquinone derivative were synthesized, but found not to be stereospecific.

Protection against inactivation by these reagents was provided by CoA; however, NAD^+, NADH, $NADP^+$, NADPH, and choline provided no protection, even at high concentration. 3'-Dephospho CoA, lacking the phosphate essential for the tight binding of CoA to ChA, provided less protection against arginine reagents than CoA. These results are compatible with the presence of an arginine residue within the CoA binding site of ChA capable of interacting with the 3'-phosphate of the cofactor. The partial protection by dephospho CoA can presumably be attributed to an interaction, although a relatively inefficient one, between the arginine residue of the enzyme and the pyrophosphate residue of the cofactor.

Partial protection against ChA inactivation by arginine reagents was provided by salt. It seems likely that the 3'-phosphate of CoA and salt anions can compete for the same active-site arginine. As already noted, only at low salt concentration is the departure of CoA from its complex with ChA rate-limiting (Hersh, et al., 1978a), whereas the binding of dephospho CoA is only minimally salt-sensitive. We found sodium chloride, bromide, and iodide, as well as cesium chloride, to be equiactive in protecting ChA against arginine inactivation (Mautner et al., 1981b). This agrees with the lack of ion specificity seen in studies of the effect of salt on the binding of the substrates and products of the ChA reaction (Hersh and Peet, 1978), and disagrees with the claim of Rossier et al. (1977a) that chloride is a specific activator of ChA.

7.6. Hydrophobic Binding Sites

The first suggestion that ChA contains lipophilic binding sites was provided by the tendency of this enzyme to be retained by alkyl- and aryl-substituted columns (Mautner and Currier, 1976; Mautner, 1977; Rossier, 1977b). Lengthening of alkyl chains leads to tighter binding until elution becomes almost impossible. "Hydrophobic columns," as already discussed, provide a useful purification step.

At this time, little is known about the details of the lipophilic interactions of ChA. The elution patterns of ChA bound to columns carrying aromatic dyes found to be competitive antagonists of CoA, as well as the elution of ChA bound to CoA columns, suggest the presence of a lipophilic component in the binding of

CoA to ChA. However, lipophilic interactions outside the CoA binding site also seem likely. Elucidation of this question may throw light on the possibility that ChA, generally thought of as a soluble cytoplasmic enzyme, may interact with intracellular membranes during its physiological functions. Normally, highly lipophilic enzymes tend to be membrane-associated. If this were so in the case of ChA, the well-known fact that the properties of membrane-bound enzymes often differ from their properties in solution, "allotopy" (Racker, 1968), should lead to caution in interpreting studies of ChA in solution.

8. Physical Properties of Choline Acetyltransferase

In contrast to the other biopolymers of the Ach cycle, acetylcholinesterase and the AchR, ChA, is not retained by any lectin column investigated (Mautner, unpublished data). This suggests that ChA is not a glycoprotein, and indicates that if ChA were membrane-bound, this interaction would involve intracellular membranes.

The molecular weight and subunit composition of ChA have been studied for many years. Chao and Wolfgram (1973) reported a molecular mass of 120 kdaltons and the presence of subunits of 51 and 69 kdaltons. Later they reported that ammonium sulfate precipitation yielded aggregates with a molecular mass of 1600 kdaltons (Chao and Wolfgram, 1974), followed by a claim by Chao (1975) that this enzyme had a molecular weight of 84 kdaltons and was made up of six identical subunits.

In recent years, most reported molecular masses for ChA, isolated from different sources, have been in the range of 63 kdaltons for ChA from *Torpedo* electric organ (Brandon and Wu, 1978) to 69 kdaltons for ChA from bovine caudate nucleus (Malthe-Sørenssen et al., 1978; Ryan and McClure, 1979) or *Drosophila* heads (Driskell et al., 1978). The close agreement in the molecular mass of highly purified ChA from mammalian sources with that from *Drosophila* heads should be noted. For the latter enzymes, reported to have an activity of 500 μM Ach/min/mg protein (Slemmon et al., 1982), the close agreement between the molecular mass obtained by molecular exclusion chromatography and velocity sedimentation suggests that ChA is a globular protein with little axial symmetry (Slemmon et al., 1984).

The *Drosophila* enzyme contains two major peptides of 67 and 54 kdaltons; however, only the former peptide had catalytic activity. Polyacrylamide-gel electrophoresis showed the presence of a

13-kdalton peptide, probably generated by limited proteolysis. The 54- and 13-kdalton peptides were not linked by disulfide bonds and the peptide pattern was not changed when enzyme purification was carried out in the presence of an array of protease inhibitors (Slemmon et al., 1984).

Relatively high molecular weights for ChA from bovine caudate nuclei were reported by Cozzari and Hartman (1983a), who found 72- and 76-kdalton forms of the enzyme, both with catalytic activity. Similarly, Hersh et al. (1984) reported a molecular weight of 73 kdaltons for bovine brain enzyme if purification was carried out in the presence of a series of proteolytic inhibitors (EDTA, EGTA, phenylmethylsulfonyl fluoride, L-1-tosylamido-2-phenylethylchloromethyl ketone, *N-p*-tosyl-L-lysyl-chloromethyl ketone, and leupeptin). A molecular weight of 71 kdaltons was reported for nematode *(C. elegans)* ChA (Rand and Russell, 1985). Clearly, the importance of proteolysis and the difficulty of inhibiting it has been underestimated by workers in this field.

Another difficulty is the tendency of ChA to interact with other proteins. The many reports that ChA is fractionated into multiple peaks on isoelectric focusing (Malthe-Sørenssen and Fonnum, 1972; Malthe-Sørenssen, 1976b), ion-exchange chromatography (Malthe-Sørenssen et al., 1978; Ryan and McClure, 1979), gel filtration (Chao and Wolfgram, 1974; Banns, 1976), or dye-ligand chromatography (Dietz and Salvaterra, 1980) led to the belief that ChA exists in the form of multiple isozymes. It was suggested recently that some of these multiple forms are caused by the interaction of ChA with other proteins and that such interactions can be minimized by the presence of dilute ethanol during chromatography (Cozzari and Hartman, 1983a). Whatever the cause—proteolysis or protein interactions—of the heterogeneity of ChA, it does not seem to be due to posttranslational modification of the enzyme or distinct genes for the different enzyme forms (Hersh et al., 1984). Reexamination of the many reported "isozymes" of ChA should be of interest.

Another area of ambiguity is presented by the membrane and protein interactions of ChA. That they can take place and are favored by low pH and low ionic strength is well known (Fonnum, 1968). The ability of ChA to interact with tubulin has been noted (Fonnum and Malthe-Sørenssen, 1973), and it was shown recently that colchicine, which interferes with tubulin assembly, increases ChA activity in spinal cord cell cultures (Ishida and Deguchi, 1984).

In recent years, it has been reported that some ChA may be associated with neuronal or vesicular membranes and that this "membrane-bound" ChA differs in its specificity for substrates and inhibitors from that of soluble ChA. It also differs in susceptibility to denaturation and in pH profile (Nelson et al., 1980; Benishin and Carroll, 1981, 1983; Luqmani and Richardson, 1982). It has been claimed that, during neonatal development, membrane-bound and cytoplasmic ChA appear at different times (Benishin and Carroll, 1984).

It is not yet clear whether membrane interactions are an important component of the physiological functions of ChA. It would not be surprising if such interactions played a part in the axonal transport of ChA or in the regulation of ChA synthesis. The report that, in the presence of colchicine, ChA synthesis is increased, fits this hypothesis.

9. Inhibitors of Choline Acetyltransferase

9.1. Derivatives of Choline

Because the binding of choline (or Ach) is much looser than the binding of acetyl-CoA (or CoA), it is not surprising that few analogs of choline proved to be potent inhibitors of this enzyme. Aminocholine is a substrate. Methoxycholine and methylthiocholine do not inhibit, even at 50 m*M* concentration. Choline disulfide and diselenide are weak inhibitors (Currier and Mautner, 1974). The stibonium analog, in which nitrogen is replaced with antimony, was prepared and found to be a substrate for ChA (Meyer et al., 1981).

Several attempts were made to prepare analogs of choline capable of alkylating ChA. 3-Bromoacetonyltrimethylammonium bromide could be attached irreversibly to ChA; however, even $5 \times 10^{-3} M$ solutions caused only 95% inhibition of the placental enzyme. The compound did not inhibit acetylcholinesterase or pseudocholinesterase (Persson et al., 1967).

The fluoro-, chloro-, bromo-, and iodoacetyl derivatives of choline were synthesized and tested as ChA inhibitors (Chiou and Rama-Sastry, 1968; Morris and Grewaal, 1969). These compounds proved to be rather weak, reversible inhibitors of ChA. They also show little selectivity, being able to inactivate acetylcholinesterase and the AchR.

Acryloylcholine inactivates brain ChA irreversibly. Preincubation with choline, acetyl-CoA, or sodium chloride does not protect the enzyme. Methylation of the α- or β-carbons and saturation of the double bond greatly reduces inhibitory potency (Malthe-Sørenssen et al., 1974). The authors postulated that resonance should render the β-carbon of acryloylcholine an effective electrophile, enabling it to react with mercapto, hydroxy, and amino groups.

```
                    O –
                   /
+ CH2 – CH = C
                   \
                    OR
```

The dimethylamino analog of acryloylcholine is a much poorer inhibitor of ChA than the onium analog and rapidly reacts in aqueous solution to form the β-hydroxy derivative (Rowell and Chiou, 1976).

A group of keto analogs of acetyl- and benzoylcholine were synthesized and evaluated as ChA inhibitors (Chaturvedi et al., 1978). The most potent of these, 2-benzoylethyltrimethylammonium chloride, caused 50% inhibition of ChA at $3 \times 10^{-6}M$ concentration. Later this and related derivatives were used to inhibit sperm motility, an effect attributed to inhibition of sperm ChA (Rama-Sastry et al., 1981). Because, as already noted, sperm contains a great deal of carnitine acetyltransferase and very little ChA, one must assume that these derivatives show little selectivity in inhibiting acyl transferases.

9.2. Derivatives of Coenzyme A

During an early study of the specificity of ChA, the Nachmansohn group found that acetyl- and propionyl-CoA were equiactive substrates of ChA and that benzoylcholine was an inhibitor (Berman et al., 1953). Recently, another acyl derivative, phenylacetyl-CoA, was found to be a potent inhibitor with a K_i of $3 \times 10^{-7}M$ (Potempska et al., 1984).

Alkyldisulfides of CoA are effective, reversible inhibitors of acetyl-CoA attachment to ChA, as is the fluorescent $1,N^6$-etheno derivative (Mautner and Currier, 1976, Currier and Mautner, 1977). The similarity in the inhibitory activities of the methyl, ethyl, and propyl disulfides suggests that the region with which the alkyl chains interact must be quite small and vicinal to the sulfur atom. CoA alkyldisulfides also inhibit carnitine acetyltransferase (Currier and Mautner, 1977) and spermidine/spermine N^1-acetyl transferase (Della Raggione and Pegg, 1983).

Methoxycarbonyl-CoA disulfide was introduced as an active site-directed reagent by Hersh et al. (1979). In contrast to the CoA

Table 2
Inhibitory Activity of Derivatives of Coenzyme A

Derivative	Inhibitory potency, K_i^a
CoA	$75 \times 10^{-6}M$
Dethia-CoA	$78 \times 10^{-6}M$
$1,N^6$-Etheno-CoA	$400 \times 10^{-6}M$
3'-Dephospho-CoA	$7000 \times 10^{-6}M$
CoA methyldisulfide	$2 \times 10^{-6}M$
CoA ethyldisulfide	$2 \times 10^{-6}M$
CoA propyldisulfide	$5 \times 10^{-6}M$
$1,N^6$-Etheno-CoA methyldisulfide	$4 \times 10^{-6}M$
Methoxycarbonyl-CoA disulfide	$20 \times 10^{-6}M$ (more than 98% inhibition)

[a]Except for methoxycarbonyl-CoA disulfide, the K_i values for the various derivatives were determined using squid enzyme (Currier and Mautner, 1977). The inhibitory potency of methoxycarbonyl-CoA disulfide was determined using human placental ChA (Hersh et al., 1979). All the compounds, except for the last, were competitive inhibitors with respect to acetyl-CoA.

alkyldisulfides, which are attached tightly to the CoA binding site, but don't react with it, methoxycarbonyl-CoA disulfide attaches CoA to the enzyme. Because inhibition of ChA by this reagent is not complete (95% with $2 \times 10^{-5}M$ reagent), it is assumed that CoA is attached by a disulfide link near—but not at—the CoA binding site (Hersh et al., 1979) (*see* Table 2).

9.3. Aromatic Dyes

During a study of the inhibitory action of CoA derivatives, we found that a simple, fluorescent aromatic dye, 1-anilinonaphthalene-8-sulfonic acid (1,8-ANS), inhibited ChA. Inhibition was competitive with respect to CoA and noncompetitive with respect to choline (Currier and Mautner, 1977). This observation, coupled

Table 3
The Inhibition of Choline Acetyltransferase From Squid Ganglia by Aromatic Dyes

Dye	Inhibition constant, K_i
1-Anilinonaphthalene-8-sulfonic acid	$200 \times 10^{-6} M$
2-Anilinonaphthalene-8-sulfonic acid	$150 \times 10^{-6} M$
1-Anilinonaphthalene-2-sulfonic acid	NI[a]
1-Anilinonaphthalene-5-sulfonic acid	$1500 \times 10^{-6} M$
Bis-(1,8-ANS)	$3 \times 10^{-6} M$
Cibacron blue (*para*-isomer)	$0.8 \times 10^{-6} M$
Cibacron blue (*meta*-isomer)	$0.8 \times 10^{-6} M$
Cibacron blue (mixture, *meta*- and *para*-isomers)	$1.2 \times 10^{-6} M$
Hydroxy Cibacron blue	$1.8 \times 10^{-6} M$
Congo red	$0.1 \times 10^{-6} M$
Procion red	$0.3 \times 10^{-6} M$
Tetraiodofluorescein	$8 \times 10^{-6} M$

[a]NI = No inhibition at limit of solubility ($4.4\ 10^{-3} M$). All the dyes were competitive inhibitors with respect to acetyl-CoA, except for Procion red, 1,5-ANS and 2,6-ANS, which were noncompetitive inhibitors (Mautner et al., 1981a).

with the fact that ChA was retained on agarose-Cibacron blue columns, led to a systematic study of the inhibition of ChA and related enzymes by aromatic dyes.

Table 3 summarizes the inhibitory potencies of several dyes (Fig. 3). The activity of 1,8-ANS was compared with that of several positional isomers. The inhibitory action of the dimer of 1,8-ANS, as in other systems (York et al., 1978; Bohnert et al., 1982) greatly exceeded that of the monomer. The most potent inhibitory dyes were Cibacron blue F3GA, Procion red, and Congo red. All the dyes showed competitive inhibition with respect to acetyl-CoA, except Procion red, 1,5-ANS, and 2,6-ANS, which were noncompetitive inhibitors.

The possibility that Cibacron blue F3GA, with its chlorotriazino group, might be reacting with ChA was eliminated by showing that replacing the chlorine with a hydroxy group (Moe and Piszkiewicz, 1979) did not lower inhibitory activity (Mautner et al., 1981a).

It was claimed that Cibacron blue and related dyes bind to the "dinucleotide fold" of enzymes, i.e., the region to which the aden-

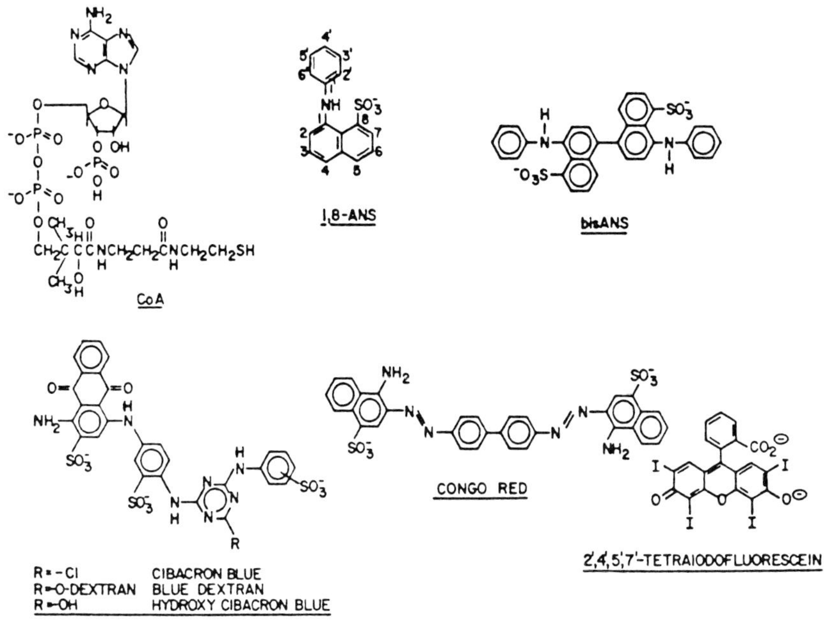

Fig. 3. The structures of coenzyme A and of aromatic dyes capable of inhibiting choline acetyltransferase (from Mautner et al., 1981).

osyl diphosphoryl portion of cofactors is attached, and that such binding is diagnostic for the presence of the dinucleotide fold (Thompson and Stellwagen, 1976; Stellwagen, 1977). Numerous exceptions to this postulate have been found (Turner, 1981; Subramanian, 1984). As first noted by Glazer, however, (1970a), it is true that aromatic dyes with no obvious similarity to normally attached ligands can bind tightly to globular proteins, predominantly to areas overlapping with cofactor and substrate binding sites. Hydrophobic or electrostatic interactions or both can be involved in the binding of dyes to enzymes (Subramanian, 1984). Similarly, the elution behavior of ChA bound to dye columns suggests that both electrostatic and lipophilic forces are involved in the tight binding of aromatic dyes to this enzyme (Mautner et al., 1981a).

The affinity of aromatic dyes for ChA is useful because such dyes can serve as probes of their binding sites, as well as probes of conformational alterations brought about by the binding of choline or of salts. Several of the dyes are highly fluorescent, whereas others, such as Congo red or Cibacron blue, show induced circular dichroism spectra when introduced into the asymmetric binding environment of enzyme binding sites (Edwards and Woody, 1977, 1979). The topography of dye-binding sites can also be studied with difference spectroscopy (Subramanian, 1982a,b). Organic dyes also bind tightly to carnitine acetyltransferase (Mautner et al., unpublished data) and to spermine/spermidine acetyltransferase (Della Regione et al., 1983; Della Ragione and Pegg, 1983). A systematic study of these dye interactions should be useful for clarifying the relationships of the binding sites of these enzymes.

9.4. Arylvinyl Pyridines and Related Compounds

Arylvinyl pyridines were found to be effective inhibitors of ChA (Smith et al., 1967). The literature dealing with these compounds is voluminous and has been reviewed (Mautner, 1977; Ryan and McClure, 1981). Hundreds of analogs have been synthesized, the most potent of which have the general structure:

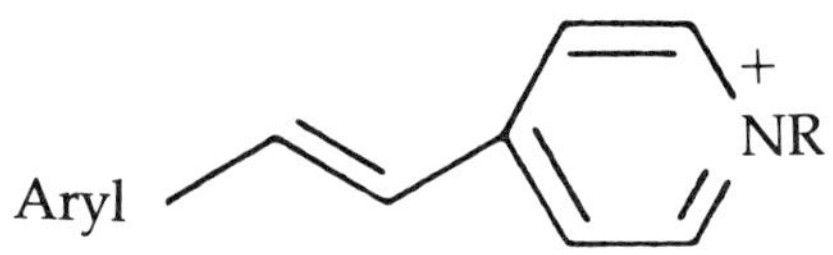

R = alkyl

Quaternization of the ring nitrogen tends to increase inhibitory potency, but lowers the specificity for inhibiting ChA as related to acetylcholinesterase. In general, the attachment of nonpolar electron-releasing groups on the aryl substituent favors inhibition of ChA and lessens inhibition of acetylcholinesterase (Baker and Gibson, 1971). Two theories have been proposed to explain the inhibitory action of these compounds: (1) The coplanar molecules form a charge–transfer complex with the enzyme. For high activity, a π electron-deficient pyridinium group is important (Cavallito et al., 1970, 1971), and (2) The bridge vinyl group is the point of binding to ChA, either through π bond interactions or through a nucleophilic attack of an enzyme group on the charge-separated double bond (Baker and Gibson, 1972).

The action of these derivatives is complicated by their ability to undergo *trans-cis* photoisomerization, with the *trans* compounds being much better inhibitors. The double bond can undergo photohydration (Cavallito et al., 1970) similar to the behavior of acryloylcholine.

The crystal structures of *cis*- and *trans*-analogs of 4-(β-1-naphthylvinyl) pyridine were determined recently (Chweh et al., 1984). The *trans*-compound is almost coplanar; the *cis*-compound is not. The authors concluded that the greater inhibitory potency of the *trans*-, as compared to the *cis*-derivative, is a result of the ability of both aromatic groups of the former to interact simultaneously with the ChA binding site. Ethyl mercaptan, but not cysteine or dithiothreitol, could react with the vinyl group of the inhibitor.

White and Cavallito (1970b) proposed that arylvinyl pyridines inhibit ChA by interfering with the catalytic action of an active-site histidine. This postulate seems plausible. As already noted, ChA carries an essential histidine residue and arylvinyl pyridines block the transfer of the acyl group from acetyl-CoA to choline without interfering with the attachment of either substrate. In view of the precise alignment needed for the active-site histidine to activate ChA-bound choline for reacting with ChA-bound acetyl-CoA, distortion of this arrangement by binding a bulky inhibitor in the vicinity of the substrates would be expected to block enzyme activity.

Azo-analogs of arylvinyl pyridines and *p*-phenylazo trimethylammonium chloride also have some ability to inhibit ChA (Adams and Mautner, unpublished data). Since *cis*- and *trans*-azo compounds can be interconverted readily by irradiation at defined wavelengths, such compounds can be used as photochromic reagents, i.e., compounds capable of undergoing defined, reversible conformational changes when bound to a biopolymer (Erlanger, 1976). Photochromic reagents related to arylvinyl pyridines should be useful tools for studying the mechanism of action of ChA.

10. Factors Controlling Acetylcholine Formation

An important factor in the control of ChA activity is the fact that choline and Ach are bound much more loosely than CoA or acetyl-CoA. Because there is evidence that acetyl-CoA is the first substrate to interact with ChA, and CoA is the last product to depart, and because these molecules are bound relatively very tightly, it seems likely that acetyl-CoA and CoA, rather than Ach or choline, are regulating Ach synthesis.

Lefresne et al. (1973) provided evidence that in brain tissue, in vitro, Ach synthesis depended on acetyl-CoA availability. Řičny and Tuček (1980) showed that the release of Ach from brain slices was proportional to acetyl-CoA content. Excellent, thoughtful reviews (Tuček, 1984, 1985) appeared recently, in which the origins of the acetyl-CoA used for Ach synthesis are discussed in detail. This substance is synthesized from pyruvate in intraterminal mitochondria. Some of the acetyl groups used for Ach synthesis are transferred to the cytosol, in the form of citrate, from which acetyl-CoA is formed by ATP citrate lyase. Other acetyl groups are transferred in the form of acetylcarnitine.

Choline used for Ach synthesis can be derived from several sources, including free choline in the plasma, the hydrolysis of Ach, or from phosphatidylcholine. In view of the cholinergic deficiency seen in Alzheimer's dementia, attempts have been made to elevate cortical Ach by feeding massive amounts of choline (Hirsch and Wurtman, 1978). Whether this approach actually elevates cortical Ach remains controversial (Wecker and Dettbarn, 1979; Brunello et al., 1982). The situation is complicated by the ability of brain neurons to synthesize their own choline by the methylation of phosphatidylethanolamine, followed by the hydrolysis of phosphatidyl choline (Blusztajn and Wurtman, 1983).

It should be noted that inhibition of acetylcholinesterase, leading to increases in Ach levels, does not seem to shut down the ChA reaction. This is emphasized by the observation that blocking ChA leads to a delay in the death of animals whose acetylcholinesterase was inhibited by large doses of organophosphates (Hopff et al., 1984).

As already noted, it has been proposed that choline transport and choline acetylation are coupled. High-affinity, sodium-dependent choline transport was demonstrated by several laboratories (Yamamura and Snyder, 1972, 1973; Dowdall and Simon, 1973; Haga and Noda, 1973; Kuhar et al., 1973). Research to determine whether choline transport is the rate-limiting step in Ach synthesis and whether choline transport and Ach synthesis are functionally, or even structurally, linked, has led to a voluminous body of literature reviewed lucidly by Tuček (1984, 1985). Most available data suggest that choline transport and Ach synthesis are not coupled (O'Regan and Collier, 1981; Marchbanks and Kessler, 1982), and that neither choline nor Ach levels regulate Ach synthesis in the brain. On the other hand, there is fairly good evidence that, at least in vitro, Ach synthesis is regulated by acetyl-CoA availability.

The possibility that membrane-bound forms of ChA might exist and behave differently from soluble ChA cannot be ignored, nor can the possible existence of regulatory proteins (Ryan and McClure, 1981) or inhibitory proteins (Cozzari and Hartman, 1983c). The claim that colchicine increases the levels of ChA in tissue culture (Ishida and Deguchi, 1984) raises the possibility that ChA anabolism is regulated by interaction with microtubules.

On the basis of information summarized in section 7 of this review, the Fig. 4 model for the substrate interactions of ChA can be proposed (Mautner, 1985).

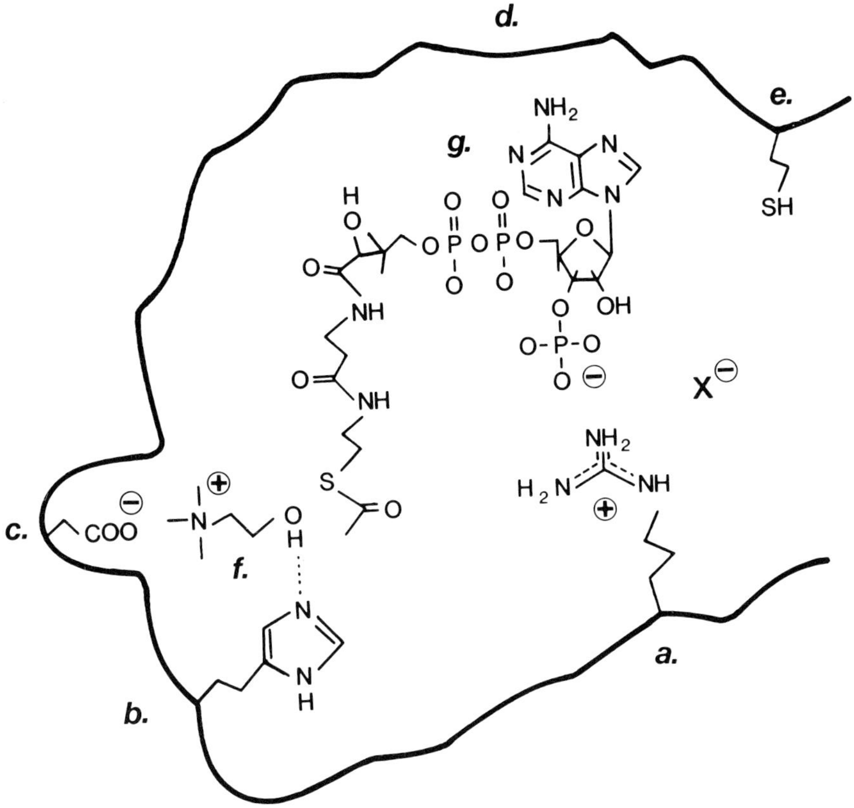

Fig. 4. Active site interactions of choline acetyltransferase. a, active-site arginine; b, active-site histidine; c, anionic binding site; d, hydrophobic binding site; e, cysteine; f, choline; g, acetyl-CoA.

The leading substrate, acetyl-CoA, is held to its binding site by a coulombic interaction between its 3'-phosphate and the guanidinium group of an active-site arginine, as well as by lipophilic interactions. The second substrate, choline, is held relatively loosely by a coulombic interaction between its trimethylammonium group and the enzyme's anionic binding site that presumably carries a carboxyl group. This attachment is likely to result in a conformational change of choline from the constrained *gauche* to the fully extended *trans* configuration. Conformational changes within the enzyme, following substrate attachment, are likely as well. General-base catalysis by the imidazole group of an active-site histidine permits the hydroxy group of ChA-bound choline to attack the thiolester group of acetyl-CoA, leading to transacetylation. The resulting ester, Ach, being held relatively loosely, is released first, followed by the departure of CoA.

An increase in salt concentration can have a regulatory effect by shifting the rate-limiting step from the departure of CoA to the interconversion of the ternary complexes ChA:AcCoA:Ch and ChA:CoA:AcCh. Salt anions can compete with the 3'-phosphate of CoA for attachment to the active-site arginine. Whether this effect is of physiological importance remains to be proved.

The role of cysteine near—but not within—the active site of ChA remains to be defined. The considerable lipophilicity of ChA makes interactions with intracytoplasmic membranes likely. Whether such interactions play a physiological role remains as poorly defined as the question of the possible interactions of ChA with regulatory proteins (Ryan and McClure, 1981; Cozzari and Hartman, 1983c).

More than 40 years have passed since the discovery of ChA. It is astonishing how many questions regarding the mechanism and control of Ach synthesis remain to be answered.

Acknowledgment

Work from our laboratory was supported by a grant from the National Science Foundation (BNS-81-04175).

References

Alpert A., Kisliuk R. L., and Shuster L. (1966) A simple radioisotopic assay for choline acetyltransferase and its application in the *Lactobacillus plantarum* system. *Biochem. Pharmacol.* **15,** 465–473.

Amicon Corp. (1980) *Dye-Ligand Chromatography. Applications, Method, Theory of Matrex Gel Media.* Amicon Corp. p. C-11, Massachusetts.

Baker B. R. and Gibson R. E. (1971) Irreversible enzyme inhibitors. 181. Inhibition of brain choline acetyltransferase by derivatives of 4-stilbazole. *J. Med. Chem.* **14,** 315–322.

Baker B. R. and Gibson R. E. (1972) Irreversible enzyme inhibitors. 193. Inhibition of choline acetyltransferase. Mode of binding by 4-stilbazoles. *J. Med. Chem.* **15,** 639–642.

Banns H. E. (1976) A study of the molecular sizes and stability of choline acetyltransferase. *J. Neurochem.* **26,** 967–971.

Barlow R. B. and Dixon R. O. D. (1973) Choline acetyltransferase in the nettle *Urtica dioica. Biochem. J.* **132,** 15–18.

Bartus R. T., Dean R. L., Beer B., and Lippa S. (1982) The cholinergic hypothesis of geriatric memory dysfunction. *Science* **217,** 408–417.

Becket A. H., Harper N. J., and Clitherow J. W. (1963) The absolute configuration of α- and β-methyl choline isomers and their acetyl and succinyl esters. *J. Pharm. Pharmacol.* **15,** 349–361.

Behr J. P. and Lehn J. M. (1972) Molecular dynamics of acetylcholine and choline. *Biochem. Biophys. Res. Commun.* **49,** 1573–1579.

Bender M. L. (1971) *Mechanisms of Homogeneous Catalysis from Protons to Proteins,* p. 511, John Wiley & Sons, New York.

Bender M. L., Pollock E. J., and Neveu M. (1962) Deuterium oxide solvent isotope effects in the nucleophilic reactions of phenyl esters. *J. Am. Chem. Soc.* **84,** 595–600.

Benishin C. G. and Carroll P. T. (1981) Acetylation of choline and homocholine by membrane-bound choline-*O*-acetyltransferase in mouse forebrain nerve endings. *J. Neurochem.* **36,** 732–740.

Benishin C. G. and Carroll P. T. (1983). Multiple forms of choline-*O*-acetyltransferase in mouse and rat brain: Solubilization and characterization. *J. Neurochem.* **41,** 1030–1039.

Benishin C. G. and Carroll P. T. (1984) Developmental differences between soluble and membrane-bound fractions of choline-*O*-acetyltransferase in neonatal mouse brain. *J. Neurochem.* **43,** 885–887.

Berman R., Wilson I. B., and Nachmansohn D. (1953) Choline acetylase specificity in relation to biological function. *Biochim. Biophys. Acta* **12,** 315–324.

Berman-Reisberg R. (1957) Sulfhydryl groups of choline acetylase. *Biochim. Biophys. Acta* **14,** 442–443.

Berman-Reisberg R. (1974) Stimulation of choline acetyltransferase activity by ethanol in *in vitro* preparations of rat cerebrum. *Life Sci.* **14,** 1965–1973.

Biellmann J. F., Samama J. P., Bränden C. I., and Eklund H. (1979) X-ray studies of the binding of Cibacron Blue F3GA to liver alcohol dehydrogenase. *Eur. J. Biochem.* **102,** 107–110.

Birman S. (1985) Determination of acetylcholinesterase activity by a new chemiluminescence assay with a natural substrate. *Biochem. J.* **225,** 825–828.

Bishop M. R., Rama-Sastry B. V., Schmidt D. E., and Harbison R. D. (1976) Occurrence of choline acetyltransferase and acetylcholine and other quaternary ammonium compounds in mammalian spermatozoa. *Biochem. Pharmacol.* **25,** 1617–1622.

Blow D. M. (1976) Structure and mechanism of chymotrypsin. *Accts. Chem. Res.* **9,** 145–152.

Blusztajn J. K. and Wurtman R. J. (1983) Choline and cholinergic neurons. *Science* **221,** 614–620.

Bohnert J. L., Malencik D. A., Anderson S. R., Teller D., and Fischer E. H. (1982) Binding of 5,5'-*bis*[8-(phenylamino)-1-naphthalene sulfonate] by the regulatory subunits of adenosine cyclic 3',5'-phosphate-dependent protein kinase. *Biochemistry* **21,** 5570–5576.

Bowen D. M., Smith G. B., White P., and Davison A. N. (1976) Neurotransmitter-related enzymes and indices of hypoxia in senile dementia and other abiotrophics. *Brain* **99,** 459–496.

Brandon C. and Wu J. Y. (1978) Purification and properties of choline acetyltransferase from *Torpedo californica. J. Neurochem.* **30,** 791–797.

Brunello N., Cheney D. L., and Costa E. (1982) Increase in exogenous choline fails to elevate the content or turnover rate of cortical, striatal or hippocampal acetylcholine. *J. Neurochem.* **38,** 1160–1163.

Bueding E. (1952) Acetylcholinesterase activity of *Schistosoma mansoni. Br. J. Pharmacol.* **7,** 563–567.

Bulbring E., Lourie E. M., and Pardoe U. (1949) The presence of acetylcholine in *Trypanosoma rhodesiense* and its absence from *Plasmodium gallinaceum. Br. J. Pharmacol.* **4,** 290–294.

Canepa F. G., Pauling P., and Sörum H. (1966) Structure of acetylcholine and other substrates of cholinergic systems. *Nature* **210,** 907–911.

Cavallito C. J., Yun H. S., Smith J. C., and Foldes F. F. (1970) Choline acetyltransferase inhibitors. Dimensional and substituent effects among styrylpyridine analogs. *J. Med. Chem.* **13,** 221–244.

Cavallito C. J., Yun H. S., Edwards M. L., and Foldes F. F. (1971) Choline acetyltransferase inhibitors. Styrylpyridine analogs with nitrogen atom modifications. *J. Med. Chem.* **14,** 130–133.

Chao L. P. (1975) Subunits of choline acetyltransferase. *J. Neurochem.* **25,** 261–266.

Chao L. P. and Wolfgram F. (1973) Purification and some properties of choline acetyltransferase (EC 2.3.1.6) from bovine brain. *J. Neurochem.* **20,** 1075–1081.

Chao L. P. and Wolfgram F. (1974) Activation, inhibition and aggregation of choline acetyltransferase. *J. Neurochem.* **23,** 697–701.

Chao L. P., Wolfgram F. J., and Eng L. F. (1977) Immunohistochemical localization of choline acetyltransferase: Real or artifact? A reply. *Neurochem. Res.* **2,** 323–325.

Chaturvedi A. K., Rowell P. P., and Rama-Sastry B. V. (1978) Relationship between chemical structure and inhibition of human placental choline acetyltransferase by keto analogs of acetylcholine. *J. Pharm. Sci.* **67,** 657–660.

Chibata I., Tosa T., and Matuo Y. (1974) Coenzyme A. *Methods Enzymol.* **34,** 267–272.

Chiou C. Y. and Rama-Sastry B. V. (1968) Acetylcholinesterase hydrolysis of halogen-substituted acetylcholines. *Biochem. Pharmacol.* **17,** 805–815.

Chu S. H. and Mautner H. G. (1966). Analogs of neuroeffectors. V. Neighboring-group effects in the reactions of esters, thiolesters and selenolesters. The hydrolysis and aminolysis of benzoylcholine, benzoylthiolcholine, benzoylselenolcholine and of their dimethylamino analogs. *J. Org. Chem.* **31,** 308–312.

Chweh A. Y., DeBernardis J. F., Siuda J. F., Rondan N. G., Abola J. E., and Abraham D. J. (1984) Structural correlations of choline acetyltransferase inhibitors: *trans-N*-(carboxymethyl)-4-β-1-naphthylvinyl pyridinium bromide and *cis-N*-(2-aminoethyl)-4-(β-1-naphthylvinyl)-3-methylpyridinium bromide hydrobromide. *J. Med. Chem.* **27,** 825–830.

Connors K. A. and Bender M. L. (1961) The kinetics of alkaline hydrolysis and *n*-butylaminolysis of ethyl *p*-nitrobenzoate and ethyl *p*-nitrothiolbenzoate. *J. Org. Chem.* **26,** 2498–2504.

Cozzari C. and Hartman B. K. (1983a) Choline acetyltransferase and factors affecting chromatographic properties and enzyme stability. *J. Biol. Chem.* **258,** 10013–10019.

Cozzari C. and Hartman B. K. (1983b) Synthesis of a naphthylpyridine derivative and its use for affinity chromatography of choline acetyltransferase. *Anal. Biochem.* **133,** 120–125.

Cozzari C. and Hartman B. K. (1983c) An endogenous inhibitory factor for choline acetyltransferase. *Brain Res.* **276,** 109–117.

Crawford G. D., Correa L., and Salvaterra P. M. (1982) Interaction of monoclonal antibodies with mammalian choline acetyltransferase. *Proc. Natl. Acad. Sci. USA* **79,** 7031–7035.

Cuatrecasas P. (1970) Protein purification by affinity chromatography: Derivatizations of agarose and polyacrylonitrile beads. *J. Biol. Chem.* **245,** 3059–3065.

Culvenor C. C. J. and Ham N. S. (1966) The proton magnetic resonance spectrum and conformation of acetylcholine. *Chem. Commun.* **537–539.**

Currier S. F. and Mautner H. G. (1974) On the mechanism of choline acetyltransferase. *Proc. Natl. Acad. Sci. USA* **71,** 3355–3358.

Currier S. F. and Mautner H. G. (1976) Evidence for a thiol reagent inhibiting choline acetyltransferase by reacting with the thiol group of coenzyme A, forming a potent inhibitor. *Biochem. Biophys. Res. Commun.* **69,** 431–436.

Currier S. F. and Mautner H. G. (1977) The interaction of analogs of coenzyme A with choline acetyltransferase. *Biochemistry* **16,** 1944–1948.

Cushley R. J. and Mautner H. G. (1970) NMR studies on the conformation of acetylcholine isologues. *Tetrahedron,* **26,** 2151–2159.

Davies P. and Maloney A. J. R. (1976) Selective loss of central cholinergic neurons in Alzheimer's disease. *Lancet* **2,** 1403–1408.

Dean P. D. G. and Watson D. H. (1979) Protein purification using immobilized triazine dyes. *J. Chromatogr.* **165,** 301–319.

Della Raggione F. and Pegg A. E. (1983) Studies of the specificity and kinetics of rat liver spermidine/spermine N^1-acetyltransferase. *Biochem. J.* **213,** 701–706.

Della Raggione F., Erwin B. G., and Pegg A. E. (1983) Studies of the acetyl coenzyme A binding site of rat liver spermidine/spermine N'-acetyltransferase. *Biochem. J.* **213,** 707–712.

DeLorenzo A. J., Dettbarn W. D., and Brzin M. (1969) Fine structural localization of acetylcholinesterase in single axons. *J. Ultrastruct. Res.* **28,** 27–40.

Denburg J. L. (1972) An axon plasma preparation from the walking legs of the lobster *Homarus americanus. Biochim. Biophys. Acta* **282,** 453–458.

Dietz G. W. and Salvaterra P. M. (1980) Purification and peptide mapping of rat brain choline acetyltransferase. *J. Biol. Chem.* **255,** 10612–10617.

Doetsch R. N. (1972) A unified theory of bacterial motile behavior. *J. Theor. Biol.* **35,** 55–60.

Dowdall M. J. and Simon E. J. (1973) Comparative studies on synaptosomes: Uptake of (*N*-Me-^{3}H)choline by synaptosomes from squid optic lobes. *J. Neurochem.* **21,** 969–982.

Driskell W. J., Weber B. H., and Roberts E. (1978) Purification of choline acetyltransferase from *Drosophila melanogaster. J. Neurochem.* **30,** 1135–1141.

Eckenstein F. and Thoenen H. (1982) Production of specific antisera and monoclonal antibodies to choline acetyltransferase: Characterization and use for identification of cholinergic neurons. *EMBO J.* **1,** 363–368.

Eckenstein F., Barde Y. A., and Thoenen H. (1981) Production of specific antibodies to choline acetyltransferase purified from pig brain. *Neuroscience* **6,** 993–1000.

Edwards R. A. and Woody R. W. (1977) Induced circular dichroism as a probe of Cibacron Blue and Congo Red bound to dehydrogenases. *Biochem. Biophys. Res. Commun.* **79,** 470–476.

Edwards R. A. and Woody R. W. (1979) Spectroscopic studies of Cibacron blue and Congo red bound to dehydrogenases and kinases. Evaluation of dyes as probes of the dinucleotide fold. *Biochemistry* **18,** 5197–5204.

Erlanger B. F. (1976) Photoregulation of biologically active macromolecules. *Ann. Rev. Biochem.* **45,** 267–283.

Fonnum F. (1968) Choline acetyltransferase binding to and release from membranes. *Biochem. J.* **103,** 389–398.

Fonnum F. (1969) Isolation of choline esters from aqueous solutions by extraction with sodium tetraphenyl boron in organic solvents. *Biochem. J.* **113,** 291–298.

Fonnum F. (1975) A rapid radiochemical method for the determination of choline acetyltransferase. *J. Neurochem.* **24,** 407–409.

Fonnum F. and Malthe-Sørenssen D. (1973) Membrane affinities and subcellular distribution of the different molecular forms of choline acetyltransferase from rat. *J. Neurochem.* **20,** 1351–1359.

Frankenberg L., Heimbüger G., Nilsson C., and Sörbo B. (1973) Biochemical and pharmacological studies on the sulfonium analogues of choline and acetylcholine. *Eur. J. Pharmacol.* **23,** 37–46.

Garrison-Gund C. K. and Neff N. H. (1976) Deuterium-labeled choline: Lack of isotope effects during the synthesis and catabolism of acetylcholine *in vivo*. *Neurochem. Res.* **1,** 679–682.

Glazer A. N. (1970a) On the prevalence of "nonspecific binding" at the specific binding sites of globular proteins. *Proc. Natl. Acad. Sci. USA* **65,** 1057–1063.

Glazer A. N. (1970b) Specific chemical modification of proteins. *Ann. Rev. Biochem.* **39,** 110–112.

Goodman D. R., Adatsi K., and Harbison R. D. (1984) Evidence for the extreme overestimation of choline acetyltransferase in human sperm, human seminal plasma and rat heart: A case of mistaking carnitine acetyltransferase for choline acetyltransferase. *Chem. Biol. Interacts.* **49,** 39–53.

Haga T. and Noda H. (1973) Choline uptake systems of rat brain synaptosomes. *Biochim. Biophys. Acta* **291,** 564–575.

Hamprecht B. and Amano T. (1974) Differential assay for choline acetyltransferase. *Anal. Biochem.* **57,** 162–172.

Harbison R. D., Dwivedi C., and Evans M. A. (1976) A proposed mechanism for trimethylphosphate-induced sterility. *Toxicol. Appl. Pharmacol.* **35,** 481–490.

Hartmann E. and Kilbinger H. (1974a) Gas-liquid chromatographic determination of light-dependent acetylcholine concentration in moss callus. *Biochem. J.* **137,** 249–252.

Hartmann E. and Kilbinger H. (1974b) Occurrence of light-dependent acetylcholine concentrations in higher plants. *Experientia* **30,** 1387–1388.

Hawkins P. J. and Tarbell D. S. (1953) Studies on model compounds of coenzyme A. Studies of the aminolysis and hydrolysis of ethyl thioacetate and β-acetaminoethylthioacetate in aqueous solution. *J. Am. Chem. Soc.* **75,** 2982–2985.

Hebb C. (1972) Biosynthesis of acetylcholine in nervous tissue. *Physiol. Rev.* **52,** 918–957.

Hemsworth B. A. and Smith J. C. (1970) Enzymic acetylation of the stereoisomers of α- and β-methyl choline. *Physiol. Rev.* **19,** 2925–2927.

Hersh L. B. (1979) The lack of specificity towards salts in the activation of choline acetyltransferase from human placenta. *J. Neurochem.* **32,** 991–996.

Hersh L. B. (1982) Kinetic studies of the choline acetyltransferase reaction using isotope exchange at equilibrium. *J. Biol. Chem.* **257,** 12820–12825.

Hersh L. B. and Peet M. (1977) Re-evaluation of the kinetic mechanism of the choline acetyltransferase reaction *J. Biol. Chem.* **252,** 4796–4802.

Hersh L. B. and Peet M. (1978) Effect of salts on the physical and kinetic properties of human placental choline acetyltransferase. *J. Neurochem.* **30,** 1087–1093.

Hersh L. B., Barker L. A., and Rush B. (1978a) Effect of sodium chloride on changing the rate-limiting step in the human placental choline acetyltransferase reaction. *J. Biol. Chem.* **253,** 4966–4970.

Hersh L. B., Coe B., and Casey L. (1978b) A fluorometric assay for choline acetyltransferase and its use in the purification of the enzyme from human placenta. *J. Neurochem.* **30,** 1077–1085.

Hersh L. B., Nair R. V., and Smith D. J. (1979) The reaction of choline acetyltransferase with sulfhydryl reagents. Methoxycarbonyl-CoA-disulfide as an active site-directed reagent. *J. Biol. Chem.* **254,** 11988–11992.

Hersh L. B., Wainer B. H., and Andrews L. P. (1984) Multiple isoelectric and molecular weight variants of choline acetyltransferase. Artifact or real? *J. Biol. Chem.* **259,** 1253–1258.

Hestrin S. (1949) The reaction of acetylcholine and other carboxylic acid derivatives with hydroxylamine and its analytical application. *J. Biol. Chem.* **180,** 249–261.

Hirsch M. J. and Wurtman R. J. (1978) Lecithin consumption elevates acetylcholine concentrations in rat brain and adrenal gland. *Science* **202,** 223–225.

Hopff W. H., Riggio G., and Waser P. G. (1984) Blockade of acetylcholine synthesis in organophosphate poisoning. *Toxicol. Appl. Pharmacol.* **72,** 513–518.

Husain S. S. and Mautner H. G. (1973) The purification of choline acetyltransferase of squid head ganglia. *Proc. Natl. Acad. Sci. USA* **70,** 3749–3753.

Ichikawa T., Ishida I., and Deyuchi T. (1983) Monoclonal antibodies to choline acetyltransferase of rat brain. *FEBS Lett.* **155,** 306–310.

Israël M. and Lesbats B. (1980) Détection continue de la libération d'acétylcholine de l'organe électrique de la Torpille à l'aide d'une réaction de chimiluminescence. *Couptes Rendus Acad. Sci. Paris* **291,** 791–796.

Israël M. and Lesbats B. (1981) Continuous determination by a chemiluminescent method of acetylcholine release and compartmentation in *Torpedo* electric organ synaptosomes. *J. Neurochem.* **37,** 1475–1483.

Israël M. and Lesbats B. (1982) Application to mammalian tissues of the chemiluminescent method for detecting acetylcholine. *J. Neurochem.* **39,** 248–250.

Ishida I. and Deguchi T. (1984) Increase of choline acetyltransferase by colchicine in primary cell cultures of spinal cord. *J. Neurochem.* **43,** 42–48.

Jaenicke L. and Lynen F. (1960) Coenzyme A, in *The Enzymes,* Vol. 3B, 2nd Ed., (Boyer P. D., Lardy H., and Myrback K., eds.), pp. 1–103, Academic, New York.

Jaffe M. J. (1970) Evidence for the regulation of phytochrome-mediated processes in bean roots by the neurohumor, acetylcholine. *Plant Physiol.* **46,** 768–777.

Jencks W. P. (1969) *Catalysis in Chemistry and Enzymology* McGraw-Hill, New York.

Jumblatt J. E., Marquis J. K., and Mautner H. G. (1981) On the specificity of ^{125}I-α-bungarotoxin binding to axonal membranes. *J. Neurochem.* **37,** 392–400.

Kuhar M. J., Sethy V. H., Roth R. H., and Aghajanian G. K. (1973) Choline: Selective accumulation by central cholinergic neurons. *J. Neurochem.* **20,** 581–583.

Lefresne P. L., Guyenet P., and Glowinski J. (1973) Acetylcholine synthesis from (2-^{14}C)pyruvate in rat striatal slices. *J. Neurochem.* **20,** 1083–1097.

Levey A. I., Rye D. B., and Wainer B. H. (1982) Immunochemical studies of bovine and human choline-*O*-acetyltransferase using monoclonal antibodies. *J. Neurochem.* **39,** 1652–1659.

Levey A. I. and Wainer B. H. (1982) Cross-species and intra-species reactivities of monoclonal antibodies against choline acetyltransferase. *Brain Res.* **234,** 469–473.

Levey A. I., Aoki M., Fitch F. W., and Wainer B. H. (1981) The production of monoclonal antibodies reactive with bovine choline acetyltransferase. *Brain Res.* **218,** 383–387.

Levey A. I., Armstrong D. M., Atweh S. F., Terry R. D., and Wainer B. H. (1983) Monoclonal antibodies to choline acetyltransferase: Production, specificity and immunohistochemistry. *J. Neurosci.* **3,** 1–9.

Lienhard G. E. and Jencks W. P. (1965) The reaction of carbanions with *N,S*-diacetylcysteamine. A model for enzymatic carbon-carbon condensation. *J. Am. Chem. Soc.* **87,** 3863–3874.

Lipmann F. (1945) Acetylation of sulfanilamide by liver homogenates and extracts. *J. Biol. Chem.* **160,** 173–190.

Loewi O. (1921) Über humorale Übertragbarkeit der Herznervenwirkung. *Pflügers Arch. ges. Physiol.* **189,** 239–242.

Luqmani Y. A. and Richardson P. J. (1982) Homocholine and short-chain *N*-alkyl choline analogues as substrates for *Torpedo* choline acetyltransferase. *J. Neurochem.* **38,** 368–374.

Makriyannis A. (1974) Acetylcholine and choline. *Pharmacologist* **16,** 285–287.

Makriyannis A. and Mautner H. G. (1972) Conformational relationships between analogs of acetylcholine and those of local anesthetics in solution. *Proc. Natl. Acad. Sci. USA* **69,** 3416–3419.

Makriyannis A., Theard J. M., and Mautner H. G. (1979) Direct approach to the determination of the rotational barriers of the acetylcholine and choline molecules. *Biochem. Pharmacol.* **29,** 1911–1915.

Malthe-Sørenssen D. (1976a) Choline acetyltransferase—Evidence for acetyl transfer by a histidine residue. *J. Neurochem.* **27,** 873–881.

Malthe-Sørenssen D. (1976b) Molecular properties of choline acetyltransferase from different species: Investigation by isoelectric focusing and ion exchange adsorption. *J. Neurochem.* **26,** 861–865.

Malthe-Sørenssen D. and Fonnum F. (1972) Multiple forms of choline acetyltransferase in several species demonstrated by isoelectric focusing. *Biochem. J.* **127,** 229–236.

Malthe-Sørenssen D., Andersen R. A., and Fonnum F. (1974) Biochemical and pharmacological properties of acryloylcholine, an inhibitor of choline acetyltransferase. *Biochem. Pharmacol.* **28,** 577–586.

Malthe-Sørenssen D., Lea T., Fonnum F., and Eskeland T. (1978) Molecular characterization of choline acetyltransferase from bovine brain caudate nucleus and some immunological properties of the highly purified enzyme. *J. Neurochem.* **30,** 35–46.

Mannervik B. and Sörbo B. (1970) Inhibition of choline acetyltransferase from bovine caudate nucleus by sulfhydryl reagents—reactivation of the inhibited enzyme. *Biochem. Pharmacol.* **19,** 2509–2516.

Marchbanks R. M. and Kessler P. D. (1982) The independency of choline transport and acetylcholine synthesis. *J. Neurochem.* **30,** 1424–1433.

Mautner H. G. (1974) Studies of the Conformation, in Solution, of Molecules Related to Acetylcholine, in *Molecular and Quantum Pharmacology,* (Bergman E. and Pullman B., eds.), pp. 119–129, Reidel, Dordrecht, Holland.

Mautner H. G. (1977) Choline acetyltransferase. *CRC Crit. Rev. Biochem.* **4,** 341–370.

Mautner H. G. (1980) Acetylcholine and choline acetyltransferase. *Neurochem. Int.* **2,** 123–134.

Mautner H. G. (1985) The Mechanism of Action of Choline Acetyltransferase, in *Molecular Basis of Nerve Activity.* (Changeux J.-P., Hucho F., Maelicke A., and Neumann E., eds.), Walter de Gruyter, Berlin.

Mautner H. G. and Currier S. F. (1976) Studies on the Mechanism of Action of Choline Acetyltransferase, in *Cholinergic Mechanisms and Psychopharmacology.* (Jenden D. J., ed.), pp. 197–206, Plenum, New York.

Mautner H. G., Dexter D. D., and Low B. W. (1972) Conformational requirements of acetylcholine analogues and local anaesthetics. *Nature* **238,** 87–91.

Mautner H. G., Merrill R. E., Currier S. F., and Harvey G. (1981a) Interaction of aromatic dyes with the coenzyme A binding site of choline acetyltransferase. *J. Med. Chem.* **24,** 1534–1537.

Mautner H. G., Pakula A. A., and Merrill R. E. (1981b) Evidence for the presence of an arginine residue in the coenzyme A binding site of choline acetyltransferase. *Proc. Natl. Acad. Sci. USA* **78,** 7449–7452.

McCaman R. E. and Hunt J. M. (1965) Microdetermination of choline acetylase in nervous tissue. *J. Neurochem.* **12,** 253–259.

McGeer P. L., McGeer E. G., and Peng J. H. (1984) Minireview. Choline acetyltransferase: Purification and immunohistochemical localization. *Life Sci.* **34,** 2319–2338.

Meyer E. M., White D. L., Andrews B. B., Barrnett R. J., and Cooper J. R. (1981) Synthesis and radioenzymic assay of an antimony analog of choline. *Biochem. Pharmacol.* **30,** 3003–3005.

Miura G. A. and Shih T. M. (1984) Cholinergic constituents in plants: Characterization and distribution of acetylcholine and choline. *Physiol. Plant* **61,** 417–421.

Moe J. G. and Piszkiewicz D. (1979) Isoleucyl transfer ribonucleic acid synthetase. Competitive inhibition with respect to transfer ribonucleic acid by Blue Dextran. *Biochemistry* **13,** 2810–2814.

Morris D. (1966) The choline acetyltransferase of human placenta *Biochem. J.* **98,** 754–762.

Morris D. and Grewaal D. S. (1969) Halogen-substituted derivatives of acetylcholine as inhibitors of choline acetyltransferase. *Life Sci.* **8,** 511–515.

Morris D. and Grewaal D. S. (1971) Human placental choline acetyltransferase: Radiometric assay, inhibition by analogues of choline and isotopic exchange between choline and acetylcholine. *Eur. J. Biochem.* **22,** 563–572.

Morris D., Maneckjee A., and Hebb C. (1971). The kinetic properties of human placental choline acetyltransferase. *Biochem. J.* **125,** 857–863.

Nachmansohn D. (1954) Metabolism and function of the nerve cell. Harvey Lectures, 1953–1954, 57–99.

Nachmansohn D. (1959) *Chemical and Molecular Basis of Nerve Activity,* Academic, New York.

Nachmansohn D. (1972) Biochemistry as part of my life. *Ann. Rev. Biochem.* **41,** 1–28.

Nachmansohn D. and Machado A. L. (1943) The formation of acetylcholine, a new enzyme "choline acetylase." *J. Neurophysiol.* **6,** 397–403.

Nachmansohn D. and Weiss M. S. (1948) Studies on choline acetylase. IV. Effect of citric acid. *J. Biol. Chem.* **172,** 677–687.

Nachmansohn D., John H. M., and Waelsch H. (1943) Effect of glutamic acid on the formation of acetylcholine. *J. Biol. Chem.* **150,** 485–486.

Nelson S. H., Benishin C. G., and Carroll P. T. (1980) Accumulation and metabolism of choline and homocholine by mouse brain subcellular fractions. *Biochem. Pharmacol.* **29,** 1949–1957.

O'Regan S. and Collier B. (1981) Factors affecting choline transport by the cat superior cervical ganglion and the relationship between choline uptake and acetylcholine synthesis. *Neuroscience* **6,** 511–520.

Peng J. H., McGeer P. L., and McGeer E. G. (1983) Anti-human choline acetyltransferase fragments: Antigen binding (Fab) Sepharose chromatography for enzyme purification. *Neurochem. Res.* **8,** 1481–1486.

Persson B. O., Larsson L., Schuberth J., and Sörbo B. (1967) 3-Bromoacetonyltrimethylammonium bromide, a choline acetylase inhibitor. *Acta Chem. Scand.* **21,** 2283–2284.

Pieklik J. R. and Guynn R. W. (1975) Equilibrium constants of the reactions of choline acetyltransferase, carnitine acetyltransferase and acetylcholinesterase under physiological conditions. *J. Biol. Chem.* **250,** 4445–4450.

Potempska A., Loo Y. H., and Wisniewski H. M. (1984) On the possible mechanism of phenylacetate neurotoxicity: Inhibition of choline acetyltransferase by phenylacetyl coenzyme A. *J. Neurochem.* **42,** 1499–1501.

Potter L. T., Glover V. A. S., and Saelens J. K. (1968) Choline acetyltransferase from rat brain. *J. Biol. Chem.* **243,** 3864–3870.

Racker E. (1968) The membrane of the mitochondrion. *Sci. Am.* **218,** 32–41.

Rama-Sastry B. V. and Henderson G. I. (1972) Kinetic mechanisms of human placental choline acetyltransferase. *Biochem. Pharmacol.* **21,** 787–802.

Rama-Sastry B. V. and Sadavongvivad C. (1979) Cholinergic systems in non-nervous tissue. *Pharmacol. Rev.* **30,** 65–132.

Rama-Sastry B. V., Janson V. E., and Chaturvedi A. V. (1981) Inhibition of human sperm motility by inhibitors of choline acetyltransferase. *J. Pharmacol. Exp. Ther.* **216,** 378–384.

Rand J. B. and Johnson C. D. (1981) A single-vial biphasic liquid extraction assay for choline acetyltransferase using [^{3}H] choline. *Anal. Biochem.* **116,** 361–371.

Rand J. B. and Russell R. L. (1985) Properties and partial purification of choline acetyltransferase from the nematode *Caenorhabditis elegans. J. Neurochem.* **44,** 189–200.

Řičny J. and Tuček S. (1980) Relation between the content of acetyl coenzyme A and acetylcholine in brain slices. *Biochem. J.* **188,** 683–688.

Roskoski R. (1973) Choline acetyltransferase. Evidence for an acetyl-enzyme reaction intermediate. *Biochemistry* **12,** 3709–3714.

Roskoski R. (1974a) Choline acetyltransferase. Inhibition by thiol reagents. *J. Biol. Chem.* **249,** 2156–2159.

Roskoski R. (1974b) Choline acetyltransferase and acetylcholinesterase: Evidence for essential histidine residues. *Biochemistry* **13,** 5141–5144.

Roskoski R., Lim C. T., and Roskoski L. M. (1975) Human brain and placental choline acetyltransferase: Purification and properties. *Biochemistry* **14,** 5105–5110.

Rossier J. (1975) Immunohistochemical localization of choline acetyltransferase: Real or artifact? *Brain Res.* **98,** 619–622.

Rossier J. (1977a) Choline acetyltransferase: A review with special reference to its cellular and subcellular localization. *Int. Rev. Neurobiol.* **20,** 283–330.

Rossier J. (1977b) Acetyl coenzyme A and coenzyme A analogues. Their effect on rat brain choline acetyltransferase. *Biochem. J.* **165,** 321–326.

Rossier J. (1981) Serum monospecificity: A prerequisite for reliable immunohistochemical localization of neuronal markers including choline acetyltransferase. *Neuroscience* **6,** 989–991.

Rossier J., Bauman A., and Benda P. (1973) Improved purification of rat brain choline acetyltransferase by using an immunoabsorbent. *FEBS Lett.* **32,** 231–234.

Rossier J., Spantidakis Y., and Benda P. (1977) The effect of Cl^- on choline acetyltransferase kinetic parameters and a proposed role for Cl^- in the regulation of acetylcholine synthesis. *J. Neurochem.* **29,** 1007–1012.

Roth J., Le Roith D., Shiloach J., Rosenzweig J. L., Lesniak B. A., and Havrankova J. (1982) The evolutionary origins of hormones, neurotransmitters and other extracellular chemical messengers. *New Eng. J. Med.* **306,** 523–527.

Rowell P. P. and Chiou C. Y. (1976) *In vivo* effects of some tertiary alkylaminoethyl esters with choline acetyltransferase inhibitory properties. *Eur. J. Pharmacol.* **40,** 83–91.

Ryan R. L. and McClure W. O. (1979) Purification of choline acetyltransferase from rat and cow brain. *Biochemistry* **24,** 5357–5365.

Ryan R. L. and McClure W. O. (1981) Inhibition by styrylpyridines of purified rat and bovine brain choline acetyltransferase. *Neurochem. Res.* **6,** 163–173.

Schaap A. P., Thayer A. L., Blossey E. C., and Neckers D. C. (1975) Polymer-based sensitizers for photo-oxidations. *J. Am. Chem. Soc.* **97,** 3741–3745.

Schrier B. K. and Shuster L. (1967) A simplified radiochemical assay for choline acetyltransferase. *J. Neurochem.* **14,** 977–985.

Shaltiel S. (1974) Hydrophobic chromatography. *Methods Enzymol.* **34,** 126–132.

Shuster L. and O'Toole C. (1974) Inactivation of choline acetyltransferase by an antiserum to the enzyme from mouse brain. *Life Sci.* **15,** 645–656.

Slemmon J. R., Salvaterra P. M., Crawford G. D., and Roberts E. (1982) Purification of choline acetyltransferase from *Drosophila melanogaster*. *J. Biol. Chem.* **257,** 3847–3852.

Slemmon J. R., Salvaterra P. M., and Roberts E. (1984) Molecular characterization of choline acetyltransferase from *Drosophila melanogaster*. *Neurochem. Int.* **6,** 519–525.

Sluyterman L. A. and Wijdenes J. (1970) An agarose mercurial column for the separation of mercaptopapain from nonmercaptopapain *Biochim. Biophys. Acta* **200,** 593–596.

Smallman B. N. and Maneckjee A. (1981) The synthesis of acetylcholine by plants. *Biochem. J.* **194,** 361–364.

Smith D. J., Maggio E. T., and Kenyon G. L. (1975) Simple alkane thiol groups for temporary blocking of sulfhydryl groups of enzymes. *Biochemistry* **14,** 755–771.

Smith J. C., Cavallito C. J., and Foldes F. F. (1967) Choline acetyltransferase inhibitors: A group of styryl-pyridine analogs. *Biochem. Pharmacol.* **16,** 2438–2441.

Stellwagen E. (1977) Use of blue dextran as a probe for the nicotinamide adenine dinucleotide domain in proteins. *Accts. Chem Res.* **10,** 92–98.

Stephens-Newsham L. G., Hebb C., Mann S. P., and Banns H. (1979) Choline acetyltransferase in the equine spleen. *Gen. Pharmacol.* **10,** 385–388.

Stephenson M. and Rowatt E. (1947) The production of acetylcholine by a strain of *Lactobacillus plantarum. J. Gen. Microbiol.* **1,** 279–298.

Subramanian S. (1982a) Spectral changes induced in Cibacron Blue F3GA by salts, organic solvents, and polypeptides: Implications for Blue Dye interaction with proteins. *Arch. Biochem. Biophys.* **216,** 116–125.

Subramanian S. (1982b) Specific interaction of spermine with Cibacron Blue F3GA. *Arch. Biochem. Biophys.* **217,** 388–391.

Subramanian S. (1984) Dye ligand affinity chromatography: The interaction of Cibacron Blue F3GA with proteins and enzymes. *CRC Crit. Revs. Biochem.* **16,** 169–205.

Takahashi K. (1968) The reaction of phenylglyoxal with arginine residues in proteins. *J. Biol. Chem.* **243,** 6171–6179.

Terry R. D. and Katzman R. (1983) Senile dementia of the Alzheimer type. *Ann. Neurol.* **14,** 497–506.

Thompson S. T. and Stellwagen E. (1976) Binding of Cibacron Blue F3GA to proteins containing the dinucleotide fold. *Proc. Natl. Acad. Sci. USA* **73,** 361–365.

Tsuchida Y., Yokomori K., Morikawa J., Mori K., and Saito S. (1982) Differential assay of choline acetyltransferase activity in human neuroblastomas and cloned adrenergic neuroblastoma of mouse. *Gann* **73,** 21–26.

Tuček S. (1984) Problems in the organization and control of acetylcholine synthesis in brain neurons. *Prog. Biophys. Molec. Biol.* **44,** 1–46.

Tuček S. (1985) Regulation of acetylcholine synthesis in the brain. *J. Neurochem.* **44,** 11–24.

Turner A. J. (1981) Scope and applications of dye ligand chromatography. *Trends Biochem. Sci.* 171–173.

Wainer B. H., Levey A. I., Mutson E. J., and Mesulam M. M. (1984) Cholinergic systems in mammalian brain identified with antibodies against choline acetyltransferase. *Neurochem. Int.* **6,** 163–182.

Webb G. D. and Mautner H. G. (1966) Sulfur and selenium compounds related to acetylcholine and choline. VI. Effects of homocholine derivatives on the electroplax preparation. *Biochem. Pharmacol.* **15,** 2105–2111.

Wecker L. N. and Dettbarn W. D. (1979) Relationship between choline availability and acetylcholine synthesis in discrete regions of rat brain. *J. Neurochem.* **32,** 961–967.

Westhead E. W. (1965) Photooxidation with rose bengal of a critical histidine residue in yeast enolase. *Biochemistry* **4,** 2139–2144.

White H. L. and Cavallito C. J. (1970a) Inhibition of bacterial and mammalian choline acetyltransferase by styrylpyridine analogues. *J. Neurochem.* **17,** 1579–1589.

White H. L. and Cavallito C. J. (1970b) Choline acetyltransferase. Enzyme mechanism and mode of inhibition by a styrylpyridine analogue. *Biochim. Biophys. Acta* **206,** 343–358.

Yamamura H. I. and Snyder S. H. (1972) Choline: High affinity uptake by rat brain synaptosomes. *Science* **178,** 626–628.

Yamamura H. I. and Snyder S. H. (1973) High affinity transport of choline into synaptosomes of rat brain. *J. Neurochem.* **21,** 1355–1374.

York S. S., Lawson R. C., and Worah D. M. (1978) Binding of recrystallized and chromatographically purified 8-anilino-1-napthalenesulfonate to *Escherichia coli lac* repressor. *Biochemistry* **17,** 4480–4486.

Chapter 9
Mammalian Brain Acetylcholinesterase

Zoltán Rakonczay

1. Introduction

Acetylcholinesterase (acetylcholine acetylhydrolase, AChE: EC 3.1.1.7) has been the focus of much attention since it was first suggested that it plays an important role in the rapid destruction of acetylcholine (ACh) in a living organ (Dale, 1914). In the past decade there has been considerable progress regarding the nature, chemistry, and physiology of AChE. The catalytic properties, and their occurrence, histochemical localization, and molecular heterogeneity in the different tissues of different animal species have been extensively studied (Silver, 1974; Rosenberry, 1975; Trevor et al., 1978; Massoulié and Bon, 1982; Brimijoin, 1983). The advances have mainly been a result of the application of affinity chromatography and immunochemical techniques for enzyme purification.

AChE is widely distributed; it occurs in the central (CNS) and peripheral nervous systems (PNS), and the motor end-plates of the skeletal muscle and the electric organ, but it is also found in many other tissues and in erythrocytes.

It is the intention of this chapter to summarize our recent knowledge on mammalian brain AChE, first because it is one of the best characterized enzymes, and second because of its possible relevance to human brain enzyme. Special attention has been paid to the newly developed enzyme activity measurements, to the purification and characterization of AChE, to its possible biological function in the brain, and to its relationship to mental disorders.

2. Some Enzymological Aspects

AChE plays an essential role in the mechanism of cholinergic neurotransmission (Dale, 1914; Koelle, 1963; Silver, 1974), by rapidly splitting acetylcholine into acetate and choline. The natural substrate for brain enzyme is assumed to be ACh, but other possible functions have recently been suggested (*see* section 7).

2.1. Reaction and Substrate Specificity

AChE catalyzes the hydrolysis of ACh and other acetic acid esters by means of a two-step process. The enzyme–substrate complex first gives an acylated enzyme, which is then hydrolyzed (Wilson and Bergmann, 1950; Wilson et al., 1950). Kinetic studies have distinguished two subsites of the enzyme-active site; a negatively charged anionic subsite and an esteric subsite, with the latter component involving an "active" Ser moiety (Schaffer et al., 1953; 1973). The positively charged quaternary ammonium group of substrates or inhibitors binds to the negatively charged anionic site through a combination of coulombic and Van der Waals' forces. The esteric site determines the substrate specificity and catalytic functions. The two principal subsites are conveniently visualized as a "two-pocket" model (Wilson, 1960), though more elaborate models are frequently discussed (e.g., Krupka, 1966; Beckett, 1967; Michelson and Zeimal, 1973).

These considerations on the enzyme mechanism have developed from studies mainly on electric eel tissue AChE, but the general mechanism may be similar for the enzyme from mammalian brain tissue. This conclusion is strengthened by the fact that the kinetic constants of brain AChE isolated from different mammalian species are quite similar (Table 1).

The enzyme activity toward butyrylcholine (BuCh) or butyrylthiocholine (BuThCh) is low compared with that toward ACh. It rapidly splits the substrate acetylthiocholine (AThCh, developed for spectrophotometric assay). The rate of hydrolysis of propionylcholine (PrCh) is only slightly less than that of the natural substrate ACh (Table 1).

2.2. Specific Inhibitors

Thousands of agents inhibit cholinesterases (*see* Saunders, 1957; Holmstedt, 1951, 1959; Heath, 1961; Koelle, 1963; O'Brien, 1967; Karczmar, 1967; Main, 1976). The compounds considered here are those most commonly used in physiology or in biochemical or histochemical experiments (Table 2).

Table 1
Michaelis Constants and Rates of Hydrolysis of Various Choline Esters by Mammalian Brain AChE

Source of enzyme	Substrate[a]	$K_m \times 10^{-4}$, mmol/L	Rate[b]	References
Calf caudate nucleus	ACh	1.38	100.00	Jackson and Aprison (1966a)
Guinea pig	ACh	1.47	—	Yamamura et al. (1973)
Human	ACh	4.3–4.4	—	Patocka and Bajgar (1969)
Human	ACh	4.9	—	Zech and Domagk (1975)
Calf caudate nucleus	AThCh	4.17	170.00	Jackson and Aprison (1966a)
Mouse	AThCh	3–5	—	Adamson et al. (1975)
Bovine	AThCh	1.4–2.3	—	Chan et al. (1972b)
Rat	AThCh	1.02–1.05	—	Ho and Ellman (1969)
Human	AThCh	4.16	—	Patocka and Bajgar (1969)
Human	AThCh	2.1	—	Zech and Domagh (1975)
Calf caudate nucleus	BuCh	—	2.69	Jackson and Aprison (1966a)
Human	BuCh	1.72	—	Patocka and Bajgar (1969)
Rat	BuThCh	—	3.00	Ellman et al. (1961)
Calf caudate nucleus	PrCh	—	85.4	Jackson and Aprison (1966a)

[a]Substrate concentration, 0.75 mmol/L.
[b]Rate expressed as percentage of ACh hydrolysis.

Table 2
Most Commonly Used Antiacetylcholinesterase Agents

Trivial name and chemical nomenclature	Mol wt	Properties, remarks, references
Reversible inhibitors		
Ambenonium chloride (*N,N'*-bis-(2-chlorobenzyldiethylammoniummethyl)oxamide)	609	"Selective" inhibitor of AChE. Slowly reversible, poor lipid solubility (Silver, 1974). Used in studies of extracellular AChE (Koelle, 1957)
BW284C51 (1,5-bis-(4-allyldimethylammoniumphenyl)-pentan-3-one)	566	"Selective" inhibitor of AChE. Reversible, poor lipid solubility (Fulton and Mogey, 1954; Silver, 1974). Used to protect external AChE, while inactivating cytoplasmic activity with DFP (Taylor et al., 1981)
Edrophonium bromide (ethyl(*m*-hydroxyphenyl)dimethylammonium bromide)	246	A quaternary ammonium compound that inhibits reversibly of both AChE and BuChE (Long, 1963; Silver, 1974). May interfere with the binding of the substrate to the enzyme by occupying the anionic site. Recently used as a component of affinity chromatography material to purify the secretory form of soluble AChE (Hodgson and Chubb, 1983)

Irreversible inhibitors		
Echothiophate or phospholine iodide (*O,O*'-diethyl *S*-ethyltrimethylamine phosphorothiolate iodide)	382	A quaternary quasi-irreversible cationic, organophosphate inhibitor of AChE with low lipid solubility (McIsaac and Koelle, 1959)
217-AO (*O,O*'-diethyl *S*-ethyldimethylamine phosphorothiolate acid-oxalate	331	A tertiary analog of echothiophate, used in combination with echothiophate in localization of external and internal AChE (McIssac and Koelle, 1959; Brimijoin et al., 1978)
DFP (diisopropylfluorophosphate)	184	Highly soluble in lipids. Differential irreversible inhibitor: at low concentrations (10^{-7}–10^{-8} mol/L) inhibits BuChE, at higher concentrations (10^{-5}–10^{-6} mol/L) inhibits AChE (Silver, 1974)
Eserine, physostigmine (1'-methylpyrrolidino (2',3', 2,3)1,3-dimethylindolin-5-yl *N*-methylcarbamate)	275	Very potent irreversible inhibitor; 10^{-5} mol/L is adequate for total inhibition of both AChE and BuChE (Silver, 1974)
Nerve gases, war gases		These organophosphorus compounds are lipid-soluble, irreversible inhibitors that form stable phosphorylated enzyme complexes. They are extremely toxic for both AChE and BuChE (Grob, 1963; Saunders, 1957)
Sarin (isopropylmethylphosphonofluoridate)	140	
Soman (pinacolyl methylphosphonofluoridate)	182	
Tabun (ethyl *N*-dimethylphosphoramidocyanidate)	162	

AChE inhibitors are traditionally divided into two groups, depending on their mode of action. These are reversible and irreversible inhibitors. Reversible inhibitors form only noncovalent bonds, and almost all contain at least one positively charged nitrogen group; they therefore inhibit the enzyme, at least in part, by occupation of the anionic site. There are two major classes of irreversible inhibitors, organophosphates and carbamates. They form both covalent and noncovalent bonds. In this case the phosphorylated or carbamylated enzymes are relatively stable to hydrolysis compared to the very unstable acetyl-enzyme intermediate. It should be noted, however, that the term specific inhibitors rarely means absolute selectivity toward AChE: They will usually inhibit butyrylcholinesterase (BuChE) if the concentration is increased. Moreover, the response of the agents also depends greatly on the species (Silver, 1974).

2.3. Assays

Several methods have been developed for quantitative estimation of AChE (for reviews, *see* Augustinsson, 1971; Silver, 1974). Here we have collected some of the most convenient and common methods, to consider some of the advantages and disadvantages associated with each of these approaches.

2.3.1. Spectrophotometric Methods

Probably the most widely used AChE assay is the Ellman method (Ellman et al., 1961). The thiocholine (ThCh) formed during the hydrolysis of AThCh rapidly reacts with 5,5'-dithio-bis-(2-nitrobenzoic acid) (DTNB) and releases a yellow 5-thio-2-nitrobenzoic acid anion. The rate of production of the colored ion is recorded at 412 nm on a spectrophotometer. The main advantages of the method are simplicity; continuous increase in color density as a function of incubation time; accuracy; and relatively low expense. The following are disadvantages: detection limit is in the order of milligrams of brain tissue; SH groups in the sample may react with DTNB; the substrate used, AThCh, and the natural substrate are not identical from a kinetic point of view.

The AChE activity is usually expressed in international units, IU = μmol of substrate hydrolyzed/min/mg protein.

$$\mathrm{IU} = \frac{A}{13.6} \times \frac{V}{S \times C}$$

where A = the difference in absorbance/min between the sample and a blank (without substrate or in the presence of a specific AChE inhibitor); V = total volume (mL); S = sample volume (mL); C = concentration of enzyme preparation (mg protein/mL). The molar extinction coefficient of the yellow anion is 1.36×10^4 (Ellman, 1959).

2.3.2. Fluorometric

Parvari et al. (1983) recently developed a highly sensitive microfluorometric assay for cholinesterases. The method is based on the principle that the ThCh produced by specific enzymic hydrolysis of AThCh is reacted with the fluorogenic compound *N*-[4-(7-diethylamino-4-methylcoumarin-3-yl)phenyl] maleimide (CPM) to yield an intensely blue fluorescent product.

Reagents: Buffer solution; 50 mmol/L Tris-HCl, pH 7.5, 0.12 mol/L NaCl, 1 mmol/L EDTA. Substrate; 30 mmol/L AThCh (10-fold concentrated). Fluorogenic reagent; 0.4 mmol/L CPM in isopropanol (CPM is from Molecular Probes, Inc.).

The standard assay mixture contains 0.09 mL buffer, 0.01 mL AThCh substrate, enzyme, and H_2O to give a final volume of 0.15 mL. After a suitable incubation time the enzyme reaction is stopped by the addition of 1 vol of CPM-containing isopropanol to the aliquot of the samples diluted 1:200 in 1% Triton X-100. The intensity of fluorescence emission at 473 nm is determined with a spectrofluorimeter. Excitation is at 390 nm. To standardize the procedure, 0.005 mL 0.4 mmol/L CPM in isopropanol is added to 0.005 mL samples containing different amounts of ThCh (produced from AThCh by complete hydrolysis, using AChE purified from the electric organ of *E. electricus* as described by Parvari et al., 1983). One milliliter of 1% Triton X-100 is then added to each sample and the intensity of fluorescence is measured as described above.

The assay is linear over a range extending from a few picomoles to nanomoles of ThCh. The advantages of this assay are the wide linear range of accuracy, the sensitivity (100-fold higher than that of radiometric methods), and the relative lack of expense. Disadvantages are: the substrate used (AThCh) is not a natural substrate, and CPM may interact with the SH group and nonthiol nucleophilic groups of proteins to produce nonspecific zero-time fluorescence. However, the background values are comparable to those in other methods.

2.3.3. Radiometric

The two outlined radioisotopic assays for AChE differ only in the procedure used to separate the labeled acetate product from unhydrolyzed labeled substrate (ACh): (1) organic solvent extraction method (Johnson and Russel, 1975); (2) DEAE anion-exchange disk method (Gordon et al., 1982).

2.3.3.1. Organic Solvent Extraction Method (Johnson and Russel, 1975). Reagents: Buffer solution; 0.05 mol/L potassium phosphate, pH 7.0. Substrate; 10 mmol/L (^{3}H-acetyl)-choline iodide (Specific activity, 1 mCi/mmol or 37 MBq/mmol) (fivefold concentrated). Stopping mixture: 1 mol/L chloroacetic acid, 0.5 mol/L NaOH, and 2 mol/L NaCl. Scintillation mixture; 0.5% PPO + 0.03% POPOP in toluene: isoamyl alcohol (9:1, v/v).

The standard assay mixture contains 0.05 mL 0.05 mol/L potassium phosphate buffer, pH 7.0, 0.02 mL [^{3}H]-acetylcholine substrate, enzyme, and H_2O to give a final volume of 0.1 mL. After a suitable incubation time (usually 10 min), the reaction is stopped by adding 0.1 mL of stopping mixture, followed immediately by 4.0 mL of scintillation mixture. The vials are capped and shaken to extract [^{3}H]-acetate into the nonaqueous scintillation cocktail. The radioactivity is measured in a liquid scintillation counter.

2.3.3.2. DEAE Anion-Exchange Disk Method (Gordon et al., 1982). Reagents: Buffer solution; 0.04 mol/L Tris-HCl, pH 7.4, containing 0.5% Triton X-100. Substrate; 2.66 mmol/L acetylcholine iodide containing 0.8 μCi [^{14}C]-acetylcholine (10-fold concentrated). The standard assay mixture contains: 0.22 mL 0.04 mol/L Tris-HCl buffer, pH 7.4, 0.5% Triton X-100, 0.03 mL [^{14}C]-acetylcholine substrate, enzyme, and H_2O to give a final volume of 0.3 mL. Blanks contain the above and 10^{-5} mol/L eserine sulfate, incubated in the above mixture for 10 min prior to the addition of substrate. After a suitable incubation period, a 0.05-mL aliquot of the assay mixture is placed on four stacked DE81 filter disks (Whatman) held in a filtration manifold, and 0.1 mL of ice-chilled water is added. Fifteen seconds later, the disks are washed with 15 mL water under suction. After drying, they are placed in vials containing scintillation mixture and the radioactivity is counted.

The enzyme activities are determined from the slopes of the kinetic curves, according to the following formula. AChE activity = μmol acetylcholine hydrolyzed/min = slope (in dpm/min) × μmol ACh/dpm after total hydrolysis.

The advantages of both radiometric assays are that they are simple, reproducible, quite sensitive (μg of tissue per sample), suitable for application in multiple samples, and they use natural substrate. An additional advantage of the anion-exchange disk method (Gordon et al., 1982) is the requirement for relatively little manipulation. The only disadvantage of the radiometric methods is that they are relatively expensive.

3. Isolation

Various attempts at the solubilization and purification of mammalian brain AChE have been made. The main difficulty is that AChE is largely membrane-bound and therefore the bulk of the enzyme is insoluble in the usual aqueous media.

3.1. Solubilization and Extraction

In early attempts, a number of workers prepared soluble AChE by pretreating brain tissue with snake venoms, lipases, and proteases (Ord and Thompson, 1951; Ellman, 1962; Lawler, 1964; Ho and Ellman, 1969). Morton (1950) and Jackson and Aprison (1966a) first removed lipids with *n*-butanol and then extracted AChE into aqueous solution. Several nonionic detergents were tried (Ord and Thompson, 1951, Zittle et al., 1954), including Triton X-100 (Kremzner et al., 1967; Ho and Ellman, 1969). Jackson and Aprison (1966b) studied the effects of 30 commercially available surface-active agents, including anionic, cationic, and nonionic types, on the activity of AChE in homogenates of caudate nuclei. Sucrose-EDTA extraction of bovine caudate nuclei released 70% of the AChE activity (Chan et al., 1972a). Similarly, EDTA-tetracaine extraction of calf or mouse brain tissue led to the solubilization of 80–90% of the enzyme activity (Hollunger and Niklasson, 1973; Adamson et al., 1975). Kaplay and Jagannathan (1970) used pancreatic elastase, Lawler (1964) used ultrasonication, and Bernsohn and Barron (1964) used freezing and thawing to bring brain AChE into solution.

The most widely employed solubilization procedures have involved the combination of a high-salt concentration (0.4–1.0 mol/L NaCl) and the nonionic detergent Triton X-100 (0.5–1%) in the homogenation media. These result in the best solubilization of the AChE activity of brain tissue (Rieger and Vigny, 1976; Ruess et al., 1976; Rakonczay et al., 1981a; Sørensen et al., 1982).

Depending on the solubility, two main classes of AChE can be distinguished in mammalian brain: part of the enzyme activity is soluble without detergent, whereas the remainder is soluble only in the presence of micellar concentrations of detergents (Hollunger and Niklasson, 1973; Wenthold et al., 1974a; Rieger and Vigny, 1976; Rakonczay et al., 1981a, 1981b; Sørensen et al., 1982). It is noteworthy that in brain the ratio between salt-extractable and detergent-solubilized AChE is approximately 1:9 and does not depend on the species source.

3.2. Purification

The structural characterization of AChE begins, as with any other enzyme, with its purification. However, obtaining AChE in pure form from such a heterogenous material as brain is not an easy task. Another major problem is the solubilization of the enzyme, because of its relatively firm association with membrane compounds (see previous section).

3.2.1. Conventional Methods

The initial purification of AChE from brain tissue involved conventional purification methods (Jackson and Aprison, 1966a; Lawler, 1964). As starting material, acetone or *n*-butanol-extracted powder of bovine caudate nuclei was used. After laborious and low-yielding chromatography procedures, an overall purification of 30–35-fold with a 4–6% yield was attained (Jackson and Aprison, 1966a). Chan et al., (1972a) achieved a better purification by ion-exchange and molecular-sieve chromatography, with a specific activity of 13 mmol AThCh/mg protein/h. Their preparation was still not pure, since the most active AChE peak showed one contaminating band of protein on polyacrylamide gel electrophoresis.

3.2.2. Affinity Chromatography

Application of the selective separation technique of affinity chromatography has led to dramatic progress in the methods of purification of mammalian brain AChE. In this method, a substrate-analog, competitive, reversible inhibitor is covalently attached to a suitable matrix, a solid support thereby being obtained that is capable of binding the enzyme specifically and reversibly.

The first affinity chromatography procedure described was based on the cholinergic antagonist phenyltrimethylammonium and was used to purify AChE from the electric fish *Electrophorus electricus* (Kalderon et al., 1970). Subsequently, Chan et al. (1972b) published a five-step purification procedure for brain AChE. Bovine caudate nucleus tissue was extracted with sucrose-EDTA solution. After ammonium sulfate fractionation and agarose affinity gel chromatography (affigel 202-*m*-trimethylammonium aniline), the complete procedure represented an approximate 23,000-fold purification of the enzyme, with an average specific activity of 575 mmol AThCh hydrolyzed/mg protein/h. However, the purity of the enzyme cannot be assessed, since the electrophoretic photographs are not given.

Yamamura et al. (1973) reported the first case of affinity chromatography of AChE solubilized by Triton X-100. The enzyme was partially purified from guinea pig brain on a column of *m*-carboxyphenyltrimethylammonium iodide attached through 1,4-diaminobutane to Sepharose 4B. They achieved a 1000-fold purification with a specific activity of 10 mmol ACh hydrolyzed/mg protein/h.

Highly purified AChE was obtained from rat brain cortex by Wenthold et al. (1974a) using a similar method to that of Chan et al. (1972b); the specific activity was 270 mmol AThCh hydrolyzed/mg protein/h.

Partial purification of AChE from mouse brain (Adamson, 1977), rat brain synaptosomal membranes (Goodkin and Howard, 1974), and pig cerebral cortex (Reavill and Plummer, 1978) was achieved by affinity chromatography on a column of *N*-methyl-3-aminopyridinium iodide linked to agarose (MAP-agarose). The purified enzyme preparation from mouse brain was later used to produce antibodies (Adamson, 1977). Affinity column chromatography, based on the affinity ligand *N*-methylacridinium (MAC-agarose), was employed in the purification of AChE from pig cerebral cortex (Reavill and Plummer, 1978) and calf, rat, and human brain (Valette et al., 1983). The latter authors pointed out that this type of chromatography depends critically on the synthesis of the ligand and on its coupling to Sepharose. They underlined the necessity of establishing the optimum MAC concentrations for the binding of enzyme from different sources.

Phenylammonium ligands have recently been successfully employed in the purification to homogeneity of the soluble (Hodgson and Chubb, 1983) and the membrane-bound (Rakonczay et al., 1981a; Sørensen et al., 1982) AChE from mammalian brain. In the following sections a detailed protocol is given for the extraction and purification of these enzymes.

3.2.2.1. SECRETORY FORM OF SOLUBLE ACETYLCHOLINESTERASE It is well known that AChE is largely membrane-bound, especially in brain tissue. However, part of the enzyme can be solubilized in detergent-free buffers simply by homogenization and centrifugation (at 100,000*g* for 1 h).

There is general agreement that brain tissue contains several molecular forms (*see* section 4.1), but the released activity is usually attributable to the presence of one special form. This soluble form has been called secretory AChE (sAChE) (Chubb and Hodgson, 1982). Hodgson and Chubb (1983) reported a simple method for the selective isolation of this type of AChE from a variety of mammalian tissues, including the brain, adrenal glands, cerebrospinal fluid, and serum. The method involves a one-step purification based on adsorption of the enzyme to an affinity matrix of edrophonium coupled to epoxy-activated Sepharose 4B.

Preparation of extracts containing soluble AChE (Hodgson and Chubb, 1983) were performed as follows. The tissue was homogenized (10%, w/v) in either 0.32*M* sucrose or a medium containing 5 mmol/L Tris-HCl, 83 mmol/L glycine, and 0.15 mol/L NaCl, pH 8.2, in a Potter-Elvehjem homogenizer. The homogenates were frozen and thawed, and then centrifuged at 100,000*g* for 60 min to obtain the soluble AChE in the high-speed supernatant fraction. Samples were filtered through a range of microporous filters (final pore size, 0.22 μm) before application to the column. They were not dialyzed, and thus the salt concentration in each sample was in excess of 0.15 mol/L. All preparative work was carried out at 4°C.

Absorption and elution of AChE (Hodgson and Chubb, 1983) were performed as follows. Synthesis of the affinity gel (edrophonium-Sepharose-4B) was performed as described by Hodgson and Chubb (1983). Extracts were applied to a column with a bed volume of 12 mL, at a flow rate of 12 mL/h. The gel was washed with 50 mmol/L phosphate buffer, pH 8.0, followed by a salt wash using the same buffer containing 0.5 mol/L NaCl. The gel was then eluted with 12 mmol/L edrophonium chloride dissolved in phosphate buffer containing 0.5 mol/L NaCl. Fractions of 2 mL were collected. Prior to use, edrophonium and salts were removed from the enzyme preparations either by centrifugation through a dry bed of hydrated Sephadex G25 (Neal and Florini, 1973) or by chromatography on CM-Sephadex. The purity of the enzyme was analyzed by electrophoresis. The results of the purification are summarized in Table 3.

3.2.2.2. Membrane-Bound Acetylcholinesterase in Rat Brain. The membrane-bound AChE from adult rat brain was purified to homogenity using sequential affinity chromatography on Concanavalin A-Sepharose 4B and *m*-carboxyphenyldimethylethylammonium (*m*CPA)-Sepharose 4B, followed by DEAE-cellulose chromatography (Rakonczay et al., 1981a).

Preparation of an extract containing membrane-bound AChE was performed as follows. Rat brains were homogenized (10% w/v) in saline buffer (12.5 mmol/L sodium phosphate buffer, pH 7.2, 0.4 mol/L NaCl) in a glass homogenizer with a Teflon pestle, and the suspension was centrifuged at 53,000*g* for 2 h. The supernatant (S_1) was discarded and the pellet (P_1) was homogenized in 10 vol of cold distilled water. The suspension was added slowly to 90 vol of saline buffer containing 0.5% Triton X-100 and stirred for 30 min. The suspension was then centrifuged as above. The pellet (P_2) was

Table 3
Preparative Isolation of Secretory Acetylcholinesterase[a]

	AChE, U	Protein, mg	Specific activity, U/mg	Yield, %
Guinea pig brain soluble fraction	140	4,410	0.032	100
Unbound	8.5	4,320	0.002	6.1
Low-salt (0.15*M*)	11.9	300	0.040	8.5
Medium-salt (0.3*M*)	16.8	64	0.262	12.1
High-salt (1.0*M*)	21.0	2	9.500	15.0
Edrophonium eluate	39.2	nd		28.0

[a]Soluble fraction (1L) from guinea pig brain was applied to a 12-mL column of edrophonium-agarose. This was washed with sodium chloride in buffer in three steps: (a) low-salt (0.15*M*), (b) medium-salt (0.3*M*), and (c) high-salt (1.0*M*) before elution with edrophonium chloride (12 m*M*). Abbreviations: U, μmol AThCh hydrolyzed/min; nd, not detectable (from Hodgson and Chubb, 1983).

discarded and the supernatant (S_2) was used as starting material for purification of membrane-bound AChE. All operations were carried out at 4°C.

Adsorption and Elution of AChE were performed as follows. Concanavalin A-Sepharose 4B chromatography: The S_2 fraction was applied to a previously equilibrated (saline buffer containing 0.5% Triton X-100) 40 × 2 cm Concanavalin A-Sepharose 4B column with a flow rate of 70 mL/h. The column was then washed with 10 column vol of buffer. The enzyme was eluted with 6–8 column vol of the same buffer containing 0.5 mol/L α-methyl-D-mannoside (α-MM). The Concanavalin A-Sepharose chromatography step is important because the efficiency of the second affinity column is greatly reduced when it is omitted.

Affinity chromatography on *m*CPA-Sepharose 4B: The *m*CPA-Sepharose was prepared by the method of Massoulié and Bon (1976). The α-MM eluent from the Concanavalin A column was applied to a *m*CPA-Sepharose 4B column (40 × 2 cm) previously equilibrated in saline-Triton X-100 buffer and washed with 10 column vol of the same buffer. Bound AChE was eluted with 1 column vol of the washing buffer [containing 0.2 mol/L tetraethylammonium bromide (TEA)] at a flow rate of 70 mL/h. This fraction was dialyzed against large volumes of 12.5 mmol/L phosphate buffer, pH 7.2, containing 0.5% Triton X-100, for 24 h, with two changes to remove NaCl, α-MM, and TEA.

DEAE-cellulose chromatography: The dialyzed TEA eluate was applied to a 15 × 1 cm DEAE-cellulose column previously equilibrated with 12.5 mmol/L phosphate buffer containing 0.5% Triton X-100. The elution was performed with a 300-mL linear gradient of NaCl (0–0.4 mol/L) at a flow rate of 15 mL/h.

Second affinity chromatography on mPCA-Sepharose: The most active fractions were pooled and applied to a small (10 × 1 cm) column equilibrated as described above. The column was then washed with 10 column vol of saline buffer containing 0.05% Triton X-100 and eluted with the smallest volume of the same buffer (containing 0.2 mol/L TEA) at a flow rate of 15 mL/h. The purified enzyme was dialyzed against 0.05% Triton X-100 and stored at –20°C.

The results of the purification are summarized in Table 4. After the last step the purity of the enzyme was at least 95–98%, as judged by means of SDS-PAGE.

3.2.2.3. MEMBRANE-BOUND ACETYLCHOLINESTERASE IN HUMAN CAUDATE NUCLEUS Very recently, Sørensen et al. (1982) purified to homogenity a detergent-extractable AChE form from human brain caudate nucleus.

Extraction of caudate nucleus (Sørensen et al., 1982) was performed as follows. Caudate nuclei were extracted in the following sequence. First, the tissue was homogenized in a Waring-type blender, with 15 vol of 10 mmol/L Tris-HCl, pH 7.4, containing 1 mol/L NaCl and 50 mmol/L $MgCl_2$. After stirring for 3 h, the homogenate was centrifuged at 100,000*g* for 1 h. The supernatant (S_1) was removed and the pellet was resuspended in 15 vol of 10 mmol/L Tris-HCl, pH 7.4, containing 1% Triton X-100 and 144 mmol/L NaCl. After stirring for at least 3 h, the suspension was centrifuged as above. The supernatant (S_2) was removed and used in the purification procedure. All steps were performed at 4°C.

Enzyme purification (Sørensen et al., 1982) was performed as follows. The synthesis of the affinity gel (*N,N,N*-trimethylammonium *m*-phenylenediamine-Sepharose 4B) was performed as described by Brodbeck et al. (1981). The S_2 fraction was applied to a column (2.5 × 25 cm), which was washed with about 500 column vol of 20 mmol/L Tris-HCl buffer, pH 7.4, containing 1 mol/L NaCl, 30 mmol/L EDTA, and 1% Triton X-100, until the protein content of the eluate was below the detection limit. The enzyme was then specifically eluted with 20 mmol/L edrophonium. Enzyme activity appeared in one peak, which was collected and dialyzed to remove inhibitor. The results of the purification are summarized in Table 5.

Table 4
Purification of Membrane-Bound Rat Brain Acetylcholinesterase[a]

Fraction	Total protein, mg	Total activity, U	Specific activity, U/mg protein	Purification, X-fold	Yield, % of S_2 activity
Homogenate	1179	206.8	0.18	1.00	—
S_2 solubilized only by Triton X-100	559	159.0	0.29	1.64	100
Concanavalin A-Sepharose 4B chromatography	25.9	130.4	5.03	28.8	82
Affinity chromatography on *m*CPA-Sepharose 4B	0.096	120.8	1254.6	7176.6	76
DEAE-cellulose chromatography	nd	84.3	nd	nd	53
Affinity chromatography on *m*CPA-Sepharose 4B	0.025	81.2	3068.0	17551	51

[a]This typical control preparation was obtained from 12 rat brains. The solubilization and chromatographic procedures are given in section 3.2.2.2. Abbreviations: nd, not determined (because of the very dilute protein solution in 0.5% Triton X-100); U, μmol AThCh hydrolyzed/min (from Rakonczay et al., 1981a).

Table 5
Summary of Purification of the Amphiphile-Dependent Form of AChE From Human Brain Caudate Nucleus[a]

	Activity, IU/mL	Protein, mg/mL	Specific activity, IU/mg prot.	Yield %	Purification, X-fold
Homogenate	1.16	3.4	0.34	100	1
Supernatant (S_1)	0.154	2.2	0.07	18	—
Supernatant (S_2)	1.32	1.7	0.73	80	2
Affinity-Sepharose eluate, combined fractions	11.0	bld	—	77	—
Peak fraction	35.4	0.0078	4500	(43)	27,000

[a]Enzyme activity was measured by the Ellman assay (Ellman et al., 1961) and protein content by the method of Wang and Smith (1975), which excluded interference by Triton X-100. Abbreviations: IU, μmol AThCh hydrolyzed/min; bld, below limit of detection (from Sørensen et al., 1982).

4. Structure and Properties

Most information about the structure and properties of AChE has come from studies of the electric organ of Torpedo and of the electric eel *(Electrophorus electricus)*, which is one of the richest sources of the enzyme available (Nachmansohn and Lederer, 1939). Much of the work aimed at purifying AChE has used this tissue (e.g., Rothenberg and Nachmansohn, 1947; Lawler, 1959; Kremzner and Wilson, 1963; Leuzinger and Baker, 1967; Kalderon et al., 1970; Berman and Young, 1971; Dudai et al., 1972; Massoulié and Bon, 1976). Characterization of the physical properties of AChE from mammalian brain has progressed more slowly because of the relatively small quantities of enzyme available.

4.1. Types of Molecular Forms

The polymorphism of AChE in brain tissue has been well documented by electrophoretic analysis (Bernsohn et al., 1962, 1963; Davis and Agranoff, 1968; Srinivasan et al., 1972; McIntosh and Plummer, 1973, 1976; Vijayan and Brownson, 1974) and by sedimentation and gel chromatography (Hollunger and Niklasson, 1973; Adamson et al., 1975; Rieger and Vigny, 1976; Adamson, 1977; Rakonczay et al., 1981a, 1981b; Grassi et al., 1982; Sørensen et al., 1982). The heterogenity of AChE in neuroblastoma cells has been described by many authors (Rieger et al., 1976; Chang and Blume, 1976; Vimard et al., 1976; Lazar and Vigny, 1980).

Bernsohn et al. (1962, 1963) first suggested that in human brain the charge-isomers of AChE are aggregates. A similar observation has been made for size-isomers in calf brain by Hollunger and Niklasson (1973), and also those in mouse brain by Adamson et al. (1975). Later it was discovered that there is not a one-to-one relationship between the charge-isomers and the size-isomers (Gisiger et al., 1978; Rakonczay et al., 1982; Kása and Rakonczay, 1982).

Characterization of the physical properties of AChE progressed considerably when Massoulié and Rieger (1969) introduced density gradient ultracentrifugation to separate the different stable size-isomers of AChE. Later, the term size-isomers was frequently replaced by molecular forms, since sedimentation coefficients gives more accurate physicochemical information than electrophoresis in a nondenaturing system.

Recent studies have shown that the AChE molecular forms from mammalian tissue are homologous in their quaternary structure with those of the electric organ (Bon et al., 1979). Generally, two sets of AChE have been distinguished: the asymmetric forms (A_4, A_8, and A_{12}), in which 1–3 tetramers of catalytic subunits are associated with a three-stranded collagen-like tail, and the globular (G_1, G_2, and G_4) forms, including 1, 2, and 4 catalytic subunits (Massoulié et al., 1971, 1980; Silman, 1976; Rosenberry and Richardson, 1977; Anglister and Silman, 1978; Massoulié and Bon, 1982).

Sedimentation analysis of mammalian brain AChE shows that the enzyme is present predominantly in the G_4 (10S) form (70–90%) (Rieger and Vigny, 1976; Rakonczay et al., 1981a, 1981b; Rodriguez-Borrajo et al., 1982). In the tetrameric (G_4) form, the catalytic chains are associated in dimers, by disulfide bonds (Vigny et al., 1979). However, small proportions of G_1 (4S) (10–30%) and a detectable level of A_{12} (16S) (2–5%) are also present (Rieger et al., 1980; Rakonczay et al., 1981b; Rodriguez-Borrajo et al., 1982). It worth noting that it is possible to increase the yield of the solubilized G_1 form to 40–90% by repeated extractions of calf or bovine caudate nuclei in a detergent-free buffer (Hollunger and Niklasson, 1973; Grassi et al., 1982). The G_2 (6.5S) form might be considered an intermediate transitional form that quickly transforms to G_4; it can therefore be revealed under special circumstances (e.g., during ontogenesis) (Wade and Timiras, 1980). This hypothesis is supported by the experiments of Vigny et al. (1979), who have shown that the dimeric form (G_2) is readily reduced into the monomer (G_1), whereas the tetramer (G_4) is very stable, being dissociated only by a combination of reduction and proteolysis with a high concentration of trypsin. For this reason, it is useful to work in the presence of antiproteolytic agents (EDTA, bacitracin, benzamidine, and so on) (*see* Silman et al., 1978) during the extraction and the separation of the molecular forms. Finally, the A_4 and A_8 forms seem to be present only as traces (0.002%) in brain tissue (Grassi et al., 1982).

It should be pointed out, however, that differential solubilization can be achieved from the brain using a high-salt buffer in the absence and presence of detergent (Rieger and Vigny, 1976; Rakonczay et al., 1981a,b; Grassi et al., 1982). Hydrophilic AChE molecular forms are preferentially extracted in detergent-free medium (G_1, A_{12}, and a small part of G_4), whereas a second treatment in detergent-containing medium extracts the hydrophobic form (G_4). It should also be noted that the detergent-solubilized forms aggregate upon the removal of detergent.

On the other hand, the 10S form may possess a short hydrophobic polypeptide that is probably digested by pronase

(Grassi et al., 1982). This polypeptide might act as an anchor of the enzyme in the lipid bilayer of membranes (Lazar and Vigny, 1980; Sørensen et al., 1982; Grassi et al., 1982).

One of the interesting properties of the asymmetric forms is that they are poorly soluble at low ionic strength (Bon et al., 1978; Cartaud et al., 1978). This property makes it possible to separate them from globular forms. If the tissue is initially extracted in high-salt buffer and then dialyzed against a low-salt buffer, the asymmetric forms are precipitated (Bon et al., 1979; Gisiger and Stephens, 1983).

Another approach to separate globular and asymmetric forms is to homogenize the tissue initially in detergent-containing low-salt buffer (solubilized globular forms) and to follow this by rehomogenization of the pellet in high-salt buffer, releasing the asymmetric forms (Grassi et al., 1982).

However, if EDTA is added to the homogenization buffer, a relatively high amount of AChE will be recovered in the supernatant (*see* section 3.1), which may mean that Ca^{2+} ions play a role in the attachment of the enzyme to the membranes.

It is worth emphasizing that in mammalian brain the globular forms constitute the major part of AChE, and only a small percentage of the enzyme is asymmetric. Although these asymmetric forms are collagenase-sensitive (Rodriguez-Borrajo et al., 1982), there is no direct evidence (electron microscopic pictures) that the purified forms contain a collagen-like tail similar to that of the electric eel enzyme (Cartaud et al., 1978).

Grossman et al. (1979) have determined the hydroxyproline content of highly purified bovine caudate nucleus AChE. They have shown that the amount of hydroxyproline is more than 20 times less in the brain enzyme than that of the 14S/18S forms of electric organ AChE with similar specific activity.

Further experiments are needed to determine whether the asymmetric forms in the brain really possess a collagen-like tail. There must be some other way(s) of anchoring AChE to the membranes than that in the electric organ. The hydrophobic globular forms (mostly G_4) probably have similar roles in the brain as the asymmetric forms have in the neuromuscular junctions or in the electric organ. The hydrophobic peptide(s) (see above) might help to bind the G_4 form to the membraneous structure of the cholinergic synapse.

For more details of the nature, chemistry, and dynamics of AChE forms, the reader is referred to the excellent reviews of Massoulié and Bon (1982) and Brimijoin (1983).

4.1.1. Methods for Separation

The existence of multiple molecular forms of brain AChE can be demonstrated by sucrose density gradient ultracentrifugation or by application of other separation procedures, including molecular sieve (gel) chromatography and gel electrophoresis. The molecular forms may also be separated by analytical or preparative isoelectric focusing (Wenthold et al., 1974a; Rakonczay et al., 1981a), but they give a very complex pattern even when almost pure brain AChE is used (Rakonczay et al., 1981a), suggesting artifacts caused by the interactions between the enzyme and ampholytes during electrofocusing.

4.1.1.1. DENSITY GRADIENT ULTRACENTRIFUGATION. Sucrose density gradient ultracentrifugation has been found to be a suitable method for determining the sedimentation coefficients of the enzyme in protein mixtures (Martin and Ames, 1961). A linear 5–20% sucrose gradient, which is that most commonly used, permits nearly complete separation of the different molecular forms of AChE, assuming a linear relationship between the migration and sedimentation coefficients. Moreover, it also permits analytical measurements to be made (Massoulié and Rieger, 1969; Bon et al., 1973; Hall, 1973; Dudai and Silman, 1974). The 5–20% sucrose gradients are formed prior to centrifugation, either by the hand-layering process or with gradient-former devices. The former begins with a half vol of 20% sucrose solution in the tube. The second half vol of 5% sucrose solution is carefully overlayered and allowed to diffuse. The sample (0.2–0.5 mL, depending on the volume of the gradient) is mixed with marker enzymes *E. coli* β-galactosidase (EC 3.2.1. 23, 16S), beef liver catalase (EC 1.11.1.6, 11.3S), and calf intestine alkaline phosphatase (EC 3.1.3.1, 6.1S), and centrifuged (BECKMAN SW 55 rotor, 105,000*g*, 33,000 rpm for 15–17 h; SW 28.1 rotor, 105,000*g*, 27,700 rpm for 32–34 h; SW 41 rotor, 178,000*g*, 38,000 rpm for 14–16 h). The gradients are emptied from the bottom with a glass capillary by means of a peristaltic pump, or the bottom of the cellulose nitrate centrifuge tube is simply punctured with a needle. Fractions are collected and assayed for AChE activity, and the sedimentation coefficients are estimated by comparison with enzyme markers. The activities of the marker enzymes are measured according to the following methods: β-galactosidase (Quarles and Brady, 1970), catalase (Beers and Sizer, 1952), and alkaline phosphatase (Bessey et al., 1946).

The relative proportions of the different forms can be established by integrating the areas under the centrifugation profiles and dividing by the total area.

4.1.1.2. GEL FILTRATION. Calibrated molecular-sieve chromatography is a suitable tool with which to estimate the approximate molecular weights of brain AChE forms. Although this separation method is less convenient and requires more time than gradient centrifugation, its main advantage is the high capacity. It is a useful technique if large quantities of one specific form are needed for detailed analysis.

The best separations of the molecular forms can be obtained with a column of Biogel A 1.5 m (Bio-Rad) (10–1500 kdaltons) or with Ultrogel AcA 34 (LKB) (20–750 kdaltons) (Bon et al., 1973; Rakonczay et al., 1981b). For analytical work, columns 90 cm high and 1 cm in diameter are employed, whereas for preparative purposes, 90 cm × 2 cm columns are routinely used. After equilibration of the column, samples (1–3 mL) of enzyme solution containing molecular weight markers are always employed and eluted with buffer in the presence or absence of 0.5–1% Triton X-100. Fractions of 0.5–1 mL are collected.

Standards with the following molecular weights and Stokes' radii (R_e) are used: cytochrome c, 12,000 R_e 1.74 nm; ovalbumin, 43,000 R_e 3.05 nm; aldolase, 158,000 R_e 4.81 nm; catalase, 232,000 R_e 5.22 nm; β-galactosidase, 540,000 R_e 8.2 nm; thyroglobulin, 750,000 R_e 8.5 nm; dextran blue, 2,000,000. The positions of the markers are determined via absorbance (610 nm for blue dextran, 527 nm for cytochrome c; 280 nm for all proteins) or enzyme activity (*see* section 4.1.1.1).

The elution parameter K_d is defined as $V_e - V_o / V_t - V_o$ (V_o = void volume, V_t = total volume, determined with potassium ferricyanide, V_e = elution volume of the protein or AChE). Apparent molecular weights can be estimated by comparison with a graph produced by plotting K_d values of marker proteins against the logarithm of molecular weight.

The Stokes' radii of the molecular forms can be calculated from the gel filtration parameters by means of two different equations (Ackers, 1964; Siegel and Monty, 1966), as described by Rakonczay et al. (1981b).

4.1.1.3. GEL ELECTROPHORESIS. Electrophoretic techniques depending on both charge differences and molecular sieving usually give much more complex patterns than do density gradient ultracentrifugation or gel filtration. AChE variants may arise from covalent modifications of the enzyme (such as phosphorylation or glycosylation, and so on); they therefore differ in their electrostatic charges. According to Wenthold et al. (1974a), a small difference in unit charge could be sufficient to alter the electrophoretic mobility.

Electrophoretic analysis of soluble or membrane-bound AChE

is carried out on 6% polyacrylamide disk or slab gels, by means of the simplified system of Clarke (1964). When Triton extracts are examined, 0.05% (v/v) Triton X-100 should be present in both the electrophoresis buffer and the gels. The advantage of using gradient gels (4–30%) is that the enzyme-active bands become sharper. The demonstration of enzyme activity in gels is based on the histochemical method of Koelle (1951) using substrate AThCh, as modified by Chubb and Smith (1975), where copper is stained by dithiooxamide reagent. After densitometry, the relative specific activity of each charge-isomer may be calculated from the percentage distribution of each peak relative to the total AChE activity.

The subunit molecular weight of purified AChE (or the purity of the enzyme) can be estimated via polyacrylamide gel electrophoresis in the presence of sodium dodecyl sulfate and β-mercaptoethanol in a discontinuous buffer system, according to Leammli (1970).

AChE samples and standard proteins are denatured prior to electrophoresis by heating for 3 min at 100°C in a solution of 10% v/v glycerol and 60 mmol/L Tris-HCl buffer (pH 6.8) containing 2% w/v sodium dodecyl sulfate and 5% v/v β-mercaptoethanol. Polypeptides are visualized by any protein staining method. The molecular weights of the subunits can be estimated from the calibration curve, which is determined by the method of Shapiro et al. (1967) from the relative mobilities of standard proteins.

4.2. Biochemical Properties

As already discussed in section 4.1, the multiple forms of brain AChE may be related to the aggregation of a single low-molecular-weight species. The biochemical properties of enzyme forms from different mammalian species are similar (Table 6). Subunit molecular weights are in the 60–78 kdalton range and several association states of these subunits exist, giving rise to a multiplicity of molecular forms (Table 6). These forms are stable entities and can be characterized through the Stokes' radii and sedimentation coefficients. They are all glycoproteins, containing about 15% carbohydrate, since they have a great affinity for lectins (Wenthold et al., 1974a; Gurd, 1976; Rieger and Vigny, 1976; Rakonczay et al., 1981a).

It seems extremely interesting to examine whether the different forms are catalytically equivalent and insensitive to subunit interactions. Vigny et al. (1978) have used a stable, irreversible phosphorylating inhibitor to titrate the active site of the different forms of AChE from *Electrophorus, Torpedo,* rat, and chicken brain.

They found that in all cases the molecular forms were equivalent in their catalytic activity per active site. Moreover, in the case of rat brain enzyme, the activity per active site of solubilized (with or without Triton X-100) and membrane-bound enzyme were identical within experimental error.

4.3. Immunochemical Properties

Immunochemical studies with mammalian brain AChE depend on the availability of highly purified enzyme in sufficient quantities to elicit antibodies. Only a few papers have been reported on the production of specific polyclonal (Adamson, 1977; Greenberg et al., 1977; Zanetta et al., 1981) or monoclonal (Brimijoin et al., 1983) antibodies against brain enzyme.

An interesting property of the specific antibodies is that they inhibit the activity of AChE from mouse brain, bovine caudate nucleus, or rat brain: up to 50–60% of the activity is lost during precipitation with antibody (Adamson, 1977; Greenberg et al., 1977; Zanetta et al., 1981). The remaining activity is important in the detection of antibody–enzyme complexes. Such inhibition may be caused by certain of the antigenic determinants located close to the active site of the enzyme and restricting access of the substrate to the active site region.

Immunochemical methods can detect subtle structural differences between AChE isolated from different species. In cross-over experiments using both eel AChE and bovine caudate nucleus AChE polyclonal antisera, there was a complete absence of cross-reactivity between the mammalian brain enzyme and either the globular or the elongated forms of the electric tissue enzyme (Greenberg et al., 1977).

Immunodiffusion revealed multiple precipitation bands for mouse brain polyclonal antisera, which might reflect the presence of distinct antibodies to the different molecular forms of the enzyme (Adamson, 1977). Antibody against purified rat brain membrane-bound AChE does not react on the species of AChE that are soluble in the absence of detergent. These results suggest that membrane-bound and soluble brain enzyme do not share similar antigenic determinants (Zanetta et al., 1981). The difference in immunological reactivity could be related to the presence of a short hydrophobic peptide(s) on the catalytic subunits (*see* section 4.1).

One of the most suitable tools for determination of the molecular basis for the heterogeneity of AChE of different tissue origins is the use of the hybridome technique. Two recent reports indicate that the forms of mammalian enzyme are immunologically similar

Table 6
Biochemical Properties of Mammalian Brain AChE

Source	Subunit, mol wt	"Native," mol wt	Stokes radius, nm	$s_{20,w}$, S	References
Human caudate nucleus (membrane-bound)	78,000	320,000	8.5	10.0	Sørensen et al., 1982
Mouse	74,000	74,000		3.5	Adamson et al., 1975
		155,000			Rieger et al., 1980
		360,000		9.5	
		720,000		16.0	
Rat	74,500 (membrane-bound)	328,000	8.1	9.8	Rakonczay et al., 1981a,b,c
		106,000	5.4	4.4	
	70,000	304,000	6.8	10.7	
	(soluble)	608,000	10.9	16.0	
Bovine caudate nucleus	60,000	60,000			Ruess et al., 1976
	72,000	120,000			Gordon et al., 1976
		230,000			
		330,000			
Pig	60,000	60,000			McIntosh and Plummer, 1976
		270,000			
		350,000			
Mouse neuroblastoma		116,000		4.0	Chang and Blume, 1976
		186,000		6.0	
		284,000		9.6	

in that they all cross-react with monoclonal antibodies to AChE (Fambrough et al., 1982; Brimijoin et al., 1983). Fambrough et al. (1982) have concluded that there is a high level of homology between the human AChE of erythrocytes and neuromuscular junctions, although they were unable to stain AChE-positive neurons in human caudate nucleus and spinal cord with monoclonal antibodies. On the other hand, Brimijoin et al. (1983) measured identical dissociation constants for the monoclonal antibody complexes with human AChE from red blood cell (G_2) and brain (G_4), which means that they are similar with regard to the recognized epitopes.

There are as yet no data on whether the molecular forms in the brain differ from each other in primary structure and sites of antigenicity. However, it was shown by Doctor et al. (1983) that monoclonal antibodies with a high affinity for the dimeric form cross-react poorly with the tail-containing asymmetric tetrameric enzyme from *Torpedo californica*.

5. Occurrence and Distribution

Biochemical examination of the specific activity of AChE in different brain areas indicates that in cat, rat, rabbit, and human (Ord and Thompson, 1952; Thomas et al., 1971; Yaksh et al., 1975; Hetnarski et al., 1980; Kása et al., 1982a) the highest levels are to be found in the caudate nuclei, and the lowest in the cortex. A comprehensive survey of the enzyme in different brain parts in mammalians was given earlier by Silver (1974).

Histochemically (Koelle and Friedenwald, 1949), the enzyme is present in the perikaryon of the neurons. Ultrastructurally, the reaction end-product caused by the enzyme is present in the rough endoplasmic reticulum (RER) and the smooth endoplasmic reticulum (SER), as well as on the outer surface of the axon and the dendrite membrane (Kása and Csillik, 1966, 1968; Eränkö et al., 1967; Kása, 1968).

More recently, it has been demonstrated that the different molecular forms of AChE (4S, 6S, 10S, and 16S) are present not only in the PNS, but also in the CNS (Rieger and Vigny, 1976; Rieger et al., 1980; Rakonczay et al., 1981b; Rodriguez Borrajo et al., 1982). Histochemically, these forms of the enzyme in the PNS can be localized in the different structures of cholinergic and cholinoceptive neurons: e.g., 4S and 6S in the perikaryon, 10S (B and B′) in the axon (Kása and Rakonczay, 1982), and 16S in the synaptic region (Hall, 1973). Since the enzyme is present in similar structures in the CNS, we can suppose that the different molecular forms of AChE are also present in the RER, SER, nuclear envelope, axonal and dendrite surface, and synaptic region.

6. Ontogenesis and Biosynthesis of the Enzyme

During the development of the rat brain, the G_1 (4S) form is predominant before birth, and is converted rapidly to the G_4 (10S) form. In adult brain, 90% of the total AChE activity is a result of the G_4 form (Rieger and Vigny, 1976). This observation suggests that the light form is a precursor of the heavier forms. This hypothesis has found support from other results: e.g., after complete inhibition of AChE activity, neuroblastoma cells synthesize the G_1 form first; the G_4 form appears only later (Rieger et al., 1976), or during the induction of AChE activity in murine neuroblastoma × sympathetic ganglion hybrid cell cultures, there is an extensive and rapid conversion of the G_1 to the G_4 form (Lazar and Vigny, 1980). During the postnatal development of AChE in rat cerebellum, there is an enormous increase (400%) in the enzyme activity, which is closely connected with certain histogenic events in the brain. Moreover, the ratio 4S/10S is high at birth and low in the adult (Kása and Lehotai, 1979; Wade and Timiras, 1980; Kása et al., 1982b). Meanwhile, the relative percentage of the 16S form does not change (Rakonczay and Kása, unpublished results). These experiments strongly suggest that the molecular forms are metabolically related.

The AChE molecule has a fairly rapid turnover. In rat brain cortex, the half-life of the enzyme was estimated by two different methods to be about 2.8–3.6 d (Wenthold et al., 1974b; Yaksh et al., 1975). These experiments present good evidence that the half-life of AChE is considerably lower than the average half-life of about 20 d for membrane proteins in rat brain (von Hungen et al., 1968).

A promising new technique has been developed by Soreq et al. (1982) for monitoring the level of the mRNA species, poly(A)-containing RNA from whole rat brain or rat cerebellum, that direct the synthesis of AChE. Microinjected *Xenopus* oocytes are used as a translation system. This in vitro method may open new avenues in the study of the development and regulation of AChE.

7. Possible Biological Function in the Brain

This is perhaps the most speculative part of the present chapter. There is no doubt that the main function of brain AChE is to hydrolyze AChE in cholinergic and cholinoceptive neurons. Although AChE is a widely distributed enzyme, there remains a long-standing question as to whether AChE has other functions, especially when cholinergic transmission is not operating (Silver, 1974). The following results may suggest that AChE does have other functions than to destroy ACh.

For example, AChE is found in the paraventricular and supraoptic nuclei, where the hormones oxytocin and vasopressin are synthesized (Cottle and Silver, 1970). The occurrence of the enzyme with neurosecretory systems in the same morphological region is evident, but the relationship between them has not yet been satisfactorily explained. It has also been shown that AChE has the ability to degrade substance P (Chubb et al., 1980) and enkephalins (Chubb et al., 1983). The authors interpret their results to mean that one physiological function of the enzyme may be to hydrolyze these neuropeptides in order to terminate their biological activity. This hypothesis is strongly supported by Altschuler et al. (1983) and Salipan et al. (1983), who have demonstrated a coincidence of AChE and neuropeptides in certain brain areas. Kása and Csillik (1966) showed that the enzyme is present in the basement membrane of capillaries of cat cerebellar cortex. On the basis of these electron microscopic studies, Joó and Várkonyi (1969) concluded that AChE and BuChE may play a role in the maintenance of the blood–brain barrier. Kása and coworkers (Kása and Csillik, 1964; Kása et al., 1966) made the interesting observation in the adult rat that, when lesions were produced in the archicerebellar white matter, transient AChE activity was demonstrable histochemically in the normally nonreactive Purkinje cells. The authors suggested that the enzyme could be involved in protein synthesis initiated in the cells in an attempt to repair the damage (metabolic function of AChE). In recent years it has been demonstrated that AChE also possesses amidase activity (Moore and Hess, 1975) and an aryl acylamidase activity (Oommen and Balasubramanian, 1979), though their physiological substrates are still not known.

8. Relationship to Mental Disorders

Cholinergic processes have been implicated in many neurological disorders, such as mania, depression, schizophrenia, tardive dyskinesia, Huntington's disease, and senile dementia (see in comprehensive books; Usdin et al., 1977; Lipton et al., 1978; Davis and Berger, 1979).

Janowsky et al. (1972a, 1972b) proposed the possibility that the affective disorders, mania and depression, in which a disorder of mood or affect predominates, might be caused by a complex balance or interaction between adrenergic and cholinergic factors. For example, parasympathetic suppression or reversal of manic symptoms has been elicited with physostigmine, a centrally active AChE inhibitor (Janowsky et al., 1973). Gershon and Shaw (1961) studied a number of patients who had been poisoned by organophosphate cholinesterase inhibitor insecticides, and reported an increased incidence of depression and peripheral parasympathetic toxicity as sequalae.

The evidence for cholinergic disturbances in schizophrenia is weak (Domino, 1979). Nevertheless, it appears that further research is needed to test this possibility.

Tardive dyskinesia comprises a group of involuntary movements. Dopaminergic hypersensitivity results from cholinergic hypoactivity in the striatum, and this state gives rise to the symptomatology (Gerlach et al., 1974; Klawans and Rubovits, 1974). The same authors have also reported that physostigmine, which increases brain ACh, produces some improvement in tardive dyskinesia.

Huntington's chorea is an extrapyramidal genetic disorder, transmitted as an autosomal dominant trait and characterized by the loss of small neurons in the caudate and putamen, with associated astrocytic proliferation (McMenemy, 1958; Bruyn, 1968). Biochemical studies suggest that the most consistent lesion in Huntington's disease may be a degeneration of small GABAergic striatonigral neurons, with a loss of CAT activity, muscarinic receptors, and striatal cholinergic neurons (Perry et al., 1973; McGeer et al., 1973; Bird and Iversen, 1974; Stahl and Swanson, 1974; Hiley and Bird, 1974).

Although changes in AChE activity have been reported that differ from normal, their biological significance may be questionable. It is well known, however, that one can inhibit AChE up to 80–90% before functional impairment results. At what level an increased or decreased enzyme activity will result in functional alteration is not known.

Human and animal cognitive studies, as well as neurochemical investigations, have led to the cholinergic deficiency hypothesis of senile dementia. Evidence has recently accumulated that AChE may be directly involved in this mental disorder.

8.1. Dementia of Alzheimer Type

Alzheimer's disease is a progressive cerebral degeneration that continues without remission until death; it usually involves profound dementia. Neither the cause of the disease nor any effective treatment are yet known. The principal neuropathological features are characterized by a large number of cortical senile plaques and neurofibrillary tangles (Kidd, 1964; Terry, 1963; Terry et al., 1964).

With respect to a cholinergic abnormality, many authors (Davies and Malloney, 1976; Perry et al., 1977, 1978; Davies, 1979) have noticed marked decreases in CAT and AChE activities. Davies and Malloney (1976) have shown maximum reductions in those areas of the cerebral cortex that contain the greatest density of neurofibrillary tangles. The same authors have also demonstrated that, in all areas of Alzheimer's disease brain studied, the activity of glutamic acid decarboxylase appears to be well within the normal range, which supports the notion that a selective cholinergic degeneration process has occurred.

Using density gradient centrifugation, Atack et al. (1983) have distinguished three molecular forms of AChE (4S, 10S, and 16S) in the postmortem neocortex in both normal cases and senile dementia of Alzheimer type. Although the levels of the 4S and 16S forms were unaltered in Alzheimer's disease, there was a selective and extensive loss of the 10S form. In light of the above-mentioned data, they concluded that the selective loss of the tetrameric form in the cerebral cortex in this disease may reflect degeneration of the cholinergic axonal process.

9. Conclusions

In recent years there has been significant progress in the field of AChE research, although many questions concerning brain AChE are far from being answered, in spite of the steady accumulation of detailed knowledge. The aim here has been to identify the principal themes, but the limitations of space have prevented anything approaching a comprehensive survey.

Whatever the animal source, brain AChE is heterogenous and consists of oligomers of basic subunit(s); both the quaternary structure and the subunit composition are reasonably well understood, though several questions remained unanswered.

There is no clear evidence that the molecular structures of asymmetric brain forms include a tail component similar to that existing in the "native" forms of the electric organ enzyme. The primary structures (both amino acid sequence and carbohydrate composition) of the subunits have not yet been determined.

The functional roles of the molecular forms and their precise localization are still unsolved, especially for the asymmetric forms, which are present in surprisingly low amounts in the brain compared to the muscle and some parts of the PNS. Furthermore, it is not known whether the molecular heterogeneity (e.g., charge-isomers) of the enzyme is posttranslational in origin or whether different AChE genes operate in different tissues, nor how these determine the distribution among the molecular forms. There are no data on the number of genes encoding this enzyme, and the possible physiological roles of the multiple molecular weight forms, and the assembly processes of the larger forms still remain largely unclear.

In spite of these facts, many problems have already been solved, and if AChE research goes forward as quickly as it has during the last ten years, many of the above remaining questions will soon be answered.

Acknowledgments

The author is especially indebted to Prof. P. Kása for helpful discussion and for critical reading of this manuscript. He would also like to extend his appreciation to Prof. U. Brodbeck, Dr. R. K. Gordon, Dr. A. J. Hodgson, and Dr. H. Soreq for permission to reproduce parts of their published work, as well as to Mrs. Gyöngyi Ugrin for the secretarial work.

Abbreviations

ACh	Acetylcholine
AChE	Acetylcholinesterase
αMM	α-Methyl-*D*-mannoside
AThCh	Acetylthiocholine
BuCh	Butyrylcholine
BuChE	Butyrylcholinesterase
BuThCh	Butyrylthiocholine
CAT	Choline acetyltransferase
CNS	Central nervous system
CPM	*N*-(4-(7-Diethylamino-4-methylcoumarin-3-yl)phenyl)maleimide
DFP	Diisopropylfluorophosphate
DTNB	5,5'-Dithiobis-(2-nitrobenzoic acid)
EDTA	Ethylenediaminetetraacetate
MAC	*N*-Methylacridinium
*m*CPA	*m*-Carboxyphenyldimethylethylammonium

MAP	*N*-Methyl-3-aminopyridinium iodide
Mol wt	Molecular weight
PNS	Peripheral nervous system
POPOP	1,4-Di-2-(5-phenyloxazolyl)benzene
PPO	2,5-Diphenyloxazole
PrCh	Propionylcholine
R_e	Stokes' radius
RER	Rough endoplasmic reticulum
SER	Smooth endoplasmic reticulum
TEA	Tetraethylammonium
ThCh	Thiocholine
Tris	2-Amino-hydroxymethylpropane-1,3-diol

References

Ackers G. K. (1964) Molecular exclusion and restricted diffusion processes in molecular-sieve chromatography. *Biochemistry* **3,** 723–730.

Adamson E. D. (1977) Acetylcholinesterase in mouse brain erythrocytes and muscle. *J. Neurochem.* **28,** 605–615.

Adamson E. D., Ayers S. E., Deussen Z. A., and Graham C. F. (1975) Analysis of the forms of acetylcholinesterase from adult mouse brain. *Biochem. J.* **147,** 205–214.

Altschuler R. A., Parakkal M. H., and Fex J. (1983) Localization of enkephalin-like immunoreactivity in acetylcholinesterase positive cells in the guinea-pig lateral superior olivary complex that project to the cochlear. *Neuroscience* **9,** 621–630.

Anglister L. and Silman I. (1978) Molecular structure of elongated forms of electric eel acetylcholinesterase. *J. Mol. Biol.* **125,** 293–311.

Atack J. R., Perry E. K., Bonham J. R., Perry R. H., Tomlinson B. E., Blessed G., and Fairbairn A. (1983) Molecular forms of acetylcholinesterase in senile dementia of Alzheimer type: Selective loss of the intermediate (10S) form. *Neurosci. Lett.* **40,** 199–204.

Augustinsson K.-B. (1971) Determination of Activity of Cholinesterases, in *Methods of Biochemical Analysis,* suppl. vol. *Analysis of Biogenic Amines and their Related Enzymes,* (Glick, D., ed.), Interscience Publishers, New York.

Beckett A. H. (1967) Stereospecificity in the reactions of cholinesterase and the cholinergic receptor. *Ann. NY Acad. Sci.* **144,** 675–688.

Beers R. F., Jr. and Sizer I. W. (1952) A spectrophotometric method for measuring the breakdown of hydrogen peroxide by catalase. *J. Biol. Chem.* **195,** 133–140.

Berman J. D. and Young M. (1971) Rapid and complete purification of acetylcholinesterases of electric eel and erythrocyte by affinity chromatography. *Proc. Natl. Acad. Sci. USA* **68,** 395–398.

Bernsohn J. and Barron K. D. (1964) Multiple molecular forms of brain hydrolases. *Int. Rev. Neurobiol.* **7,** 297–344.

Bernsohn J., Barron K. D., and Hedrick M. T. (1963) Some properties of isozymes of brain acetylcholinesterase. *Biochem. Pharmacol.* **12,** 761–763.

Bernsohn J., Barron K. D., and Hess A. R. (1962) Multiple nature of acetylcholinesterase in nerve tissue. *Nature* **195,** 285–286.

Bessey D. A., Lowry H. O., and Broch M. I. (1946) A method for the rapid determination of alkaline phosphatase with five cubic millimeters of serum. *J. Biol. Chem.* **164,** 321–329.

Bird E. D. and Iversen L. L. (1974) Huntington's chorea: Postmortem measurement of glutamic acid decarboxylase, choline acetyltransferase and dopamine in the basal ganglia. *Brain* **97,** 457–472.

Bon S., Cartaud J., and Massoulié J. (1978) The dependence of acetylcholinesterase aggregation at low ionic strength upon a polyanionic component. *Eur. J. Biochem.* **85,** 1–14.

Bon S., Rieger F., and Massoulié J. (1973) Propriétés des formes allongées de l' acétylcholinestérase en solution. Rayon de Stokes, densité et masse. *Eur. J. Biochem.* **35,** 372–379.

Bon S., Vigny M., and Massoulié J. (1979) Asymmetric and globular forms of acetylcholinesterase in mammals and birds. *Proc. Natl. Acad. USA* **76,** 2546–2550.

Brimijoin S. (1983) Molecular forms of acetylcholinesterase in brain, nerve and muscle: Nature, localization, and dynamics. *Prog. Neurobiol.* **21,** 291–322.

Brimijoin S., Mintz K. P., and Alley M. C. (1983) Production and characterization of separate monoclonal antibodies to human acetylcholinesterase and butyrylcholinesterase. *Mol. Pharmacol.* **24,** 513–520.

Brimijoin S., Skau K., and Wiermaa M. J. (1978) On the origin and fate of external acetylcholinesterase in peripheral nerve. *J. Physiol.* (Lond.) **285,** 143–158.

Brodbeck U., Gentinetta R., and Ott P. (1981) Purification by Affinity Chromatography of Red Cell Membrane Acetylcholinesterase, in *Membrane Proteins* (Azzi A., Brodbeck U., and Zahler P., eds.), Springer-Verlag, Berlin.

Bruyn G. W., ed. (1968) Huntington's Chorea: Historical Clinical and Laboratory Synopsis, in *Vicken and Bruyn Handbook of Clinical Neurology, Disease of Basal Ganglia,* vol. 6, North Holland, Amsterdam.

Cartaud J., Bon S., and Massoulié J. (1978) *Electrophorus* acetylcholinesterase; biochemical and electron microscope characterization of low ionic strength aggregates. *J. Cell. Biol.* **77,** 315–322.

Chang C. H. and Blume A. J. (1976) Heterogeneity of acetylcholinesterase in neuroblastoma. *J. Neurochem.* **27,** 1427–1435.

Chan S. L., Shirachi D. Y., and Trevor A. J. (1972a) Purification and properties of brain acetylcholinesterase (EC 3.1.1.7). *J. Neurochem.* **19,** 437–447.

Chan S. L., Shirachi D. Y., Bhargava H. N., Gardner E., and Trevor A. J. (1972b) Purification and properties of multiple forms of brain acetylcholinesterase (EC 3.1.1.7). *J. Neurochem.* **19,** 2747–2758.

Chubb I. W. and Hodgson A. J. (1982) Release of Non-Transmitter Substances: Especially Acetylcholinesterase, in *Trends in Autonomic Pharmacology,* 2, (Kalsner A., ed.), Urban and Schwarzenberg, Baltimore and Munich.

Chubb I. W. and Smith A. D. (1975) Isoenzymes of soluble and membrane-bound acetylcholinesterase in bovine splanchnic nerve and adrenal medulla. *Proc. Roy. Soc. Lond. B.* **191,** 245–261.

Chubb I. W., Hodgson A. J., and White G. H. (1980) Acetylcholinesterase hydrolyzes substance P. *Neuroscience* **5,** 2065–2072.

Chubb I. W., Ranieri R., White G. H., and Hodgson A. J. (1983) The enkephalins are amongst the peptides hydrolyzed by purified acetylcholinesterase. *Neuroscience* **10,** 1369–1377.

Clarke J. T. (1964) Simplified "disc" (polyacrylamide gel) electrophoresis. *Ann. NY Acad. Sci.* **121,** 428–436.

Cottle M. K. W. and Silver A. (1970) Histochemical demonstration of acetylcholinesterase in the hypothalamus of the female guinea-pig. *Z. Zellforsch. Mikroskop. Anat.* **103,** 570–588.

Dale H. H. (1914) The action of certain esters of choline and their relation to muscarine. *J. Pharmacol. Exp. Ther.* **6,** 147–190.

Davies P. (1979) Neurotransmitter-related enzymes in senile dementia of the Alzheimer type. *Brain Res.* **171,** 319–327.

Davies P. and Maloney A. J. F. (1976) Selective loss of central cholinergic neurons in Alzheimer's disease. *Lancet* **II,** 1403.

Davis G. A. and Agranoff B. W. (1968) Metabolic behaviour of isozymes of acetylcholinesterase. *Nature* **220,** 277–280.

Davis K. L. and Berger P. A., eds. (1979) *Brain Acetylcholine and Neuropsychiatric Disease.* Plenum, New York.

Doctor B. P., Camp S., Gentry M. K., Taylor S. S., and Taylor P. (1983) Antigenic and structural differences in the catalytic subunits of the molecular forms of acetylcholinesterase. *Proc. Natl. Acad. Sci. USA* **80,** 5767–5771.

Domino E. F. (1979) Brain Cholinergic Enzymes in Schizophrenia, in *Brain Acetylcholine and Neuropsychiatric Disease,* (Davis K. L. and Berger P. A., eds.), Plenum, New York.

Dudai Y. and Silman I. (1974) The molecular weight and subunit structure of acetylcholinesterase preparations from the electric organ of the electric eel. *Biochem. Biophys. Res. Commun.* **59,** 117–124.

Dudai Y., Silman I., Kalderon N., and Blumberg S. (1972) Purification by affinity chromatography of acetylcholinesterase from electric organ tissue of the electric eel subsequent to tryptic treatment. *Biochim. Biophys. Acta* **268,** 138–157.

Ellman G. L. (1959) Tissue sulfhydryl groups. *Arch. Biochem. Biophys.* **82,** 70–77.

Ellman G. L. (1962) Solubilization of brain proteins. *Fed. Proc.* **21,** 402.

Ellman G. L., Courtney D. K., Andres V., and Featherstone R. M. (1961) A new and rapid colorimetric determination of acetylcholinesterase activity. *Biochem. Pharmacol.* **7,** 88–95.

Eränko O., Rechardt L., and Hänninen L. (1967) Electron microscopic demonstration of cholinesterases in nervous tissue. *Histochemie* **8,** 369–376.

Fambrough D. M., Engel A. G., and Rosenberry T. L. (1982) Acetylcholinesterase of human erythrocytes and neuromuscular junctions: Homologies revealed by monoclonal antibodies. *Proc. Natl. Acad. Sci. USA* **79,** 1078–1082.

Fulton M. P. and Mogey G. A. (1954) Some selective inhibitors of true cholinesterase. *Brit. J. Pharmacol.* **9,** 138–144.

Gerlach J., Reisby N., and Randrup A. (1974) Dopaminergic hypersensitivity and cholinergic hypofunction in the pathophysiology of tardive dyskinesia. *Psychopharmacologia* **34,** 21–25.

Gershon S. and Shaw F. H. (1961) Psychiatric sequelae of chronic exposure to organophosphorus insecticides. *Lancet* **I,** 1371–1374.

Gisiger V. and Stephens H. R. (1983) Asymmetric and globular forms of AChE in slow and fast muscles of 129/ReJ normal and dystrophic mice. *J. Neurochem.* **41,** 919–929.

Gisiger V., Vigny M., Gautron J., and Rieger F. (1978) Acetylcholinesterase of rat sympathetic ganglion: Molecular forms, localization and effects of denervation. *J. Neurochem.* **30,** 501–516.

Goodkin P. and Howard B. D. (1974) Studies on acetylcholinesterase of rat brain synaptosomal plasma membranes. *J. Neurochem.* **22,** 129–136.

Gordon M. A., Chan S.-L., and Trevor A. J. (1976) Active-site determination on forms of mammalian brain and eel acetylcholinesterase. *Biochem. J.* **157,** 69–76.

Gordon R. K., Doctor B. P., and Chiang P. K. (1982) A radioactive assay for acetylcholinesterase using anion-exchange disk. *Anal. Biochem.* **124,** 333–337.

Grassi J., Vigny M., and Massoulié J. (1982) Molecular forms of acetylcholinesterase in bovine caudate nucleus and superior cervical ganglion: Solubility properties and hydrophobic character. *J. Neurochem.* **38,** 457–469.

Greenberg A. J., Parker K. K., and Trevor A. J. (1977) Immunochemical studies of mammalian brain acetylcholinesterase. *J. Neurochem.* **29,** 911–917.

Grob D. (1963) Anticholinesterase Intoxication in Man and its Treatment, in *Cholinesterases and Anticholinesterase Agents,* (Koelle, G. B., ed.), Springer, Berlin.

Grossmann H., Ruess K.-P., and Liefländer M. (1979) Concerning the anchoring of acetylcholinesterase in biomembranes. *Experientia* **35,** 1545.

Gurd J. W. (1976) Fractionation of rat brain acetylcholinesterase by lectin affinity chromatography. *J. Neurochem.* **27,** 1257–1259.

Hall Z. W. (1973) Multiple forms of acetylcholinesterase and their distribution in endplate and non-endplate regions of rat diaphragm muscle. *J. Neurobiol.* **4,** 343–361.

Heath D. F., ed. (1961) *Organophosphorus Poisons. Anticholinesterases and Related Compounds.* Pergamon, Oxford.

Hetnarski B., Wisniewski H. M., Iqbal K., Dziedzic J. D., and Lajtha A. (1980) Regional distribution of choline acetyltransferase and acetylcholinesterase in rabbit brain. *Neurochem. Res.* **5,** 385–392.

Hiley C. R. and Bird E. D. (1974) Decreased muscarinic receptor concentration in postmortem brain in Huntington's chorea. *Brain Res.* **80,** 355–358.

Ho I. K. and Ellman G. L. (1969) Triton-solubilized acetylcholinesterase of brain. *J. Neurochem.* **16,** 1505–1513.

Hodgson A. J. and Chubb I. W. (1983) Isolation of the secretory form of soluble acetylcholinesterase by using affinity chromatography on edrophonium-Sepharose. *J. Neurochem.* **41,** 654–662.

Hollunger E. G. and Niklasson B. H. (1973) The release and molecular state of mammalian brain acetylcholinesterase. *J. Neurochem.* **20,** 821–836.

Holmstedt B. (1951) Synthesis and pharmacology of dimethylamidoethoxyphosphoryl cyanide (Tabun) together with a description of some allied anticholinesterase compounds containing the N–P bond. *Acta Physiol. Scand.* **25,** Suppl. 90, 7–120.

Holmstedt B. (1959) Pharmacology of organophosphorus cholinesterase inhibitors. *Pharmacol. Rev.* **11,** 567–688.

Hungen von K., Mahler H. R., and Moore W. J. (1968) Turnover of protein and ribonucleic acid in synaptic subcellular fractions from rat brain. *J. Biol. Chem.* **243,** 1415–1423.

Jackson R. L. and Aprison M. H. (1966a) Mammalian brain acetylcholinesterase. Purification and properties. *J. Neurochem.* **13,** 1351–1365.

Jackson R. L. and Aprison M. H. (1966b) Mammalian brain acetylcholinesterase. Effects of surface active agents. *J. Neurochem.* **13,** 1367–1371.

Janowsky D. S., El-Yousef M. K., Davis J. M., Sekerke H. J., and Hubbard B. J. (1972a) Cholinergic antagonism of manic symptoms. *Lancet* **I,** 1236–1237.

Janowsky D. S., El-Yousef M. K., Davis J. M., and Sekerke H. J. (1972b) A cholinergic-adrenergic hypothesis of mania and depression. *Lancet* **II,** 632–535.

Janowsky D. S., El-Yousef M. K., Davis J. M., and Sekerke H. J. (1973) Parasympathetic suppression of manic symptoms by physostigmine. *Arch. Gen. Psychiatry* **28,** 542–547.

Johnson C. D. and Russell R. L. (1975) A rapid, simple radiometric assay for cholinesterase, suitable for multiple determinations. *Anal. Biochem.* **64,** 229–238.

Joó F. and Várkonyi T. (1969) Correlation between the cholinesterase activity of capillaries and the blood–brain barrier in the rat. *Acta Biol. Acad. Sci. Hung.* **20,** 359–372.

Kalderon N., Silman I., Blumberg S., and Dudai Y. (1970) A method for the purification of acetylcholinesterase by affinity chromatography. *Biochim. Biophys. Acta* **207,** 560–562.

Kaplay S. S. and Jagannathan V. (1970) Purification and properties of ox brain acetylcholinesterase. *Arch. Biochem. Biophys.* **138,** 48–57.

Karczmar A. G. (1967) Pharmacological, Toxicologic and Therapeutic Properties of Anticholinesterase Agents, in *Physiological Pharmacology,* vol. III, *The Nervous System*-Part C (Root W. S. and Hofmann, F. G., eds.), Academic, New York.

Kása P. (1968) Ultrastructural localization of acetylcholinesterase in the cerebellar cortex with special reference to the intersynaptic organelles. *Histochemie* **14,** 161–167.

Kása P. and Csillik B. (1964) Histochemical studies on the effect of nerve degeneration in the cerebellar cortex. *II. International Congress of Histochemistry and Cytochemistry,* pp. 195–196. Springer, Berlin.

Kása P. and Csillik B. (1966) Electron microscopic localization of cholinesterase by a copper-lead-thiocholine technique. *J. Neurochem.* **13,** 1345–1349.

Kása P. and Csillik B. (1968) AChE synthesis in cholinergic neurons: Electron histochemistry of enzyme translocation. *Histochemie* **12,** 175–183.

Kása P., Csillik B., Joó F., and Knyihár E. (1966) Histochemical and ultrastructural alterations in the isolated archicerebellum of the rat. *J. Neurochem.* **13,** 173–178.

Kása P. and Lehotai L. (1979) Multiple Molecular Forms of Acetylcholinesterase, in *The Cholinergic Synapse* (Tucek S., ed.), Elsevier Scientific, Amsterdam, New York, Oxford.

Kása P. and Rakonczay Z. (1982) Histochemical and biochemical demonstration of the molecular forms of acetylcholinesterase in peripheral nerve of rat. *Acta Histochem.* **70,** 244–257.

Kása P., Szepesy G., Gulya K., Bánsághy K., and Rakonczay Z. (1982a) The effect of 4-(1-naphthylvinyl)-pyridine on the acetylcholine system and on the number of synaptic vesicles in the central nervous system of the rat. *Neurochem. Int.* **4,** 185–193.

Kása P., Bánsághy K., Rakonczay Z., and Gulya K. (1982b) Postnatal development of the acetylcholine system in different parts of the rat cerebellum. *J. Neurochem.* **39,** 1726–1732.

Kidd M. (1964) Alzheimer's disease—an electron microscopical study. *Brain* **87,** 307–320.

Klawans H. L., Jr. and Rubovits R. (1974) Effect of cholinergic and anticholinergic agents on tardive dyskinesia. *J. Neurol. Neurosurg. Psychiatr.* **37,** 941–947.

Koelle G. B. (1951) The elimination of enzymatic diffusion artifacts in the histochemical localization of cholinesterases and survey of their cellular distribution. *J. Pharmacol. Exp. Ther.* **103,** 153–171.

Koelle G. B. (1957) Histochemical demonstration of reversible anticholinesterase action at selective cellular sites *in vivo. J. Pharmacol. Exp. Ther.* **120,** 488–503.

Koelle G. B. (1963) *Cholinesterases and Anticholinesterase Agents.* Springer, Berlin.

Koelle G. B. and Friedenwald J. S. (1949) A histochemical method for localizing cholinesterase activity. *Proc. Soc. Exptl. Biol. Med.* **70,** 617–622.

Kremzner L. T. and Wilson I. B. (1963) A chromatographic procedure for the purification of acetylcholinesterase. *J. Biol. Chem.* **238,** 1714–1717.

Kremzner L. T., Kitz R. J., and Ginsburg S. (1967) A partial purification and characterization of the acetylcholinesterase of human brain. *Fed. Proc.* **26,** 296.

Krupka R. M. (1966) Chemical structure and function of the active center of acetylcholinesterase. *Biochemistry* **5,** 1988–1998.

Laemmli J. K. (1970) Cleavage of structural proteins during the assembly of the head of bacteriophage T4. *Nature* **227,** 680–685.

Lawler H. C. (1959) A simplified procedure for the partial purification of acetylcholinesterase from electric tissue. *J. Biol. Chem.* **234,** 799–801.

Lawler H. C. (1964) The preparation of a soluble acetylcholinesterase from brain. *Biochim. Biophys. Acta* **81,** 280–288.

Lazar M. and Vigny M. (1980) Modulation of the distribution of acetylcholinesterase molecular forms in a murine neuroblastoma × sympathetic ganglion cell hybrid cell line. *J. Neurochem.* **35,** 1067–1079.

Leuzinger W. and Baker A. L. (1967) Acetylcholinesterase. I. Large-scale purification, homogeneity and amino acid analysis. *Proc. Natl. Acad. Sci. USA* **57,** 446–451.

Lipton M. A., DiMascio A., and Killam F. K. (1978) *Psychopharmacology: A Generation of Progress.* Raven, New York.

Long J. P. (1963) Structure–Activity Relationships of the Reversible Anticholinesterase Agents, in *Cholinesterases and Anticholinesterase Agents* (Koelle G. B., ed.), Springer, Berlin.

Main A. R. (1976) Structure and Inhibitors of Cholinesterase, in *Biology of*

Cholinergic Function (Goldberg A. M. and Hanin I., ed.), Raven, New York.

Martin R. G. and Ames B. N. (1961) A method for determining the sedimentation behavior of enzymes: Application to protein mixtures. *J. Biol. Chem.* **236,** 1372–1379.

Massoulié J. and Bon S. (1976) Affinity chromatography of acetylcholinesterase; The importance of hydrophobic interactions. *Eur. J. Biochem.* **68,** 531–539.

Massoulié J. and Bon S. (1982) The molecular forms of cholinesterase and acetylcholinesterase in vertebrates. *Ann. Rev. Neurosci.* **5,** 57–106.

Massoulié J. and Rieger F. (1969) L'acétylcholinestérase des organes électriques de poissons (*Torpille* et *Gymnote*); complex membranainres. *Eur. J. Biochem.* **11,** 441–455.

Massoulié J., Bon S., and Vigny M. (1980) The polymorphism of cholinesterase in vertebrates. *Neurochem. Int.* **2,** 161–184.

Massoulié J., Rieger F., and Bon S. (1971) Espéces acétylcholinesterasiques globulaires et allongées des organes électriques de possions. *Eur. J. Biochem.* **21,** 542–551.

McGeer P. L., McGeer E. G., and Fibiger H. C. (1973) Choline acetylase and glutamic acid decarboxylase in Huntington's chorea. *Neurology* (Minneap.) **23,** 912–917.

McIntosh C. H. S. and Plummer D. T. (1973) Multiple forms of acetylcholinesterase from pig brain. *Biochem. J.* **133,** 655–665.

McIntosh C. H. S. and Plummer D. T. (1976) The subcellular localization of acetylcholinesterase and its molecular forms in pig cerebral cortex. *J. Neurochem.* **27,** 449–457.

McIsaac R. J. and Koelle G. B. (1959) Comparison of the effects of inhibition of external, internal and total acetylcholinesterase upon ganglionic transmission. *J. Pharmacol. Exp. Ther.* **126,** 9–20.

McMenemy W. H. (1958) Huntington's Chorea, in *Neuropathology,* (Grenfield J. G., McMenemy W. H., Norman R. M., Blackwood W., and Meyer A., eds.), Arnold, London.

Michelson M. J. and Zeimal E. V., eds. (1973) *Acetylcholine.* Pergamon, Oxford.

Moore D. E. and Hess G. P. (1975) Acetylcholinesterase-catalyzed hydrolysis of an amide. *Biochemistry* **14,** 2386–2389.

Morton R. K. (1950) Separation and purification of enzymes associated with insoluble particles. *Nature* **166,** 1092–1095.

Nachmansohn D. and Lederer E. (1939) Sur la biochimie de la cholinesterase. Préparation de l' enzyme, role de groupements SH. *Bull. Soc. Chim. Biol.* **21,** 797–808.

Neal M. W. and Florini J. R. (1973) A rapid method for desalting small volumes of solution. *Anal. Biochem.* **55,** 328–330.

O'Brien R. D., ed. (1967) *Insecticides: Action and Metabolism.* Academic, New York.

Oommen A. and Balasubramanian A. S. (1979) The association of the serotonin-sensitive aryl acylamidase with acetylcholinesterase in the monkey brain. *Eur. J. Biochem.* **94,** 135–143.

Ord M. G. and Thompson R. H. S. (1951) The preparation of soluble cholinesterases from mammalian heart and brain. *Biochem. J.* **49,** 191–199.

Ord M. G. and Thompson R. H. S. (1952) Pseudo-cholinesterase activity in the central nervous system. *Biochem. J.* **51,** 245–251.

Parvari R., Pecht I., and Soreq H. (1983) A microfluorometric assay for cholinesterases, suitable for multiple kinetic determinations of picomoles of released thiocholine. *Anal. Biochem.* **133,** 450–456.

Patocka J. and Bajgar J. (1969) Some enzymatic properties of human brain acetylcholinesterase. *FEBS Lett.* **2,** 195–197.

Perry T. L., Hansen S., and Kloster M. (1973) Huntington's chorea: Deficiency of γ-aminobutyric acid in brain. *N. Eng. J. Med.* **288,** 337–342.

Perry E. K., Perry R. H., Blessed G., and Tomlinson B. E. (1977) Necropsy evidence of central cholinergic deficits in senile dementia. *Lancet* **I,** 189.

Perry E. K., Tomlinson B. E., Blessed G., Bergmann K., Gibson P. H., and Perry R. H. (1978) Correlation of cholinergic abnormalities with senile plaques and mental test scores in senile dementia. *Brit. Med. J.* **2,** 1457–1459.

Quarles R. H. and Brady H. O. (1970) Sialoglycoproteins and several glycosidases in developing rat brain. *J. Neurochem.* **17,** 801–807.

Rakonczay Z., Mallol J., Schenk H., Vincendon G., and Zanetta J.-P. (1981a) Purification and properties of the membrane-bound acetylcholinesterase from adult rat brain. *Biochim. Biophys. Acta* **657,** 243–256.

Rakonczay Z., Vincendon G., and Zanetta J.-P. (1981b) Heterogeneity of rat brain acetylcholinesterase: A study by gel filtration and gradient centrifugation. *J. Neurochem.* **37,** 662–669.

Rakonczay Z., Zanetta J.-P., Vincendon G., and Kása P. (1981c) Purification and characterization of soluble acetylcholinesterase from rat brain. *Neurosci. Lett.* suppl. **7,** S429.

Rakonczay Z., Zanetta J.-P., Reeber A., Vincendon G., and Kása P. (1982) Characterization of soluble acetylcholinesterase molecular forms in rat brain, spinal ganglia and sciatic nerve. *Neuroscience* suppl. **7,** S174.

Reavill C. A. and Plummer D. T. (1978) Purification of acetylcholinesterase from pig cerebral cortex by affinity chromatography. *J. Chromatogr.* **157,** 141–151.

Rieger F. and Vigny M. (1976) Solubilization and physicochemical characterization of rat brain acetylcholinesterase: Development and maturation of its molecular forms. *J. Neurochem.* **27,** 121–129.

Rieger F., Chételat R., Nicolet M., Kamal L., and Poullet M. (1980)

Presence of tailed, asymmetric forms of acetylcholinesterase in the central nervous system of vertebrates. *FEBS Lett.* **121,** 169–174.

Rieger F., Faivre-Bauman A., Benda P., and Vigny M. (1976) Molecular forms of acetylcholinesterase: Their *de novo* synthesis in mouse neuroblastoma cells. *J. Neurochem.* **27,** 1059–1063.

Rodriguez-Borrajo C., Barat A., and Ramirez G. (1982) Solubilization of collagen-tailed molecular forms of acetylcholinesterase from several brain areas in different vertebrate species. *Neurochem. Int.* **4,** 563–568.

Rosenberry T. L. (1975) Acetylcholinesterase. *Adv. Enzymol.* **43,** 103–218.

Rosenberry T. L. and Richardson J. M. (1977) Structure of 18S and 14S acetylcholinesterase. Identification of collagen-like subunits that are linked by disulfide bonds to catalytic subunits. *Biochemistry* **16,** 3550–3558.

Rothenberg M. A. and Nachmansohn D. (1947) Studies on cholinesterase. III. Purification of the enzyme from electric tissue by fractional ammonium sulfate precipitation. *J. Biol. Chem.* **168,** 223–231.

Ruess K.-P., Weinert M., and Liefländer M. (1976) Affinitätschromatographische reinigung der acetylcholinesterase aus nucleus caudatus vom rind im grossmassstab. *Hoppe-Seyler's Z. Physiol. Chem.* **357,** 783–793.

Salipan N., Morgan I. G., and Chubb I. W. (1983) Decreased concentrations of acetylcholinesterase and leu-enkephalin in dark-adapted chick retinae. *Neurosci. Lett.* suppl. **11,** S73.

Saunders B. C., ed. (1957) *Some Aspects of the Chemistry and Toxic Action of Organic Compounds Containing Phosphorus and Fluorine.* Cambridge Univ. Press, Cambridge.

Schaffer N. K., May S. C., and Summerson W. H. (1953) Serine phosphoric acid from diisopropylphosphoryl chymotrypsin. *J. Biol. Chem.* **202,** 67–76.

Schaffer N. K., Michel H. O., and Bridges A. F. (1973) Amino acid sequence in the region of the reactive serine residue of eel acetylcholinesterase. *Biochemistry* **12,** 2946–2950.

Shapiro A. L., Vinuela E., and Maizel J. V. (1967) Molecular weight estimation of polypeptide chains by electrophoresis in SDS-polyacrylamide gels. *Biochem. Biophys. Res. Commun.* **28,** 815–820.

Siegel L. M. and Monty K. J. (1966) Determination of molecular weights and frictional ratios of proteins in impure systems by use of gel filtration and density gradient centrifugation. Application to crude preparations of sulfite and hydroxylamine reductases. *Biochim. Biophys. Acta* **112,** 346–362.

Silman I. (1976) Molecular structure of acetylcholinesterase. *Trends Biochem. Sci.* **1,** 225–227.

Silman I., Lyles J. M., and Barnard E. A. (1978) Intrinsic forms of acetylcholinesterase in skeletal muscle. *FEBS Lett.* **94,** 166–170.

Silver A. (1974) *The Biology of Cholinesterases.* North Holland, Amsterdam.

Sørensen K., Gentinetta R., and Brodbeck U. (1982) An amphiphile-dependent form of human brain caudate nucleus acetylcholinesterase: Purification and properties. *J. Neurochem.* **39,** 1050–1060.

Soreq H., Parvari R., and Silman I. (1982) Biosynthesis and secretion of catalytically active acetylcholinesterase in *Xenopus* oocytes microinjected with mRNA from rat brain and from *Torpedo* electric organ. *Proc. Natl. Acad. Sci. USA* **79,** 830–834.

Srinivasan R., Karczmar A., and Bernsohn J. (1972) Activation of acetylcholinesterase by Triton X-100. *Biochim. Biophys. Acta* **284,** 349–354.

Stahl W. L. and Swanson P. D. (1974) Biochemical abnormalities in Huntington's chorea brains. *Neurology* (Minneap.) **24,** 813–819.

Taylor P. B., Rieger F., Shelanski M. L., and Greene L. A. (1981) Cellular localization of the multiple molecular forms of acetylcholinesterase in cultured neuronal cells. *J. Biol. Chem.* **256,** 3827–3830.

Terry R. D. (1963) The fine structure of neurofibrillary tangles in Alzheimer's disease. *J. Neuropath. Exp. Neurol.* **22,** 629–642.

Terry R. D., Gonatas N. K., and Weiss M. (1964) Ultrastructural studies in Alzheimer's presenile dementia. *Am. J. Pathol.* **44,** 269–297.

Thomas N. C., Fleisher J. H., and Harris L. W. (1971) Utilization of (^{32}P)soman for measurement of acetylcholinesterase in brain tissues. *Biochim. Biophys. Acta* **235,** 542–547.

Trevor A. J., Gordon M. A., and Parker K. K. (1978) Acetylcholinesterase. *Life Sci.* **23,** 1209–1220.

Usdin E., Hamburg D., and Barchas J., eds. (1977) *Neuroregulators and Psychiatric Disorders.* Oxford University Press, New York.

Vallette F. M., March D. J., Muller F., and Massoulié J. (1983) Comparative affinity chromatography of acetylcholinesterases from five vertebrate species. *J. Chromatogr.* **257,** 285–296.

Vigny M., Bon S., Massoulié J., and Gisiger V. (1979) The subunit structure of mammalian acetylcholinesterase: Catalytic subunits, dissociating effect of proteolysis and disulfide reduction of the polymeric forms. *J. Neurochem.* **33,** 559–565.

Vigny M., Bon S., Massoulié J., and Leterrier F. (1978) Active-site catalytic efficiency of acetylcholinesterase molecular forms in *Electrophorus, Torpedo,* rat and chicken. *Eur. J. Biochem.* **85,** 317–323.

Vijayan V. K. and Brownson R. H. (1974) Polyacrylamide gel electrophoresis of rat brain acetylcholinesterase: Isoenzymes of normal rat brain. *J. Neurochem.* **23,** 47–53.

Vimard C., Jeantet C., Netter Y., and Gros F. (1976) Changes in the sedimentation properties of acetylcholinesterase during neuroblastoma differentiation. *Biochimie* **58,** 473–478.

Wade P. D. and Timiras P. S. (1980) A regional study of the molecular forms of acetylcholinesterase in the brain of developing and adult rats. *Dev. Neurosci.* **3,** 101–108.

Wang C.-S. and Smith R. L. (1975) Lowry determination of protein in the presence of Triton X-100. *Anal. Biochem.* **63,** 414–417.

Wenthold R. J., Mahler H. R., and Moore W. J. (1974a) Properties of acetylcholinesterase from rat brain. *J. Neurochem.* **22,** 945–949.

Wenthold R. J., Mahler H. R., and Moore W. J. (1974b) The half-life of acetylcholinesterase in mature rat brain. *J. Neurochem.* **22,** 941–943.

Wilson I. B. (1960) Acetylcholinesterase. *Enzymes* **4,** 501–520.

Wilson I. B. and Bergmann F. (1950) Acetylcholinesterase. VIII. Dissociation constants of the active groups. *J. Biol. Chem.* **186,** 683–692.

Wilson I. B., Bergmann F., and Nachmansohn D. (1950) Acetylcholinesterase. X. Mechanism of catalysis of acylation reactions. *J. Biol. Chem.* **186,** 781–790.

Yaksh T. L., Filbert M. G., Harris L. W., and Yamamura H. I. (1975) Acetylcholinesterase turnover in brain, cerebrospinal fluid and plasma. *J. Neurochem.* **25,** 853–860.

Yamamura H. I., Reichard D. W., Gardner T. L., Morrisett J. D., and Broomfield C. A. (1973) Guinea pig brain acetylcholinesterase: Partial purification by affinity chromatography. *Biochim. Biophys. Acta* **302,** 305–315.

Zanetta J.-P., Rakonczay Z., Reeber A., Kása P., and Vincendon G. (1981) Antibodies against the membrane-bound acetylcholinesterase from adult rat brain. *FEBS Lett.* **129,** 293–296.

Zech R. and Domagk G. F. (1975) Puromycin and cycloheximide as inhibitors of human brain acetylcholinesterase. *Brain Res.* **86,** 339–342.

Zittle C. A., DellaMonica E. S., and Custer J. H. (1954) Purification of human red cell acetylcholinesterase. *Arch. Biochem. Biophys.* **48,** 43–49.

Chapter 10
Brain Glutamate Decarboxylase

David L. Martin

1. Introduction

Glutamate decarboxylase (glutamate 1-carboxylyase, EC 4.1.1.15, GAD)* is the major, rate-limiting enzyme in brain for synthesizing gamma-aminobutyric acid (GABA). Total GAD activity in brain is 10–20 times greater than the observed rate of GABA synthesis (Collins, 1972; Matsui and Deguchi, 1977; Casu and Gale, 1981), indicating that GAD operates at only a fraction of its capacity. Although other routes of synthesis have been suggested, they appear to be minor (Baxter, 1976). GAD has a distinct regional distribution in brain that parallels the regional distribution of GABA (Collins, 1972; Fahn, 1976), as would be expected of a rate-limiting biosynthetic enzyme. Because GAD appears to be the major biosynthetic enzyme for GABA, it is used as one of the principal markers for GABA neurons and presynaptic terminals; it does not appear to be present in appreciable amounts in glial cells (Schousboe et al., 1977; Wu et al., 1979). GAD is concentrated in the presynaptic endings of GABA neurons, but is also present in the cell bodies, as shown by subcellular fractionation (Salganicoff and DeRobertis, 1965; Van Kempen et al., 1965; Fonnum, 1968) and by immunocytochemistry (Saito et al., 1974; Wood et al., 1976; Ribak et al., 1978; Oertel et al., 1981c).

Lesioning and immunocytochemical studies using GAD as a marker have provided evidence of long-axon, GABA neurons that project between brain regions. For example, GAD-containing Purkinje cells project from cerebellar cortex to deep cerebellar nuclei

*Abbreviations used: AET, 2-aminoethylisothisuronium bromide; AOAA, aminooxyacetic acid; GAD, glutamate decarboxylase; pyridoxal-P, pyridoxal 5'-phosphate; pyridoxamine-P, pyridoxamine 5'-phosphate; SSA, succinic semialdehyde.

(Saito et al., 1974; Oertel et al., 1981a), and some GAD-containing cells project from the striatum to the substantia nigra (McGeer et al., 1973a; Hattori et al., 1973; Gale and Casu, 1981), and others project from the substantia nigra to the thalamus (Di Chiara et al., 1979; Kilpatrick et al., 1980). In contrast, GAD-containing nerve endings in hippocampus arise principally from intrinsic interneurons involved in local circuits (Storm-Mathisen, 1972, 1976), although there is evidence that septal neurons containing GAD-like immunoreactivity project to the hippocampus (Kohler et al., 1984). Immunochemically identified, GAD-containing interneurons have been studied in some detail in cortex and hippocampus (Houser et al., 1983; Ribak et al., 1978) and comprise a morphologically diverse group of cells.

Because GABA is one of the major inhibitory neurotransmitters in the central nervous system, GAD has been the focus of an enormous variety of studies in humans and animals concerned with various disease states, drug effects, and correlations between neurochemistry and behavior. Such studies have shown, for example, that GAD activity is markedly reduced in the basal ganglia of patients with Huntington's disease (Bird et al., 1973; McGeer et al., 1973b; Iversen et al., 1979). This reduction is apparently the result of a loss of GAD-containing interneurons from these regions. GAD has also been implicated in seizure mechanisms and epilepsy. Numerous convulsive agents are potent inhibitors of GAD (Baxter, 1970; Tapia, 1975), and tissue GAD activity is reduced in experimental epileptic foci in animals (Emson and Joseph, 1975; Ribak et al., 1979, 1981; Bakay and Harris, 1981) and in some epileptic foci in humans (Lloyd et al., 1981). Studies of the involvement of GAD in many other processes have been summarized in recent volumes of symposia and meetings (Roberts et al., 1976; Krogsgaard-Larsen et al., 1979; Lal et al., 1980; Hertz et al., 1983).

GAD has been reported to be present outside central nervous tissue in heart, kidney, oviduct, myenteric plexus, and cerebral blood vessels (Wu, 1977; Kanazawa et al., 1976; del Rio, 1981; Jessen et al., 1979; Hamel et al., 1982). GAD in heart and kidney, termed nonneuronal GAD, appears to differ in kinetic and immunological properties from brain GAD, but is only partially characterized (Wu, 1977). In some tissues, this GAD may be associated with GABA neurons of the peripheral nervous system (Beart et al., 1974; Jessen et al., 1979).

Almost all neurobiological studies of GAD involve one of three kinds of measurements: (i) determination of total GAD activity, (ii) determination of the relative amounts of holo- and apoenzyme, or (iii) localization of GAD with immunocytochemical methods to identify specific GABA neurons and map their connections. In many respects, measuring GAD activity is straightforward. However, not all procedures are accurate and reliable, and a number of pitfalls and necessary precautions have been recognized. Inappropriate procedures have sometimes been used, so evaluating the literature on GAD requires an understanding of the properties of the enzyme and the principles and limitations of the methods. This paper will review the properties of GAD, describe the principles and procedures for reliably measuring GAD activity, and also describe the newer methods of purifying the enzyme, especially those for resolving the multiple forms of GAD. Immunocytochemical methods *per se* are beyond the scope of this paper, although the occurrence of multiple forms of GAD in brain is obviously important and relevant to studies of the localization of GAD.

2. Characteristics of GAD

2.1. Physical Properties of GAD

Mammalian GAD has been purified to various degrees from mouse (Susz et al., 1966; Wu et al., 1973; Wu, 1976), rat (Maitre et al., 1978; Oertel et al., 1981a, b), human (Blindermann et al., 1978; Spink and Martin, 1983), porcine (Spink et al., 1983), and bovine (Wu, 1982; Heinamaki et al., 1983) brain. As might be expected, the physical and kinetic characteristics of these enzymes are similar, although some differences have been reported. For example, the values of K_m for glutamate range from 0.5 to 7.9 m*M*. Such differences could be a result of the intrinsic properties of the enzymes, or to differences in assay conditions, particularly pH (Blindermann et al., 1978; Spink and Martin, 1983). In all cases the enzyme is anionic at neutral pH (p*I* values ranged from 5.3 to 5.8) and chromatographs readily on anion exchange matrices. The enzyme from human and rat brain has been reported to be a dimer of 67,000 dalton subunits (Blindermann et al., 1978; Maitre et al., 1978). The enzyme from mouse brain was initially reported to be a dimer of 44,000 dalton subunits (Wu et al., 1973), although a more complex subunit structure was subsequently proposed (Matsuda et al., 1973). A number of studies have shown that GAD can adhere strongly to membranes, particularly in the presence of Ca^{2+}, but this characteristic has not been extensively studied with purified GAD (Salganicoff and DeRobertis, 1965; Fonnum, 1968; Covarrubias and Tapia, 1978).

Recent studies have revealed that hog, rat, and human brain

each contain multiple forms of GAD that differ in physical and kinetic properties (Spink et al., 1983, 1985; Spink and Martin, 1983). Although three forms of hog and rat GAD were separated, only two forms of human GAD could be resolved. It is not clear at present whether the smaller number of forms of human GAD is because of a species difference, insufficient resolving power of the chromatographic methods used, or a selective loss of one of the forms during the unavoidably long postmortem period before the human tissue could be frozen.

The three forms of GAD from hog brain (termed α-, β-, and γ-GAD) have been studied in greatest detail. They can be separated by hydrophobic interaction chromatography on phenyl-Sepharose, by isoelectric focusing, and by polyacrylamide gel electrophoresis under nondenaturing conditions. They have the same approximate molecular weights, as determined by chromatography on Sephadex G-200, indicating that they are not merely different sized aggregates. They also differ in isoelectric point, K_m for glutamate, rate of substrate-promoted inactivation, and pH dependence. The amount of γ-GAD appears to be somewhat lower in synaptosomes than in whole brain extracts.

The relationship of these multiple forms to the previously purified enzymes remains to be established. However, the presence of multiple forms in brain has many obvious implications for both metabolic and anatomic studies.

2.2. Reactions Catalyzed by GAD

In addition to its substrate, glutamate, GAD also requires the cofactor pyridoxal 5'-phosphate (pyridoxal-P), which participates in catalysis through its ability to form Schiff base adducts. Recent studies of the reactions catalyzed by GAD have clarified the actions of a number of inhibitors and effectors of GAD and suggested mechanisms by which GAD might be regulated in vivo (Miller et al., 1977, 1978; Seligmann et al., 1978; Martin and Martin, 1979, 1982; Meeley and Martin, 1983a, b; Porter and Martin, 1984). The reaction mechanism proposed for GAD is shown in Fig. 1 (Martin et al., 1980; Meeley and Martin, 1983a, b; Porter and Martin, 1984; Porter et al., 1985; Spink et al., 1985). An important feature is the branch that occurs at the step immediately following the release of CO_2. The principal path from this branch leads to the production of GABA. The alternative path, which is termed the transamination pathway, leads to the production of succinic semialdehyde and pyridoxamine 5'-phosphate (pyridoxamine-P) and results in the conversion of the holoenzyme (active GAD with bound pyridoxal-

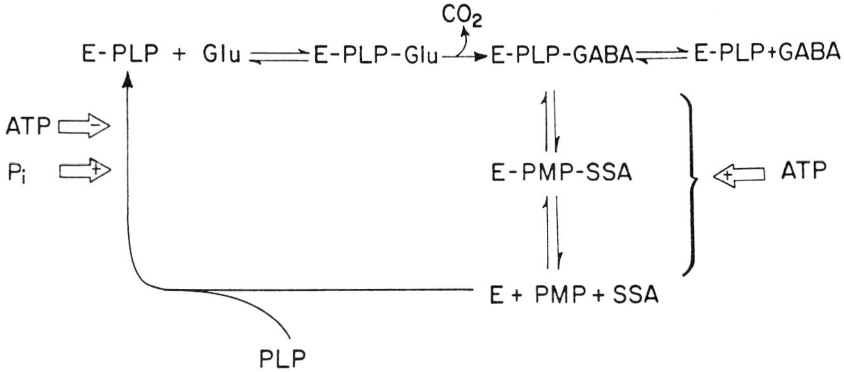

Fig. 1. Outline of the reaction mechanism of brain glutamate decarboxylase, including the interconversion of apo- and holoGAD and the effects of ATP and Pi. E = apoGAD; PLP = pyridoxal-P; E-PLP = holoGAD; Glu = glutamate; PMP = pyridoxamine-P; SSA = succinic semialdehyde (modified from Porter and Martin, 1984).

P) to the apoenzyme (inactive GAD without pyridoxal-P). Although the absolute rate of the transamination pathway is far slower than the rate of GABA production, the transamination pathway plays a key role in the interaction of GAD with its cofactor because it is the major mechanism by which holoGAD is converted to apoGAD under most conditions. ApoGAD is not readily formed by simple dissociation of pyridoxal-P from holoGAD, as shown by dialysis experiments with GAD from rat and hog brain (Miller et al., 1978; Meeley and Martin, 1983a, b).

Also shown in Fig. 1 is the conversion of apoGAD to holoGAD by reaction with pyridoxal-P. Together the activation reaction and the transamination pathway form a cycle that interconverts apo- and holoGAD and is sometimes called the cycle of inactivation and reactivation. Thus, under normal reaction conditions (in the presence of glutamate and pyridoxal-P), GAD continuously passes between the apo- and holoenzyme forms.

Several important aspects of the behavior of GAD are explained by this mechanism. First, GAD remains active even after exhaustive dialysis in the absence of pyridoxal-P, because the cofactor does not dissociate readily from holoGAD. However, GAD is rapidly inactivated when incubated with glutamate in the absence of pyridoxal-P, because of the formation of apoGAD. When saturating pyridoxal-P is present there is little net inactivation by glutamate, because apoGAD is rapidly reconverted to holoGAD. Because the transamination pathway is a branch of the normal reaction mechanism, the rate of inactivation increases with the glutamate concentration in the same way as the rate of decarboxylation. The rates of inactivation also differ among the multiple forms of the enzyme (Spink and Martin, 1983). Nucleoside triphosphates and certain other inhibitors appear to act, at least in part, on the cycle of inactivation and reactivation, so several minutes are required for these agents to have their full effect, in agreement with the observed time dependence of the steps of the cycle (Meeley and Martin, 1983a, b; Martin and Martin, 1982).

The reactions comprising the GABA release and transamination paths are simple Schiff base reactions, so these paths are readily reversible. Consequently apoGAD is formed and GAD is inactivated when holoGAD is incubated with GABA in the absence of pyridoxal-P (Porter and Martin, 1984).

As a decarboxylase, GAD appears to have good substrate specificity; glutamate is decarboxylated about 20 times more rapidly than aspartate (Wu, 1976). Two other acidic amino acids, cysteiate and cysteine sulfinate, also appear to be decarboxylated at appreciable rates (Oertel et al., 1981b; Wu, 1982; Spears and Martin, 1982).

2.3. Inhibitors and Effectors of GAD Activity

GAD activity is affected by a wide variety of compounds, including simple competitive inhibitors, carbonyl trapping agents, compounds that affect the interconversion of apo- and holoGAD, suicide substrates, and metal ions. Several studies in which substantial numbers of inhibitors were tested have been reported (Roberts and Simonsen, 1963; Tursky, 1970; Wu and Roberts, 1974; Wu, 1976; Taberner et al., 1977; Blindermann et al., 1978; Martin and Martin, 1982).

2.3.1. Competitive Inhibitors

The list of competitive inhibitors of GAD is extensive. Acidic amino acids, such as aspartate and D-glutamate, and simple dicarboxylic acid analogs of glutamate, such as glutarate, malate, and other Krebs' cycle intermediates, are competitive inhibitors with K_is in the low millimolar range. Some glutamate analogs with restricted conformation are much more potent inhibitors, having K_is in the micromolar range (Endo and Kitahara, 1981; Porter et al., 1982). GABA inhibits competitively, as expected of the last product released by the normal enzymatic mechanism [K_i >10 m*M*; (Blindermann et al., 1978; Porter and Martin, 1984)]. Other potent competitive inhibitors are 3-mercaptopropionate, thiomalate, and the sulfhydryl reagents 5,5'-dithiobis(2-nitrobenzoate) and *para*-chloromercuribenzoate (Wu, 1976; Taberner et al., 1977). It is not clear whether a reactive sulfhydryl group on GAD is involved in the competitive inhibition by the sulfhydryl reagents because these reagents react covalently with sulfhydryl groups and would, therefore, be expected to inhibit GAD irreversibly in a time-dependent manner. Finally, the inorganic anion Cl^- is a competitive inhibitor of GAD [K_i = 17–22 m*M*, (Wu, 1976; Blindermann et al., 1978)]. Although the K_i for Cl^- is quite high, this inhibition can be important in purification procedures in which high concentrations of Cl^- may be present. Other inorganic anions (SO_4^{2-}, SCN^-, I^-, Br^-) also inhibit GAD (Wu, 1976). The presence of substantial amounts of these ions in assay media should be avoided.

2.3.2. Carbonyl Trapping Agents

GAD is highly sensitive to a variety of carbonyl trapping agents (Baxter, 1970; Tapia, 1975). This group of compounds, which includes hydroxylamine aminooxyacetate (AOAA), certain hydrazines, semicarbazide, and isoniazid, reacts with pyridoxal-P to form Schiff base adducts. In vitro they inhibit GAD by forming adducts with pyridoxal-P in the active site. Aminooxyacetate is one of the most potent of these agents, giving almost complete inhibition at a concentration of 10 μ*M* (Roberts and Simonsen, 1963). Once formed, the adduct can dissociate from the active site. Thus, hydroxylamine and AOAA have been used to prepare apoGAD by treating the enzyme with one of the compounds and then removing the excess trapping agent and released adduct by dialysis or chromatography on Sephadex G-25 (Ryan and Roskoski, 1976; Martin and Martin, 1979).

Primary amines, including alpha-amino acids, can also form Schiff base adducts with pyridoxal-P and can be considered weak carbonyl-trapping agents. If adduct formation occurs in the active site, the amines would inhibit GAD directly, like the more potent carbonyl agents discussed above. However, the amines can also inhibit GAD indirectly by forming a Schiff base with free pyridoxal-P in solution, thereby depleting the cofactor available to reactivate any apoGAD formed by the transamination branch of the enzyme mechanism (*see* Fig. 1) and thereby leading to a reduction in GAD activity. Thus primary amine buffers, such as glycine or Tris, should be avoided in GAD assays.

2.3.3. Compounds That Affect the Interconversion of Apo- and HoloGAD

The cycle that interconverts apo- and holoGAD (*see* Sec. 2.2, Fig. 1) appears to be an important point of action of compounds, such as ATP and inorganic phosphate (P_i), that may regulate GAD activity. ATP appears to inhibit reactivation of apoGAD by pyridoxal-P and to enhance the rate of apoGAD formation by the transamination branch of the mechanism (Seligmann et al., 1978; Meeley and Martin, 1983a, b). Thus, its effect is to decrease GAD activity by causing a net reduction in the amount of holoGAD. Similar effects have been observed with other nucleoside triphosphates, nucleoside diphosphates, and polyanions (Martin and Martin, 1983). In procedures designed to measure tissue levels of holoGAD activity, ATP is included in the tissue homogenizing medium to prevent the rapid conversion of apoGAD to holoGAD by endogenous pyridoxal-P (Miller et al., 1977, 1980). ATP appears to act at an allosteric site rather than the glutamate or pyridoxal-P binding sites (Meeley and Martin, 1983a,b; Martin and Martin, 1982; Wu and Martin, 1984). Inorganic phosphate opposes the inactivating effects of ATP and enhances the activation of apoGAD prepared in the presence of ATP or by treatment with aminooxyacetate (Martin and Martin, 1979; Meeley and Martin, 1983b). Thus, the inhibitory effects of low concentrations of ATP can be overcome by pyridoxal-P and P_i (Tursky, 1970; Seligmann et al., 1978; Meeley and Martin, 1983a,b).

Because turnover of the cycle requires minutes, a compound that acts on it does not have full effect immediately, but requires a correspondingly long period of time to exert its full effect. Thus the effects of ATP are highly time-dependent. In contrast, competitive inhibitors such as glutarate act immediately (Martin and Martin, 1982).

2.3.4. Enzyme-Activated Irreversible Inhibitors

An enzyme-activated, irreversible inhibitor or suicide substrate, is a compound that is converted to a reactive intermediate in the active site of the target enzyme where it reacts with nearby amino acids or the cofactor to irreversibly inactivate the enzyme. In principle, a suicide substrate is highly specific because it will only act on an enzyme that accepts it as a substrate.

Three suicide substrates for GAD have been reported—γ-acetylenicGABA (Metcalf et al., 1979), trifluoromethylGABA (Rando, 1979), and 2-methyl-3,4-didehydroglutamic acid (Chrystal et al., 1979). γ-AcetylenicGABA is not specific for GAD, but inhibits GABA-transaminase (Metcalf et al., 1979). Radioactive γ-acetylenicGABA has been used to label GAD in homogenates and in vivo after GABA-transaminase was inactivated by gabaculine (Rando, 1981). γ-AcetylenicGABA is also produced in vivo by metabolism of 5-hexane-1,4-diamine, a suicide substrate for ornithine decarboxylase (Danzin et al., 1983).

2.3.5. Metal Ions

Several divalent metal ions inhibit GAD. The most potent is Zn^{2+} (approximately 50% inhibition at 10 μM) followed by Cd^{2+}, Hg^{2+}, and Cu^{2+} (Wu, 1976). Other divalent ions tested, including transition and akaline earth metal ions, are much less effective as inhibitors. Alkali metal ions (Li^+, Na^+, K^+, Rb^+, and Cs^+) inhibit little, if at all.

GAD adheres tightly to membranes and liposomes in the presence of millimolar concentrations of Ca^{2+} (Salganicoff and DeRobertis, 1965; Fonnum, 1968; Covarrubias and Tapia, 1978). Thus, EDTA is often included in homogenizing media to minimize the binding of GAD to membranes and any possible inhibition by divalent metal ions.

3. Determination of GAD Activity

3.1. General Considerations

3.1.1. Choice of Assay

Although GAD activity has been determined by manometric or chemical measurement of the reaction products (Tursky, 1970; MacDonnell and Greengard, 1975; Holdiness and Justice, 1981; Pahuja and Reid, 1983; Wu et al., 1979), the two most convenient and widely used methods are radiometric. The simplest is the so-called CO_2-trapping assay in which $^{14}CO_2$ from [1-^{14}C]-glutamate is collected and measured (Albers and Brady, 1959). The second method is the GABA production assay in which labeled GABA is measured. The GABA production assay is somewhat less convenient than the CO_2-trapping assay because labeled GABA must first be separated by chromatography from labeled glutamate and other possible products, such as glutamine, before its radioactivity can be measured. For both assays it is important to purify the labeled substrate before use (*see* section 3.2.1), because commercial labeled glutamate contains impurities that can result in overestimation of GAD activity by either the CO_2 assay or the GABA production assay (Molinoff and Kravitz, 1968; Miller and Martin, 1973; Gonnard and Wicker, 1974; Morin and Wasterlain, 1978). The GABA assay is more specific for GAD than the CO_2-trapping assay and is the method of choice when measuring GAD in tissues, because CO_2 can be produced from glutamate by other metabolic pathways (Drummond and Phillips, 1974; MacDonnell and Greengard, 1975; White, 1981). However, it is possible to underestimate the GAD content with the GABA-production assay if the tissue contains little endogenous GABA and a high level of GABA-transaminase (Wu et al., 1979). This particular problem could be minimized by adding a specific inhibitor of GABA-transaminase, such as gabaculine or γ-vinylGABA to the preparation. In general, these difficulties are most serious in tissues with low GAD activity, because the alternative routes of CO_2 production or GABA degradation become relatively more important.

3.1.2. Sensitivity

The sensitivity of the CO_2 and GABA assay procedures are determined by the ability to detect the radioactivity in the labeled product above blank values. In our experience, blank values are generally lower with the CO_2 method, although low blanks can be achieved with either method if the labeled glutamate is carefully purified. Most of the factors (time of assay, recovery of labeled product, amount of enzyme) that control the amount of label incorporated into the products are similar for the two assay procedures. The amount of label in the product also depends on the specific radioactivity of the substrate. This is usually controlled by the experimenter by mixing labeled and unlabeled glutamate. The amount of product, and hence the enzyme activity, is calculated from the radioactivity in the product and the specific radioactivity of the substrate. When GAD is measured in tissues, the endogenous glutamate contributes to the total unlabeled glutamate and reduces the actual specific radioactivity of the substrate. The glutamate concentration in brain is about 10 μmol/g, so the concentration of endogenous glutamate in a 10% brain homogenate is about 1 m*M*. Thus, to minimize the diluting effect of endogenous glutamate on the specific activity of the substrate, it is necessary to add a high concentration of substrate to the assay mixture. Because the specific activity of commercially available [^{3}H]-glutamate is about 1000-fold greater than that of [^{14}C]-glutamate, the GABA production assay can, in principle, be made more sensitive using [^{3}H]-glutamate as substrate instead of [^{14}C]-glutamate. However, the specific activity of [^{14}C]-glutamate is sufficiently high for most tissues, especially because a substantial amount of unlabeled glutamate must be added to obviate the contribution of the endogenous glutamate. Furthermore, the radioactivity in blanks also increases as the label increases, so frequently little or no advantage is gained by adding large amounts of label.

3.1.3. Buffers and Reaction Conditions

Phosphate or Hepes buffers are suitable for many purposes, such as assaying fractions from column chromatography. Tris buffer should be avoided because it reacts with the cofactor, pyridoxal-P (*see* section 2.3.2). Phosphate buffer may be inappropriate in some studies of the enzyme mechanism or action of inhibitors because it strongly affects the action of some effectors, such as ATP and pyridoxal-P (Tursky, 1970; Martin and Martin, 1979; Meeley and Martin, 1983a, b).

A thiol protecting agent, such as 2-aminoethylisothiouronium bromide (AET) or reduced glutathione, should be used. Dithiothreitol and high concentrations of 2-mercaptoethanol should be avoided because these agents somewhat inhibit GAD. Some workers have recommended that the assays be conducted under anaerobic conditions. However, we have found this to be unnecessary in short (<30 min) assays.

3.1.4. Postmortem Stability and Premortem State

Postmortem stability of GAD is not a serious problem for measurement of GAD activity in either human or animal brain. Experiments with rats and mice have shown that total GAD activity is reasonably stable for up to 3 h postmortem, even in brains held at 37°C or cooled slowly (Bowen et al., 1976; Fahn and Coté, 1976; Iversen et al., 1979; Puymirat et al., 1979; Spokes and Koch, 1978; Spokes et al., 1979). Because GAD is stable for long periods in frozen tissue, postmortem changes in activity present no problem in animal studies. In studies with human tissue, losses in GAD activity appear to be small if the postmortem time before autopsy is 1 d or less. Substantial losses in activity have been reported when autopsies were delayed for longer periods (Perry et al., 1977; Crow et al., 1978), although others have reported that GAD activity remains constant for more than 48 h postmortem (Iversen et al., 1979).

Although postmortem changes do not present a serious impediment to studies of GAD in humans, the premortem state of a patient can have a profound influence on GAD activity. Levels of GAD activity have been found to be markedly lower in patients likely to have suffered from prolonged cerebral hypoxia before death than in patients who died suddenly. McGeer et al. (1971) first commented on the unusually low levels of GAD activity in patients comatose for 24 h or longer. Subsequent studies showed that GAD was greatly reduced in patients dying after prolonged hospitalization or, for example, with bronchopneumonia or signs of oxygen deprivation (Bowen et al., 1976; Perry et al., 1977; Spokes et al., 1979).

3.2. Procedures for Measuring GAD

3.2.1. Purification of Labeled Glutamate

Regardless of the assay method, purification of the radioactive substrate is necessary to achieve low blanks and reliable estimates of CO_2 or GABA production by GAD (Molinoff and Kravitz, 1968; Miller and Martin, 1973; Gonnard and Wicker, 1974; Morin and Wasterlain, 1978). At least seven impurities have been detected in commercially available [1-^{14}C]-glutamate and [U-^{14}C]-glutamate, and one of these releases labeled CO_2 when reacted with aminooxyacetate (Miller and Martin, 1973).

For the CO_2-trapping assay, labeled glutamate can be purified conveniently by thin-layer chromatography. Thin-layer plates (20 × 20 cm) are prepared by spreading a 250 μm layer of cellulose powder MN-300 (Sigma or Machery and Nagle) with any standard thin layer spreader. Commercially prepared cellulose plates contain a binder that is extracted from the cellulose with the glutamate and have not proved as satisfactory. About 500 μCi of labeled glutamate is applied as a streak and the plate developed in a mixture of isopropanol: 2-butanone: 1*M* HCl (60:15:25 by vol) until the solvent is 2–3 cm below the top of the plate (3–4 hr). In some cases it may be useful to reduce the vol of the labeled glutamate from the supplier before applying it to the plate. The plate is allowed to dry thoroughly at room temperature and then exposed for about 1–2 h to Kodak XAR-2 X-ray film. Overexposure of the film may obscure the location of impurities that migrate close to glutamate. The film is developed in Kodak GBX developer and replenisher and fixed with Kodak Fixer according to the manufacturer's instructions, washed thoroughly in distilled water, and dried. To mark the location of the glutamate, the film is aligned on the plate and the glutamate band outlined by pressing on the film with a stylus. This embosses the outline of the band in the cellulose, so the glutamate band can be accurately scraped off. The cellulose is collected in a centrifuge tube and extracted with 2 mL of hot (90°C) water. The cellulose suspension is centrifuged and the supernatant collected, allowed to cool, and filtered through a Millex disposable filter (Millipore 45 μm). The extraction, centrifugation, and filtration is repeated twice. The vol of the combined supernatants is reduced to about 1 mL with a stream of nitrogen. To avoid the formation of degradation products, the temperature of the solution is maintained at <37°C during vol reduction.

$[^3H]$-Glutamate can be purified by the same procedure, except that the band of $[^3H]$-glutamate cannot be detected directly by autoradiography. To locate the band, small amounts of $[^{14}C]$-glutamate are chromatographed at each end of the plate and located by autoradiography.

An alternative purification procedure can be used when GAD activity is to be determined by measuring the GABA produced (Bird, 1976). The pH of the labeled glutamate is adjusted to approximately neutral with NaOH. The labeled glutamate (500 μCi) is placed on a 0.5 × 2.5 cm column of Dowex 1X8 anion exchange resin (200 mesh, acetate form prepared as described below). The column is washed extensively with water until no more radioactivity is present in the eluate (approximately 10 mL) and then eluted with 3 mL of 1*M* acetic acid. The acetic acid eluate containing the labeled glutamate is loaded on a 3 mL column of Dowex 50-X8 (200 mesh, H^+ form prepared as described below). The column is washed with 20 mL of water and the glutamate eluted with 15 mL of 2*M* ammonium hydroxide. The samples containing the radioactive glutamate are combined and freeze-dried thoroughly to remove the ammonia. The labeled glutamate is dissolved and stored in a small vol of 0.01*M* HCl.

The Dowex resins (Sigma Chemical Co.) are prepared for use by repeated washing with NaOH and HCl. To do this, the resin is repeatedly suspended in 1*M* NaOH and filtered on a Buchner funnel until the filtrate is colorless. The resin is then washed with 1*M* HCl by the same procedure, followed by another cycle of washes with NaOH and HCl. Dowex 1 anion exchange resin is then converted to the acetate form by suspending it in NaOH, collecting it by filtration, washing with water, and resuspending it in 2*M* acetic acid. Each resin is finally washed exhaustively with water until the pH of the filtrate is greater than 5.

3.2.2. Measuring GAD by CO_2 Assay

In this assay, $^{14}CO_2$ released from the labeled glutamate is trapped and counted. Because of its simplicity, the CO_2 assay is the most commonly used method with purified GAD. It can also be used with preparations, such as dialyzed supernatants, in which other routes for producing CO_2 from glutamate are unimportant. D,L-$[1\text{-}^{14}C]$-glutamate is acceptable as substrate because the D isomer is generally not present in high enough concentration to interfere with the assay. This form of labeled glutamate is also less expensive than either L-$[1\text{-}^{14}C]$-glutamate or uniformly labeled glutamate. Commercially available $[^{14}C]$-glutamate should be purified before it is used for the CO_2 assay (*see* section 3.2.1).

To assay GAD, 80 μL of enzyme and 10 μL of 0.5*M* Hepes buffer (pH 7.0) containing 1 m*M* AET are placed in a 12-mL glass conical centrifuge tube. After warming the tube in a water bath for 3 min, the reaction is started by adding 10 μL of 0.1*M* L-glutamate containing 1 × 10^5 dpm of [1-^{14}C]-glutamate. The tube is capped immediately with a serum stopper to which a plastic center well (Kontes Glass Co., Vineland, NJ) containing 50 μL of 1*M* methylbenzethonium hydroxide (Hyamine hydroxide) in methanol has been attached. After 20 min, the reaction is stopped by injecting 0.1 mL of 7*M* sulfuric acid through the serum stopper. The tube is incubated for an additional 90 min to trap the CO_2. The plastic center well is then cut off into a scintillation vial, 10 mL of scintillant added, and the radioactivity determined. If L-[1-^{14}C]-glutamate is used, the moles of CO_2 produced are calculated by dividing the radioactivity of the CO_2 by the specific radioactivity of the labeled glutamate after correction for counting efficiency. If racemic labeled glutamate is used after dilution with unlabeled L-glutamate, the specific radioactivity must first be divided by 2, and if uniformly labeled glutamate is used, the specific radioactivity must be divided by 5, to account for the numbers of labeled atoms in the molecule.

3.2.3. Measuring GAD by the Production of GABA

Either uniformly labeled [^{14}C]-glutamate or [^{3}H]-glutamate labeled on carbon 3 or 4 can be used to measure the production of GABA by GAD. In most cases [^{14}C]-glutamate is preferable to [^{3}H]-glutamate because ^{14}C is counted with higher efficiency. [^{3}H]-Glutamate may be advantageous if higher specific radioactivity is required. Regardless of the form of labeled glutamate used, it should be purified (*see* section 3.2.1).

A number of procedures for separating labeled GABA from glutamate have been published. Some involve two small ion-exchange columns, the first to separate neutral amino acids (GABA) from glutamate and the second to separate GABA from other amino acids, such as glutamine, that might be produced from labeled glutamate. In most assays, production of glutamine should be minimal because glutamine synthesis also requires ATP, NH_3, and Mg^{2+} and these species are unlikely to be present in sufficient amounts unless added to the incubation medium. None of these procedures are designed to measure the production of GABA from labeled glutamate in intact cells where many routes of glutamate metabolism are possible. Hamel et al. (1982) have developed a simple, single column procedure with improved sensitivity.

To measure GAD, 80 μL of enzyme or tissue extract and 10 μL

of 0.5*M* sodium phosphate buffer, pH 7.2, containing 10 m*M* AET and 200 μ*M* pyridoxal-P are placed in a 1.5 mL conical plastic centrifuge tube for an Eppendorf clinical microcentrifuge and warmed in a 37°C water bath for 3 min. The pyridoxal-P and AET can be omitted from the buffer when included in the homogenizing medium for preparing tissue extracts. The reaction is started by adding 10 μL of 0.1*M* glutamate containing 2×10^5 dpm of labeled glutamate. After incubating for 60 min, the reaction is stopped by adding 500 μL of fresh cupric phosphate suspension (prepared as described below). The tubes are placed on ice for 10 min and then centrifuged for 2 min in an Eppendorf microcentrifuge. A measured volume (400 μL) of the supernatant is placed on a small column (0.5 × 2.5 cm) of Dowex 1-X8 anion exchange resin (200 mesh, acetate form prepared as described in section 3.2.1). The column is eluted with 2 mL of water directly into a scintillation vial and the radioactivity determined.

The moles of GABA produced are calculated using the equation

$$\frac{1.5 \times (\text{sample dpm})}{E \times SA}$$

where E is the counting efficiency, SA is the specific radioactivity of the labeled GABA in dpm/mol, and the factor 1.5 corrects for the fraction of the incubation mixture placed on the column. With [^{3}H]-glutamate, the SA of GABA is assumed to be equal to the SA of the glutamate added to the tube. When [U-^{14}C]-glutamate is used, the SA of GABA is obtained by multiplying the SA of the glutamate by 0.8 to account for the difference in the number of labeled carbon atoms in GABA and glutamate.

The cupric phosphate suspension is prepared fresh before each experiment by mixing solutions of cupric chloride (27.3 g/L), trisodium phosphate (64.5 g/L), and sodium borate (28.6 g/L containing 50 mL of 1*M* HCl) in a volume ratio of 1:2:2.

3.2.4. Measuring Total GAD in Tissues

Measurement of total GAD activity is a simple matter for adult brain and brain regions. As mentioned previously, the labeled glutamate should be purified and GAD activity should be determined by measuring GABA rather than CO_2 production. These are particularly important considerations with immature brain or with tissues in which GAD activity is low, because impurities in the labeled glutamate, or production of CO_2 from labeled glutamate, may result in an inaccurate estimation of GAD activity. Triton X-100 is generally added to the homogenizing medium to disrupt membranes and ensure complete exposure of GAD to the substrate (van Kempen et al., 1965).

The following procedure has been used to measure GAD activity in hippocampus of rats as young as 4 d old. At this age, GAD activity is about 4% of the adult level. The tissue is dissected, weighed, and homogenized in 9 mL/g tissue of 0.1*M* sodium phosphate buffer, pH 7.0, containing 20 μ*M* pyridoxal-P, 0.1 % Triton X-100 (w/v), and 1 m*M* AET. Ninety μL of homogenate are placed in a 1.5-mL plastic conical centrifuge tube (for a high-speed clinical microcentrifuge, such as an Eppendorf) and warmed to 37°C for 3 min. The reaction is started by adding 10 μL of 0.1*M* L-glutamate containing 3×10^5 dpm of L-[U-^{14}C]-glutamate. The reaction is allowed to proceed for 60 min and the amount of GABA produced is measured as described in section 3.2.3.

3.2.5. Measuring HoloGAD in Tissues

To measure holoGAD, the general approach has been to determine GAD activity in homogenates or supernatants in the absence of added pyridoxal-P. To obtain accurate results, however, several factors must be taken into account in the experimental design (Miller et al., 1977, 1980). First, assay conditions must be selected to minimize the inactivation of GAD via the transamination branch of the enzyme mechanism (*see* section 2.2). Because the rate of inactivation depends strongly on the substrate concentration, this can be achieved by carrying out short assays (≤5 min) at low concentrations of glutamate. Levels of holoGAD undoubtedly have been underestimated in studies that involved long assays with high concentrations of glutamate in the absence of pyridoxal-P. Second, there is a very rapid, postmortem activation of apoGAD by endogenous pyridoxal-P (Miller et al., 1977, 1980). To minimize this, brains must be frozen immediately when the animals are killed and homogenates prepared in the presence of ATP to inhibit activation of apoGAD by pyridoxal-P. Finally, the extracts contain several factors that can influence the relative amounts of apo- and holoGAD during the assay (e.g., endogenous glutamate, pyridoxal-P, Pi, and added ATP). Thus these factors must be removed to obtain accurate measurements of GAD activity.

The amount of holoGAD (measured in the absence of pyridoxal-P) is generally expressed as a percentage of the total GAD activity (measured in the presence of added pyridoxal-P). The amount of apoGAD (inactive GAD without bound pyridoxal-P) is calculated by substracting the holoGAD activity from total GAD activity. It should be noted that the term apoenzyme has also been used to refer to total GAD protein. In one study, total GAD protein was measured immunochemically (Blindermann et al., 1979), but in other studies total GAD protein was measured by determining GAD activity in the presence of added pyridoxal-P (Sze and Lovell, 1970). Thus, in the latter cases the term "apoenzyme" is equivalent to the term "total GAD activity" that is used here.

To measure holoGAD activity, the animals are killed and the brains frozen by the most rapid means available. With rats, the freeze-blowing technique, in which the brains are removed and frozen against liquid nitrogen cooled plates within 1 s, is the most suitable, but decapitation into liquid nitrogen can also be used (Miller et al., 1977). Samples of brain are removed and ground into a fine powder while frozen, homogenized in 9 mL of 5 m*M* ATP (pH 6.5)/g brain, and immediately centrifuged at 16,000*g* for 30 min at 4°C. The supernatant is diluted with an equal volume of 200 m*M* imidazole-acetate buffer, pH 6.5, containing 1 m*M* AET. Low molecular weight compounds (ATP, glutamate, pyridoxal-P, and Pi) are then removed by chromatographing 2 mL of the diluted supernatant on a column of Sephadex G-25 (bed vol, 70 mL), which has been equilibrated with 100 m*M* imidazole-acetate buffer, pH 6.5, containing 1 m*M* AET, and kept at 4°C. The absorbance at 280 nm is monitored to identify protein-containing fractions, which are pooled and used for determining enzyme activity. GAD activity is measured by a 5-min assay in the absence and presence of 100 μ*M* pyridoxal-P, using 100 μ*M* glutamate as substrate. Either CO_2 or GABA production can be measured, because they are produced in equimolar amounts in these experiments (Miller et al., 1977).

4. Separation of Multiple Forms of GAD

4.1. Hydrophobic Interaction Chromatography of GAD

The preparative separation of multiple forms of GAD is best achieved by chromatography on phenyl-Sepharose. The exact elution conditions required to separate the multiple forms depend on the source of the enzyme, so procedures developed for GAD from one species cannot be applied directly to another. For example, the three forms of GAD from hog brain can be resolved by a simultaneously decreasing phosphate and increasing glycerol gradient, but this same gradient only elutes two forms of rat brain GAD and the third form must be eluted with Triton X-100 (Spink and Martin, 1983).

Hog brains from a local abattoir are obtained shortly after slaughter and stored frozen or transported to the laboratory in ice-cold 0.32*M* sucrose containing 2.5 m*M* EDTA (pH 7). Rat brains are removed within 2 min of death and immediately frozen on dry ice. Brains can be stored frozen for at least 3 mo before preparing the enzyme.

In preparative procedures, the best resolution of the multiple forms on phenyl-Sepharose chromatography is obtained if the enzyme is first partially purified by chromatography on DEAE-Sephacel and hydroxylapatite. The procedure requires a DEAE-Sephacel column (2.6 × 20 cm) equilibrated with 5 m*M* sodium phosphate buffer (pH 7.0), a hydroxylapatite column (1.6 × 20 cm) equilibrated with 5 m*M* sodium phosphate buffer, and a phenyl-Sepharose column (2.6 × 15 cm) equilibrated with 200 m*M* potassium phosphate buffer, pH 7.0. All chromatography buffers also contain 1 m*M* AET and 20 μ*M* pyridoxal-P. DEAE-Sephacel and phenyl-Sepharose are obtained from Pharmacia Fine Chemicals, and hydroxylapatite (Bio-Gel HT) is obtained from Bio-Rad Laboratories. The elution of protein is followed by measuring A_{280} with an absorbance monitor equipped with a flow cell. The following procedure is described for 150 g of brain (2 hog brains or approximately 100 rat brains); the scale can be doubled by running all columns in tandem. The frozen brains are cut up, allowed to thaw for about 1 h at room temperature, and then homogenized in 850 mL of ice-cold deionized water containing 2.5 m*M* EDTA, 100 μ*M* phenylmethylsulfonylfluoride, 1 m*M* AET, 20 μ*M* pyridoxal-P, and 1% polyethylene glycol (~300 mol wt). The homogenate is then transferred to a 2-L stainless steel beaker and the pH is adjusted to 5.4 with 7% acetic acid. The beaker is then placed in an 80°C water bath and heated to 48°C for 5 min with constant stirring. After cooling in an ice bath (with occasional stirring), the pH is readjusted to 7.0 with 1*M* NaOH and the homogenate is centrifuged at 1200*g* for 45 min in a Sorvall GSA rotor. The supernatant (~700 mL) is loaded at 35 mL/h onto the DEAE-Sephacel column, which is then washed with 100 mL of 5 m*M* sodium phosphate buffer (pH 7.0) and eluted with a 0–0.6*M* linear NaCl gradient in the same buffer. The total vol of the gradient is 700 mL. The fractions (11 mL) are assayed for GAD and protein (Bradford, 1976), the peak fractions are pooled and applied directly to the hydroxylapatite column, which is then eluted with a 5–225 m*M* linear potassium phosphate gradient (pH 7.0). The columns are run in reverse flow at about 10 mL/h. The fractions (4 mL) are assayed and those containing peak activity are pooled and applied to the phenyl-Sepharose column. The column is then washed with 150 mL of 200 m*M* potassium phosphate buffer, pH 7.0, and eluted with a linear increasing glycerol gradient

(0–50%, v/v) and simulataneously linear decreasing phosphate gradient (200–10 m*M*) over 700 mL. The column is washed with 250 mL of 50% glycerol in 10 m*M* potassium phosphate buffer. With hog brain, this procedure separates the three forms of GAD. With rat brain, only two forms are obtained by this procedure; the third is eluted by washing the column with 250 mL of Triton X-100 (10 mL/L of 10 m*M* phosphate buffer). The fractions comprising the peak of activity of each form are pooled and dialyzed against 30 m*M* sodium phosphate buffer containing 1 m*M* AET and 20 μ*M* pyridoxal-P. This results in at least a twofold increase in the vol of the solutions of β- and γ-GAD, because of the high concentrations of glycerol. The dialyzed solution of each form is then concentrated to 0.12 units of activity per milliliter by ultrafiltration through an Amicon PM-10 membrane, mixed with an equal vol of 80% (w/v) sucrose solution, containing 20 μ*M* pyridoxal-P, 1 m*M* AET, and 30 m*M* phosphate, and stored frozen in 0.5 mL samples at –20°C.

4.2. Separation of α-, β-, and γ-GAD by Isoelectric Focusing

The three forms of GAD have different isoelectric points and can be partially resolved by isoelectric focusing in a flat bed of granulated gel, provided a suitably shallow pH gradient is used (Spink et al., 1983). The procedure is carried out with an LKB isoelectric focusing apparatus (LKB 2117 Multiphor). A flat bed of granulated gel is prepared from a suspension (100 mL total vol) of 5 g of Ultrodex (LKB) in a solution of 1% polyethylene glycol (approximate mol wt, 300) and 2% ampholytes (consisting of a mixture of BIO-Lytes 4/6, 5/7, and 3/10 in a 3 : 2 : 2 ratio by volume; BIO-Lytes are obtained from Bio-Rad Laboratories). To prepare for focusing, GAD is dialyzed against 10 m*M* imidazole-acetate buffer, pH 7.0, containing 1 m*M* AET and 20 μ*M* pyridoxal-P. Approximately 3 mL of enzyme solution is mixed with about 20 mL of gel that is removed from the bed with the sample applicator. The gel is reapplied to the bed, allowed to hydrostatically equilibrate for about 20 min, and isoelectric focusing is carried out at 8 W constant power for 10 h at 6°C. At the end of the focusing period, zones of gel are removed from the bed and transferred to small columns. Each sample of gel is eluted with distilled water, and the GAD activity, pH, and amount of protein in each eluate is determined.

4.3. Separation of α-, β-, and γ-GAD by Nondenaturing Polyacrylamide Gel Electrophoresis

The three forms of GAD can be separated by discontinuous polyacrylamide gel electrophoresis (Spink et al., 1983). Gels are prepared in tubes or slabs essentially by the method developed by

Ornstein (1964) and Davis (1964). Resolving gels are prepared with a solution containing 6% (w/v) acrylamide, 0.3% (w/v) *N,N'*-methylene-bis-acrylamide, in 0.375*M* Tris-HCl buffer, pH 8.8. Stacking gels are formed from a solution of 2.4% acrylamide, 0.6% *N,N'*-methylene-bis-acrylamide, and 0.06*M* Tris-HCl buffer, pH 6.8. The electrode buffer contains 0.025*M* Tris and 0.192*M* glycine, pH 8.3. Tube gels are run at 5 mA/tube for 80 min. The gels are immediately sliced into 3 mm sections, and each slice homogenized in 0.5 mL of 75 m*M* phosphate buffer, pH 6.2. The homogenates are allowed to stand for at least 30 min and then assayed for GAD by the CO_2 trapping assay. The pH of the homogenizing buffer is selected to compensate for the relatively high pH of the electrophoresis buffer and achieve a final value near the pH optimum of the enzyme.

4.4. Separation of α-GAD From β- and γ-GAD With ATP-Agarose

ATP-agarose provides a convenient means of separating α-GAD from β- or γ-GAD when incomplete resolution is achieved by other means (Wu and Martin, 1984). A 5-mL column of ATP-agarose [agarose-N^6(aminohexyl) carbamoyl-methyl-ATP from Sigma Chemical Co.] is equilibrated with 50 m*M* Hepes buffer, pH 7.2, containing 1 m*M* AET. GAD (to 0.1 units) is loaded in 1 mL of the same buffer. α-GAD is eluted by washing the column with 10 mL of buffer. The other forms are eluted with 20 mL of 200 m*M* potassium phosphate buffer, pH 7.2.

References

Albers R. W. and Brady R. O. (1959) The distribution of glutamic decarboxylase in the nervous system of the monkey. *J. Biol. Chem.* **234,** 926–928.

Bakay R. E. and Harris A. B. (1981) Neurotransmitter, receptor and biochemical changes in monkey cortical epileptic foci. *Brain Res.* **206,** 387–404.

Baxter C. F. (1970) The Nature of γ-Aminobutyric Acid, in *Handbook of Neurochemistry* Vol. III (Lajtha A., ed.), Plenum, New York.

Baxter C. F. (1976) Some Recent Advances in Studies of GABA Metabolism and Compartmentation, in *GABA* in *Nervous System Function* (Roberts E., Chase T. N., and Tower D. B., eds.), Raven, New York.

Beart P. M., Kelly J. S., and Schon F. (1974) γ-Aminobutyric acid in the peripheral nervous system, pineal, and posterior pituitary. *Biochem. Soc. Trans.* **2,** 266–268.

Bird T. D. (1976) Normal glutamic acid decarboxylase activity in kidney tissue from patients with Huntington's disease. *J. Neurochem.* **27,** 1555–1557.

Bird E. D., MacKay A. V. P., Rayner C. N., and Iversen L. L. (1973) Reduced glutamic-acid decarboxylase activity of postmortem brain in Huntington's Chorea. *Lancet* **ii,** 1090–1092.

Blindermann J.-M., Maitre M., Ossola L., and Mandel P. (1978) Purification and some properties of L-glutamate decarboxylase from human brain. *Eur. J. Biochem.* **86,** 143–152.

Blindermann J.-M., Maitre M., and Mandel P. (1979) Apoenzyme concentration and turnover number of L-glutamate decarboxylase in some regions of rat brain. *J. Neurochem.* **32,** 245–246.

Bowen D. M., Smith C. R., White P., and Davison A. N. (1976) Neurotransmitter-related enzymes and indices of hypoxia in senile dementia and other abiotrophies. *Brain* **99,** 459–496.

Bradford M. M. (1976) A rapid and sensitive method for the quantitation of microgram quantities of protein using the principle of protein-dye binding. *Anal. Biochem.* **72,** 248–254.

Casu M. and Gale K. (1981) Intracerebral injection of gamma-vinylGABA: Method for measuring rates of GABA synthesis in specific brain regions in vivo. *Life Sci.* **29,** 681–688.

Chrystal E., Bey P., and Rando R. R. (1979) The irreversible inhibition of brain L-glutamate-1-decarboxylase by (2RS,3E)-2-methyl-3,4-didehydroglutamic acid. *J. Neurochem.* **32,** 1501–1507.

Collins C. G. S. (1972) GABA-2-oxoglutarate transaminase, glutamate decarboxylase and the half-life of GABA in different areas of rat brain. *Biochem. Pharmacol.* **21,** 2849–2858.

Covarrubias M. and Tapia R. (1978) Calcium-dependent binding of brain glutamate decarboxylase to phospholipid vesicles. *J. Neurochem.* **31,** 1209–1214.

Crow T. J., Owen F., Cross A. J., Lefthouse R., and Longden A. (1978) Brain biochemistry in schizophrenia. *Lancet,* **i,** 36–37.

Danzin C., Casara P., Claverie N., and Grove J. (1983) Effects of the enantiomers of 5-hexyne-1,4-diamine on ODC, GAD and GABA-T activities in the rat. *Biochem. Pharmacol.* **32,** 941–942.

Davis B. J. (1964) Disc-electrophoresis. II. Methods and application to human serum proteins. *Ann. NY Acad. Sci.* **121,** 404–427.

del Rio R. M. (1981) γ-Aminobutyric acid system in rat oviduct. *J. Biol. Chem.* **256,** 9816–9819.

Di Chiara G., Porceddu M. L., Morelli M., Mulas M. L., and Gessa G. L. (1979) Evidence for a GABAergic projection from the substantia nigra to the ventromedial thalamus and to the superior colliculus of the rat. *Brain Res.* **176,** 273–284.

Drummond R. J. and Phillips A. T. (1974) L-glutamic acid decarboxylase in non-neural tissues of the mouse. *J. Neurochem.* **23,** 1207–1213.

Emson P. C. and Joseph M. H. (1975) Neurochemical and morphological changes during the development of cobalt-induced epilepsy in the rat. *Brain Res.* **93,** 91–110.

Endo A. and Kitahara N. (1981) Inhibition of brain glutamate decarboxylase by 4,5-dihydroxyisophthalic acid and related compounds. *J. Antibiot.* **34,** 1351–1354.

Fahn S. (1976) Regional Distribution Studies of GABA and Other Putative Neurotransmitters and Their Enzymes, in *GABA* in *Nervous System Function* (Roberts E., Chase T. N., and Tower D. B., eds.), Raven, New York.

Fahn S. and Coté L. J. (1976) Stability of enzymes in postmortem rat brain. *J. Neurochem.* **26,** 1039–1042.

Fonnum F. (1968) The distribution of glutamate decarboxylase and aspartate transaminase in subcellular fractions of rat and guinea-pig brain. *Biochem. J.* **106,** 401–412.

Gale K. and Casu M. (1981) Dynamic utilization of GABA in the substantia nigra: Regulation by dopamine and GABA in the striatum and its clinical and behavioral implications. *Mol. Cell. Biochem.* **39,** 369–405.

Gonnard P. and Wicker A. (1974) A propos de la glutamate decarboxylase mitochondriale cerebrale. *Biochimie* **56,** 1437–1438.

Hamel E., Krause D. N., and Roberts E. (1982) Characterization of glutamic acid decarboxylase activity in cerebral blood vessels. *J. Neurochem.* **39,** 842–849.

Hattori T., McGeer P. L., Fibiger H. C., and McGeer E. G. (1973) On the source of GABA-containing terminals in the substantia nigra. Electron microscopic autoradiographic and biochemical studies. *Brain Res.* **54,** 103–114.

Heinamaki A. A., Malila S. I., Tolonen K. M., Valkonen K. H., and Piha R. S. (1983) Resolution and purification of taurine- and GABA-synthesizing decarboxylases from calf brain. *Neurochem. Res.* **8,** 207–218.

Hertz L., Kvamme E., McGeer E. G., and Schousboe A., eds. (1983) *Glutamine, Glutamate, and GABA in the Central Nervous System.* Alan R. Liss, New York.

Holdiness M. R. and Justice J. B. (1981) Gas chromatographic-mass spectrometric determination of glutamic acid decarboxylase activity in subregions of rat brain. *J. Chromatogr.* **225,** 283–290.

Houser C. R., Hendry S. H. C., Jones E. G., and Vaughn J. E. (1983) Morphological diversity of immunocytochemically identified GABA neurons in the monkey sensory-motor cortex. *J. Neurocytol.* **12,** 617–638.

Iversen L. L., Bird E., Spokes E., Nicholson S. H., and Suckling C. J. (1979) Agonist Specificity of GABA Binding Sites in Human Brain and GABA in Huntington's Disease and Schizophrenia, in *GABA-Neurotransmitters: Pharmacochemical, Biochemical, and Pharmacological*

Aspects (Krogsgaard-Larsen P., Scheel-Kruger J., and Kofod H., eds.), Academic, New York.

Jessen K. R., Mirsky R., Dennison M. E., and Burnstock G. (1979) GABA may be a neurotransmitter in the vertebrate peripheral nervous system. *Nature* **281,** 71–74.

Kanazawa I., Iversen L. L., and Kelly J. S. (1976) Glutamate decarboxylase activity in the rat posterior pituitary, pineal gland, dorsal root ganglion, and superior cervical ganglion. *J. Neurochem.* **27,** 1267–1269.

Kilpatrick I. C., Starr M. S., Fletcher A., James T. A., and MacLeod M. K. (1980) Evidence for a GABAergic nigrothalamic pathway in the rat. I. Behavioural and biochemical studies. *Exp. Brain Res.* **40,** 45–54.

Kohler C., Chan-Palay V., and Wu J.-Y. (1984) Septal neurons containing glutamic acid decarboxylase immunoreactivity project to the hippocampal region in the rat brain. *Anat. Embryol.* **169,** 41–44.

Krogsgaard-Larsen P., Scheel-Kruger J., and Kofod H., eds. (1979) *GABA-Neurotransmitters.* Academic, New York.

Lal H., Fielding S., Malick J., Roberts E., Shah N., and Usdin E., eds. (1980) *GABA Neurotransmission: Current Developments in Physiology and Neurochemistry. Brain Res. Bull.* **5,** suppl. 2, 1–946.

Lloyd K. G., Munari C., Bossi L., Stoeffels C., Talairach J., and Morselli P. L. (1981) Biochemical Evidence for the Alterations of GABA-Mediated Synaptic Transmission in Pathological Brain Tissue (Stereo EEG or Morphological Definition) From Epileptic Patients, in *Neurotransmitters, Seizures and Epilepsy.* (Morselli P. L., Lloyd K. G., Löscher W., Meldrum B., and Reynolds E. H., eds.), Raven, New York.

MacDonnell P. and Greengard O. (1975) The distribution of glutamate decarboxylase in rat tissues: Isotopic vs fluorimetric assays. *J. Neurochem.* **24,** 615–618.

Maitre M., Blindermann J. M., Ossola L., and Mandel P. (1978) Comparison of the structures of L-glutamate decarboxylases from human and rat brains. *Biochem. Biophys. Res. Commun.* **85,** 885–890.

Martin D. L. and Martin S. B. (1982) Effect of nucleotides and other inhibitors on the inactivation of glutamate decarboxylase. *J. Neurochem.* **39,** 1001–1008.

Martin S. B. and Martin D. L. (1979) Stimulation by phosphate of the activation of glutamate apodecarboxylase by pyridoxyl-5'-phosphate and its implications for the control of GABA synthesis. *J. Neurochem.* **33,** 1275–1283.

Martin D. L., Meeley M. P., Martin S. B., and Pedersen S. (1980) Factors influencing the activation and inactivation of glutamate decarboxylase. *Brain Res. Bull.* **5,** suppl. 2, 57–61.

Matsuda T., Wu J.-Y., and Roberts E. (1973) Sodium dodecyl sulfate acrylamide gel electrophoresis of glutamic acid decarboxylase from mouse brain. *J. Neurochem.* **21,** 167–172.

Matsui Y. and Deguchi T. (1977) Effects of gabaculine, a new potent inhibitor of gamma-aminobutyrate transaminase on the brain gamma-aminobutyrate content and convulsions in mice. *Life Sci.* **20,** 1291–1296.

McGeer E. G., Fibiger H. C., McGeer P. L., and Brooke S. (1973a) Temporal changes in amine-synthesizing enzymes of rat extrapyramidal structures after hemitranssection or 6-hydroxydopamine administration. *Brain Res.* **52,** 289–300.

McGeer P. L., McGeer E. G., and Fibiger H. C. (1973b) Choline acetylase and glutamic acid decarboxylase in Huntington's chorea. *Neurology* **23,** 912–917.

McGeer P. L., McGeer E. G., and Wada J. A. (1971) Glutamic acid decarboxylase in Parkinson's disease and epilepsy. *Neurology* **21,** 1000–1007.

Meeley M. P. and Martin D. L. (1983a) Inactivation of brain glutamate decarboxylase and the effects of adenosine 5'-triphosphate and inorganic phosphate. *Cell. Mol. Neurobiol.* **3,** 39–54.

Meeley M. P. and Martin D. L. (1983b) Reactivation of substrate-inactivated brain glutamate decarboxylase. *Cell. Mol. Neurobiol.* **3,** 55–68.

Metcalf B. W., Jung M. J., Lippert B., Casara P., Bohlen P., and Schechter P. J. (1979) γ-AcetylenicGABA and γ-VinylGABA—two Enzyme-Activated Irreversible Inhibitors of GABA Aminotransferase, in *GABA-Neurotransmitters* (Krogsgaard-Larsen P., Scheel-Kruger J., and Kofod H., eds.), Academic, New York.

Miller L. P. and Martin D. L. (1973) An artifact in the radiochemical assay of brain mitochondrial glutamate decarboxylase. *Life Sci.* **13,** 1023–1032.

Miller L. P., Walters J. R., and Martin D. L. (1977) Postmortem changes implicate adenine nucleotides and pyridoxal-5'-phosphate in regulation of brain glutamate decarboxylase. *Nature* **266,** 847–848.

Miller L. P., Martin D. L., Mazumder A., and Walters J. R. (1978) Studies on the regulation of GABA synthesis: Substrate-promoted dissociation of pyridoxal-5'-phosphate from GAD. *J. Neurochem.* **30,** 361–369.

Miller L. P., Walters J. R., Eng N., and Martin D. L. (1980) Glutamate holodecarboxylase levels and the regulation of GABA synthesis. *Brain Res. Bull.* **5,** suppl. 2, 89–94.

Molinoff P. B. and Kravitz E. A. (1968) The metabolism of γ-aminobutyric acid (GABA) in the lobster nervous system—glutamic decarboxylase. *J. Neurochem.* **15,** 391–409.

Morin C. W. and Wasterlain C. G. (1978) Stoichiometry of GABA and CO_2 formation in glutamate decarboxylase assays: Alteration by an impurity in L-[U-^{14}C]-glutamate. *J. Neurochem.* **31,** 371–373.

Oertel W. H., Schmechel D. E., Tappaz M. L., and Kopin I. J. (1981a) Production of a specific antiserum to rat brain glutamic acid decarboxylase by injection of an antigen–antibody complex. *Neuroscience* **6,** 2689–2700.

Oertel W. H., Schmechel D. E., Weise V. K., Ransom D. H., Tappaz M. L., Krutzsch H. C., and Kopin I. J. (1981b) Comparison of cysteine sulphinic acid decarboxylase isoenzymes and glutamic acid decarboxylase in rat liver and brain. *Neuroscience* **6,** 2701–2714.

Oertel W. H., Schmechel D. E., Mugnaini E., Tappaz M. L., and Kopin I. J. (1981c) Immunocytochemical localization of glutamate decarboxylase in rat cerebellum with a new antiserum. *Neuroscience* **6,** 2715–2735.

Ornstein L. (1964) Disc electrophoresis. I. Background and Theory. *Ann. NY Acad. Sci.* **121,** 321–349.

Pahuja S. L. and Reid T. W. (1983) A simple microassay for glutamic acid decarboxylase by ion exchange thin-layer chromatography. *J. Liquid. Chromatogr.* **6,** 127–137.

Perry E. K., Gibson G. B., Perry R. H., and Tomlinson B. E. (1977) Neurotransmitter enzyme abnormalities in senile dementia. *J. Neurol. Sci.* **34,** 247–265.

Perry T. L., Hansen S., and Kloster M. (1973) Huntington's Chorea—Deficiency of gamma-aminobutyric acid in brain. *N. Engl. J. Med.* **288,** 337–342.

Porter T. G. and Martin D. L. (1984) Evidence for feedback regulation of glutamate decarboxylase by γ-aminobutyric acid. *J. Neurochem.* **43,** 1464–1467.

Porter T. G., Martin S. B., and Martin D. L. (1982) Inhibitors and inactivation promoters of glutamate decarboxylase. *Soc. Neurosci. Abstr.* **8,** 819.

Porter T. G., Spink D. C., Martin S. B., and Martin D. L. (1985) Transaminations catalyzed by brain glutamate decarboxylase. *Biochem. J.* **231,** 705–712.

Puymirat J., Javoy-Agid F., Gaspar P., Ploska A., Prochiantz A., and Agid Y. (1979) Post mortem stability and storage in the cold of brain enzymes. *J. Neurochem.* **32,** 449–454.

Rando R. R. (1979) Specific Inhibitors of GABA Metabolism, in *GABA-Neurotransmitters* (Krogsgaard-Larsen P., Scheel-Kruger J., and Kofod H., eds.), Academic, New York.

Rando R. R. (1981) The chemical labeling of glutamate decarboxylase in vivo. *J. Biol. Chem.* **256,** 1111–1114.

Ribak C. E., Harris A. B., Vaughn J. E., and Roberts E. (1981) Immunocytochemical Changes in Cortical GABA Neurons in a Monkey Model of Epilepsy, in *Neurotransmitters, Seizures and Epilepsy.* (Morselli P. L., Lloyd K. G., Löscher W., Meldrum B., and Reynolds E. H., eds.), Raven, New York.

Ribak C. E., Vaughn J. E., and Saito K. (1978) Immunocytochemical localization of glutamic acid decarboxylase in neuronal somata following colchicine inhibition of axonal transport. *Brain Res.* **140,** 315–332.

Ribak C. E., Harris A. B., Vaughn J. E., and Roberts E. (1979) Inhibitory, GABAergic nerve terminals decrease at sites of focal epilepsey. *Science* **205,** 211–214.

Roberts E. and Simonsen D. G. (1963) Some properties of L-glutamic decarboxylase in mouse brain. *Biochem. Pharmacol.* **12,** 113–134.

Roberts E., Chase T. N., and Tower D. B., eds. (1976) *GABA in Nervous System Function.* Raven, New York.

Ryan L. D. and Roskoski R. (1976) Resolution and reconstitution of glutamate decarboxylase from cerebellum. *Neurochem. Res.* **1,** 37–45.

Saito K., Barber R., Wu J.-Y., Matsuda T., Roberts E., and Vaughn J. E. (1974) Immunohistochemical localization of glutamate decarboxylase in rat cerebellum. *Proc. Natl. Acad. Sci. USA* **71,** 269–273.

Salganicoff L. and DeRobertis E. (1965) Subcellular distribution of the enzymes of the glutamic acid, glutamine, and γ-aminobutyric acid cycles in brain. *J. Neurochem.* **12,** 287–309.

Schousboe A., Svenneby G., and Hertz L. (1977) Uptake and metabolism of glutamate in astrocytes cultured from dissociated mouse brain hemispheres. *J. Neurochem.* **29,** 999–1005.

Seligmann B., Miller L. P., Brockman D. E., and Martin D. L. (1978) Studies on the regulation of GABA synthesis: The interaction of adenine nucleotides and glutamate with brain glutamate decarboxylase. *J. Neurochem.* **30,** 371–376.

Spears R. M. and Martin D. L. (1982) Resolution and brain regional distribution of cysteine sulfinate decarboxylase isoenzymes from hog brain. *J. Neurochem.* **38,** 985–991.

Spink D. C. and Martin D. L. (1983) Multiple Forms of Glutamate Decarboxylase in Hog, Rat, and Human Brain, in *Glutamate, Glutamine and GABA in the Central Nervous System* (Hertz L., Kvamme E., McGeer E. G., and Schoushoe A., eds.), Alan R. Liss, New York.

Spink D. C., Wu S. J., and Martin D. L. (1983) Multiple forms of glutamate decarboxylase in porcine brain. *J. Neurochem.* **40,** 1113–1119.

Spink D. C., Porter T. G., Wu S. J., and Martin D. L. (1985) Characterization of three kinetically distinct forms of glutamate decarboxylase from pig brain. *Biochem. J.* **231,** 695–703.

Spokes E. G. S. and Koch D. J. (1978) Postmortem stability of dopamine, glutamate decarboxylase and choline acetyltransferase in the mouse brain under conditions simulating the handling of human autopsy material. *J. Neurochem.* **31,** 381–383.

Spokes E. G. S., Garrett N. J., and Iversen L. L. (1979) Differential effects of agonal status on measurements of GABA and glutamate decarboxylase in human postmortem tissue from control and Huntington's chorea subjects. *J. Neurochem.* **33,** 773–778.

Storm-Mathisen J. (1972) Glutamate decarboxylase in the rat hippocampal region after lesions of the afferent fiber systems. Evidence that the enzyme is localized in intrinsic neurones. *Brain Res.* **40,** 215–235.

Storm-Mathisen J. (1976) Distribution of the Components of the GABA System in Neuronal Tissue: Cerebellum and Hippocampus—Effects of Axotomy, in *GABA in Nervous System Function* (Roberts E., Chase T. N., and Tower D. B., eds.), Raven, New York.

Susz J. P., Haber B., and Roberts E. (1966) Purification and some properties of mouse brain L-glutamic decarboxylase. *Biochemistry* **5,** 2870–2876.

Sze P. Y. and Lovell R. A. (1970) Reduction of level of L-glutamic acid decarboxylase by γ-aminobutyric acid in mouse brain. *J. Neurochem.* **17,** 1657–1664.

Taberner P. V., Pearce M. J., and Watkins J. C. (1977) The inhibition of mouse brain glutamate decarboxylase by some structural analogues of L-glutamic acid. *Biochem. Pharmacol.* **26,** 345–349.

Tapia R. (1975) Biochemical Pharmacology of GABA in the CNS, in *Handbook of Psychopharmacology,* Vol. 4 (Iversen L. L., Iversen S. D., and Snyder S. H., eds.), Plenum, New York.

Tursky T. (1970) Inhibition of brain glutamate decarboxylase by adenosine triphosphate. *Eur. J. Biochem.* **12,** 544–549.

van Kempen G. M. J., van den Berg C. J., van der Helm H. J., and Veldstra H. (1965) Intracellular localization of glutamate decarboxylase, gamma-aminobutyrate transaminase and some other enzymes in brain tissue. *J. Neurochem.* **12,** 581–587.

White H. L. (1981) Glutamate as a precursor of GABA in rat brain and peripheral tissues. *Mol. Cell. Biochem.* **39,** 253–259.

Wood J. G., McLaughlin B. J., and Vaughn J. E. (1976) Immunocytochemical Localization of GAD in Electronmicroscopic Preparations of Rodent CNS, in *GABA in Nervous System Function* (Roberts E., Chase T. N., and Tower D. B., eds.), Raven, New York.

Wu J.-Y. (1976) Purification, Characterization and Kinetic Studies of GAD and GABA-T from Mouse Brain, in *GABA in Nervous System Function* (Roberts E., Chase, T. N. and Tower D. B., eds.), Raven, New York.

Wu J.-Y. (1977) A comparative study of L-glutamate decarboxylase from mouse brain and bovine heart with purified preparations. *J. Neurochem.* **28,** 1359–1367.

Wu J.-Y. (1982) Purification and characterization of cysteic acid and cysteine sulfinic acid decarboxylase and L-glutamate decarboxylase from bovine brain. *Proc. Natl. Acad. Sci. USA* **79,** 4270–4274.

Wu J.-Y. and Roberts E. (1974) Properties of brain L-glutamate decarboxylase: Inhibition studies. *J. Neurochem.* **23,** 759–767.

Wu J.-Y., Matsuda T., and Roberts E. (1973) Purification and characterization of glutamate decarboxylase from mouse brain. *J. Biol. Chem.* **248,** 3029–3034.

Wu P. H., Durden D. A., and Hertz L. (1979) Net production of γ-aminobutyric acid in astrocytes in primary cultures determined by a sensitive mass spectrometric method. *J. Neurochem.* **32,** 379–390.

Wu S. J. and Martin D. L. (1984) Binding of ATP to brain glutamate decarboxylase as studied by affinity chromatography. *J. Neurochem.* **42,** 1607–1612.

Chapter 11
4-Aminobutyrate Transaminase

Godfrey Tunnicliff

1. Introduction

1.1. Role of GABA in the Nervous System

γ-Aminobutyric acid (GABA) is present in many tissues of the mammalian body at low concentrations, and we believe its main function to be that of an inhibitory chemical transmitter in the central nervous system, in which it is present in comparatively large amounts (Davidson, 1976). It has been estimated that as a transmitter, GABA is involved at over one third of the synapses in the brain (Fonnum and Storm-Mathisen, 1978).

Until recently it was understood that the inhibitory action of GABA was the result of its binding to one of two types of receptors. One of these is located presynaptically, and the second is located postsynaptically. In both cases, Cl^- conductance is increased when the receptor is activated (Alger and Nicoll, 1982; Thampy and Barnes, 1984). Occupancy of the postsynaptic receptor by an agonist will give rise to a *hyperpolarization.* This can be antagonized by bicuculline (Curtis et al., 1971). An agonist at the presynaptic receptor causes a *depolarization* of the neuron, and this leads to an inhibition of transmitter release. This is known as presynaptic inhibition and has been demonstrated on 1a afferent cells in the spinal cord (Eccles et al., 1963; Davidson and Southwick, 1971). This receptor is probably also sensitive to antagonism by bicuculline (Bowery, 1983; Desarmenien et al., 1984).

Evidence is now available that a second type of presynaptic GABA receptor exists. This is bicuculline-insensitive, but is activated by baclofen. Ca^{2+} and Na^+ ions are required for agonist action. It appears that this receptor has no effect on Cl^- conductance, but could instead be linked to Ca^{2+} flux (Dunlap, 1981). This baclofen-sensitive $GABA_B$ receptor is apparently coupled to adenylate cyclase (Wojick and Neff, 1984; Hill et al., 1984). The $GABA_B$ receptor has been shown to occur on autonomic terminals (Bowery et al., 1981; Anwar and Mason, 1982; Kaplita et al., 1982; Muhyaddin et al., 1982; Desarmenien et al., 1984) and on central nerve terminals (Bowery et al., 1980; Ault and Nadler, 1982).

1.2. Function of GABA Transaminase

The release of GABA by nerve terminals and its subsequent binding to any of its receptors must be followed by a rapid inactivation of the transmitter in order that the sensitivity of the mechanism be retained. This inactivation process is, in all probability, the active uptake system for GABA that has been demonstrated in synaptosomes (Martin, 1973), brain slices (Iversen and Neal, 1968), glial fractions (Schon and Kelly, 1974), and retina (Tunnicliff et al., 1975). Evidently, GABA in the synaptic cleft is taken up into three morphologically distinct sites: the presynaptic neuron, postsynaptic neuron, and surrounding glial cells (Martin, 1976). The metabolic degradation of GABA is brought about by the action of γ-aminobutyrate α-ketoglutarate aminotransferase (GABA transaminase; GABA-T; EC 2.6.1.19). This enzyme is located in the mitochondria, where it is localized in the inner matrix (Schousboe et al., 1977).

1.3. Description of the Reaction

The enzyme catalyzes the transfer of the amino group of GABA to the alpha carbon of α-ketoglutaric acid (α-KG). The reaction depends on the presence of pyridoxal 5'-phosphate at the active site, where it is tightly bound. Succinic semialdehyde and glutamic acid are the products of the reaction. Because of the presence of an excess of succinic semialdehyde dehydrogenase over GABA-T in the mitochondria, succinic semialdehyde is rapidly converted to succinic acid in the presence of NAD^+ as proton acceptor (Fig. 1).

The rapid metabolism of succinic semialdehyde ensures that for all practical purposes the reaction is irreversible. Succinic acid can enter the tricarboxylic acid cycle. The route from GABA to succinic acid is part of the so-called GABA shunt pathway, and it has been proposed that this enables GABA to be utilized as an energy source (Baxter, 1970). Thus, GABA-T does not terminate the action of GABA, but instead, merely metabolically degrades the transmitter.

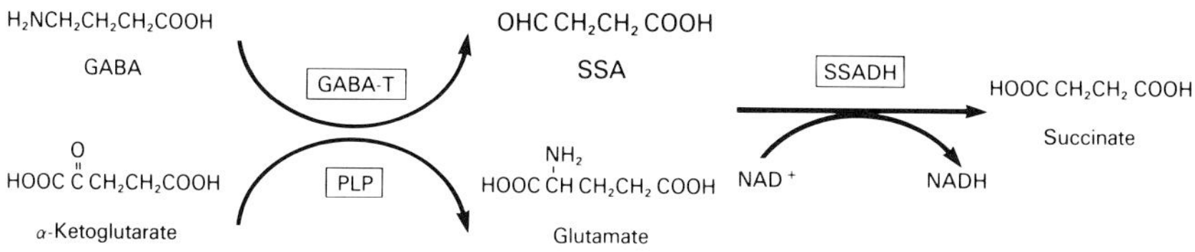

Fig. 1. Metabolic degradation of GABA. Abbreviations: PLP, pyridoxal 5'-phosphate; SSA, succinic semialdehyde; SSADH, succinic semialdehyde dehydrogenase.

1.4. Distribution

1.4.1. Phylogenic Distribution

GABA and its metabolizing enzymes are widely distributed in nature. For example, many bacteria and unicellular plants possess this amino acid. Higher plants are also a good source of GABA. The function of this amino acid in organisms with no nervous system is not understood. In the animal kingdom, GABA is mainly associated with the nervous system, in which it is thought to function as an inhibitory neurotransmitter.

In invertebrates, GABA has been shown to occur in nerve cells. For example, inhibitory neurons in the lobster possess 200 times as much GABA as do excitatory neurons (Kravitz et al., 1965). GABA-T from lobster has been well studied and does not differ significantly from the mammalian enzyme (Hall and Kravitz, 1967). In the mollusc *Aplysia*, GABA has been detected in ganglia and individual neurons (McCaman et al., 1979).

The detection of GABA has been reported in insects. Ray (1964) observed the amino acid in the nerve cord of the cockroach. Nerve tissue from honey bee, locust, and *Drosophila* have all been demonstrated to be able to synthesize GABA (Frontali, 1961; Huggins et al., 1967; Bradford et al., 1969; Tunnicliff et al., 1969). Naturally, in higher animals GABA and GABA-T have received a lot of attention (Davidson, 1976; Roberts et al., 1976; Krogsgaard-Larsen et al., 1979).

1.4.2. Tissue Distribution

When GABA and its metabolizing enzymes were first discovered in the brain, it was thought that they were confined to this tissue. Further studies, however, usually employing more sensitive techniques, demonstrated that GABA and GABA-T can be found in many parts of the mammalian body; for example, liver, kidney, heart, blood vessels, blood platelets, and various structures of the urogenital tract (White, 1979; Zachman et al., 1966; Vasilev et al., 1973; Lancaster et al, 1973; Martin del Rio, 1981). Indeed, the GABA content of the rat oviduct is over twice that of whole brain, and GABA-T activity is comparable to that measured in brain (Martin del Rio, 1981).

1.4.3. Regional Distribution in the Nervous System

The regional localization of both GABA and GABA-T has been extensively studied. Interestingly, their distribution patterns are not similar. As an example, GABA levels are greatest in the substantia nigra, globus pallidus, and hypothalamus, whereas GABA-T activity is greatest in the inferior olivary nucleus, the dentate nucleus, and the inferior and superior colliculus. On the other hand, the transaminase activity is comparatively low in the globus pallidus. Both GABA and GABA-T are present at much higher levels in the grey matter than the white matter (Fahn and Cote, 1968; Okada et al., 1971; Salvador and Albers, 1959).

1.4.4. Subcellular Distribution

The enzyme is associated primarily with the P_2 fraction described by Gray and Whittaker (1962). This fraction is enriched with mitochondria and synaptosomes. Kuriyama et al. (1968a) reported that over half the enzyme activity was recovered from this fraction. Further fractionation into myelin-, synaptosome-, and mitochondria-enriched fractions revealed that GABA-T was associated mainly with the latter. It appears, therefore, that the mitochondrial-bound enzyme originates largely from postsynaptic neurons, endothial cells, and glia. The comparatively low GABA-T activity measured in synaptosomes probably reflects the mitochondria populations at nerve terminals. Schousboe et al. (1977) have evidence that GABA-T is associated with the inner matrix of mitochondria.

Immunochemical techniques have been employed to more accurately establish the cellular distribution of the enzyme. These techniques have been made possible by the availability of purified mouse brain GABA-T (Schousboe et al., 1973). Antisera to the enzyme have been prepared in rabbits by administration of GABA-T in complete Freund's adjuvant (Saito et al., 1974). This method has led to the direct visualization of GABA-T in cerebellar neurons, for example. Under the light microscope the enzyme has been observed in Purkinje cell bodies and dendrites, as well as in stellate cell bodies (Barber and Saito, 1976).

1.5. GABA Transaminase in CNS Disorders

GABA has been implicated in certain types of behaviors and disorders of the central nervous system in animals and humans. The activity of GABA-T, however, has rarely been determined in these same models and, consequently, little is known of the enzyme's involvement.

GABA levels and/or glutamic acid decarboxylase (GAD) activity have been reported to be reduced in postmortem brain tissue from parkinsonian, Huntington's chorea, and schizophrenic patients (McGeer et al., 1971; Bird and Iversen, 1974; Perry et al., 1979). GABA concentrations are decreased in the cerebrospinal fluid of epileptics (Wood et al., 1980). Of course, in many models of epilepsy either GAD activity or GABA levels or both have been shown to be reduced. Furthermore, anticonvulsant effects are seen if GABA-T is inhibited (Wood, 1975). In a study on postmortem brains from Huntington's disease patients, GABA-T was measured, but no alterations were observed even though GABA levels were significantly reduced (Urquhart et al., 1975). This is, no doubt, indicative of the fact that GABA-T activity is not a good marker for GABA function.

It is noteworthy that GABA-T activity (and that of GAD) is markedly lower than control levels in the cerebellum of the two neurological mice mutants known as Staggerer and Weaver (Beart and Lee, 1977).

1.6. Development

The GABA system undergoes sharp increases during development. Both GABA-T and GAD increase eightfold in rat brain from birth to maturity (Van den Berg et al., 1965). The most rapid increases occur between 10 and 20 d. Bloch-Tardy et al. (1971) and Sims et al. (1968) found 10–15-fold increases in enzyme activity between birth and adulthood. The development of GABA enzymes and the growth of synaptic structures in chick cerebellum show a remarkably good correlation (Kuriyama et al., 1968b). Similarly, Sisken et al. (1961) observed parallel increases in the development of the GABA system in chicken optic lobes and increases in the size of the optic tectum.

Nervous tissue culture can provide a convenient way to study development, both morphological and biochemical. In a dissociated cell culture containing both neurons and glia from rat cerebral cortex, GABA-T activity was shown to increase steadily with time in culture over a 20-d period (Hauser and Bernasconi,

1980). These authors ascribed most of the increase in enzyme activity to an increase in *neuronal* GABA-T. In dissociated cultures from mouse cerebral hemispheres, GABA-T increases steadily over a 3-wk period (Tardy et al., 1979). Surprisingly, these workers found that the increase in activity was significantly more pronounced if GABA was added to the culture medium. It was speculated that the GABA brought about an induction of the enzyme in the culture.

Histochemically, Hyde and Robinson (1974) noticed that by 10 d in the rat central nervous system GABA-T is fully developed, yet in glial cells a rapid increase occurs between 10 and 15 d. A conclusion to be drawn from this observation is that some of the sharp increases found by other workers in GABA-T activity during development are caused by the enzyme located in glial cells.

2. Purification, Assay, and Properties—Discussion

2.1. Purification

One of the earliest reports of the purification of GABA-T appeared about 20 yr ago (Waksman and Roberts, 1965). This concerned the enzyme from mouse brain. These workers employed acetone powder extraction followed by ammonium sulfate fractionation and gradient elution. In the final step the enzyme was subjected to calcium phosphate gel fractionation. The final specific activity was 0.5 μmol/min/mg protein at 37°C. This represented a 150-fold increase in activity over the original extract. This particular preparation gave a single band in the ultracentrifuge and was homogeneous after carboxymethylcellulose column chromatography. Later work from the same laboratory led to the isolation of an enzyme with a 10-fold greater specific activity and an overall purification of 1200 times the original sucrose homogenate (Schousboe et al., 1973). This time a more rigorous testing of the enzyme's purity was carried out. By both PAGE and high speed sedimentation runs, the GABA-T preparation was judged to be homogeneous. The enzyme was purified from a sucrose homogenate by ammonium sulfate fractionation followed by column chromatography employing Sephadex, calcium gel, and DEAE-Sephadex.

The enzyme has also been purified from pig brain (Bloch-Tardy et al., 1974). It was subjected to acetone powder extraction, heat precipitation at acid pH, and ammonium sulfate fractionation. The preparation was then applied to three columns in succession—a CM cellulose column, a DEAE cellulose column, and a hydroxyapatite column. The final specific activity was 8.3 μmol/min/mg protein. This represented a 1700-fold purification over the original extract.

At about the same time GABA-T was purified from human brain (Cash et al., 1974). An ammonium sulfate fractionation was carried out after an initial homogenization in sucrose and preparation of a crude mitochondrial fraction. Gel filtration chromatography was followed by DEAE-cellulose and DEAE-Sephadex chromatography. The resulting specific activity was 4 μmol/min/mg protein (1440-fold purification). PAGE established that the enzyme was homogeneous. The authors felt that the specific activity of their enzyme was rather low compared to that of rat brain GABA-T (Maitre et al., 1975) and suggested that a contributing factor could have been the extended time between the death of the patient and start of the purification procedures.

Maitre and colleagues (1975) have extracted and purified GABA-T from rat brain. They started with a sucrose homogenate and undertook an ammonium sulfate fractionation after a crude mitochondrial extract was prepared. Chromatography using a Sephadex G-200 column, a DEAE-cellulose column, an affinity column of 4-aminobutyric acid, and a final DEAE Sephadex column was carried out and an enzyme of specific activity 10 μmol/min/mg protein was obtained. The purification was 2400-fold over the initial homogenate. Homogeneity was established by SDS-PAGE.

One of the most straightforward, and indeed, most effective procedures for the purification of GABA-T has been published by John and Fowler (1976). These workers extracted the enzyme from rabbit brain and obtained a specific activity of 17.3 μmol/min/mg protein. This was a 1169-fold increase in activity compared to the initial homogenate. This enzyme was subjected to heat and low pH treatment, ammonium sulfate fractionation, and two column chromatography steps—DEAE-cellulose and CM-cellulose. A starch gel electrophoresis step demonstrated homogeneity.

It is noteworthy that the method that produced the enzyme of highest specific activity (John and Fowler, 1976) also involved the least degree of purification (Table 1). This is probably a reflection of the activity in the initial homogenate. In those methods in which the brain was homogenized in sucrose, the GABA-T activity was three or four times less than that reported by John and Fowler. The latter authors simply homogenized their tissue in a Waring blender using a 10 m*M* sodium phosphate buffer that also contained the detergent 0.5% Triton X-100. These conditions would have favored liberation of the mitochondrial bound GABA-T. In assays carried

Table 1
Comparison of the Specific Activities of GABA Transaminase Isolated From Different Species[a]

Species	Initial homogenate	Initial activity	Final activity	Purification, X-fold
Rabbit	10 m*M* Buffer	0.015	17.3	1169
Rat	0.25*M* Sucrose	0.042	10.0	2400
Pig	—	0.005	8.3	1700
Mouse	0.25*M* Sucrose	0.0042	5	1200
Human	0.25*M* Sucrose	0.0027	4	1440

[a]Activity is μmol/min/mg protein. Rabbit data from John and Fowler (1976); rat data from Maitre et al. (1975); pig data from Bloch-Tardy et al. (1974); mouse data from Schousboe et al. (1973); human data from Cash et al. (1974).

out on sucrose homogenates (Table 1), not all the enzyme may have been available during the incubation period. Thus, the low activity may not have accurately reflected the amount of enzyme present.

The reason for the different specific activities measured in different laboratories can be attributed to such factors as (1) the presence of inactive enzyme in the final extract (*see* Cash et al., 1974); (2) unrecognized problems with the assay of the enzyme; and (3) the possibility that the mechanism of the enzyme includes a slow step.

2.2. Assay

In principle there are a number of ways to measure the forward reaction of GABA-T (*see* Fig. 1). For instance, the rate of disappearance of either GABA or α-KG could be monitored. In general, however, the sensitivity of such an approach leaves much to be desired since a decrease in substrate concentration of the order of only 5% would have to be detectable. It is almost always far superior to determine the rate of increase in the formation of product. In the case of GABA-T, this means monitoring the production of either succinic semialdehyde (SSA) or glutamic acid. This can be done directly or indirectly by coupling to another reaction.

2.2.1. Nonradiometric Methods

Examples of direct measurement of succinic semialdehyde are those published by Salvador and Albers (1959), in which a fluorescent product is formed with 3,5-diaminobenzoic acid and can be detected in a fluorometer. The condensation product is optimally activated at 405 nm and read at 505 nm. This sensitive method can detect the presence of SSA down to 10^{-10} mol. Salganicoff and DeRobertis (1965) have modified this procedure.

An early procedure (Baxter and Roberts, 1958) takes advantage of the fact that alpha-amino acids chelate copper, whereas gamma-amino acids do not. At the termination of the GABA-T catalyzed reaction, a copper-glutamic acid chelate is formed that can be detected by a phenanthroline reagent. The product absorbs at 454 nm and can thus be read in a spectrophotometer.

Beeler and Churchich (1978) described a method for the measurement of SSA by first reacting the aldehyde product of the enzymic reaction with cyclohexane-1-3-dione. The condensation product is fluorescent and can be read at 460 nm after excitation at 365 nm. Another method to determine SSA synthesis is based on that of Paz et al. (1965), who obtained a colored product with 3-methylbenzothiazolione-2-hydrazone. Tabener et al. (1972) exploited this observation to develop an assay for mouse brain GABA-T. These workers found that the colored complex absorbed at about 640 nm with an extinction coefficient of 3.4×10^7 cm^2/mol and that SSA could conveniently be assayed over the concentration range of 0.01–0.1 μmol.

Other methods exist that employ coupling the reaction catalyzed by GABA-T to another enzymic reaction. Such a method was described by Jacoby (1962) for the determination of bacterial GABA-T. The procedure relied on the conversion of SSA to succinic acid. The concomitant formation of NADP was monitored in a spectrophotometer at 340 nm. Pitts and coworkers (1965) utilized this concept to assay brain GABA-T. The assay depends on the addition of guinea pig kidney succinic semialdehyde dehydrogenase (SSADH) to ensure adequate conversion of SSA to succinic acid. At the same time, NAD^+ is converted to NADH. This product is then converted to a highly fluorescent derivative by hydrogen peroxide and read in a fluorometer at 470 nm. This method allows detection of 10^{-11} mol of NADH, which makes it far more sensitive than the original procedure of Jacoby. A modification of the previous method has been made by De Boer and Bruinvels (1977). Other methods have been reported that are very similar to those mentioned (Rengachary and Yang, 1980; Churchich, 1982).

Another method that makes use of a coupled reaction is that of Schousboe et al. (1973), in which the Glu product of GABA-T is converted to α-KG by the addition of glutamic dehydrogenase and acetylpyridine-NAD^+. The rate of production of acetylpyridine-NADH is measured in a spectrophotometer at 363 nm. The sensitivity of the assay is greatly enhanced with the use of the synthetic analog of NAD.

2.2.2. Radiometric Methods

The majority of these procedures depend on the use of either radioactive α-KG and the isolation of labeled glutamic acid, or they utilize radioactive GABA and rely on the isolation of labeled SSA or succinic acid.

Waksman and Roberts (1963) described a procedure in which 2-keto-5-^{14}C-glutaric acid was converted to [^{14}C]-Glu by GABA-T. The radioactive product was separated from the unmetabolized [^{14}C]-α-ketglutarate by means of a Dowex-50 (H^+ form) ion exchange column. The substrate was washed out with water while the glutamate remained on the column. A final elution with 2*N* NH_4OH recovered the labeled product. A modification of this method has been reported by Clifford et al. (1973).

The use of [^{14}C]-GABA as substrate was an idea exploited by Hall and Kravitz (1967). Radiolabeled SSA and succinic acid were products of the reaction. These were separated from [^{14}C]-GABA by use of ion exchange chromatography. GABA remained on the column while SSA and succinic acid were eluted with water. This method has also been used, with slight modification, by White and Sato (1978).

Sterri and Fonnum (1978) also made use of radiolabeled GABA in their procedure for the assay of GABA-T. These authors incubated their enzyme with [^{14}C]-GABA and relied on endogenous SSADH and the addition of NAD^+ to metabolize [^{14}C]-SSA to [^{14}C]-succinic acid. The latter was extracted into chloroform containing tri-*n*-octylamine, a liquid anion exchanger. The unmetabolized radioactive GABA remained behind in the aqueous layer.

Lees and Weiner (1975) measured the activity of GABA-T from rat brain by incubating with [^{14}C]-GABA and reacting the radiolabeled SSA with 2,4-dinitrophenylhydrazine. The resulting [^{14}C]-dinitrophenylhydrazone was much more efficiently extracted into ethyl acetate than the unmodified SSA. Unmetabolized [^{14}C]-GABA remained in the aqueous layer.

A novel method to indirectly measure GABA-T activity is one described by Gonnard and coworkers (1973). This depends on the use of 1-[^{14}C]-α-ketoglutarate and glutamate decarboxylase. The latter will decarboxylate the labeled product of the reaction and thus liberate $^{14}CO_2$, which can be trapped and counted.

2.3. Properties

2.3.1. Kinetics

The K_m values for GABA vary considerably, depending upon the source of the enzyme (Table 2). The enzyme with the highest affinity for the substrate was that isolated from human brain (White and Sato, 1978). This was reported to have a K_m of 0.31 mM. On the other hand, the enzyme with the least affinity for GABA has been that extracted from rabbit brain by John and Fowler (1976), who reported a K_m of 18 mM. The enzyme from human kidney has the greatest affinity for α-KG (White and Sato, 1978) (K_m = 0.09 mM), whereas the lowest affinity has been shown by rat brain GABA-T, which was reported to have a K_m of 5.5 mM (Maitre et al., 1975). The pH optimum for the GABA-T reaction varies between 7.95 for mouse brain to 8.6 for both rat and human brain enzyme (Maitre et al., 1975; Cash et al., 1974).

When one substrate is varied in the presence of a fixed concentration of the other, a family of parallel lines is observed (Schousboe et al., 1974; Maitre et al., 1975; White and Sato, 1978). According to the ideas of Cleland (1963), this suggests a ping pong bi bi mechanism. In the case of GABA-T, this would involve two stable forms of the enzyme. One form would be a complex of enzyme and cofactor (Enz:PLP) to which GABA can bind. The second form would be a complex between enzyme and the pyridoxamine form of the cofactor (Enz:PAM). The pyridoxamine would be formed at the active site of the enzyme when GABA gives up its amino group. This scheme is represented in Fig. 2.

2.3.2. Turnover Rates

Reports from at least two laboratories (Waksman and Roberts, 1965; De Boer and Bruinvels, 1977) have shown that the specific activity of GABA-T can vary with the concentration of the enzyme. In neither of these studies was pure enzyme used. In fact, in a study using pure enzyme, no such variation in specific activity was detected (Schousboe et al., 1973). One possible explanation for a decrease in specific activity with increasing amounts of enzyme is the existence of modulators present in the brain. If these acted as inhibitors or activators, they could apparently alter the specific activity of GABA-T. Ossola and colleagues (1980) have studied the turnover number of GABA-T from different regions of rat brain, and found a substantial variation from 3.5/s in the cerebellum cortex to 30.6/s in the internal capsule. The turnover number was defined as the number of mol converted/mol of enzyme/s at 37°C and pH 8.6. Activity of the enzyme was measured by separating

Table 2
Comparison of Substrate Affinities and Optimum pH Values for GABA Transaminase From Different Species

Species	Tissue	Substrate K_m, mM		pH	Reference
		GABA	α-KG		
Human	Brain	4	1.0	8.6	Cash et al., 1974
Human	Brain	0.31	0.16	—	White and Sato, 1978
Human	Kidney	0.35	0.09	—	White and Sato, 1978
Mouse	Brain	12	2	7.95	Waksman and Roberts, 1965
Mouse	Brain	1.1	0.25	8.05	Schousboe et al., 1973
Rabbit	Brain	18	5	—	John and Fowler, 1975
Rat	Brain	0.38	0.17	—	White and Sato, 1978
Rat	Brain	4	5.5	8.6	Maitre et al., 1975
Pig	Liver	1.1	0.7	8.65	Buzenet et al., 1978

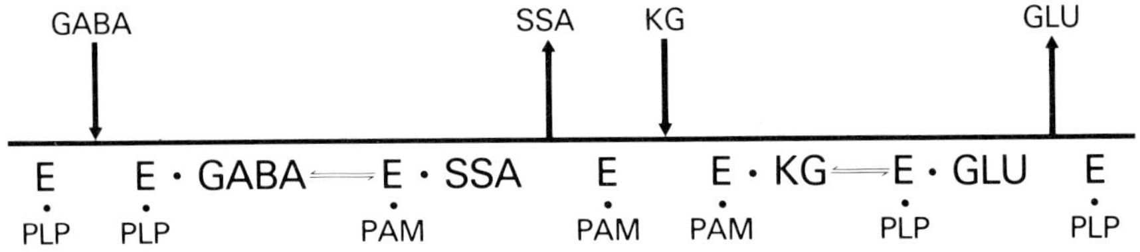

Fig. 2. Representation of the ping pong bi bi mechanism of GABA-T. Abbreviations: E:PLP, enzyme pyridoxal 5'-phosphate complex; E:PAM, enzyme pyridoxamine 5'-phosphate complex; GLU, Glu.

[^{3}H]-SSA from Glu on a Dowex ion exchange column. The amount of enzyme was determined in the same homogenate by the use of a radioimmunoassay.

John and Fowler (1976) measured the turnover rate (K_{cat}) of rabbit brain GABA-T by a method developed by Velick and Vavra (1962). This is a steady-state kinetic method in which the molecular weight of the subunits of GABA-T and the protein concentration are used. The K_{obs} is determined from the V_{max} and plotted against substrate concentration, and the K_{cat} is obtained from the intercept.

The turnover rate of GABA-T can be monitored by using a method published by Gale (1984). Gamma-vinyl-GABA is known to irreversibly inhibit GABA-T by binding at the GABA substrate site of the enzyme (Lippert et al., 1977). Tritiated gamma-vinyl GABA is incubated with rat brain homogenates. The specific binding of the label is proportional to enzyme activity. This method offers an extremely convenient way to estimate the turnover rate of GABA-T.

2.3.3. Molecular Weight

Waksman and Roberts (1965) observed that partially purified mouse brain GABA-T gave rise to a single boundary upon high-speed centrifugation. The molecular weight was calculated to be of the order of 100,000. When the enzyme was later purified to homogeniety, a value of 109,000 was obtained (Schousboe et al., 1973). This protein was found to dissociate into two subunits of unequal size, one of 53,000 molecular weight, and the other, 58,000 (Schousboe et al., 1974). Purified GABA-T from rat brain was subjected to gel filtration and SDS-PAGE (Maitre et al., 1975). The molecular weight was estimated to be 105,000. In the presence of SDS, a single band was noted that had a molecular weight of 57,000. This study provides strong evidence that rat brain GABA-T is a dimer made up of two identical subunits.

John and Fowler (1976) obtained a molecular weight of 120,000 for purified rabbit brain GABA-T when the enzyme was applied to a Sephadex G-200 column. In the presence of SDS and starch gel electrophoresis, a single band of 58,000 was found.

Pig brain enzyme also appears to be a dimer of identical subunits of 50,000 molecular weight (Beeler and Churchich, 1978). White and Sato (1978) studied GABA-T from human brain, kidney, and liver and calculated that the enzyme from all three sources had a molecular weight of only 70,000. They presented no evidence for the existence of subunits.

The reason for two unequal subunits of mouse brain GABA-T is not obvious (Schousboe et al., 1974). However, it has been suggested (John and Fowler, 1976) that the smaller 53,000 subunit is a product of proteolytic degradation of the larger subunit.

2.3.4. Isozymes

Whether or not GABA-T exists in different forms is a matter not yet satisfactorily resolved. Both mouse and rat brain enzyme have been reported to exist in four distinct forms (Waksman and Bloch, 1968). However, other workers, using highly purified preparations of the enzyme, found no evidence for isozymes (Schousboe et al., 1973; Maitre et al., 1975; John and Fowler, 1976). Yet Vasilev et al. (1973) obtained three peaks of GABA-T activity after chromatography of pig liver and kidney enzyme on carboxymethylcellulose. Moreover, two peaks of activity were noted by Ho and colleagues (1975) after DEAE-cellulose chromatography. In addition, Cash et al. (1974) reported the existence of a shoulder on peaks from PAGE *and* on a pH activity curve. This led to the speculation that two isozymes could be present with pH optima of 8.0 and 8.6, respectively.

Indirect evidence, also, is available suggesting GABA-T may exist in different forms. For instance, using different mouse brain fractions, Buu and Van Gelder (1974) found that both the K_m for GABA and the pH optimum differed considerably between synaptosomal enzyme and cytoplasmic enzyme. In a similar study using rat brain, Tunnicliff et al. (1977b) discovered marked differences in K_m values and in the potency of the inhibitor aminooxyacetic acid between GABA-T from nerve endings and cytoplasmic enzyme.

2.3.5. Amino Acid Composition

The amino acid composition of GABA-T purified from the brain of the pig, rat, and human has been determined and is illustrated in Table 3.

There are no great differences among the three species. Indeed, Glu, Thr, Ser, Val, and Ala yielded extremely similar values.

Table 3
Amino Acid Composition of Brain GABA Transaminase From Various Species (Residues per Molecule)[a]

Amino acid	Rat enzyme	Human enzyme	Pig enzyme
Glu	75	75	79
Asp	81	110	80
Thr	30	31	33
Ser	60	69	61
Gly	84	57	96
Cys	12	4	6
Met	8	8	15
Val	42	46	42
Pro	30	50	47
Arg	72	37	37
His	39	43	20
Lys	90	89	52
Phe	27	43	43
Tyr	12	30	17
Leu	63	108	72
Isoleucine	30	47	39
Ala	78	75	76
Trp	—	—	—

[a]Rat and human data are from Maitre et al. (1978); pig enzyme data from Kim and Churchich (1981).

Asp, Gly, Pro, Phe, and isoleucine also produced values that varied little.

2.3.6. Binding Sites

Little is known of those amino acid residues at the active site that are important for substrate and cofactor binding and for catalytic activity. *p*-Chloromercurybenzoate and 5,5'-dithiobis(2-nitrobenzoate), both sulfhydryl reagents, have been demonstrated to inhibit GABA-T activity, although the presence of either GABA or α-KG did not afford protection (Schousboe et al., 1974). Similar results were obtained by Moses and Churchich (1980). They concluded that the loss of one SH group per dimer of the enzyme is responsible for the loss of almost all the activity after exposure to either 5,5'-dithiobis-2-nitrobenzoic acid or *N*-iodoacetylaminoethyl 5-naphthylamine-1-sulfonic acid. Since GABA afforded no protection, the idea was proposed that reaction of an essential SH group effects a conformational change in the enzyme, reducing its catalytic activity.

An essential Lys residue has been implicated in the binding of α-KG to the enzyme. Kim and Churchich (1981) have reported that *o*-phthalaldehyde modifies six Lys residues per dimer to account for about 90% of the total activity loss. These workers found that α-KG, but not pyridoxal phosphate (PLP), protected against the inhibition and concluded that the former binds to a Lys at or near the active site. For some inexplicable reason, GABA was not tested as a protector.

Evidence is available that Arg residues are critical for the binding of PLP at the active site. The Arg-specific reagent phenylglyoxal inactivates the enzyme and the cofactor can prevent this effect. On the other hand, neither of the substrates affords any protection (Tunnicliff, 1980). There is a good correlation between the rate of inactivation of the enzyme and the rate of incorporation of ^{14}C from 1-^{14}C-phenylglyoxal.

Recently, Churchich and Moses (1981) presented evidence that purified GABA-T contains 1 mol of PLP per dimer. A second molecule of cofactor binds to the enzyme with a dissociation constant 3000 times greater. Therefore, two nonequivalent binding sites exist for the cofactor. Further study revealed that the enzyme contains 1 mol and 1.8 mol PLP residues per dimer (Churchich, 1982). A distance of 2.7 nm separating the two binding sites was calculated. Results from another study (Kim et al., 1981) prompted the conclusion that the second PLP binding site becomes functional only after chemical modification.

2.3.7. Inhibitors

The action of inhibitors can tell us much about the active site of GABA-T, as well as provide a basis for the development of more potent drugs for the future. Such drugs have the potential to be useful antiepileptics, antihypertensives, or even analgesics (see below).

Three major categories of drugs exist in principle: (1) those that interfere with GABA as substrate; (2) those that interfere with α-KG as substrate; and (3) those that interfere with the cofactor. This interference would, generally, but not always, be binding to the active site. In theory, the most effective approach is through GABA structural analogs, since α-KG is used in so many other reactions in the body. Similar arguments can be made in the case of PLP. In practice this appears to hold true because almost all of the interesting inhibitors are those that interfere with GABA binding to the enzyme.

The most effective structural analog of GABA is BW 357U (1-(*N*-decyl)-3-pyrazolidinone). This compound has an IC_{50} of 0.02 μM with respect to GABA (White et al., 1983). The drug passes the blood–brain barrier and inhibits the enzyme in vivo. There is evidence, however, that a metabolite of the analog is, in fact, the actual inhibitor.

Amino-oxyacetic acid has been known for some time to inhibit GABA-T in vivo as well as in vitro (Wallach, 1961). Wallach obtained a K_i of 7.5 μM for inhibition of the enzyme from *E. coli*. This inhibitor is over 100 times more powerful against mammalian GABA-T (Tunnicliff et al., 1977b; Ngo and Tunnicliff, 1978). Hydrazinopropionic acid is also a potent competitive inhibitor with respect to GABA (Van Gelder, 1968). A K_i of 0.235 μM has been calculated for the mouse brain enzyme. Like the previous inhibitor, hydrazinopropionic acid penetrates the blood–brain barrier and elevates brain GABA levels by virtue of an in vivo inhibition of GABA-T. Fowler and John (1972) reported that ethanolamine-*O*-sulfate is an effective inhibitor of rabbit brain GABA-T (K_i = 0.44 μM). This compound in reality acts as an inactivator of the enzyme since it binds irreversibly at the GABA binding site. GABA cannot displace the inhibitor but it can protect against its initial binding to the enzyme. Fowler and John propose that the inhibitor acts as a substrate for the enzyme and, in so doing, undergoes a β-elimination to produce aminoethylene, which may be the actual inactivator. This type of inhibitor is known as a K_{cat} (Rando, 1975) or suicide inactivator (Abeles and Maycock, 1976). Ethanolamine-*O*-sulfate does not penetrate the blood–brain barrier, but intracisternal administration will give rise to an in vivo inactivation of GABA-T. This drug has been found to possess anticonvulsant properties (Anlezark et al., 1976).

Another powerful K_{cat} inhibitor is gabaculline, which occurs naturally in *Streptomyces toyocaensis*. This compound irreversibly inactivates mouse brain GABA-T (Rando and Bangerter, 1976). The K_i has been reported as 0.6 μM. This inhibitor is taken up into the brain, where it increases GABA concentrations by inactivating GABA-T in vivo. Gamma-acetylenic-GABA (Jung and Metcalf, 1975) and gamma-vinyl GABA (Lippert et al., 1977) are also suicide inhibitors of GABA-T, albeit not as potent as gabaculline or ethanolamine-*O*-sulfate. Their K_i values have been calculated as 50 and 350 μM, respectively.

An interesting competitive inhibitor of GABA-T is *N*-(5'-phosphopyridoxyl)-4-aminobutyric acid (PLP-GABA). This compound is a stable complex of GABA and cofactor. Like the above inhibitors, it binds to the GABA site of the enzyme (Tunnicliff et al.,

1977a). The K_i has been reported as 1.2 μM. The GABA agonist imidazoleacetic acid is also an inhibitor of rat brain GABA-T. Interestingly, although structurally related to GABA, imidazoleacetic acid does not compete for the GABA binding site of GABA-T. Instead this drug is a noncompetitive inhibitor of the enzyme (Clifford et al., 1973). In principle, this compound can act as a GABA-mimetic by both activating the GABA receptor and inhibiting GABA degradation.

Dipropylacetic acid is a clinically useful drug for the treatment of certain types of epilepsy. It has been claimed that this compound is a weak competitive inhibitor of GABA-T with respect to GABA (Simler et al., 1973). The antiepileptic action of dipropylacetic acid may well be the result of an elevation of brain GABA concentrations; however, this has yet to be established.

Very few inhibitors have been discovered that act at the α-KG binding site of the enzyme. Ciesielski et al. (1979) claimed that 1-propyl-3,5-dicyano-4-phenyl-6-hydroxy-2-pyridone inactivates GABA-T, and that its effects can be reduced in the presence of α-KG. These authors believe this drug acts as a K_{cat} inhibitor.

Although inhibitors that act by interfering with PLP are not particularly specific, it is noteworthy that the most potent inhibitor of GABA-T is such a compound. This is *N*-(5'-phosphopyridoxyl)-aminooxyacetic acid (PLP-AOAA), which competes for PLP binding with a K_i of 0.01 μM (Churchich, 1982). Another structural analog of the cofactor is 4-vinylpyridoxal-5-phosphate, with a K_i of 0.2 μM as a competitive inhibitor of PLP (Churchich, 1982).

The neurochemist's interest in discovering inhibitors of GABA-T is twofold. First, the structure of an inhibitor and its mechanism of action may shed light on the nature of the enzyme. Second, there is definite evidence that, in principle, drugs that elevate brain GABA levels can be important clinically. For example, antiepileptic action often stems from inhibition of GABA-T (Wood, 1975; Anlezark et al., 1976). Furthermore, evidence is now accumulating that points to the involvement of GABA in such physiological processes as pain threshold and the regulation of blood pressure (DeFeudis, 1982; 1983). Drugs that inhibit GABA-T may be of potential use in analgesia and the control of hypertension.

2.3.8. Regulation

The rate-limiting step and therefore the key enzyme in GABA metabolism is GAD. The activity of GABA-T in the cell is in excess of that of GAD. Fine control of the degradative enzyme would not seem as important as that for the enzyme responsible for GABA synthesis. Nevertheless, certain observations have been made that indicate GABA-T may be under the control of endogenous factors. First, the turnover number for the enzyme differs in different brain regions (Ossola et al., 1980). This could be interpreted to mean that activators or inhibitors exist that somehow regulate GABA-T activity. Second, it has recently been demonstrated that GABA-T can undergo phosphorylation in the presence of radioactive ATP and a protein kinase (Spielholz et al., 1984). It is well established that phosphorylation of enzymes can be an important regulatory mechanism (Wilson, 1980). The significance of the apparent phosphorylation of GABA-T in its regulation, however, remains to be established. Third, the availability of cofactor has been suggested as a means of controlling enzyme activity in vivo (White et al., 1983).

3. Methods

3.1. Purification of Rat Brain GABA Transaminase

A purification method that we have employed in this laboratory is one based on that reported by John and Fowler (1976). Typically, 25 adult male Sprague-Dawley rats are killed by a blow to the base of the skull and their brains disrupted in 9 vol of ice-cold 0.32*M* sucrose in a Potter-Elvehjem homogenizer. The homogenate is centrifuged in a Beckman J21 centrifuge at 4°C for 15 min at 1500*g* (3000 rpm using the JA20 rotor). The supernatant is carefully removed and recentrifuged at 20,000 rpm for 60 min at 4°C. The pellet, known as the crude mitochondrial fraction, is subjected to hypotonic shock by suspending in ice-cold water and stirring for 15 min. The mixture is centrifuged at 4°C for 60 min at 100,000*g*. The resulting pellet is suspended in buffer to a total volume of 44 mL. The buffer is 10 m*M* sodium phosphate, pH 7.0, containing 100 m*M* PLP, 1 m*M* 2-aminoethylisothiouronium bromide hydrobromide (AET), 1 m*M* EDTA, and 0.5% Triton X-100. A Waring blender is used to obtain a homogeneous mixture.

Solid ammonium sulfate is added gradually until 40% saturation is reached. During this time the extract is stirred at 4°C. Stirring is continued for a further 15 min after the addition of the salt. The mixture is then centrifuged in the cold for 20 min at 20,000 rpm. The supernatant is again subjected to ammonium sulfate fractionation. This time solid $(NH_4)_2SO_4$ is added until 70% saturation is attained.

After centrifugation, the pellet is dissolved in the original homogenization buffer and dialyzed against 3 L of 10 m*M* sodium acetate and 10 m*M* HCl (adjusted to pH 7.5 with Tris). The pH of the extract is adjusted to 7.5 by the addition of solid Tris. A column of DEAE-cellulose (1.5 × 15 cm) is equilibrated with the dialysis buffer and the enzyme solution applied. A linear gradient composed of equal volumes of equilibration buffer and 50 m*M* sodium acetate is used to elute the column. Fractions of 0.5 mL are collected, and those that contain enzyme are pooled and dialyzed against 3 L of 40 m*M* sodium acetate-acetic acid buffer (pH 5.4). The fraction is applied to a CM-cellulose column (1.5 × 15 cm) that has been equilibrated with the previous dialysis buffer. Elution is with a linear gradient consisting of 40 m*M* sodium acetate buffer and the same volume of 80 m*M* sodium acetate buffer, both at pH 5.4. Fractions containing GABA-T activity are pooled and subjected to ammonium sulfate precipitation (70% saturation). The pellet after centrifugation is dissolved in 2 mL of the original homogenization buffer, but without the Triton X-100, and dialyzed against 3 L of the same buffer. The enzyme is stored at 4°C. The purity of the enzyme is checked by PAGE. A summary of the purification procedure is shown in Table 4.

3.2. Assay of the Enzyme

An extremely sensitive and convenient way to measure GABA-T activity is by the micromethod of Sterri and Fonnum (1978). A 50-m*M* Tris-HCl buffer, pH 8.6, contains 5 m*M* [^{14}C]-GABA, 5 m*M* α-KG, 3 m*M* NAD^+, 0.1 m*M* PLP, and 1 m*M* AET. The total volume is 19 μL. Addition of 1 μL of enzyme initiates the reaction. Incubation is usually for 30 min at 37°C. The reaction is stopped by the addition of 100 μL of chloroform containing 0.2*M* tri-*n*-octylamine, a liquid anion exchanger. This tertiary amine has to be treated with 0.25*M* phosphoric acid before its use in the assay. The entire solution is thoroughly mixed to extract the radioactive product into the chloroform layer. The unmetabolized GABA remains in the aqueous layer. An aliquot of the chloroform layer is taken for scintillation counting. Blanks are tubes in which either no enzyme is added or in which the α-KG has been omitted. The radioactive GABA can be labeled in any position other than the amino group. Furthermore, tritium or carbon-14 can be used. The assay, as developed, was for the measurement of comparatively crude enzyme preparations in which SSADH was present in excess over GABA-T. For more highly purified forms of GABA-T, the

Table 4
Purification of Rat Brain GABA Transaminase

	Volume, mL	Protein, mg	Specific activity[a]	Yield, %	Purification
Sucrose homogenate	260	3740	0.006	100	1
Mitochondrial fraction	44	33	0.156	24	27
Ammonium sulfate (40–70%)	10	8	0.62	22.8	105
DEAE-cellulose	20	1.5	2.4	16.6	414
CM-cellulose	12	0.2	11.2	10.1	1931
Ammonium sulfate (0–85%)	2	0.2	11.2	10.1	1931

[a]Specific activity is defined as μmol/min/mg protein at 37°C and pH 8.6.

addition of SSADH to the assay system is required. The SSADH can be purified from guinea pig liver by the procedure of Pitts et al. (1965), in which the homogenate is subjected to ammonium sulfate fractionation (33–50%) and dialyzed against 3 m*M* phosphate buffer, pH 7.5. An alternative source of SSADH is from the commercial enzyme preparation known as Gabase that contains GABA-T and SSADH from *Pseudomonas fluorescens*. The SSADH can be separated from the mixture by means of a Sephadex G-25-40 column (Rengachary and Yang, 1980).

The assay for GABA-T described by Sterri and Fonnum (1978) can also make use of [^{14}C]-α-KG instead of labeled GABA. In fact, such a method has been used to measure enzyme activity in rat brain (Tunnicliff and Smith, 1981). The product of the reaction is labeled Glu that remains in the aqueous layer. The unmetabolized radioactive α-KG is extracted into the chloroform layer containing the tri-*n*-octylamine.

4. Summary

The first part of this chapter attempted to introduce and explain the part played by GABA and its degradative enzyme GABA transaminase in the central nervous system. This then led to a discussion of methods available for the purification and measurement of the enzyme. The section reviewing the properties of GABA-T explains how we now believe the protein to consist of two identical subunits, each with a molecular weight of the order of 53,000. The existence of isozymes is a matter of conjecture. The mechanism of GABA-T appears to be of the ping pong bi bi type. Sulfhydryl groups, Lys and Arg residues are all important for the catalytic activity of the enzyme. The most potent inhibitor is pyridoxal phosphate-aminooxyacetic acid. As a rule, inhibition of the enzyme in the intact brain leads to an anticonvulsant action.

Very little is known of the regulation of GABA-T, but the latest evidence points to the possibility that phosphorylation of the protein could be important.

The latter part of the chapter dealt with a detailed method, based on that used by John and Fowler (1976), to purify the enzyme from rat brain. A sensitive, convenient method for the assay of GABA-T, after the procedure of Sterri and Fonnum (1978), is described.

References

Abeles R. H. and Maycock A. L. (1976) Suicide enzyme inactivators. *Acc. Chem. Res.* **9,** 313–319.

Alger B. E. and Nicoll R. A. (1982) Pharmacological evidence for two kinds of GABA receptor on rat hippocampal pyramidal cells studied in vitro. *J. Physiol.* (Lond.) **328,** 125–141.

Anlezark G., Horton R. W., Meldrum B. S., and Sawaya M. C. (1976) Anticonvulsant action of ethanolamine-*O*-sulphate and di-*n*-propylacetate and the metabolism of γ-aminobutyric acid (GABA) in mice with audiogenic seizures. *Biochem. Pharmacol.* **25,** 413–417.

Anwar N. and Mason D. F. J. (1982) Two actions of γ-aminobutyric acid on the responses of the isolated basilar artery from the rabbit. *Brit. J. Pharmacol.* **75,** 177–181.

Ault B. and Nadler V. (1982) Baclofen selectively inhibits transmission at synapses made by axons of CA3 pyramidal cells in the hippocampal slice. *J. Pharmacol. Exp. Ther.* **223,** 291–297.

Barber R. P. and Saito K. (1976) Light Microscopic Visualization of GAD and GABA-T in Immunocytochemical Preparations of Rodent CNS, in *GABA in Nervous Tissue Function* (Roberts E., Chase T. N., and Tower D. B., eds.), Raven, New York.

Baxter C. F. (1970) The Nature of γ-Aminobutyric Acid, in *Handbook of Neurochemistry,* vol. 3, 1st Ed. (Lajtha A., ed.), Plenum, New York.

Baxter C. F. and Roberts E. (1958) The γ-aminobutyric acid-α-ketoglutarate transaminase of beef brain. *J. Biol. Chem.* **233,** 1135–1139.

Beart P. M. and Lee V. (1977) Transmitter enzymes in Weaver and Staggerer neurological mutant mice. *Brain Res.* **124,** 171–177.

Beeler T. and Churchich J. E. (1978) 4-Aminobutyrate aminotransferase fluorescence studies. *Eur. J. Biochem.* **85,** 365–371.

Bird E. D. and Iversen L. L. (1974) Huntington's chorea. Postmortem measurement of glutamic acid decarboxylase, choline acetyltransferase and dopamine in basal ganglia. *Brain* **97,** 457–472.

Bloch-Tardy M., Rolland B., and Gonnard P. (1971) Ontogenetic evolution of the molecular forms of 4-aminobutyrate 2-oxoglutarate aminotransferase in rat brain and liver. *J. Neurochem.* **18,** 1779–1781.

Bloch-Tardy M., Rolland B., and Gonnard P. (1974) Pig brain 4-aminobutyrate-2-ketoglutarate transaminase. Purification, kinetics and physical properties. *Biochimie* **56,** 823–832.

Bowery N. G. (1983) Classification of GABA Receptors, in *The GABA Receptors* (Enna S., ed.), Humana, Clifton, New Jersey.

Bowery N. G., Doble A., Hill D. R., Hudson A. L., Middlemiss D. N., Shaw J., and Turnbull M. J. (1980) (–)Baclofen decreases neurotransmitter release in the mammalian CNS by an action at a novel GABA receptor. *Nature* **284,** 92–94.

Bowery N. G., Doble A., Hill D. R., Hudson A. L., Shaw J. S., Turnbull M. J., and Warrington R. (1981) Bicuculline-insensitive GABA receptors on peripheral autonomic nerve terminals. *Eur. J. Pharmacol.* **71,** 53–70.

Bradford H. F., Chain E. B., Cory H. T., and Rose S. P. R. (1969) Glucose and amino acid metabolism in some invertebrate nervous systems. *J. Neurochem.* **16,** 969–978.

Buu N. T. and Van Gelder N. M. (1974) Differences in biochemical properties of γ-aminobutyric acid aminotransferase from synaptosome-enriched and cytoplasmic mitochondria-enriched subcellular fractions of mouse brain. *Can. J. Physiol. Pharmacol.* **52,** 674–680.

Buzenet A. M., Fages C., Bloch-Tardy M., and Gonnard P. (1978) Purification and properties of 4-aminobutyrate 2-ketoglutarate aminotransferase from pig liver. *Biochim. Biophys. Acta* **522,** 400–411.

Cash C., Maitre M., Ciesielski L., and Mandel P. (1974) Purification and partial characterisation of 4-aminobutyrate 2-ketoglutarate transaminase from human brain. *FEBS Letts.* **47,** 199–203.

Churchich J. E. (1982) 4-Aminobutyrate aminotransferase: Different susceptibilities to inhibitors, microenvironment of the cofactor binding site and distance of the catalytic sites. *Eur. J. Biochem.* **126,** 507–511.

Churchich J. E. and Moses U. (1981) 4-Aminobutyrate aminotransferase. The presence of nonequivalent binding sites. *J. Biol. Chem.* **256,** 1101–1104.

Ciesielski L., Simler S., Gensburger C., Mandel P., Taillandier G., Benoit-Guyod J. L., Boucherle A., Cohen-Addad C., and Lajzerowicz J. (1979) GABA transaminase inhibitors. *Adv. Exp. Med. Biol.* **123,** 21–41.

Cleland W. W. (1963) The kinetics of enzyme-catalyzed reactions with two or more substrates or products. *Biochim. Biophys. Acta* **67,** 104–137.

Clifford J. M., Tabener P. V., Tunnicliff G., Rick J. T., and Kerkut G. A. (1973) Biochemical and pharmacological actions of imidazoleacetic acid. *Biochem. Pharmacol.* **22,** 535–542.

Curtis D. R., Duggan A. W., Felix D., and Johnston G. A. R. (1971) Bicuculline, an antagonist of GABA and synaptic inhibition in the spinal cord of the cat. *Brain Res.* **32,** 69–96.

Davidson N. (1976) *Neurotransmitter Amino Acids.* Academic, London.

Davidson N. and Southwick C. A. P. (1971) Amino acids and presynaptic inhibition in the rat cuneate nucleus. *J. Physiol.* (Lond.) **219,** 689–708.

De Boer T. and Bruinvels J. (1977) Assay and properties of 4-aminobutyric-2-oxoglutaric acid transaminase and succinic semialdehyde dehydrogenase in rat brain tissue. *J. Neurochem.* **28,** 471–478.

DeFeudis F. V. (1982) γ-Aminobutyric acid and analgesia. *Trends Pharmacol. Sci.* **3,** 444–446.

DeFeudis F. V. (1983) γ-Aminobutyric acid and cardiovascular function. *Experientia* **39,** 845–849.

Desarmenien M., Feltz P., Occhipinti G., Santangelo F., and Schlichter R. (1984) Coexistence of $GABA_A$ and $GABA_B$ receptors on Aδ and C primary afferents. *Br. J. Pharmacol.* **81,** 327–333.

Dunlap K. (1981) Two types of γ-aminobutyric acid receptor on embryonic sensory neurones. *Br. J. Pharmacol.* **74,** 579–586.

Eccles J. C., Schmidt R. F., and Willis W. D. (1963) Pharmacological studies on presynaptic inhibition. *J. Physiol.* (Lond.) **168,** 500–530.

Fahn S. and Cote L. J. (1968) Regional distribution of gamma-aminobutyric acid (GABA) in brain of the rhesus monkey. *J. Neurochem.* **15,** 209–213.

Fonnum F. and Storm-Mathisen J. (1978) Localization of GABAergic Neurons in the CNS, in *Handbook of Psychopharmacology,* Vol. 9 (Iversen L. L., Iversen S. D., and Snyder S. H., eds.), Plenum, New York.

Fowler L. J. and John R. A. (1972) Active-site-directed irreversible inhibition of rat brain 4-aminobutyrate aminotransferase by ethanolamine *O*-sulphate in vitro and in vivo. *Biochem. J.* **130,** 569–573.

Frontali N. (1961) Activity of glutamic acid decarboxylase in insect nerve tissue. *Nature* **191,** 178–179.

Gale K. (1984) Binding of ^{3}H-*gamma*-vinyl-GABA to GABA transaminase in brain homogenates. *Life Sci.* **34,** 701–706.

Gonnard P. A., Wicker A., Kouyoumdjian J.-C, and Bloch-Tardy M. (1973) Methode radio-isotopique rapide de mesure de l'activite 4-aminobutyrate:2-oxoglutarate aminotransferase (GABA-transaminase). *Biochimie* **55,** 509–510.

Gray E. G. and Whittaker V. P. (1962) The isolation of nerve endings from brain. An electron microscopic study of cell fragments derived from homogenisation and centrifugation. *J. Anat.* (Lond.) **96,** 79–88.

Hall Z. W. and Kravitz E. A. (1967) The metabolism of γ-aminobutyric acid (GABA) in lobster nervous systems. I. GABA-glutamate transaminase. *J. Neurochem.* **14,** 45–54.

Hauser K. and Bernasconi R. (1980) Rat Cortical Neurons in Dissociated Cell Culture: Changes in GABA and Guanyl Cyclase Activity During Development, in *Tissue Culture In Neurobiology* (Giacobini E. et al., eds.), Raven, New York.

Hill D. R., Bowery N. G., and Hudson A. L. (1984) Inhibition of $GABA_B$ receptor binding by guanyl nucleotides. *J. Neurochem.* **42,** 652–657.

Ho P. P. K., Young A. L., and Walters P. C. (1975) Two forms of 4-aminobutyrate transaminase in guinea pig brain. *Enzyme* **19,** 244–255.

Huggins A. K., Rick J. T., and Kerkut G. A. (1967) A comparative study of the intermediary metabolism of L-glutamate in muscle and nervous tissue. *Comp. Biochem. Physiol.* **21,** 23–30.

Hyde J. C. and Robinson N. (1974) Appearance of gamma-aminobutyrate transaminase activity in developing rat brain. *J. Neurochem.* **23,** 365–367.

Iversen L. L. and Neal M. J. (1968) The uptake of ^{3}H-GABA by slices of rat cerebral cortex. *J. Neurochem.* **15,** 1141–1149.

Jacoby W. B. (1962) Enzymes of γ-aminobutyrate metabolism (bacterial). *Meth. Enzymol.* **5,** 765–778.

John A. J. and Fowler L. J. (1976) Kinetic and spectral properties of rabbit brain 4-aminobutyrate aminotransferase. *Biochem. J.* **155,** 645–651.

Jung M. J. and Metcalf B. W. (1975) Catalytic inhibition of γ-aminobutyric acid-α-ketoglutarate transaminase of bacterial origin by 4-aminohex-5-ynoic acid, a substrate analog. *Biochem. Biophys. Res. Commun.* **67,** 301–306.

Kaplita P. V., Waters D. H., and Triggle D. J. (1982) γ-Aminobutyric acid action in guinea-pig ileal myenteric plexus. *Eur. J. Pharmacol.* **79,** 43–51.

Kim D. S. and Churchich J. E. (1981) 4-Aminobutyrate aminotransferase, the reaction of lysine residues connected with enzymatic activity. *Biochem. Biophys. Res. Commun.* **99,** 1333–1340.

Kim D. S., Moses U., and Churchich J. E. (1981) Gabaculline and *m*-carboxyphenyl-pyridoxamine-5-phosphate as probes of the catalytic binding sites of 4-aminobutyrate aminotransferase. *Eur. J. Biochem.* **118,** 303–308.

Kravitz E. A., Molinoff P. B., and Hall Z. W. (1965) A comparison of the enzymes and substrates of gamma-aminobutyric acid metabolism in lobster excitatory and inhibitory axons. *Proc. Natl. Acad. Sci. USA* **54,** 778–782.

Krogsgaard-Larsen P., Scheel-Kruger J., and Kofod H., eds. (1979) *GABA-Neurotransmitters.* Academic, New York.

Kuriyama K., Roberts E., and Kakefuda T. (1968a) Association of the γ-aminobutyric acid system with a synaptic vesicle fraction from mouse brain. *Brain Res.* **8,** 132–152.

Kuriyama K., Sisken B., Ito J., Simonsen D. G., Haber B., and Roberts E. (1968b) The γ-aminobutyric acid system in the developing chick embryo cerebellum. *Brain Res.* **11,** 412–430.

Lancaster G., Mohyuddin F., Shriver C. R., and Whelan D. T. (1973) A γ-aminobutyrate pathway in mammalian kidney cortex. *Biochim. Biophys. Acta* **297,** 229–240.

Lees G. J. and Weiner N. (1975) An examination of the catalytic properties of several aminotransferases in brain and liver by means of an improved radiometric assay with dinitrophenylhydrazine. *J. Neurochem.* **25,** 315–322.

Lippert B., Metcalf B. W., Jung M. J., and Casara P. (1977) 4-Amino-hex-5-enoic acid, a selective catalytic inhibitor of 4-aminobutyric acid aminotransferase in mammalian brain. *Eur. J. Biochem.* **74,** 441–445.

Maitre M., Ciesielski L., Cash C., and Mandel P. (1975) Purification and studies on some properties of the 4-aminobutyrate:2-oxoglutarate transaminase from rat brain. *Eur. J. Biochem.* **52,** 157–169.

Maitre M., Ciesielski L., Cash C., and Mandel P. (1978) Comparison of the structural characteristics of the 4-aminobutyrate:2-oxoglutarate transaminases from rat and human brain, and their affinities for certain inhibitors. *Biochim. Biophys. Acta* **522,** 385–399.

Martin D. L. (1973) Kinetics of sodium-dependent transport of gamma-aminobutyric acid by synaptosomes. *J. Neurochem.* **21,** 345–356.

Martin D. L. (1976) Carrier-Mediated Transport and Removal of GABA From Synaptic Regions, in *GABA in Nervous System Function* (Roberts E., Chase T. N., and Tower D. B., eds.), Raven, New York.

Martin del Rio R. (1981) γ-Aminobutyric acid system in rat oviduct. *J. Biol. Chem.* **256,** 9816–9819.

McCaman W. W., Colby B. N., and McCaman R. E. (1979) Gamma-aminobutyric acid in ganglia and in single neurons from *Aplysia* and lobster. *J. Neurochem.* **33,** 967–971.

McGeer P. L., McGeer E. G., and Jung E. (1971) Effects of globus pallidus lesions and Parkinson's disease on brain glutamic acid decarboxylase. *Brain Res.* **32,** 425–431.

Moses U. and Churchich J. E. (1980) Reactivity of one thiol group in the dimeric protein, 4-aminobutyrate aminotransferase. *Biochim. Biophys. Acta* **613,** 392–400.

Muhyaddin M., Roberts P. J., and Woodruff G. N. (1982) Presynaptic γ-aminobutyric acid receptors in the rat anococcygeous muscle and their antagonism by 5-aminovaleric acid. *Brit. J. Pharmacol.* **77,** 163–168.

Ngo T. T. and Tunnicliff G. (1978) Further evidence for the existence of isozymes of brain γ-aminobutyrate aminotransferase. *Comp. Biochem. Physiol.* **59C,** 101–104.

Okada Y., Nitsch-Hassler C., Kim J. S., Bak I. J., and Hassler R. (1971) Role of gamma-aminobutyric acid (GABA) in the extrapyamidal motor system. I. Regional distribution of GABA in rabbit, rat, guinea pig and baboon CNS. *Exp. Brain Res.* **13,** 514–518.

Ossola L., Maitre M., Blinderman J.-M., and Mandel P. (1980) Turnover numbers of γ-aminobutyrate aminotransferase in some regions of rat brain. *J. Neurochem.* **34,** 293–296.

Paz M. A., Blumenfeld O. O., Rojkind M., Henson E., Furfine C., and

Gallop P. M. (1965) Determination of carbonyl compounds with *N*-methyl benzothiazolone hydrazone. *Arch. Biochem. Biophys.* **109,** 548–559.

Perry T. L., Kish S. J., Buchanan J., and Hansen S. (1979) γ-Aminobutyric acid deficiency in brain of schizophrenic patients. *Lancet* **i,** 237–239.

Pitts F. N., Quick C., and Robins E. (1965) The enzymic measurement of γ-aminobutyric-α-oxoglutaric transaminase. *J. Neurochem.* **12,** 93–101.

Rando R. R. (1975) Mechanisms of action of naturally occurring irreversible enzyme inhibitors. *Acc. Chem. Res.* **8,** 281–288.

Rando R. R. and Bangerter F. W. (1976) The irreversible inhibition of mouse brain GABA transaminase by gabaculine. *J. Amer. Chem. Soc.* **98,** 6762–6764.

Ray J. W. (1964) Free amino acid pool of the cockroach *(Periplaneta americana)* central nervous system and the effects of insecticides. *J. Insect Physiol.* **10,** 587–597.

Rengachary S. S. and Yang B.I-Y. (1980) A convenient assay for γ-aminobutyric acid transaminase. *Brain Res. Bull.* (suppl. 2) **5,** 51–55.

Roberts E., Chase T. N., and Tower D. B., eds. (1976) *GABA in Nervous System Function.* Raven, New York.

Saito K., Schousboe A., Wu J.-Y., and Roberts E. (1974) Some immunochemical properties and species specificity of GABA-α-ketoglutarate transaminase from mouse brain. *Brain Res.* **65,** 287–296.

Salganicoff L. and DeRobertis E. (1965) Subcellular distribution of the enzymes of the glutamic acid, glutamine and γ-aminobutyric acid cycle in rat brain. *J. Neurochem.* **12,** 287–309.

Salvador R. A. and Albers R. W. (1959) The distribution of glutamic γ-aminobutyric transaminase in the nervous system of the rhesus monkey. *J. Biol. Chem.* **234,** 922–925.

Schon F. and Kelly J. S. (1974) The characterisation of ^{3}H-GABA uptake into the satellite glial cells of rat sensory ganglia. *Brain Res.* **66,** 289–300.

Schousboe I., Bro B., and Schousboe A. (1977) Intramitochondrial localization of the 4-aminobutyrate-2-oxoglutarate transaminase from ox brain. *Biochem. J.* **162,** 303–307.

Schousboe A., Wu J-Y., and Roberts E. (1973) Purification and characterization of the 4-aminobutyrate-2-ketoglutarate transaminase from mouse brain. *Biochemistry* **12,** 2868–2873.

Schousboe A., Wu J.-Y., and Roberts E. (1974) Subunit structure and kinetic properties of 4-aminobutyrate-2-ketoglutarate transaminase purified from mouse brain. *J. Neurochem.* **23,** 1189–1195.

Simler S., Ciesielski L., Maitre M., Randriianarisoa H., and Mandel P. (1973) Effect of sodium *n*-dipropylacetate on audiogenic seizures and brain γ-aminobutyric acid levels. *Biochem. Pharmacol.* **22,** 1701–1708.

Sims K. L., Witztum J., Quick C., and Pitts F. N. (1968) Brain 4-aminobutyrate 2-oxoglutarate aminotransferase. *J. Neurochem.* **15,** 667–672.

Sisken B., Sano K., and Roberts E. (1961) γ-Aminobutyric acid content and glutamic decarboxylase and γ-aminobutyrate transaminase activities in the optic lobe of the developing chick. *J. Biol. Chem.* **236,** 203–207.

Spielholz C., Carr K., Schlichter D., and Wicks W. D. (1984) In Vitro Phosphorylation of Tyrosine and 4-Aminobutyrate Aminotransferases by cAMP-Dependent Protein Kinase, in *Chemical and Biological Aspects of Vitamin B_6 Catalysis: Part B* (Evangelopoulos A. E., ed.), Alan Liss, New York.

Sterri S. H. and Fonnum F. (1978) Isolation of organic anions by extraction with liquid anion exchangers and its application to micromethods for acetylcholinesterase and 4-aminobutyrate aminotransferase. *Eur. J. Biochem.* **91,** 215–222.

Tabener P. V., Barnett J. E. G., and Kerkut G. A. (1972) Preparation of succinic semialdehyde: Its colorimetric estimation as an assay for GABA aminotransferase. *J. Neurochem.* **19,** 95–99.

Tardy M., Bardakdjian J., and Gonnard P. (1979) GABA metabolism in cultured glial cells. *Adv. Exp. Med. Biol.* **123,** 177–187.

Thampy K. G. and Barnes E. M. (1984) γ-Aminobutyric acid-gated chloride channels in cultured cerebral neurons. *J. Biol. Chem.* **259,** 1753–1757.

Tunnicliff G. (1980) Essential arginine residues at the pyridoxal phosphate binding site of brain γ-aminobutyrate aminotransferase. *Biochem. Biophys. Res. Commun.* **97,** 160–165.

Tunnicliff G. and Smith J. A. (1981) The action of analogues on γ-aminobutyric acid recognition sites in rat brain. *Neurochem. Intl.* **3,** 371–376.

Tunnicliff G., Firneisz G., Ngo T. T., and Martin R. O. (1975) Developmental changes in the kinetics of γ-aminobutyric acid transport by chick retina. *J. Neurochem.* **25,** 649–652.

Tunnicliff G., Ngo T. T., and Barbeau A. (1977a) *N*-(5'-Phosphopyridoxyl)-4-aminobutyric acid: A stable bisubstrate adduct inhibitor of rat brain 4-aminobutyric acid aminotransferase. *Experientia* **33,** 20–21.

Tunnicliff G., Ngo T. T., Rojo-Ortega J. M., and Barbeau A. (1977b) The inhibition by substrate analogues of γ-aminobutyrate aminotransferase from mitochondria of different subcellular fractions of rat brain. *Can. J. Biochem.* **55,** 479–484.

Tunnicliff G., Rick J. T., and Connolly K. (1969) Locomotor activity in Drosophila—V. A comparative biochemical study of selective bred populations. *Comp. Biochem. Physiol.* **29,** 1239–1245.

Urquhart N., Perry T. L., Hansen S., and Kennedy J. (1975) GABA content and glutamic acid decarboxylase activity in brain of Huntington's chorea patients and control subjects. *J. Neurochem.* **24,** 1071–1075.

Van den Berg C. J., Van Kempen G. M. J., Schade J. P., and Veldstra H. (1965) Levels and intracellular localization of glutamate decarboxylase and γ-aminobutyrate transaminase and other enzymes during the development of the brain. *J. Neurochem.* **12,** 863–869.

Van Gelder N. M. (1968) Hydrazinopropionic acid: A new inhibitor of aminobutyrate transaminase and glutamate decarboxylase. *J. Neurochem.* **15,** 747–757.

Vasilev V. Y., Eremin V. P., Severin E. S., and Sytinskii I. A. (1973) Comparative characterization of porcine kidney and liver γ-aminobutyrate glutamate transaminase. *Biokhimiya* **38,** 355–364.

Velick S. F. and Vavra V. P. (1962) A kinetic and equilibrium analysis of the glutamic oxaloacetate transaminase mechanism. *J. Biol. Chem.* **237,** 2109–2122.

Waksman A. and Bloch M. (1968) Identification of multiple forms of aminobutyrate transaminase in mouse and rat brain: Subcellular localization. *J. Neurochem.* **15,** 99–105.

Waksman A. and Roberts E. (1963) Transaminase activity of diaphorase, phosphorylase A and several dehydrogenases. *Biochem. Biophys. Res. Commun.* **12,** 263–267.

Waksman A. and Roberts E. (1965) Purification and some properties of mouse brain γ-aminobutyric-α-ketoglutaric acid transaminase. *Biochemistry* **4,** 2132–2138.

Wallach D. P. (1961) Studies on the GABA pathway—I. The inhibition of γ-aminobutyric acid-α-ketoglutaratic acid transaminase in vitro and in vivo by U-7524 (amino-oxyacetic acid). *Biochem. Pharmacol.* **5,** 323–331.

White H. L. (1979) 4-Aminobutyrate : 2-oxoglutarate aminotransferase in blood platelets. *Science* **205,** 696–698.

White H. L. and Sato T. L. (1978) GABA-transaminase of human brain and peripheral tissues—kinetic and molecular properties. *J. Neurochem.* **31,** 41–47.

White H. L., Cooper B. R., and Howard J. L. (1983) Regulation of GABA-T: Inhibition of GABA-T by BW 357U, in *Glutamine, Glutamate, and GABA in the Central Nervous System* (Hertz L., Kvamme E., McGeer E. G., and Schousboe A., eds.), Alan R. Liss, New York.

Wilson J. E. (1980) Protein Phosphorylation—Involvement in Brain Function, in *Biochemistry of Brain* (Kumar S., ed.), Pergamon, Oxford.

Wojick W. J. and Neff N. H. (1984) γ-Aminobutyric acid B receptors are negatively coupled to adenylate cyclase in brain, and in the cerebellum these receptors may be associated with granule cells. *Mol. Pharmacol.* **25,** 24–28.

Wood J. D. (1975) The role of γ-aminobutyric acid in the mechanism of seizures. *Progr. Neurobiol.* **5,** 77–95.

Wood J H., Hare T. A., Glaeser B. S., Brooks B. R., Ballenger J. C., and Post R. M. (1980) Cerebrospinal fluid GABA variations with seizure type and cerebellar stimulation in man. *Brain Res. Bull.* **5** (suppl. 2), 747–753.

Zachman M., Tocci P., and Nyhan W. L. (1966) The occurrence of γ-aminobutyric acid in human tissues other than brain. *J. Biol. Chem.* **241,** 1355–1358.

Chapter 12
Proteinases and Peptidases

Koon-Sea Hui and Abel Lajtha

1. Introduction

Protein turnover in the brain proceeds at a high rate in all developmental stages. This high activity indicates that not only protein synthesis but also protein degradation occurs at high rates. Information is available about the rate, mechanism, and regulatory factors of protein synthesis in the nervous system, but comparable knowledge on protein breakdown is not yet available (Marks and Lajtha, 1971; Hui and Lajtha, 1983).

More recently, interest in breakdown processes has increased as it became clear that under a wide variety of conditions proteolysis plays an important role, in addition to its role in general protein turnover. Proteolysis can result in the formation (activation) of enzymes, and the formation of neuropeptides from larger inactive precursors; it also plays a role in the termination of activity of enzymes and peptides, and possibly has a similar role in the processing (activation and inactivation) of a number of other proteins, such as membrane, structural, and secretory proteins. It may also convert one active polypeptide to another active peptide with a different function. Thus, peptide bond hydrolysis has a role not only in inactivation and metabolism, but also in formation of active compounds and alteration of membrane properties or enzyme activity. Proteolytic activity changes under a number of conditions, and such changes may not be parallel to changes in protein synthesis; thus, controls of breakdown are different from those of synthesis.

In recent years much of the research on neural proteinases has been focused on peptide hormones and neuropeptide synthesis and inactivation. Peptide hormones and neuropeptides are formed from higher-molecular-weight precursors by limited proteolysis, restricted to only selected peptide bonds. This was first discovered in the conversion of proinsulin to insulin. Other neuropeptides, gastrin, somatostatin, glucagon, parathyroid hormone, β-endorphin, Met-enkephalin, and Leu-enkephalin, are also produced from larger precursors by such a postranslation process (Steiner et al., 1980).

2. Significance

Proteinases and peptidases are important for intracellular protein turnover and its control, and also for hormone and neuropeptide formation. Because the proteolytic apparatus is very complex, dysfunction of some enzymes that cannot be compensated for by the others results in pathological imbalance. Myelin basic protein has been found to be reduced and "acid proteinase" activity to be elevated in acutely developing, active plaques of multiple sclerosis (Einstein et al., 1970, 1972). Compared with the surrounding tissue and control white matter (WM), enzyme activity decreased in the order: plaque > adjacent WM > distant "normal" WM (some brain) > normal WM (control brain). Changes in protease activity have been implicated in neuronal abiotrophies (Schlaepfer, 1978), Alzheimer's disease (Bowen et al., 1973; Pope, 1968), schizophrenia (Arregui et al., 1979, 1980;), neoplasms (Fruton, 1960). Creutzfeldt-Jakob disease (Pope et al., 1964), and Guillain-Barré syndrome (Arstila et al., 1971). Thus, the study of proteinases and peptidases offers significant insights into the neurobiology and fundamental nature of the pathogenesis of neurological and mental illness. Manipulation of protease activity could lead to rational interventions that might halt or slow the progression of pathological processes. Several protease inhibitors have already shown some value as therapeutic agents in animal models of human neuropsychiatric disorders. For example, inhibitors of cathepsin D (pepstatin) and plasminogen activators (ϵ-aminocaproic acid, aminomethylcyclohexane-carboxylic acid, *p*-nitrophenyl guanidolunzoate) alleviate the symptoms of experimental allergic encephalomyelitis (Boehme et al., 1978; Brosnan et al., 1980); pepstatin and thiol protease inhibitors (leupeptin) slow the progress of murine muscular dystrophy (Chelmicka-Schorr et al., 1978; Libby and Goldberg, 1978; Sher et al., 1981); metalloendopeptidase inhibitor (thiorphan) modifies the response to nociceptive stimuli (Roques et al., 1980); peptidase inhibitors reduce opiate narcotic withdrawal signs (Pinsky et al., 1982); angiotensin-converting enzyme inhibitor (captopril) has a mood-elevating effect in depressed patients (Zubenko and Nixon, 1984);

bestatin and thiorphan, enkephalin inhibitors, are potent analgesics (Roques et al., 1980; Zhang et al., 1982; De La Baume et al., 1983); and captopril (Capoten) is an angiotensin-converting enzyme already marketed as an oral antihypertensive drug (Cushman et al., 1977; Ondetti, 1983).

Besides their potential therapeutic values, the specific inhibitors are useful tools for isolation and determination of enzymes; for example, inhibitors are good ligands for affinity chromatography. They are also useful in studies of enzyme function that measure the effects of inhibition of a specific enzyme. Based on the difference in the affinity to the enzymes, thiorphan was used to differentiate metalloendopeptidase (enkephalinase A) and angiotensin-converting enzyme (Swerts et al., 1979). Puromycin was used to differentiate arylamidase and Leu aminopeptidase (Marks et al., 1968), and aminopeptidase MI and MII (Hersh, 1981).

3. Classification and Nomenclature

The classification and hence the nomenclature of hydrolytic enzymes is complex because the overall reactions controlled by these enzymes are similar, and in most cases, the enzymes do not have a high degree of specificity. As a rule, several enzymes are capable of hydrolyzing a substrate; that is, activities usually overlap and very few substrates are metabolized by only one specific enzyme. This makes the determination of the physiologically important activity difficult. The distinction between exopeptidases and endopeptidases is not absolute. Some endopeptidases exhibit exopeptidase activity, or vice versa; for example, cathepsin B, a membrane-bound neutral metalloendopeptidase, can simulate the function of dipeptidyl carboxypeptidase or prolyl endopeptidase, depending on the amino acid residue in the penultimate position of the peptide; prolyl endopeptidase needs Pro, whereas neutral metalloendopeptidase needs a hydrophobic amino acid. Cathepsin H can act as an aminopeptidase since it hydrolyzes amino acid β-arylamides. Classification of an enzyme as an exo- or endopeptidase should be made only after the specificity of the enzyme has been defined by studying substrates representing various types. The exopeptidases are divided according to their specificity into those hydrolyzing single amino acids from the NH_2-termini of the peptide (EC 3.4.11) (International Union of Biochemistry, 1978), those splitting single residues from the COOH-terminus of the peptide (EC 3.4.16–17), those specific for dipeptide substrates (EC 3.4.13), and those splitting off dipeptide units from either the

NH_2-terminus (EC 3.4.14) or the COOH-terminus (EC 3.4.15). The COOH-terminus hydrolyzing group is subdivided into two distinct classes: one that displays maximum activity in the acid range and is organofluorophosphate-sensitive (Ser carboxypeptidases, EC 3.4.16), and one that requires divalent cations for activity (metallocarboxypeptidases, EC 3.4.17) (Table 1).

Endopeptidases are a group of enzymes that cleave peptide bonds at a distance from both terminal amino acids, in contrast to exopeptidases. Endopeptidases are grouped into Ser-(EC 3.4.21), thio-(EC 3.4.22), metallo-(EC 3.4.23), and carboxyl-(EC 3.4.24) proteinases according to their catalytic mechanisms, which can be distinctively determined from the susceptibility to inhibitors (Walsh, 1975). There are additional classification schemes for endopeptidases. The terms "acid" and "neutral" proteinases from early studies, classifying the enzymes on the basis of their pH optima, are still often used. Other properties, such as specificity, mechanism of action, localization, or function, added to these terms, are necessary to better characterize each enzyme. High specificity toward a certain peptide or certain type of peptide bond is sometimes indicated in the name of an enzyme (e.g., collagenase, prolyl endopeptidase, angiotensin-converting enzyme, aminoenkephalinase, enkephalinase, and enkephalin convertase). This type of characterization may be misleading because of the wide substrate specificity of the enzymes. For example, it was first thought that enkephalinase A is a distinct enzyme releasing the C-terminal dipeptide of the enkephalins (Schwartz et al., 1981); now it is known that it is a neutral metalloendopeptidase splitting a number of protein substrates (Almenoff et al., 1981; Relton et al., 1983).

4. Brain Proteinases and Peptidases

Peptidases and proteinases in brain have been recently reviewed by ourselves (Hui and Lajtha, 1983) and others (Orlowski, 1983; Pope and Nixon, 1984; Turner et al., 1985). Since several brain proteinases and peptidases have been identified, it is appropriate to present a short account of them here. Most of these enzymes are related to the biosynthesis and degradation of enkephalin or its related peptides.

4.1. Endopeptidases

A rat brain membrane extract has been shown to convert dynorphin B-29 to dynorphin B-29 (1–13) (Devi and Goldstein,

1984). The converting activity, a "single Arg cleavage" at Thr-Arg at positions 13 and 14 of the substrate, exhibits a pH optimum of 8. The enzyme is sensitive to thiol protease inhibitors, but is not sensitive to Ser protease inhibitors or cathepsin B inhibitors. It is inhibited by intact dynorphin A, but not by its fragments. Another dynorphin-converting enzyme has been partially characterized in extracts of the atrial gland of *Aplysia californica* (Wallace et al., 1984). The enzyme, with a pH optimum of 6.2, is also inhibited by thiol protease inhibitors and chelators, but activated by dithiothreitol and Co^{2+}. It converts dynorphin A to dynorphin (1–8). Like the one found in brain, it cleaves the bond on the amino side of the Arg residue.

4.2. Exopeptidases

4.2.1. Enkephalin Convertase

Enkephalin precursors contain multiple copies of the enkephalins flanked by pairs of basic amino acids. Biosynthesis of enkephalin is thought to involve two steps. First, a trypsin-like enzyme produces cleavage at the COOH-terminus of the basic amino acids, giving rise to the enkephalin-containing peptides with COOH-terminal basic amino acids; enkephalin convertase then removes the basic amino acids. The regional distribution of this membrane-bound enzyme in the rat brain differs from that of the soluble enzyme. Both enzymes are inhibited by chelating reagents and activated by Co^{2+}. The enzyme with an apparent molecular weight of 52,500 is inhibited by carboxypeptidase B inhibitors (Supattapone et al., 1984).

4.2.2. Aminopeptidases

Several neuropeptides and hormones, for example enkephalins, MIF, and TRH, are inactivated by aminopeptidases (*see* review by Hui and Lajtha, 1983). Furthermore, aminopeptidases can also activate neuropeptides. Membrane aminoenkephalinase converts Arg^0-Met-enkephalin, its inhibitor, to Met-enkephalin (Hui et al., 1983a). A brain enzyme, not yet fully characterized, converts *N*-Tyr-MIF-1 into MIF-1 (Marks et al., 1984).

A specific aminopeptidase for peptides with acidic NH_2-terminal amino acid residues has been purified 1000-fold (Kelly et al., 1983). The enzyme has a large molecular weight, 450,000, and a pH optimum of 7.5. The enzyme is active on dipeptides, tripeptides, and oligopeptides. Tripeptides are more rapidly hydrolyzed than dipeptides; however, activity tends to decline with increasing chain length. This metal-containing enzyme is insensitive to bestatin or amastatin. Its properties are different from

Table 1
Classification of Brain Peptide Hydrolases

Enzyme	EC Number	Other names
EC 3.4.11 Aminoacylpeptide hydrolases		
Aminopeptidase (cytosol)	3.4.11.1	Leu aminopeptidase (LAP)
Aminopeptidase (microsomal)	3.4.11.2	Particle-bound aminopeptidase
Cystyl aminopeptidase	3.4.11.3	Oxytocinase
Tripeptide aminopeptidase	3.4.11.4	—
Pro aminopeptidase	3.4.11.5	—
Arg aminopeptidase	3.4.11.6	—
Asp aminopeptidase	3.4.11.7	—
Acid aminopeptidase (cytosol)	—	—
Pyroglutamyl aminopeptidase	3.4.11.8	Pyrrolidone carboxylase
Aminopeptidase P	3.4.11.9	Aminoacylproline aminopeptidase
Tyr-MIF aminopeptidase	—	—
Aminopeptidase	3.4.11.11	Peptidase a
Soluble aminopeptidase	3.4.11.?	Aminoenkephalinase, HI, II, III
Membrane aminopeptidase	3.4.11?	Aminoenkephalinase
EC 3.4.13 Dipeptide hydrolases		
Aminoacyl-His dipeptidase	3.4.13.3	Carnosinase
Aminoacyl-Lys dipeptidase	3.4.13.4	N^2-(4-Aminobutyryl)-L-Lys hydrolase
Aminoacyl-methylhistidine dipeptidase	3.4.13.5	Anserinase
Cysteinyl-Gly dipeptidase	3.4.13.6	—
α-Glutamyl Glu dipeptidase	3.4.13.7	—

	Prolyl dipeptidase	3.4.13.8	Iminodipeptidase, prolinase, L-prolylglycine dipeptidase
	Proline dipeptidase	3.4.13.9	Prolidase, iminodipeptidase
	β-Aspartyldipeptidase	3.4.13.10	β-Aspartyl peptidase
	Dipeptidase	3.4.13.11	—
	Methionyl dipeptidase	3.4.13.12	Dipeptidase M
EC 3.4.14	Dipeptidylpeptide hydrolases		
	Dipeptidyl peptidase I	3.4.14.1	Cathepsin C, dipeptidyl-aminopeptidase I, dipeptidyl transferase
	Dipeptidyl peptidase II	3.4.14.2	Dipeptidyl aminopeptidase II
	Acylamino-acid-releasing enzyme	3.4.14.3	N-Acylaminoacyl-peptide hydrolase
	Acylamino-acid-releasing enzyme	3.4.14.4	Postproline dipeptidyl aminopeptidase
EC 3.4.15	Peptidyldipeptide hydrolases		
	Dipeptidyl carboxypeptidase	3.4.15.1	Angiotensin-converting enzyme, peptidase P, kinase II, carboxycathepsin
	Peptidyl carboxyamidase	3.4.15.2	Carboxyamidase
EC 3.4.16	Ser carboxypeptidases		
	Ser carboxypeptidase	3.4.16.1	—
	Pro carboxypeptidase	3.4.16.2	Angiotensinase C, lysosomal carboxypeptidase C
	Tyr carboxypeptidase	3.4.16.3	Thyroid peptide carboxypeptidase, thyroid peptidase
EC 3.4.17	Metallocarboxypeptidases		
	Carboxypeptidase A	3.4.17.1	Carboxypolypeptidase
	Carboxypeptidase B	3.4.17.2	Protaminase
	Enkephalin convertase	—	—
	Arginine carboxypeptidase	3.4.17.3	Carboxypeptidase N

(continued on next page)

Table 1 (*continued*)

	Enzyme	EC Number	Other names
	Aspartate carboxypeptidase	3.4.17.5	—
	Alanine carboxypeptidase	3.4.17.6	—
	Acylmuramoylalanine carboxypeptidase	3.4.17.7	Acylmuramoylalaninase
	Muramoyl-pentapeptide carboxypeptidase	3.4.17.8	—
EC 3.4.21	Ser proteinases		
	Cathepsin A	—	—
	Cathepsin G	3.4.21.?	Chymotrypsin-like enzyme
	Lysosomal elastase	3.4.21.11	—
	Plasminogen activator	3.4.21.?	—
	Prolyl endopeptidase	3.4.21.26	Postproline-cleaving enzyme, postproline endopeptidase, thyrotropin-releasing factor deamidase, kinase B
EC 3.4.22	Thiol proteinases		
	Calcium-activated proteinase	—	—
	Cathepsin B	3.4.21.1	Cathepsin B_1
	Cathepsin M	—	Cathepsin B_3
	Cathepsin L	—	—
	Cathepsin S	—	—
	Collagenolytic cathepsin	—	—
	Kinase A	—	Thiol-activated neutral endopeptidase, endooligopeptidase

EC 3.4.23 Carboxyl proteinases		
Cathepsin D	3.4.23.5	Acid proteinase not requiring thiols, Pseudorenin, isorenin
Cathepsin E	—	—
Renin	—	—
EC 3.4.24 Metalloproteinases		
Tissue collagenase	3.4.24.3	—
Lysosomal collagenase	—	—
Insulinase	—	—
Metalloendopeptidase	3.4.24.11	Enkephalinase, enkephalinase A
EC 3.4.99 Proteinases of unknown catalytic mechanism		
Cathepsin F	—	—
Pz-peptidase	—	—
"Trypsin-like" proteinase	—	—
Dynorphin-converting enzyme	—	—

those of aminopeptidase A from human serum or pig kidney, in agreement with the notion that peptidases are unique for each organ. Leu aminopeptidase in the lens, for example, is different from that in kidney.

A cytosolic aminopeptidase of rat brain has been purified (Berg and Marks, 1984) and found to release Tyr from enkephalin or enkephalin-containing peptides, leaving the metabolite intact; it is similar to the membrane aminoenkephalinase purified in our laboratory (Hui et al., 1983b). We recently compared the major cytosolic and membrane aminoenkephalinases by parallel purification with column chromatography (Hui et al., unpublished observations). They coeluted in column chromatography on ion-exchange cellulose, Sephadex G-200, and Affi-Gel Blue Cellulose. The cytosol and the membrane aminopeptidases alike can be resolved on a hydroxylapatite column, to SI and SII for the cytosol and MI and MII for the membrane. SI and MI are eluted with 0.05*M* phosphate, and SII and MII are eluted with 0.25*M* phosphate. All molecular weights are similar (102,000). Their fingerprint peptide maps are identical, but not their substrate and inhibitor specificities.

5. Inhibitors

Inhibitors are useful tools for elucidation of active sites, and for identifying enzyme mechanisms. Inhibitors may play a role in secretory mechanisms and in protein turnover. They can be classified into three categories: (1) synthetic, (2) microbial, and (3) endogenous and natural inhibitors (Table 2).

5.1. Synthetic Inhibitors

Synthetic inhibitors are usually designed on the basis of model substrates. In general, the practice is to incorporate into the inhibitor, as much as possible, the structural features that determine normal enzyme-substrate recognition. The chloromethyl ketone of the typical trypsin substrate, tosyl-Lys, inactivates trypsin by forming a covalent bond with His- at the active site (Shaw, 1970). Tosyl-Lys-CH_2Cl inhibits proteinases with Ser at the active site. Active-site-directed inhibitors have been useful in studies of enzyme mechanisms, but not in therapeutics, because the alkylation forms a covalent bond that makes the inhibition irreversible.

Recently, synthetic inhibitors of metallopeptidases were prepared by substituting the functional group of the substrates with thiol (Cushman et al., 1977; Patey et al., 1981; Chan, 1983), phosphate (Thorsett, 1982), carboxylate (Patchett et al., 1980; Byers and Wolfenden, 1973), and hydroxamate (Chan et al., 1982), so that the modified substrates become strong dentate ligands (Nishino and Powers, 1980). It is believed that the additional size of sulfur or phosphorus atoms and their conformational flexibility can effectively engage the space occupied by the peptide bond of a substrate, by ligating, for example, the divalent metal atom Zn^{2+}. D-3-Mercapto-2-methyl-propanoyl-L-Pro (SQ 14,225, captopril, Capoten) (Cushman et al., 1977) was developed for angiotensin-converting enzyme; L-leucinol and leucinthiol, for Leu aminopeptidase (Chan, 1983); D-2-benzyl-succinic acid, for carboxypeptidase A (Byers and Wolfenden, 1973); and thiorphan [*N*-(*R*,*S*)-3-mercapto-2-benzylpropanoyl-glycine], for a metalloendopeptidase that splits enkephalins (Patey et al., 1981).

5.2. Microbial Inhibitors

Bacitracin and puromycin are among the best known aminopeptidase inhibitors. Puromycin inhibits arylamidases, but not LAP (Neidle and Lajtha, 1976). Since 1965, Umezawa and his colleagues have screened peptidase and proteinase inhibitors in culture filtrates of actinomyces (Umezawa, 1978). About 20 kinds of small-molecular-size inhibitors were found. Compared to other bioactive peptides, they exhibit special characteristics, such as unusual amino acids, in their component structures. The COOH-terminal of the inhibitors is usually an aldehyde (Aoyagi, 1978). They are the inhibitors of various proteinases and peptidases, such as Ser proteinases, thiol proteinases, carboxyl proteinases, metalloendopeptidases, and aminopeptidases. Some inhibit with high specificity, whereas others have low specificity (Table 2). For example, elastinal inhibits only elastase (Aoki et al., 1975), whereas bestatin can inhibit LAP, aminopeptidase B (Nishizawa et al., 1977), arylamidases (Hui et al., 1980a), and aminoenkephalinases (Hui et al., 1983b). Bestatin [(2*S*,3*R*)-3-amino-2-hydroxy-4-phenylbutanoyl-L-Leu] was found to enhance immune responses and to cause antitumor effects in vivo; it is already being evaluated for clinical use (Saito et al., 1973, 1977).

5.3. Endogenous and Natural Inhibitors

Tissue and body fluids contain a wide range of materials inhibitory of proteinases and peptidases. Inhibitors occur in brain (Marks and Lajtha, 1963; Kopitar et al., 1983), cartilage (Dean and Woessner, 1984), serum (Travis and Salvesen, 1983; Kuehn et al.,

Table 2
Peptidase and Proteinase Inhibitors

Inhibitors	Enzymes
Synthetic	
DFP, PMSF	Serine proteases
Ac-Ala-Ala-Ala-AlaCH_2Cl	Elastases
Tosyl-Lys CH_2Cl	Trypsin, thrombin, plasmin
Tosyl-Phe CH_2Cl	Chymotrypsin
Z-Phe Ch_2Br	Cathepsin G
Thiorphan	Neutral metalloendopeptidase
Phosphoryl-Leu-Phe	Neutral metalloendopeptidase
Captopril	Angiotensin converting enzyme
Leucinthiol	Leu aminopeptidase
Microbial	
Amastatin	Aminoenkephalinase, aminopeptidase A, LAP
Antipain	Cathepsin B, cathepsin A, trypsin, papain
Bacitracin	Aminopeptidases
Bestatin	Aminoenkephalinase, aminopeptidase B, LAP
Chymostatin	Papain, cathepsin B
Elastinal	Elastase
Pepstatin	Cathepsin D, renin, pepsin
Phosphoramidone	Thermolysin, collagenase
Puromycin	Arylamidase
Endogenous	
Cystatin	Cysteine protease
α-1-Proteinase inhibitor	Trypsin, chymotrypsin, elastase, plasmin
Antithrombin	Thrombin
α_2-Antiplasmin	Plasmin
α_1-Antichymotrypsin	Chymotrypsin
Cl-Inhibitor	Clr, Cls, kallikrein, plasmin
α_2-Macroglobulin	Universal protease inhibitor
Inter-α-trypsin inhibitor	Serine proteases
α_1-Anticollagenase	Collagenase
Substance P	Aminoenkephalinase
β-Endorphin	Aminoenkephalinase
Enkephalin-containing peptides	Aminoenkepalinase
ACTH fragments	Aminoenkephalinase
Proctolin, Arg-Tyr	Aminoenkephalinase

(continued on next page)

Table 2 (*continued*)

Inhibitors	Enzymes
Phe-Ala	Neutral metalloendopeptidase
Gly-Gly-Phe-Met	Neutral metalloendopeptidase
Tyr-Tyr, Tyr-Phe	Enkephalinase B
Natural	
Kunitz inhibitor	Trypsin and chymotrypsin
Potato inhibitor (a)	Carboxypeptidase A
Potato inhibitor (b)	Cathepsin D
Lima bean trypsin inhibitor	Trypsin and chymotrypsin
Papain inhibitor	Papain and cathepsin B
Ovamucoid	Chymotrypsin
Ovoinhibitor	Trypsin and chymotrypsin

1984; Takahara and Sinohara, 1982), cerebrospinal fluid (Barrett et al., 1984), and liver (Kominami et al., 1982; Green et al., 1984; Pontremoli et al., 1983). The fact that inhibitors are usually of high molecular weight jeopardizes their clinical applications. We recently found that Arg-dipeptides, some endogenously present, are potent inhibitors of aminoenkephalinase; they are superior to the microbial inhibitors, from the therapeutic point of view, because of their potency, selectivity, and endogenous origin.

Protein and peptide degradation is a stepwise process in which several enzymes participate. The hydrolysis may be triggered by a single enzyme or by a group of enzymes. As degradation proceeds, the digested substrate becomes susceptible to cleavage by succeeding enzymes. The control mechanism of protein degradation is still unclear. There are at least four factors playing a role: (1) the availability of the substrate, controlled by the biosynthesis of the peptide and its compartmentation (binding, storage, uptake into lysosomes); (2) activation of some of the enzymes—for example, chymotrypsin is formed from chymotrypsinogen because Ca^{2+} activates the Ca^{2+}-dependent protease; (3) properties of the enzyme such as shifts in substrate specificity with pH or binding by membranes; and (4) endogenous inhibitors. Endogenous inhibitors also play a role in peptide degradation. We recently found that Arg^0-Met-enkephalin, a precursor of Met-enkephalin, is a potent inhibitor of Met-enkephalin. Several inhibitors are found in vegetables, fruits, and eggs (Stalhut and Hymowitz, 1983; Nagase and Harris, 1983; Laskowski and Kato, 1980).

6. Methods for Proteinase and Peptidase Studies

6.1. Determination of Proteinase and Peptidase Activity

With a peptide of known structure it is possible to determine the sites of action by measuring the amino acids or peptides produced in order to establish whether the enzyme is an exopeptidase or an endopeptidase. The hydrolysate can be submitted to chromatography for separation of peptides, which can be identified by acid hydrolysis followed by amino acid determination; the NH_2-terminal amino acid is determined by dansylation or by phenylthiohydration, or by comparison with synthetic standards. For better identification, optimal pH, and SH and metal ion dependence must also be known.

6.1.1. Amino Acid Determination

The amino acids released from a peptide can be separated by ion-exchange chromatography (Hamilton, 1963). The column effluent is monitored spectrofluorometrically or spectrophotometrically after it is derivatized with *o*-phthaldialdehyde (Benson and Hare, 1975), fluorescamine (Stein et al., 1973), or ninhydrin (Samajima et al., 1971). The separation of amino acids by ion-exchange chromatography has a high resolution, but it takes a long time (usually 6–8 h for a complete run). HPLC offers greater speed and maintains sensitivity. Underivatized amino acids have been assayed (Hancock et al., 1979), but precolumn derivatization and subsequent separation and detection of amino acids by HPLC (Hill et al., 1979) is now a commonly used method. Dansyl chloride (Burzynski et al., 1975), phenylthiohydantoin (Somack, 1980), and *o*-phthaldialdehyde (Hill et al., 1979) are the usual derivatizing agents.

Dansylated amino acids can be separated by two-dimensional thin-layer chromatography on polyamide layers (Hui et al., 1980a). This method is fast and highly sensitive, to the femtomolar range. Although it is excellent qualitatively, it is poor as a quantitiative method because dansyl amino acids are not stable during chromatography, and the danyslation is not stoichiometric.

6.1.2. Peptide Determination

The peptide products of cerebral hydrolases can be determined by chromatography (Kellokumpu et al., 1980), paper electrophoresis (Burbach et al., 1980), gel electrophoresis (Gráf et al., 1977), and HPLC (Stein, 1980). For small peptides, HPLC is the method of choice because of its speed, reliability, accuracy, and high sensitivity, and because the peptide can be recovered for further study. A peptide, without being modified, can be detected photometrically or fluorometrically, depending on the amino acid of interest; that is, by UV absorption at 205 nm (amide bond), 280 nm (Tyr), and 254 nm (Phe), or by fluorescence if it contains Trp (excitation, 275 nm; emission, 370 nm). The sensitivity of the UV or the fluorometric method is in the subnanomolar range. The separated peptides can be identified by comparing their chromatographic properties with those of synthetic standards. Since the peptide determined is unmodified, it can be further characterized if suitable synthetic standards are not available. Acid hydrolysis (6*N* HCl) followed by amino acid analysis is the most common method.

The NH_2-terminal or the COOH-terminal of the peptide can be identified after incubation with an aminopeptidase (aminopeptidase M) or a carboxypeptidase (carboxypeptidase Y). The NH_2-terminal amino acid can also be identified by chromatography after it is derivatized by dansyl or phenylthiohydantoin followed by acid hydrolysis.

6.2. Enzyme Assays

For purification of peptide hydrolases, several points must be considered, including yield, speed, and reproducibility; also, a proper substrate has to be selected. Natural peptides and model synthetic compounds are the two major choices for substrates. Because little is known about the major endogenous substrates of most proteinases and peptidases, the "natural substrate" may not be a significant substrate under physiological conditions. Nonetheless, determination of enzyme activity toward some selected substrates is an indispensable step in the characterization of hydrolases and the evaluation of their function. For a good characterization, a number of substrates representing various classes must be used. Many purified preparations still contain a mixture of exo- and endopeptidases. Under such conditions, characterization of a component enzyme can be improved by selecting assay conditions so that interference from other enzymes is reduced to a minimum. For example, if Leu aminopeptidase activity in the brain is assayed using 2 m*M* Mn^{2+}, at pH 8.5, the arylamidase activity (optimal pH 7.0) is suppressed (Neidle et al., 1980). The alternative method is to add specific inhibitor. In the above case, puromycin, a selective inhibitor of arylamidase, can be used. For selective purification, Leu aminopeptidase can be separated from other aminopeptidases by ion-exchange chromatography on a DEAE-cellulose column (Hui et al., unpublished observations). Leu aminopeptidase is eluted with 0.1*M* NaCl, which is too low in concentration to elute other aminopeptidases; that is, purification is centered on separating the enzyme activities even if a pure crystalline enzyme cannot be obtained.

6.2.1. Peptide Substrates

Of the peptides commonly used as substrates, Ala-Pro is a substrate for prolidase (Hui and Lajtha, 1978), Gly-Gly for Gly-Gly dipeptidase (Stern and Marks, 1979), and Ala-Ala, Lys-Ala, Tyr-Gly, and Pro-Ala for other dipeptidases (Reith and Neidle, 1979). Amino acyl-Gly-Gly is a typical substrate for aminotripeptidase (Sacks and Marks, 1982). The activity can be measured as the released amino or carboxyl groups. Such methods are applicable to general dipeptidases and tripeptidases, but not to prolidase, which can only be assayed spectrophotometrically based on the decrease in UV absorption (220 nm) as peptide bonds are split.

In the past few years, interest in the function and metabolism of neuropeptides has been growing rapidly. Inactivation can be monitored by pharmacological methods. Kininase activity on isolated guinea pig ileum has been used for brain endopeptidase activity assay for substance P and bradykinin degradation (Camargo et al., 1973). Hydrolytic activity can also be determined with the aid of radiolabels such as ^{3}H or ^{125}I, or by radioimmunoassay after separation by chromatography or by electrophoresis (Hui and Lajtha, 1983).

Activity of specific proteinases, such as renin, can be determined by measuring the hydrolytic product of natural substrates. With proteinases of broad specificity, using native proteins creates certain problems: several bonds of the protein are split, and a number of fragments are formed that are difficult to isolate and identify. SDS-PAGE may be useful for separation of complex mixtures, but it is time-consuming, and the quantitation of the separated peptides is difficult. Denatured proteins such as hemoglobin, albumin, casein, and the oxidized β-chain of insulin are the substrates used most often for endopeptidases. The activity of the enzyme can be monitored by measuring the increase of UV absorbance (280 nm) of the supernatant after trichloroacetic acid precipitation of the sample, or by measuring the release of amino acids and peptides with the ninhydrin method. Measurements of the release of ^{125}I or ^{3}H from the iodinated or tritiated proteins or the release of chromophores coupled to proteins (e.g., azocasein) are also useful methods. Besides being expensive, the natural protein substrates suffer from poor solubility, low extent of denaturation, nonuniformity, and autolysis, and they cannot be used to determine precisely the number of peptide bonds split.

6.2.2. Artificial Substrates

Synthetic substrates such as amino acids or low-molecular-weight synthetic peptides that contain a chromogenic group facilitate rapid and quantitative assays. The breakdown of the substrate releases the chromogen: diazo dye, dansyl, naphthylamine, or *p*-nitroaniline compounds. The free chromogen is measured photometrically or fluorometrically. Several enzymes have been purified by using such substrates: enkephalin convertase (Fricker and Snyder, 1982), arylamidase (Neidle and Lajtha, 1976), cathepsin B (Orlowski et al., 1981), prolyl endopeptidase (Orlowski et al., 1979), trypsin (Erlanger et al., 1961), elastase (Feinstein et al., 1973), and aminopeptidase (Hersh, 1981). The most useful synthetic substrates are specific for only one enzyme, so that in crude preparations they are free from the interference of other hydrolytic enzymes. This is not always possible, however, with enzymes of broad specificity. The specificity of an enzyme is determined by two types of interactions between the enzyme and the substrate. The "primary interaction" determines the primary specificity of the enzyme, which is controlled by the two amino acids forming the sensitive peptide bond. The "secondary interaction" determines the secondary specificity, which depends on the sequence and the number of amino acids adjacent to the sensitive bond.

Most proteinases and peptidases split more than one type of peptide bond. The active site of proteinases can probably accommodate five to seven amino acid residues. The specificity of an enzyme is also governed by the amino acid sequence near the sensitive bond of the substrate and the accommodation of this sequence to the reactive site of the enzyme. We found that the subcellular distributions of hydrolytic activities are different with enkephalins (Tyr-Gly-Gly-Phe-Met and Tyr-Gly-Gly-Phe-Leu) and Tyr-2NA* as substrates (Hui et al., 1981). With the enkephalins, the soluble fraction has the highest, and the synaptosomal fraction the lowest, activity. With Tyr-2NA, the highest activity is in the mitochondrial and the lowest is in the soluble fraction. This shows that the release of Tyr from the enkephalin and the arylamide molecule is controlled by groups of aminopeptidases with different

*Abbreviations: 2NA, 2-naphthylamide or 2-naphthylamine; *p*NA, *p*-nitroanilide or *p*-nitroaniline; DFP, diisopropylfluorophosphate; PMSF, phenylmethanesulfonylfluoride.

activities and specificities. For the major aminoenkephalinases, the kinetic constants for Met-enkephalin are different from those for Tyr-2NA or Arg-2NA (Hui et al., 1982). Furthermore, the release of Tyr from Met- and Leu-enkephalin is not parallel in rat brain regions and subcellular fractions, showing that the remote amino acid residue (Leu or Met) at a distance from the sensitive bond also contributes significant influence. Data from our laboratory and those of others (Orlowski, 1983) show that synthetic substrates are not sufficient to predict the enzyme action on large, more complex peptides, where the three-dimensional conformation of the peptide chain hinders the accessibility of the peptide bond to the enzyme action.

Several synthetic substrates are peptides with amino acid sequences similar to those of the natural substrates, so as to replicate the conformation of the sensitive peptide bond and its microenvironment. To eliminate the action of the interfering enzymes, the amino acids near the bonds susceptible to such enzymes can be blocked; e.g., the NH_2-terminal with dansyl, α-*N*-benzoyl, or benzyloxycarbonyl groups, and the COOH-terminal with *p*NA or 2NA. The endopeptidase activity can also be determined by measuring the released chromogen or the amino acid with the aid of a complementary enzyme, usually aminopeptidase M (Orlowski, 1983).

6.3. Purification of Hydrolytic Enzymes

For purification of proteinases and peptidases, the most suitable method is chromatography. Electrophoresis gives high resolution in the separation, but because its yield is low, it is not practical to use for enzyme purification.

The general procedure for purification of hydrolytic enzymes is salt precipitation, ion-exchange chromatography, gel-permeation chromatography, partition chromatography, and reverse-phase chromatography. Salt precipitation, usually by $(NH_4)_2SO_4$, is the best way to start if the preparation is large enough and contains no detergent such as Triton X, Nonidet P-40, or Brij. These detergents affect the precipitation by $(NH_4)_2SO_4$. Ion-exchange chromatography, reverse-phase chromatography, and partition chromatography are the usual methods, since they can purify a sample and concentrate it at the same time. Two chromatographic media are very useful in aminopeptidase purification. AH-Sepharose-4B ($NH_2(CH_2)_6NH$-Sepharose-4B), a supporting matrix for affinity chromatography, retains the peptidases on the column and can be eluted by NaCl (Kelly et al., 1983; Shimamura et al.,

1983). The mechanism of this chromatography, however, is still unknown. It is likely that the free NH_2 group can be a cation group for ion-exchange chromatography or a ligand for affinity chromatography. Hydroxylapatite offers convenience and resolution for aminopeptidases (Hui et al., 1981; Shimamura et al., 1983; Sacks and Marks, 1982); it can resolve two forms of the major aminoenkephalinase (Hui et al., unpublished observations).

Advanced chromatography includes affinity-chromatography and HPLC. Gel exclusion HPLC can separate aminopeptidases speedily and with high resolution, but so far only in analytical scale (Hui et al., unpublished observations). Reverse-phase HPLC (RP-HPLC) has been used to further purify trypsin from a commercial source (Waters Associates, personal communications), but it is still not applicable to enzyme purification, largely because most of the elution solvents (acetonitrile, methanol, 1-propanol, and so on) inactivate enzymes. Some success has been made in diluting the organic solvent immediately after the separation.

In successful affinity chromatography, one-step 1000-fold purification with nearly 100% recovery can usually be obtained. The speed and ease of the method have revolutionized the isolation of a large number of important enzymes, including proteinases and peptidases. Several peptidases have been purified by affinity chromatography using substrate, inhibitor, or enzyme antibody as ligand (immunoabsorbent chromatography). Dipeptidyl peptidase IV has been purified by chromatography on Gly-Pro-NH-$(CH_2)_6$-NH Sepharose (Kojima et al., 1980). With *p*-aminobenzoyl-L-arginine-Sepharose 6B affinity chromatography, purification of enkephalinase convertase from chromaffin granules was 10,000-fold, and from pituitary, 1800-fold (Fricker and Snyder, 1982). *p*-Aminobenzoyl-L-arginine is a potent inhibitor of enkephalin convertase. Monoclonal antibodies as ligands are superior to polyclonal antibodies. In the purification of neutral metalloendopeptidase (EC 2.4.11), one-step purification with the monoclonal column was 4700-fold, and with the polyclonal column, 25-fold (Relton et al., 1983).

Many factors can affect affinity chromatography because a specific selection, related to ligand binding, is involved. The basic requirements for a successful affinity chromatography are:

1. The ligand should be linked to the matrix in such a way that its binding to the enzyme of interest is not affected. The ligand should be kept away from the matrix by a "spacer arm" so that it is more accessible to the enzyme.

2. Nonspecific binding or interaction should be kept at a minimum.
3. The ligand–matrix linkage should be sufficiently stable during the chromatography and the regeneration of the column before reuse.

Affinity chromatography is not always successful, even with the many attempts to vary the process. The main problems have been the low capacity and specificity of the adsorption.

6.4. Enzyme Preservation and Storage

The sensitivity of enzymes to denaturation after removal from their natural, compartmented, or cryptic state increases as the enzyme purity increases. Influencing factors can be physical or chemical, e.g., temperature, oxidation, pH, reduction, and ions. Some, but not all, of such denaturation is reversible. Proteinases and peptidases are more labile than other enzymes because, in addition to the above-mentioned factors, they have a tendency toward autolysis. Several attempts to purify the enkephalin-processing enzyme (trypsin-like) from β-endorphin failed because of difficulties encountered in storage of the enzyme (Gráf et al., 1979). Aminoenkephalinase, which is sensitive to temperature changes, is inactivated by freezing and thawing; however, if kept in liquid condition at 4°C, it has a lifespan of 3 mo. Most peptidases can be kept for months in 10% sucrose at –80°C.

7. Procedure for Membrane Aminoenkephalinase Purification

Recently, we succeeded in purifying the major membrane aminoenkephalinase from rat brain (Hui et al., 1983b). The enzyme accounts for the membrane-bound enkephalin aminopeptidase activity (Hersh, 1985). Following is the detailed procedure for the enzyme assay, purification, and characterization.

7.1. Enkephalin Degradation Assays

Met-enkephalin (11.25 nmol) and aminopeptidase are incubated in a final volume of 150 μL of 25 m*M* Tris-HCl buffer, pH 7.5, containing 2 m*M* ethylene-bis(oxyethylenenitrilo)tetraacetic acid and 0.2 m*M* dithiothreitol (buffer A) at 37°C for 30 min. The reaction is terminated by adding 20 μL of 20% TCA. Leu-enkephalin (11.25 nmol) is added as internal standard, the sample

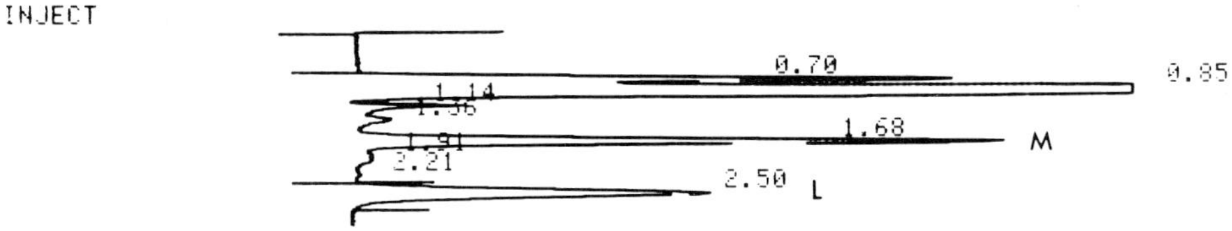

INTERNAL STANDARD QUANTITATION
IS AREA 171791

PEAK#	AMOUNT	RT	EXP RT	AREA	RF
1	0.87799	1.68		199728 F	0.395100E-2
2	INT STD	2.50		154402 L	0.582100E-2
TOTAL	0.87799				

Fig. 1. A typical HPLC chromatogram for Met-enkephalin (M) assay, using Leu-enkephalin (L) as an internal standard.

is centrifuged, and 30 μL of the supernatant is submitted to HPLC analysis. Breakdown of the Met-enkephalin is determined spectrophotometrically by measuring the disappearance of the peptide. The sample is eluted isocratically on a Radial-Pak C_8 (5 × 100 mm) column at ambient temperature with a mixture of acetonitrile and 0.1*M* phosphate buffer, pH 3.0 (20:80). The flow rate is 2 mL/min, resulting in a back-pressure of 500–1000 lb/in^2. The peptides are measured at 280 nm and/or 205 nm. Figure 1 is a typical HPLC chromatogram for Met-enkephalin. Tyr is detected by HPLC on a Lichrosorb RP_{18} column (20 × 4.6 mm, particle size 7 μm, Unimetrics, Westwood, NJ). The sample is eluted isocratically with acetonitrile and phosphate buffer (4:96) at a flow rate of 1.5 mL/min. The retention time for Tyr is 4 min, which resulted in resolution from Tyr-Gly (6.9 min) and Tyr-Gly-Gly (5.7 min). Met-enkephalin binds to the column and cannot be eluted with this solvent.

7.2. Enzyme Purification

The enzyme is prepared by homogenization of five rat brains (10 g) with 9 volumes of 0.32*M* sucrose in a glass homogenizer with a motor drive. The homogenate is centrifuged at 800*g* for 10 min, and the cell debris is discarded. After centrifugation at 30,000*g* for 20 min, the supernatant is discarded and the pellet is resuspended in 50 m*M* Tris-HCl buffer at pH 7.5, centrifuged, and washed two additional times. The membrane pellet is solubilized by suspension in 20 mL of 50 m*M* Tris-HCl buffer, pH 7.5, containing 1%

(w/v) Triton X-100, and incubated at 37°C for 45 min. The solubilized enzyme, obtained after centrifugation at 30,000*g* for 10 min, is applied to a DEAE-Sepharose CL-6B column (Pharmacia; 1.5 × 25 cm) equilibrated with buffer A. The column is washed with two bed volumes of buffer A, and the enzyme is eluted with 500 mL of a gradient of 0–0.4*M* NaCl in buffer A at a flow rate of 0.4 mL/min. Fractions of 10 mL are collected. The enzyme peak from ion-exchange chromatography is applied directly to a Bio-Gel HTP column (Bio-Rad; 0.9 × 12 cm) equilibrated with buffer A. The column is washed with two bed volumes of the same buffer, and the retained proteins are eluted at a flow rate of 0.4 mL/min with 500 mL of a 0–200 m*M* sodium phosphate gradient, pH 7.5, in buffer A. The pooled enzyme fractions from the Bio-Gel HTP chromatography are concentrated to approximately 2 mL with an Amino PM 10-membrane filter and applied to a Sephadex G-200 column (Pharmacia; 1.5 × 85 cm) equilibrated with buffer A. The protein is eluted with the same buffer at a flow rate of 0.05 mL/min. The eluate is collected in 2-mL fractions. The overall purification is 1200-fold, with 25% yield.

7.3. Characterization of the Aminoenkephalinase

The purity of an enzyme is usually tested by Ferguson analysis of putative purity, which includes (a) chromatography, (b) electrophoresis, (c) isoelectric focusing (IEF), (d) constancy of specific activity after further purification, (e) the Ouchterlony double-immunodiffusion test with antiserum, and (f) the inhibition test. The purified membrane aminoenkephalinase shows one protein band on disk gel electrophoresis. The band is also stained in parallel for enzyme activity on Met-enkephalin breakdown and detected by reduction of tetrazolium salts. The enzyme shows two bands on SDS electrophoresis with molecular weights of 62,000 and 66,000. The aminopeptidase has a pH optimum of 7.0, a K_m of 0.28 m*M*, and a V_{max} of 45 μmol/mg of protein/min for Met-enkephalin. It releases tyrosine from Met-enkephalin, but it does not split the byproduct. It does not hydrolyze γ- or β-endorphin, or dynorphin, but it does hydrolyze neutral and basic aminoacyl-2NA. The enzyme is inhibited by puromycin, bestatin, bestatin-Gly, enkephalin-containing peptides, β-endorphin, substance P, ACTH fragments, and proctolin. Its properties, such as its membrane localization, substrate specificity, pH optimum, and molecular weight, distinguish it from Leu aminopeptidase, aminopeptidase A, aminopeptidase B, aminopeptidase M, and the soluble aminopeptidase for enkephalin breakdown.

8. Conclusions

Recent studies show that proteinases and peptidases play several important roles in brain metabolism and function. Proteinases participate in general protein turnover and also in the formation of active enzymes, neuropeptides, and hormones. They may also be involved in formation and activation of other active factors from larger, inactive precursors. Although transport of peptides has been detected as a pathway of neuropeptide inactivation, peptidase activity seems to be the major pathway of inactivation and also of neuropeptide formation through the hydrolysis of larger precursors.

The heterogeneity of turnover rates indicates significant differences in synthesis rates among the various proteins. The measured half-lives vary from minutes to weeks. The metabolism of each protein may also be at different sites (neurons vs glia, cytoplasm vs membrane, and so one) and in different processes and developmental stages. This requires not only a similar heterogeneity of breakdown rates, but also interrelated parallel control of synthesis and breakdown.

In spite of some indications for such controls, our knowledge of the factors and mechanisms that regulate protein and peptide degradation is scant. Several facts indicate that enzymes seem to be in significant excess—peptidases perhaps more than proteases. This is difficult to estimate since enzyme activities in vivo may be very different from those determined in vitro under optimal conditions. It is unknown how large a role is played by the occlusion and liberation of enzymes, changes in substrate conformation, presence (and occlusion) of endogenous inhibitors, changes in intracellular pH, energy supply, changes in ions, and so on.

Although protein metabolism is fairly stable, it is altered in some pathological conditions, and it is affected by drugs, endocrine factors, temperature, and other factors. Whether such alterations are primarily mediated by changes in synthesis or breakdown is not yet established. Undoubtedly, changes in breakdown also play an important role in physiology, pharmacology, and pathology.

The number of proteases and peptidases characterized from brain is constantly increasing; however, it seems that only a limited number of enzymes with broad substrate specificity are significantly active. Their high catabolic rates result in low peptide levels in brain, thus also indicating the presence of yet-to-be-discovered control factors.

The rapid advances in the methodology in this field—

chromatography and its related separation techniques, immunology, synthesis of substrates and inhibitors, sensitive detection of activity—will be of great help in answering some of these questions in the near future.

References

Almenoff J., Wilk S., and Orlowski M. (1981) Membrane-bound pituitary metalloendopeptidase: Apparent identity to enkephalinase. *Biochem. Biophys. Res. Commun.* **102,** 206–214.

Aoki Y., Urata G., Takatsia F., and Katunuma N. (1975) A new protease inactivating α-aminolevulinic acid synthetase in mitochondria of human bone marrow cells. *Biochem. Biophys. Res. Commun.* **65,** 567–574.

Aoyagi T. (1978) Structure and Activities of Proteinase Inhibitors of Microbial Origin, in *Bioactive Peptides Produced by Microorganisms* (Umezawa H., Takita T., and Shiba T., eds.), Kodansha, Tokyo.

Arregui A., MacKay A. V. P., Iversen L. L., and Spokes E. G. (1979) Reduction of angiotensin-converting enzyme in substantia nigra in early-onset schizophrenia. *N. Engl. J. Med.* **30,** 502–503.

Arregui A., MacKay A. V. P., Spokes E. G., and Iversen L. L. (1980) Reduced activity of angiotensin-converting enzyme in basal ganglia in early onset schizophrenia. *Psychol. Med.* **10,** 307–313.

Arstila A. V., Reikkinen P. J., Rinne U. K., Pellinieme T. T., and Nevalainen T. (1971) Guillain-Barre syndrome. Neurochemical and ultrastructural study. *Eur. Neurol.* **5,** 257–269.

Barrett A. J., Elisabeth Davies M., and Grubb A. (1984) The place of human β-trace (cystatin C) amongst the cysteine proteinase inhibitors. *Biochem. Biophys. Res. Commun.* **120,** 631–636.

Benson J. R. and Hare, H. E. (1975) *o*-Phthaldialdehyde fluorogenic detection of primary amines in the picomole range comparison with fluorescamine and ninhydrin. *Proc. Natl. Acad. Sci. USA* **72,** 619–622.

Berg M. J. and Marks N. (1984) Formation of des Tyr dynorphins 5–17 by a purified cytosolic aminopeptidase of rat brain. *J. Neurosci. Res.* **11,** 313–321.

Boehme D. H., Umezawa H., Hashim G., and Marks N. (1978) Treatment of experimental allergic encephalomyelitis with an inhibitor of cathepsin D (pepstatin). *Neurochem. Res.* **3,** 185–194.

Bowen D. M., Smith C. B., and Davison A. N. (1973) Molecular changes in senile dementias. *Brain* **96,** 849–856.

Brosnan C. F., Cammer W., Norton W. T., and Bloom B. R. (1980) Proteinase inhibitors suppress the development of experimental allergic encephalomyelitis. *Nature* **285,** 235–237.

Burbach J. P. H., De Kloet E. R., and De Wied D. (1980) Oxytocin biotransformation in the rat limbic brain: Characterization of peptidase activities and significance in the formation of oxytocin fragments. *Brain Res.* **202,** 401–414.

Burzynski S. R. (1975) Quantitative analysis of amino acids and peptides in the femtomolar range. *Anal. Biochem.* **65,** 93–99.

Byers L. D. and Wolfenden R. (1973) Binding of the by-product analog benzylsuccinic acid by carboxypeptidase A. *Biochemistry* **12,** 2070–2078.

Carmago A. C. M., Shapanka R., and Greene L. J. (1973) Preparation, assay and partial characterization of a neutral endopeptidase from rabbit brain. *Biochemistry* **12,** 1838–1844.

Chan W. W. C. (1983) Leucinthiol—a potent inhibitor of leucine aminopeptidase. *Biochem. Biophys. Res. Commun.* **116,** 297–302.

Chan W. W. C., Dennis P., Demmer W., and Brand K. (1982) Inhibition of leucine aminopeptidase by amino acid hydroxamates. *J. Biol. Chem.* **257,** 7955–7957.

Chelmicka-Schorr E. E., Arnason B. E. W., Astrom K.-E., and Darzynkiewicz Z. (1978) Treatment of mouse muscular dystrophy with the protease inhibitor pepstatin. *J. Neuropathol. Exp. Neurol.* **37,** 263–268.

Cushman D. W., Cheng H. S., Sabo E. F., and Ondetti M. A. (1977) Design of potent competitive inhibitors of angiotensin-converting enzyme. Carboxyalkanoyl and mercaptoalkanoyl amino acids. *Biochemistry* **16,** 5484–5491.

Dean D. D. and Woessner F., Jr. (1984) Extracts of human articular cartilage contain an inhibitor of tissue metalloproteinases. *Biochem. J.* **218,** 277–280.

De La Baume S., Yi C. C., Schwartz J. C., Chaillet P., Marcais-Collado H., and Costentin J. (1983) Participation of both "enkephalinase" and aminopeptidase activities in the metabolism of endogenous enkephalins. *Neuroscience* **8,** 143–151.

Devi L. and Goldstein A. (1984) Dynorphin converting enzyme with unusual specificity from rat brain. *Proc. Natl. Acad. Sci. USA* **81,** 1892–1896.

Einstein E. R., Csejtey J., Dalal K. B., Adams C. W. M., Bayliss O. B., and Hallpike J. F. (1972) Proteolytic activity and basic protein loss in and around multiple sclerosis plaques. Combined biochemical and histochemical observations. *J. Neurochem.* **12,** 653–662.

Einstein E. R., Dalal K. B., and Csejtey J. (1970) Increased protease activity and changes in basic proteins and lipids in multiple sclerosis plaques. *J. Neurol. Sci.* **11,** 109–121.

Erlanger B. F., Kokowsky N., and Cohen W. (1961) The preparation and properties of two new chromogenic substrates of trypsin. *Arch. Biochem. Biophys.* **95,** 271–278.

Feinstein G., Kupfer A., and Sokolovsky M. (1973) *N*-Acetyl-(L-Ala) 3-*p*-nitroanilide as a new chromogenic substrate for elastase. *Biochem. Biophys. Res. Commun.* **50,** 1020–1026.

Fricker L. D. and Snyder S. H. (1982) Enkephalin convertase: Purification and characterization of a specific enkephalin-synthesizing carboxypeptidase localized to adrenal chromaffin granules. *Proc. Natl. Acad. Sci. USA* **79,** 3886–3890.

Fruton J. S. (1960) Cathepsins, in *The Enzymes* 2nd Ed. (Boyer P. D., Lardy H., and Myrback K., eds.), Academic, New York.

Gráf L., Cseh G., Barát E., Ronai A., Székely J. I., Kenessey A., and Bajusz S. (1977) Structure–function relationship in lipotropins. *Ann. NY Acad. Sci.* **297,** 63–83.

Gráf L., Kenessey A., Bajusz S., Patthy A., Ronai A. Z., and Berzétei I. (1979) Proteolytic Processing in the Biosynthesis and Metabolism of Endorphins, in *Endorphin in Mental Health Research* (Usdin W. E., Bunney J., and Kline N. S., eds.), MacMillan, New York.

Green G. D. J., Kembhavi A. A., Davies M. E., and Barrett A. J. (1984) Cystatin-like cysteine proteinase inhibitors from human liver. *Biochem. J.* **218,** 939–946.

Hamilton P. B. (1963) Ion exchange chromatography of amino acids, a single column, high resolving, fully automatic procedure. *Anal. Chem.* **35,** 2055–2064.

Hancock W. S., Bishop C. A., and Hearn M. T. W. (1979) The analysis of nanogram levels of free amino acids by reverse-phase high-pressure liquid chromatography. *Anal. Biochem.* **92,** 170–178.

Hersh L. B. (1981) Solubilization and characterization of two rat brain membrane-bound aminopeptidases active on Met-enkephalin. *Biochemistry* **20,** 2345–2350.

Hersh L. B. (1985) Characterization of membrane-bound aminopeptidases from rat brain: Identification of the enkephalin-degrading aminopeptidase. *J. Neurochem.* **44,** 1427–1435.

Hill D. W., Walters F. H., Wilson T. D., and Stuart J. D. (1979) High performance liquid chromatographic determination of amino acids in the picomole range. *Anal. Chem.* **51,** 1338–1341.

Hui K.-S. and Lajtha A. (1978) Prolidase activity in the brain: Comparison with other organs. *J. Neurochem.* **30,** 321–327.

Hui K.-S. and Lajtha A. (1983) Neuropeptides, in *Handbook of Neurochemistry,* 2nd Ed., Vol. 4, (Lajtha A., ed.), Plenum, New York and London.

Hui K.-S., Cheng K.-P., Wong K.-H., Salshutz M., and Lajtha A. (1980a) Degradation of melanotropin inhibiting factor by brain. *J. Neurochem.* **35,** 471–478.

Hui K.-S., Salshutz M., Davis B. A., and Lajtha A. (1980b) Separation of alkylaminonaphthylenesulfonyl peptides and amino acids: Methods for measuring melanotropin inhibiting factor breakdown. *J. Chromatogr.* **192,** 341–350.

Hui K.-S., Gráf L., and Lajtha A. (1982) β-Endorphin inhibits Met-enkephalin breakdown by a brain aminopeptidase: Structure–activity relationships. *Biochem. Biophys. Res. Commun.* **105,** 1482–1487.

Hui K.-S., Hui M., Banay-Schwartz M., DeGuzman T., Ling N., and Lajtha A. (1983a) Enkephalin-containing polypeptides are potent inhibitors of enkephalin degradation. *Peptides* **4,** 639–646.

Hui K.-S., Wang Y.-J., and Lajtha A. (1983b) Purification and characterization of an enkephalin aminopeptidase from rat brain membrane. *Biochemistry* **22,** 1062–1067.

Hui K.-S., Wang Y.-J., Wong K.-H., Tsai H., and Lajtha A. (1981) Regional and subcellular distribution of Met- and Leu-enkephalin-degrading activity in rat brain. *Neuropeptides* **1,** 167–173.

International Union of Biochemistry (1978) Recommendations of the nomenclature committee of the International Union of Biochemistry. Academic, New York.

Kellokumpu S., Vuolteenaho O., and Leppaluoto J. (1980) Behavior of rat hypothalamic and extrahypothalamic immunoreaction TRF in thin-layer chromatography (TLC) and high pressure liquid chromatography (HPLC). *Life Sci.* **26,** 475–480.

Kelly J. A., Neidle E. L., and Neidle A. (1983) An aminopeptidase from mouse brain cytosol which cleaves N-terminal acidic amino acid residues. *J. Neurochem.* **40,** 1727–1734.

Kojima K., Hama T., Kato T., and Nagatsu T. (1980) Rapid chromatographic purification of dipeptidyl peptidase IV in human submaxillary gland. *J. Chromatogr.* **189,** 233–240.

Kominami E., Uakamatsa N., and Katunuma N. (1982) Purification and characterization of thiol proteinase inhibitor from rat liver. *J. Biol. Chem.* **257,** 14648–14652.

Kopitar M., Stern F., and Marks N. (1983) Cerebrocystatin suppresses degradation of myelin basic protein by purified brain cysteine proteinase. *Biochem. Biophys. Res. Commun.* **112,** 1000–1006.

Kuehn L., Rutschmann M., Dahlmann B., and Reinauer H. (1984) Proteinase inhibitors in rat serum. *Biochem. J.* **218,** 953–959.

Laskowski M., Jr. and Kato I. (1980) Protein inhibitors of proteinases. *Ann. Rev. Biochem.* **49,** 593–626.

Libby P. and Goldberg A. L. (1978) Leupeptin, a protease inhibitor, decreases protein degradation in normal and diseased muscles. *Science* **199,** 534–536.

Marks N. and Lajtha A. (1963) Protein breakdown in brain: Subcellular distribution of neural and acid proteinases. *Biochem. J.* **89,** 438–477.

Marks N. and Lajtha A. (1971) Protein and Polypeptide Breakdown, in *Handbook of Neurochemistry,* Vol. V, Part A, (Lajtha A., ed.), Plenum, New York.

Marks N., Berg M. J., Kastin A. J., and Coy D. H. (1984) Evidence for conversion of *N*-Tyr-MIF-1 into MIF-1 by a specific brain aminopeptidase. *Neurochem. Int.* **6,** 347–353.

Marks N., Datta R. K., and Lajtha A. (1968) Partial resolution of brain arylamidases and aminopeptidases. *J. Biol. Chem.* **243,** 2882–2889.

Nagase H. and Harris E. D., Jr. (1983) Ovostatin—A novel proteinase inhibitor from chicken egg white. *J. Biol. Chem.* **258,** 7481–7489.

Neidle A. and Lajtha A. (1976) The Substrate Specificity of an Aminopeptidase (Neutral arylamidase) From Pig Brain, in *Problems of Brain Biochemistry* Vol. XI, (Buinatian H. Ch., ed.), Armenian Acad. Sci., Yerevan, CCCP.

Neidle A., Yessaian N., and Lajtha A. (1980) Degradation of prolylleucylglycinamide (MIF) by mouse brain. *Neurochem. Res.* **5,** 1011–1023.

Nishino N. and Powers J. C. (1980) *Pseudomonas aeruginosa* elastase; Development of a new substrate, inhibitors, and an affinity ligand. *J. Biol. Chem.* **255,** 3482–3486.

Nishizawa R., Saino T., Takita T., Suda H., Aoyagi T., and Umezawa H. (1977) Synthesis and structure–activity relationships of bestatin analogues, inhibitors of aminopeptidase. *J. Med. Chem.* **20,** 510–515.

Ondetti M. A. (1983) Biochemistry of the renin-angiotensin system. Introduction. *Fed. Proc.* **42,** 2722–2723.

Orlowski M. (1983) Proteinases, in *Handbook of Neurochemistry,* 2nd Ed., Vol. 4 (Lajtha A., ed.), Plenum, New York.

Orlowski M., Orlowski J., Lesser M., and Kilburn K. H. (1981) Proteolytic enzymes in bronchopulmonary lavage fluids: Cathepsin B-like activity and prolyl endopeptidase. *J. Lab. Clin. Invest.* **97,** 467–476.

Orlowski M., Wilk E., Pearce S., and Wilk S. (1979) Purification and properties of a prolyl endopeptidase from rabbit brain. *J. Neurochem.* **33,** 461–469.

Patchett A. A., Harris E., Tristram E. W., Wyrratt M. J., Wu M. T., Taub D., Paterson E. R., Ikeler T. J., Ten Broeke J., Payne L. G., Ondeyka D. L., Thorsett E. D., Greenlee W. J., Lohr N. S., Hoffsommer R. D., Joshua H., Ruyle W. V., Rothrock J. W., Aster S. D., Maycock A. L., Robinson F. M., Hirschmann R., Sweet C. S., Ulm E. H., Gross D. M., Vassil T. C., and Stone C. A. (1980) A new class of angiotensin-converting enzyme inhibitors. *Nature* **288,** 280–281.

Patey G., De La Baume S., Schwartz J.-C., Gros C., Roques B., Fournie-Zaluski M.-C., and Soroca-Lucas E. (1981) Selective protection of methionine enkephalin released from brain slices by enkephalinase inhibition. *Science* **212,** 1153–1155.

Pinsky C., Dua A. K., and LaBella F. S. (1982) Peptidase inhibitors reduce opiate narcotic withdrawal signs, including seizure activity in the rat brain. *Brain Res.* **243,** 301–307.

Pontremoli S., Melloni E., Salamino F., Sparatone B., Michett M., and Horecker B. L. (1983) Endogenous inhibitors of lysosomal proteinases. *Proc. Natl. Acad. Sci. USA* **80,** 1261–1264.

Pope A. (1968) Structural and Enzymatic Microchemistry of Human Cerebral Cortex, in *The Central Nervous System* (Bailey O. T. and Smith D. E., ed.), Williams & Wilkins, Baltimore, Maryland.

Pope A. and Nixon R. A. (1984) Proteases of human brain. *Neurochem. Res.* **9,** 291–323.

Pope A., Hess H. H., and Lewin E. (1964) Studies on the Microchemical Pathology of Human Cerebral Cortex, in *Morphological and Biochemical Correlates of Neural Activity* (Cohen M. M. and Snider R. S., eds.), Hoeber-Harper, New York.

Reith M. E. A. and Neidle A. (1979) The isolation of two dipeptide hydrolases from mouse brain cytosol. *Biochem. Biophys. Res. Commun.* **90,** 794–800.

Relton J. M., Gee N. S., Matsas R., Turner A. J., and Kenny A. J. (1983) Purification of endopeptidase-24,11 (enkephalinase) from pig brain by immunoadsorbent chromatography. *Biochem. J.* **215,** 519–523.

Roques B. P., Fournie-Zaluski M. C., Soroca E., Lecomte J. M., Malfroy B., Llorens C., and Schwartz J.-C. (1980) The enkephalinase inhibitor thiorphan shows antinociceptive activity in mice. *Nature* **288,** 286–288.

Sacks L. and Marks N. (1982) A highly specific aminotripeptidase of rat brain cytosol: Substrate specificity and effects of inhibitors. *Biochim. Biophys. Acta* **706,** 229–238.

Saito M., Aoyagi T., Umezawa H., and Nagai Y. (1977) Bestatin, A new specific inhibitor of aminopeptidase, enhances activation of small lymphocytes by concanavalin A. *Biochem. Biophys. Res. Commun.* **76,** 526–533.

Saito M., Yoshizawa T., Aoyagi T., and Nagai Y. (1973) Involvement of proteolytic activity in early events in lymphocyte transformation by phytohemagglutinin. *Biochem. Biophys. Res. Commun.* **52,** 569–575.

Samejima K., Dairman W., and Udenfriend S. (1971) Condensation of ninhydrin with aldehydes and primary amines to yield highly fluorescent ternary products. Studies on the mechanism of the reaction and some characteristics of the condensation product. *Anal. Biochem.* **42,** 222–236.

Schlaepfer W. W. (1978) Deformation of isolated neurofilaments and the pathogenesis of neurofibrillary pathology. *J. Neuropathol. Exp. Neurol.* **37,** 244–254.

Schwartz J.-C., Malfroy B., and De La Baume S. (1981) Biological inactivation of enkephalins and the role of enkephalin-dipeptidylcarboxypeptidase (enkephalinase) as neuropeptidase. *Life Sci.* **29,** 1715–1740.

Shaw E. (1970) Selective chemical modification of protein. *Physiol. Rev.* **50,** 244–296.

Sher J. H., Stracher A., Shafiq S. A., and Hardy-Stashin J. (1981) Successful treatment of murine muscular dystrophy with the proteinase inhibitor leupeptin. *Proc. Natl. Acad. Sci. USA* **78,** 7742–7744.

Shimamura M., Hazato T., and Katayama T. (1983) A membrane-bound aminopeptidase isolated from monkey and its action on enkephalin. *Biochim. Biophys. Acta* **756,** 223–229.

Somack R. (1980) Complete phenylthiohydantoin amino acid analysis by high performance chromatography on ultrasphere-octadecyltrimethyloxylsilane. *Anal. Biochem.* **104,** 464–468.

Stahlhut R. W. and Hymowitz T. (1983) Variation in the low molecular weight proteinase inhibitors of soybeans. *Crop Sci.* **23,** 766–769.

Stein S. (1980) Ultramicro Isolation and Analysis of Peptides and Proteins, in *Polypeptide Hormones* (Beers R. F. and Bassett E. G., eds.), Raven, New York.

Stein S., Bohlen P., Stone J., Dairman W., and Udenfriend S. (1973) Amino acid analysis with fluorescamine at the picomole level. *Arch. Biochem. Biophys.* **155,** 202–212.

Steiner D. F., Quinn P. S., Chan S. J., Marsh J., and Tager H. S. (1980) Processing mechanisms in the biosynthesis of proteins. *Ann. NY Acad. Sci.* **343,** 1–16.

Stern F. and Marks N. (1979) Glycyl-glycine hydrolase of rat brain: Distribution and role in cleavage of glycine-rich oligopeptides. *Brain Res. Bull.* **4,** 49–55.

Supattapone S., Fricker L. D., and Snyder S. H. (1984) Purification and characterization of a membrane-bound enkephalin-forming carboxypeptidase, "enkephalin convertase." *J. Neurochem.* **42,** 1017–1023.

Swerts J. P., Perdrisot R., Malfroy B., and Schwartz J.-C. (1979) Is "enkephalinase" identical with "angiotensin-converting enzyme?" *Eur. J. Pharmacol.* **53,** 209–210.

Takahara H. and Sinohara H. (1982) Mouse plasma trypsin inhibitor: Isolation and characterization of α-1-antitrypsin and contrapsin, a novel trypsin inhibitor. *J. Biol. Chem.* **257,** 2438–2446.

Thorsett E. D., Harris E. E., Paterson E. R., Greenlee W. J., Patchett A. A., Ulm E. H., and Vassil T. C. (1982) Phosphorus-containing inhibitors of angiotensin-converting enzyme. *Proc. Natl. Acad. Sci. USA* **79,** 2176–2180.

Travis T. and Salvesen G. S. (1983) Human plasma proteinase inhibitors. *Ann. Rev. Biochem.* **52,** 655–709.

Turner A. J., Matsas R., and Kenny A. J. (1985) Are there neuropeptide-specific peptidases? *Biochem. Pharmacol.* **34,** 1347–1356.

Umezawa H., Takita T., and Shiba T. (1978) *Bioactive Peptides Produced by Microorganisms.* Kodansha, Tokyo.

Wallace E. F., Weber E., Barchas J. D., and Evans C. J. (1984) A putative processing enzyme from Aplysia that cleaves dynorphin A at the single arginine residue. *Biochem. Biophys. Res. Commun.* **119,** 415–422.

Walsh K. A. (1975) Unifying Concepts Among Proteases, in *Proteases and Biological Control* (Reich E., Rifkin D. B., and Shaw E., eds.), Cold Spring Harbor Laboratory, New York.

Zhang A.-Z., Yang H.-Y.T., and Costa E. (1982) Nociception, enkephalin content and dipeptidyl carboxypeptidase activity in brain of mice treated with exopeptidase inhibitors. *Neuropharmacology* **21,** 625–630.

Zubenko G. S. and Nixon R. A. (1984) Mood-elevating effect of captopril in depressed patients. *Am. J. Psychiatry* **141,** 110–111.

Chapter 13
Na^+/K^+-ATPase in Nervous Tissue

P. H. Wu

1. Introduction

1.1. Historical Background

Mammalian cells are surrounded by a fluid medium that contains a high concentration of sodium and a low concentration of potassium. In contrast, the cell cytoplasm is high in potassium and low in sodium. To maintain this chemical gradient, the cell must transport excess sodium out of, and potassium into, the cell. The mechanism underlying this phenomenon was first proposed by Dean (1941), who realized that there must be active sodium excretion to control intracellular sodium levels in skeletal muscle. Schatzmann (1953) showed that the sodium and potassium fluxes across red blood cell membranes are dependent on adenosine triphosphate (ATP), and that these fluxes can be inhibited by cardiac glycosides. The idea of an "active" ion transport mechanism (ion pump) was thus formulated. Subsequent studies by Skou (1957) have indicated that there is a Mg^{2+}-dependent adenosine triphosphatase (Mg^{2+}-ATPase) in the microsomal fraction of crab nerve that is markedly activated by the concurrent presence of sodium and potassium ions. Twenty-six years have lapsed since this original observation, and it is now accepted that the enzyme system described by Skou (1957) is responsible for the active transport of sodium and potassium across the cell membrane. The enzyme system was named Na^+, K^+-activated, Mg^{2+}-dependent adenosine triphosphatase (Na^+/K^+-ATPase) (EC 3.6.1.3). The common characteristics of the ion pump and Na^+/K^+-ATPase include: localization, dependence on ATP, coupling of Na^+ and K^+, replaceability of monovalent cations, effects of cardiac glycosides, and specificity of vectral transport of ions. These facts have led to the firm belief that Na^+/K^+-ATPase and the ion transport system of the cell membrane are chemical and mechanical manifestations of a single system. There have been a number of reviews on the enzyme and the relationship between the enzyme and the ion transport (Hokin and Hokin, 1963; Albers, 1967; Bonting, 1970; Quastel, 1970; Tanaka, 1974; Schwartz et al., 1975; Schurmans-Stekhovens and Bonting, 1981). Since a detailed discussion of past observations of Na^+/K^+-ATPase is not the primary intention of this article, only a brief summary of relevant material will be presented in this chapter. Those interested in an in-depth discussion of the enzyme Na^+/K^+-ATPase should consult the review articles mentioned above.

As indicated earlier, Na^+/K^+-ATPase is associated with various cell types and is universally present in living cells. It is thus not possible to confine our discussion of this enzyme to nerve tissues alone. In this review, evidence involving Na^+/K^+-ATPase isolated from nonnervous tissue will be included. In these cases one must be prepared to accept the assumption that Na^+/K^+-ATPase is a homogeneous entity regardless of its tissue origin.

1.2. Occurrence and Distribution

From an evolutionary point of view, one of the principal functions of the cell membrane is probably to maintain appropriate sodium and potassium concentrations within the cell. Among the first enzymatic reactions to be associated with the cell membrane must have been those responsible for transporting ions such as Na^+, K^+, and Ca^{2+} across the membrane against their electrochemical gradients. It is these transport mechanisms that create ionic gradients; gradients that regulate cellular functions and determine the membrane potential of living cells. Therefore, the enzyme Na^+/K^+-ATPase, vital for these reactions, can be found in all living cell membranes, including microbial and plant cells. In nervous tissues, Na^+/K^+-ATPase has been found to be associated with both neurons and glial cells (Cummins and Hyden, 1962). Among the many tissues that have been analyzed, Na^+/K^+-ATPase is found to be most active in nervous tissue, particularly the gray matter (Bonting et al., 1961). The gray matter exhibits approximately fivefold more enzyme activity than the white matter. Other nervous tissues have also been analyzed: choroid plexus, optic nerve, sciatic nerve, superior cervical ganglion, and ocular tissue all contain Na^+/K^+-ATPase. The enzyme distribution pattern in the CNS seems to suggest that the brain regions with more electrical activity also have more Na^+/K^+-ATPase activity. This has led to the postulation that the enzyme plays a role in the generation of brain electrical potentials.

Consistent with the findings of Bonting et al. (1961), Fahn and Cote (1968) measured Na^+/K^+-ATPase activity in 21 regions of rhesus monkey brain and found that white matter has the lowest activity. The activities of the enzyme in gray matter are evenly distributed, with only a twofold difference between occipital cortex and globus pallidus. The highest activities are found in the cerebral cortex, cerebellar cortex, thalamus, and colliculi. Na^+/K^+-ATPase is also found to be highly active in invertebrate nervous systems. Grasso (1967) reported the enzyme in American cockroach nerve cords. Cheng and Cutkomp (1975) showed that the nerve cords of male cockroaches have significantly higher activity of Na^+/K^+-ATPase than those of female cockroaches. There were no explanations offered for this difference between sexes.

Heinz and Hoffman (1965) reported that Na^+ increases the amount of protein-bound ^{32}P, whereas the subsequent addition of K^+ decreases the level of phosphorylated protein. This reaction was later demonstrated to be associated with the enzyme Na^+/K^+-ATPase (Fahn et al., 1966a; Stahl et al., 1966), and was found to occur in microsomes prepared from *Electrophorus electricus* electroplex. Hosie (1965) and Albers et al. (1965) have shown that an enriched preparation of Na^+/K^+-ATPase can be obtained from a nerve ending fraction of brain, although Schwartz et al. (1962) have reported the presence of high activities of Na^+/K^+-ATPase in brain microsomal fraction. Using the preparation techniques described by Albers et al. (1965), Stahl was able to show that specific Na^+/K^+-ATPase activities are found in pinched-off nerve endings. Kurokawa et al. (1965) examined the distribution of Na^+/K^+-ATPase activity in the guinea pig brain using the sucrose gradient technique of Whittaker (1959), and confirmed that there is indeed higher specific activity in microsomal and nerve ending fractions of the brain.

In summary, Na^+/K^+-ATPase is an enzyme believed to be present in all living cell membranes. The primary function of the enzyme is to maintain a vectoral transport of cations and to establish a proper ion gradient inside and outside of cells. In the mammalian nervous tissue, the enzyme is found mainly in gray matter, and possibly in the synpatic endings where neurotransmission takes place. The enzyme is also present in significant amounts in glial cells, where it may be involved in the maintenance of the ionic gradient surrounding the neurons.

1.3. Physiological Function and Relationship to Neurological Disorder

As we have seen, the main physiological function of Na^+/K^+-ATPase is the maintenance of an electrochemical gradient that is distinguished by a high intracellular concentration of K^+ and low concentration of Na^+. The extrusion of Na^+ from the cell requires its movement against a gradient of concentration and electrical potential. Therefore, a mechanism that utilizes ATP as an energy source is evolved in the cell to maintain the Na^+ and K^+ balance and to prevent the rupture of the fragile membrane of the animal cell, which might occur if an osmotic equilibrium across the membrane is not achieved. The Na^+–K^+ gradients are involved in the propagation of the nerve impulse, reabsorption of solutes by the kidney, and various other functions, including neurotransmitter release in the CNS (Vizi, 1979).

Na^+/K^+-ATPase in the brain has been postulated to be involved in the development of seizures. The proposal originated from the observation that there is a significant difference between the Na^+/K^+-ATPase activity of DBA mice, a strain that is susceptible to audiogenic and febrile seizure, and the non-seizure-susceptible strain C-57 mice. It should be noted that this difference in enzyme activity occurred only in certain age groups of mice, specifically 28–34 d (Formby, 1975). Subsequent studies by Hertz et al. (1974) have indicated that there is a correspondence between the susceptibility of a mouse to seize and a decrease in K^+ concentration in brain tissue. The concentrations of Na^+ and Cl^- are not changed. Consistent with this finding, Hertz et al. (1974) demonstrated that there is a 50% reduction in Na^+/K^+-ATPase activity in DBA, but not in C-57, mice relative to control mice. This low Na^+/K^+-ATPase activity in the brain was suggested to account for the lower K^+ content and the diminuation of oxygen uptake. Thus, it is possible that the lower Na^+/K^+-ATPase levels contribute to the seizure susceptibility of DBA mice. However, Stefanovic et al. (1974) found a significant increase of Na^+/K^+-ATPase in cortex and hypothalamus of DBA mice compared to C-57 mice, and Reichert (1975) could not find any significant difference in the enzyme activities between these two strains of mice. This discrepancy has been suggested to be attributable to the inbreeding process in different DBA strains, in which ATPase genomes may become fixed (Rosenblatt et al., 1976). The involvement of Na^+/K^+-ATPase in the seizure susceptibility of individuals is still an unsolved issue.

Rats thyroidectomized at 1 d of age show an increase in Na^+ content and a decrease in K^+ content in the cerebral cortex at 22 d (Geel et al., 1967). This change in ionic concentration can be interpreted as a change in activity of the Na^+ pump (Valcana and Timiras, 1969). It was demonstrated that in the thyroidectomized animals there is a very significant decrease in Na^+/K^+-ATPase activity in the cortex and that this change corresponds to a marked increase in Na^+ and decrease in K^+ concentration. This hypothyroidism leads to a marked reduction in dendritic branching (Eayrs, 1955), and either to a decrease in the synthesis of Na^+/K^+-ATPase, or to a change of the enzyme's properties usually associated with the cellular membranes (Valcana and Timiras, 1969).

There are very few known cases in which a change in brain Na^+/K^+-ATPase activity is implicated in a neurological disease state. Although there is some evidence suggesting that Na^+/K^+-ATPase is abnormal in the red blood cells of some hypertensive patients, the evidence for a decrease in Na^+/K^+-ATPase activity associated with audiogenic and febrile seizures is not conclusive.

2. Properties of Na^+/K^+-ATPase

2.1. *Molecular Properties*

Although Na^+/K^+-ATPase activity was first demonstrated in crab nerve, the kidney has often been used as a source of tissue for the preparation and purification of Na^+/K^+-ATPase. Purification of the enzyme has proven to be difficult because it is imbedded in the plasma membranes. In order to remove the enzyme from its surrounding membrane structure, a number of surface-active agents (detergents) have been applied to solubilize the enzyme. Varying degrees of success have been achieved. The recovery of the enzyme, the degree of increase in specific activity of Na^+/K^+-ATPase, and the ratio between Na^+/K^+-ATPase and Mg^{2+}-ATPase activities during the successive treatments with various detergents are dependent on the source of the enzyme (Choi and Akera, 1978). Apparent turnover numbers of the brain Na^+/K^+-ATPase preparation are not significantly affected by sodium iodide treatment, but the numbers are markedly decreased by deoxycholate or glycerol treatments. Cerebral Na^+/K^+-ATPase obtained by deoxycholate treatment exhibits a lower affinity for ouabain than that exhibited by renal enzyme, suggesting that purification procedures may alter the properties of membrane-bound Na^+/K^+-ATPase and may affect the interaction with cardiac glycosides, i.e., ouabain (Choi and Akera, 1978). Jarnefelt (1964) suggested that detergents may produce an interconversion between the Mg^{2+}-ATPase and Na^+/K^+-ATPase. However, Somogyi et al. (1969) found that the interconversion probably does not occur; instead, the detergent treatments may cause a change in enzyme conformation. Later, Jorgensen and Skou (1971) concluded that deoxycholate may expose latent enzymatic sites, but suggested that the molecular activity of Na^+/K^+-ATPase is not affected by the detergent treatment. They suggested that although Na^+/K^+-ATPase activity increases severalfold when treated by detergent, the Mg^{2+}-ATPase activity is not affected: Thus, the interconversion of Mg^{2+}-ATPase to Na^+/K^+-ATPase during detergent treatment cannot be the cause of the increase in Na^+/K^+-ATPase activity. Recently, Cortas and Walser (1971) suggested that the concentration of detergents used in the treatment of membranes may affect the phospholipid composition of the enzyme preparation, since at lower concentrations of deoxycholate the enzyme–phospholipid interaction is preserved and the enzyme is able to respond to ouabain. Interestingly, they showed that Na^+/K^+-ATPase made to be insensitive to ouabain is converted to a ouabain-sensitive form upon the addition of phosphatidylserine. Triton X-100 was found to be suitable for the treatment of Na^+/K^+-ATPase (Somogyi et al., 1969; Banerjee et al., 1970). Another detergent, sodium dodecyl sulfate, has been found to bind to membrane fragments; the binding is not easily dissociated. The binding of sodium dodecyl sulfate to the membranes may induce a conformational change, and the Na^+/K^+-ATPase, whose activity is increased by this treatment, is not sensitive to ouabain, confirming what was observed by Cortas and Walser (1971).

The attempt to solubilize and purify Na^+/K^+-ATPase was carried out by Schwartz et al. (1962) with guinea pig cerebral cortex microsomes. Lubrol treatment at a concentration of 0.1% yields a very high activity in the solubilized state, but the solubilized enzyme preparation is not stimulated in the presence of Na^+. Similarly, Medzihradsky et al. (1967) also showed that Lubrol WX treatment produces a considerably higher recovery of the enzyme than does other detergents, although the solubilized enzyme loses its activity very rapidly in the absence of monovalent ions. The observation that the treatment of brain membranes with Lubrol seems to yield a solubilized enzyme containing relatively lower specific activity than the original brain homogenates indicates that either the detergent did not "purify" the enzyme, or that the detergent may have an inhibitory effect on the enzyme activity (Winter, 1972). Banerjee et al. (1970) reported that Triton X-100 is superior to Lubrol for solubilizing the renal Na^+/K^+-ATPase. The Triton X-100 treatment results in the extraction of enzyme with a higher specific activity and ouabain sensitivity than that treated with Lubrol. Variation in treatment conditions, with respect to the tissue origin, is very important in the solubilization of Na^+/K^+-ATPase from the membrane structures. One important criterion for determining whether the detergent treatment produces an increase in enzyme specific activity is to measure concomitantly whether there is an increase in partial reactions catalyzed by the Na^+/K^+-ATPase (Schwartz et al., 1975). These reactions include sodium-dependent phosphorylation, a sodium-dependent ATP–ADP exchange reaction, and the rate and maximum amount of ouabain bound. By using these criteria, one can be more certain about the increase in Na^+/K^+-ATPase specific activity during the enzyme purification.

Isolation and purification of brain Na^+/K^+-ATPase was carried out during the late 1960s by Hokin's group (Hokin, 1973). A series of publications detailed how they were able to solubilize, isolate, and partially purify Na^+/K^+-ATPase from bovine brain microsomal fractions (Medzihardsky et al., 1967; Kahlenberg et al., 1968, 1969; Uesugi et al., 1969, 1971). Procedures employed by Hokin's laboratory to isolate and purify brain Na^+/K^+-ATPase were described in a report by Uesugi et al. (1971). Bovine brain cortex microsomal fraction was treated with NaI and was frozen at a concentration of 10 mg/mL of protein in Tris-neutralized water. The frozen NaI-treated microsomal preparation was thawed and treated subsequently with Lubrol to a final composition of 20 mg/mL of protein, 20 mg/mL Lubrol, 1.25 m*M* sodium ATP (pH 7.0), and 0.5 m*M* sodium EDTA (pH 7.0). The mixture was homogenized and incubated at 0–4°C for 30 min. Following readjustment of the pH value to 6.0, the mixture was centrifuged. The supernatant was then subjected to a sucrose linear gradient zonal centrifugation. Fractions with high Na^+/K^+-ATPase activity were pooled and concentrated by ultrafiltration (under 40 to 45 psi of pressure) and subsequently concentrated to a final volume of 10–20 mL under 50 psi pressure. This fraction was designated as zonal enzyme (Uesugi et al., 1971). The zonal enzyme was further treated with 0.4*M* $(NH_4)_2SO_4$ at 30°C for 8–10 min. The pellet from a 40,000 rpm, 60 -min centrifugation was collected and resuspended in a solution containing 1 m*M* sodium EDTA and 1 m*M* sodium ATP. The enzyme fraction thus prepared was treated with Lubrol again with a Lubrol:protein ratio of 20:1 by weight. This transparent pellet was collected following a 40,000 rpm centrifugation in a Spinco No. 40 rotor. The procedure purified the enzyme by 30–50-fold over that in the starting microsomes and the enzyme was estimated to be 50% pure (Uesugi et al., 1971). Na^+/K^+-ATPase was found to be inactivated during the purification, i.e., a 35% inactivation occurred on extraction with the concentrations of Lubrol used and a 15% inactivation occurred in the ultrafiltration process. SDS-PAGE was applied to determine the molecular weight of the isolated

Na^+/K^+-ATPase from bovine brain. An estimated molecular weight of 94,000 was obtained by prelabeling the enzyme using γ-^{32}P-ATP. Negative staining of the enzyme preparation revealed that it may be in a rod-shaped form (most of the rods were 55–60 Å in diameter) and also revealed many aggregated forms in longer chains, linear aggregates, sheets, and ring-like structures. Uesugi et al. concluded from this series of experiments that Na^+/K^+-ATPase of brain consists of a principal 60 Å subunit with an estimated molecular weight of 94,000 δ for the phosphorylated subunit.

Recently, Sweadner (1979) reported that canine brain gray matter may contain two distinct molecular forms of Na^+/K^+-ATPase. These molecular forms can be resolved by gel electrophoresis in SDS gel, and can be identified by sodium-dependent, potassium-sensitive phosphorylation by γ-^{32}P-ATP. Canine brain gray matter was homogenized in 0.315*M* sucrose and 1 m*M* EDTA solution. Following centrifugation to remove nuclei and debris, it was centrifuged again at 8500*g* to yield a fraction free of mitochondria and myelin fragments. Microsomes were then isolated from this preparation and resuspended in sucrose/EDTA solution (pH 7.0). Partial purification of Na^+/K^+-ATPase was achieved by gentle extraction and SDS treatment using a modification of procedures of Jorgensen (Laemmli, 1970; Ames, 1974). The ratio of SDS to protein was set at 0.25–0.3 mg of SDS/mg of protein. SDS-PAGE showed that there are two bands associated with brain Na^+/K^+-ATPase, whereas there is only one band associated with the enzyme isolated from renal medulla, when the isolated preparation is reacted with γ-^{32}P-ATP. Once again, two closely spaced bands at molecular weights of approximately 90,000 were resolved. The higher molecular weight form α(+) has a slower mobility, equivalent to a difference in molecular weight of approximately 2000 δ. This higher molecular weight form of Na^+/K^+-ATPase was found in dog, mouse, rat, and frog brain, and eel electric organ, although there are some species-specific differences in the apparent relative amounts of the α(+) and α forms. Sweadner (1979) reported that these two forms of Na^+/K^+-ATPase have detectable structural and kinetic differences. They have different sensitivities to trypsin digestion and to Cu^{2+}-*O*-phenanthroline-catalyzed cross-linking, and differ in the number of sulfhydryl groups and in their affinities to the inhibitor strophanthidin. The possibility that α(+) and α forms of Na^+/K^+-ATPase represent isoenzymes was discussed. Sweadner (1978) suggested that they are the products of different genes. However, she cautioned that a more rigorous structural analysis of the purified forms of α(+) and α is required to prove conclusively whether the two molecular forms are isozymes.

Plasma membranes from brain gray matter show two types of Na^+/K^+-ATPase, whereas the plasma membranes from white matter contain only one type of Na^+/K^+-ATPase (Petrali and Sulakhe, 1980). The types of Na^+/K^+-ATPase seem to appear at different stages of postnatal development. When rat brain was obtained from young animals (100 g body weight), there was only one type of Na^+/K^+-ATPase identified in the gray matter. Petrali and Sulakhe suggested that since, in the rat brain, neuronal mass attains its maximum value when the animals are approximately 1 mo old, whereas the nonneuronal cells continue to grow in number for a much longer period, the second type of Na^+/K^+-ATPase originates from nonneuronal elements present in the gray matter. Although they suggested that the two types of Na^+/K^+-ATPase are localized in different neuronal and nonneuronal elements, the question of whether these two types of enzymes represent isozymes still remains unanswered.

The isolation and purification of Na^+/K^+-ATPase has been achieved from various sources of tissues, including kidney, brain, electric organ, and rectal gland (Kyte, 1971; Uesugi et al., 1971; Albers, 1967; Hokin et al., 1973). Molecular weight determinations have indicated that the enzyme may consist of two predominant proteins, as revealed by SDS gel electrophoresis, the larger one having an apparent molecular weight of 90,000 δ and the smaller one a molecular weight of approximately 47–55,000 δ (Kyte, 1971; Uesugi et al., 1971; Sweadner, 1977). Regardless of the tissue origin, Na^+/K^+-ATPase seems to consist of two major proteins (peptides), and the molar ratio of these two proteins is approximately 1.7 to 1 (Kyte, 1972). Using γ-^{32}P-ATP as a substrate, it has been demonstrated that the larger of these two proteins is phosphorylated during the course of reaction (Alexander and Rodnight, 1970; Uesugi et al., 1971; Kyte, 1971; Avruch and Fairbanks, 1972; Collins and Albers, 1972; Hart and Titus, 1973a,b; Sweadner, 1977). An analysis of the amino acid composition of the phosphorylated protein showed a high content of nonpolar amino acids. The N-terminal amino acid was found to be Ser in the enzyme prepared from electric organ of *E. electricus* (Dahl and Hokin, 1974), whereas Gly was the N-terminal amino acid in the enzyme prepared from dog kidney preparation (Kyte, 1972). The smaller protein of Na^+/K^+-ATPase appears to be a sialoglycoprotein (Kyte, 1972; Hokin et al., 1973) and has a molecular weight of 55,000 based on SDS-PAGE. This protein is also rich in hydrophobic amino acids and contains glucosamine, galactosamine, glucose, galactose, mannose, fucose, and sialic acid (Kyte, 1972; Dahl and Hokin, 1974); Ala is the N-terminal amino acid.

There are discussions regarding the function of the smaller protein (glycoprotein) component of Na^+/K^+-ATPase (Dahl and Hokin, 1974; Schwartz et al., 1975). Dahl and Hokin (1974) raised a very interesting question: Is the glycoprotein a true component of the native Na^+/K^+-ATPase? They reasoned that although Na^+/K^+-ATPase always contains glycoprotein when it is isolated and purified, the glycoprotein may be bound to Na^+/K^+-ATPase during the isolation and purification process. They suggested that if the molar ratio between the phosphorylated "subunit" to glycoprotein subunit is constant in pure enzyme from different species, this may be evidence that the glycoprotein is a true component of Na^+/K^+-ATPase. Schwartz et al. (1975) proposed that a reconstitution of the isolated components of the enzyme and the selection of the minimal requirements for Na^+/K^+-ATPase activity will provide the evidence needed to establish whether glycoprotein is a true component of Na^+/K^+-ATPase. Dahl and Hokin (1974) compared this ratio for the purified enzyme from various species and have obtained values from 2.16 to 1. Although the results are not consistent, they speculated that they did indicate the minimum molecular weight of the enzyme to be approximately 250,000, in agreement with the estimated molecular weight. Kyte (1971) estimated a mass ratio of 1.7:1 for the enzyme isolated from dog kidney. The stoichiometric ratio suggested a basic component of one large protein and two small proteins. Recent findings (Jorgensen, 1974; Kyte, 1975; Giotta, 1975; Liang and Winter, 1977; Sweadner, 1977; Huang and Askari, 1978, 1979; Askari and Huang, 1980) concerning the molecular structure of the enzyme, made by using different high-affinity ligands or by cross-linking studies, suggested that Na^+/K^+-ATPase is present in a form that contains two or more of each subunit in the native state. Askari and Huang (1981) indicated that the minimum structure of the enzyme is α-tetramer, which exhibits ligand-induced negative cooperativity across two intersubunit domains. However, a more recent study reported by Ottolenghi and Ellory (1983), using a radiation inactivation technique, demonstrated that the membrane-bound Na^+/K^+-ATPase consists of two α- and two β-peptides (a dimeric structure), and that there are subdivisions of each α-peptide by a series of domains. This study is particularly interesting in that it shows the possibility of the existence of different domains of Na^+/K^+-ATPase. However, the authors cautioned that the result may depend on the decay constants chosen. The model they presented may be reinterpreted in terms of an αβ-monomer or possibly an α-peptide alone. Therefore, it is not possible at the present time to conclude whether the glycoprotein is a true component of Na^+/K^+-ATPase.

2.2. Catalytic Properties

The primary role of the enzyme Na^+/K^+-ATPase is to transport cations in and out across the cell membranes. There has been a long debate regarding the mechanism(s) of this phenomenon. It is clear that ATP is necessary to supply the energy that is required by the enzyme to carry out its function. A number of studies using fluorescent probes to label ATP binding sites (Karlish, 1980; Moczydlowski et al., 1981a; Skou et al., 1981) have indicated that ATP is able to bind to the protein at specific sites. Using ATP analogs, Moczydlowski et al. (1981b) were able to show the active site for Na^+/K^+-ATPase and the binding-induced conformational change of the enzyme. They also proposed the stoichiometric reaction of the enzyme. Early findings (Post et al., 1965; Fahn et al., 1966a) were in favor of the idea that Na^+/K^+-ATPase catalyzed the reaction following a stepwise linear reaction sequence that is typified by an initial reaction of the enzyme with Na^+, ATP, and Mg^{2+} to form an activated phosphoenzyme (Na^+-E_1-P). Following the release of Na^+, the enzyme undergoes a conformational change to E_2-P, which reacts with a K^+ to yield an intermediate K^+-E_2-P. Finally, the intermediate enzyme undergoes a further change to release K^+, inorganic phosphate (Pi), and the original form of enzyme. This scheme has received wide acceptance because it fits in well with many other observations. However, a cyclic scheme of Na^+/K^+-ATPase catabolism has also been proposed to explain the presence of another pathway that is characterized by inhibition of Na^+/K^+-ATPase by Na^+ in simultaneous presence of Mg^{2+} and Pi (Whittam, 1962; Post et al., 1969; Tobin and Sen, 1970). There are a number of critical issues that remain unanswered, i.e., binding sites for Na^+, K^+; the binding and transporting sequence for Na^+ and K^+; and the enzyme conformational states (Whittam and Chipperfield, 1975; Dahl and Hokin, 1974). The evidence against the stepwise mechanism for the Na^+/K^+-ATPase reaction has been (a) a duality of K^+ effect on the formation of intermediates (Kanazawa et al., 1967, 1970; Fukushima and Tonomura, 1973); (b) the covalent binding of Pi to the enzyme by reversing the last step of the stepwise reaction mechanism (Lindenmeyer et al., 1968; Sen et al., 1969; Siegel et al., 1969); (c) binding of ouabain to E_2 conformations (Schwartz et al., 1968; Post et al., 1969; Tobin and Sen, 1970); and

(d) the pH dependence of ATP binding, ATP hydrolysis, and phosphorylation of the enzyme (Shamoo and Brodsky, 1972; Brodsky and Shamoo, 1973; Brodsky and Sohn, 1974).

Recently, Yamaguchi and Tonomura (1980a) showed that there are two kinds of cation binding sites on the Na^+/K^+-ATPase molecule. One of the sites is the Na^+-binding sites (3 mol/mol of active site) and the other the K^+-binding site (2 mol/mol of active site). They demonstrated that Na^+ binds to the Na^+-binding site cooperatively (Hill coefficient, 2.5–3) and the apparent dissociation constant is 0.2–0.32 m*M*. On these Na^+-binding sites, 1 mol K^+ can displace 3 mol of Na^+. Rb^+ or Na^+ is bound to the sites cooperatively (Hill coefficient, 1.5–2.0). The dissociation constants for K^+-binding sites are 0.044, 0.024, and 2.2 m*M* for K^+, Rb^+, and Na^+, respectively. To complicate the issue further, Yamaguchi and Tonomura (1980b) discovered that there are two kinds of ATP binding sites per Na^+/K^+-ATPase molecule. These two ATP binding sites represent a catalytic site (1 mol/mol phosphorylation site) with an apparent dissociation constant for ATP of 1 μ *M* and a regulatory site with a dissociation constant for ATP of greater than 0.2 m*M*. It is interesting to note that they also showed that Na^+ has no effect on the affinity of ATP for the enzyme, whereas the addition of K^+ to the incubation mixture markedly reduces the binding affinity of ATP for the catalytic site. This is consistent with the observation that ATP does not alter the affinity of the enzyme for Na^+. The binding of Na^+ to the enzyme is affected by the simultaneous presence of K^+ and Na^+. In the absence of K^+, Na^+ binds very rapidly to the Na^+ binding sites (Yamaguchi and Tonomura, 1979), whereas in the presence of K^+, Na^+ binds very slowly to the enzyme; the binding of Na^+ is markedly accelerated by the addition of ATP. This finding suggests that ATP binding to the catalytic site is not affected by K^+, nor is the binding of K^+ to K^+-binding sites affected by ATP. The binding of Na^+ to Na^+-binding sites is under the influence of the regulatory site when the enzyme is in a K^+ form (K^+-E-P state); in other words, the enzyme transition from a K^+-bound form to a Na^+-bound form is caused by the addition of ATP to a regulatory site. This finding seems to support the basic cyclic scheme of a Na^+/K^+-ATPase reaction, with the exception that native enzyme may react with K^+ and ATP to form an intermediate enzyme. Upon the presentation of the Na^+ and additional ATP at the regulatory site, the intermediate enzyme is then further transformed into a Na^+-bound form of the enzyme intermediate that undergoes conformational changes. The discovery of an ADP-insensitive phosphorylated enzyme state further strengthens the notion that the cyclic Na^+/K^+-ATPase reaction scheme is a workable model (Yamaguchi and Tonomura, 1980c). The proposed intermediate state of the enzyme Na^+/K^+-ATPase has been identified recently by Froehlich et al. (1976) and Cantley and Josephson (1976). Using a rapid mixing technique, Froehlich et al. (1976) were able to demonstrate that Na^+/K^+-ATPase is phosphorylated to reach a constant level at a rate that is dependent on ATP concentration and the presence of Na^+ and absence of K^+. The inorganic phosphate production lags during the initial phase of phosphorylation and then accumulates at a constant rate. In the presence of both Na^+ and K^+, the inorganic phosphate production can be resolved into an initial "burst" state and late steady-state reaction. The initial "burst" state of Pi production coincides with a transient decay in the level of phosphoenzyme. This observation suggests that K^+-activated turnover of the acid-stable phosphorylated enzyme gives rise to an acid-labile intermediate complex, and the breakdown of this acid-labile complex is enhanced by ATP. Although the detailed steps, intermediates, conformational states, and ion requirements of the overall Na^+/K^+-ATPase reaction have not been identified, it is clear that the reaction requires the simultaneous presence of Na^+ and K^+, and that the reaction involves the formation of phosphoenzyme intermediate. The breakdown of such phosphoenzyme intermediates gives rise to the production of Pi. With this evidence obtained by Yamaguchi and Tonomura (1980a–c), it seems reasonable to suggest that the cyclic scheme of Na^+/K^+-ATPase reaction is a suitable model to explain the catalytic properties of Na^+/K^+-ATPase.

2.3. Inhibitors and Stimulators: Na^+/K^+-ATPase Modifying Agents

2.3.1. Sulfhydryl Reagents

Na^+/K^+-ATPase is sensitive to inhibition by a variety of sulfhydryl reagents, which includes the organic mercurial diuretics, maleimide derivatives, arsenicals, ethacrynic acid, chlorpromazine, and phenothiazines. The sulfhydryl reagents inhibit Na^+/K^+-ATPase; however, the inhibition may be mediated by various mechanisms of action. *N*-ethylmaleimide (NEM) and *N*-butylmaleimide (NBM) appear to act in the same manner. Fahn et al. (1966b) found that NEM enhanced the Na^+-dependent ADP–ATP exchange in short periods of exposure of the enzyme to NEM, whereas when a long time exposure was used, NEM inhibited ADP–ATP exchange as well as Na^+-dependent phosphorylation of the enzyme molecule. The enzyme can be protected from inhibition by ATP. Post et al. (1969) determined that NEM treatment prevented conversion of E_1-P. Yamaguchi and Tonomura (1980c) indicated that 80% of all EP (phosphorylated enzyme) formed with the NEM-treated enzyme is in an ADP-sensitive form that shows little change in the affinities of the Na^+ and K^+ binding sites for monovalent cations. This is in direct contrast with the native enzyme (without prior treatment of NEM), which shows marked changes in the affinities to cations in an E-P state. Hart and Titus (1973a,b) showed that in the absence of Na^+, Mg^{2+}, K^+, or ATP, a 98 kdalton peptide containing sulfhydryl groups is protected from NEM by sodium plus ATP. In the presence of Mg^{2+}, the sulfhydryl groups on the same peptide were protected by K^+ or ouabain. However, when a phosphorylated state was achieved by the simultaneous presence of Na^+, Mg^{2+}, and ATP, additional sulfhydryl groups were exposed to the alkylation with NEM, suggesting that sulfhydryl groups can be revealed at different conformational states of Na^+/K^+-ATPase and that NEM may inhibit the enzyme reaction by interfering with different conformational states.

The organic mercurial diuretics meralluride and mercaptomerin (10^{-6}–$10^{-5}M$) markedly inhibit Na^+/K^+-ATPase; the nondiuretic mercurials *p*-chloromercuribenzoate (PCMB) and *p*-chloromercuriphenylsulfonate are effective inhibitors of Na^+/K^+-ATPase in vitro (Jones et al., 1965).

Ethacrynic acid is a relatively weak inhibitor of Na^+/K^+-ATPase activity. The inhibition of the enzyme has been proposed to be caused by the ability of ethacrynic acid to react with cysteine-sulfhydryl groups on the enzyme (Duggan and Noll, 1965; Banerjee and Sen, 1969). The potency of ethacyrnic acid in inhibiting Na^+/K^+-ATPase activity in vitro is affected by K^+ concentration (Banerjee et al. 1970). When K^+ concentration is low, the inhibitory potency of ethacrynic acid appears to increase. A mechanism for ethacrynic acid inhibition of Na^+/K^+-ATPase has been proposed: The compound apparently increases the stability of the phosphorylated intermediate and decreases the hydrolysis of this intermediate in the presence of K^+ (Banerjee et al., 1970). Inhibition of Na^+/K^+-ATPase is correlated with a decrease in Na^+-dependent dephosphorylation of the intermediate enzyme molecule.

Arsenite (Siegel and Albers, 1967) and oxophenarsine (Skou and Hilberg, 1965) are inhibitors of Na^+/K^+-ATPase, suggesting that the enzyme may have two closely associated sulfhydryl groups (Johnstone, 1963). The inhibitory effect of arsenite is enhanced in the presence of 2,3-dimercapto-1-propanol (Siegel and Albers, 1967). It has been suggested that the reaction site sulfhydryl groups on the enzyme become more accessible to an arsenite–2,3-dimercapto-1-propanol complex. However, the possibility exists that 2,3-dimercapto-1-propanol may expose more vital sulfhydryl groups that react with arsenite. It appears that arsenite may inhibit Na^+/K^+-ATPase activity via a mechanism similar to that of NEM, namely an increase in Na^+-dependent ADP–ATP exchange and an increase in the enzyme affinity for Na^+.

Chlorpromazine and other phenothiazines inhibit brain Na^+/K^+-ATPase and cation transport in this tissue (Chan and Quastel, 1970). The inhibition of Na^+/K^+-ATPase by chlorpromazine is mediated by the interaction of the chlorpromazine free radical with the sulfhydryl groups of the enzyme. The generation of the chlorpormazine free radical, as the consequence of exposure of the compound to UV light irradiation, markedly increases the ability of chlorpromazine to inhibit Na^+/K^+-ATPase (Akera and Brody, 1968, 1969, 1970). The inhibition of Na^+/K^+-ATPase by the free radicals is accompanied by a decrease in the number of free sulfhydryl groups. Complete inactivation occurs with the loss of 75% of the free sulfhydryl groups. The inhibition by chlorpromazine free radical is reversible in the presence of cysteine or 2,3-dimercapto-1-propanol (Akera and Brody, 1969). Chlorpromazine concentrations in the range of 0.1–0.5 m*M* inhibit approximately 50% of the enzyme activity (Robinson et al., 1968). The potency of chlorpromazine is enhanced when the drug is incubated with the enzyme preparation for 30 min at 37°C in the absence of Na^+ and K^+ (Squires, 1965). The monodesmethyl metabolites of chlorpromazine and promazine are more effective than the parent drugs at inhibiting Na^+/K^+-ATPase (Hackenberg and Krieglstein, 1972).

2.3.2. Cardioactive Steroid Inhibitors and Erythrophleum Alkaloids

Inhibition of Na^+/K^+-ATPase by cardioactive steroids has been the subject of intense study for over two decades. Cardioactive steroids and their glycosides, i.e., 5β,14β-card-20 (22)-enolide, 5β,14β-bufa, 20,22-dienolide, digitoxigenin (3β,14β-dihydroxy-5βk-card-20 (22)-enolide), digitoxin (digitoxigen-3β-tridigitoxoside), ouabagenin, and hellebrigenin are all naturally occurring steroids. This group of compounds is generally classified as (a) cardenolide derivatives, which have a five-member unsaturated lactone, and (b) bufadienolides, which possess a six-member doubly unsaturated lactone at the 17β position of the steroid nucleus.

Derivatives of either class that possess 14β- and 3βk-hydroxyl groups of a 3β-glycosyl substituent exhibit Na^+/K^+-ATPase inhibitory activity. The potency of these compounds as Na^+/K^+-ATPase inhibitors is dependent upon the structural modification of the basic molecule by various hydroxyl, acetoxy, and aldehyde functional groups (Yoda and Hokin, 1970). In general, the carbohydrate derivatives are more potent inhibitors than the glycone derivatives (Yoda and Hokin, 1970), and bufodienolide derivatives are more potent inhibitors of Na^+/K^+-ATPase than the corresponding cardenolides. For example, hellebrigenin is more potent than strophanthidin (a cardenolide). Furthermore, modification of the cardioactive steroid structure by saturating the 17β-lactone group results in diminished inhibitory activity (Wilson et al., 1970). The concentrations that are required to inhibit 50% of the enzyme activity by the cardioactive steroids are in the range of 10^{-7}–$10^{-6}M$. However, the inhibition of the enzyme varies with the incubation conditions, temperature, tissue preparation, and presence of ATP and/or cations (Sen et al., 1969).

Structure–activity relationship studies have indicated that these compounds possess a very rigid nonpolar concavity that exists on the surface as a result of the *cis* configuration of the ring (Dahl and Hokin, 1974). It has been suggested that the compounds interact with a complementary surface on the enzyme by hydrophobic interactions that may involve the nonpolar concavity of the steroid (Portius and Repke, 1964). This notion is supported by the observation that the bound steroid can be dissociated from the enzyme with heat or acid treatment of the steroid–enzyme complex and by the finding that the rate of inhibition is decreased by an increase in the number of hydroxyl and sugar substituent groups (Albers et al., 1968).

Ouabain, a representative of the cardioactive steroid derivatives, has been used widely to study the mechanisms of inhibition of Na^+/K^+-ATPase by cardioactive steroids. It has been observed that ouabain increases the steady-state level of ^{32}P-labeled enzyme when the incubation is carried out in the presence of Mg^{2+}, Na^+, K^+, and γ-^{32}P-ATP (Charnock et al., 1963; Whittam et al., 1964; Ahmed and Judah, 1965; Nagano et al., 1965). These findings suggested that ouabain interacts with the phosphoenzyme, resulting in a stable enzyme intermediate even in the presence of K^+. The K^+-dependent phosphate activity (part of Na^+/K^+-ATPase activity) is also inhibited by ouabain, and the finding that K^+ competes with and reverses the enzyme inhibition by ouabain further demonstrates that the K^+-dependent step of Na^+/K^+-ATPase reaction is the site sensitive to cardioactive steroid. The apparent antagonism between K^+ and low concentrations of cardioactive steroids has led many investigators to suggest that these inhibitors react reversibly with Na^+/K^+-ATPase (Glynn, 1957; Ahmed and Judah, 1964; Ahmed et al. 1966; Dunham and Glynn, 1961). However, Asano et al. (1970) were not able to reverse the inhibition of Na^+/K^+-ATPase induced by digitoxin and ouabain.

Ouabain inhibition of Na^+/K^+-ATPase occurs most rapidly in an incubation medium containing Na^+, Mg^{2+}, and ATP and lacking K^+. In order to elucidate the inhibition mechanism(s), Sen et al. (1969) derived a study to investigate the interaction of ouabain with many phosphoenzyme intermediates. They concluded that the rapid interaction of the enzyme and inhibitor is caused by the formation of a phosphorylated intermediate. Mg^{2+} enhances the rate of ouabain inhibition of Na^+/K^+-ATPase in the absence of ATP (Albers et al., 1968; Sen et al., 1969), and the inhibition under this reaction condition can be blocked by high Na^+ concentrations. Inorganic phosphate promotes the rate of Mg^{2+}-dependent inactivation of the enzyme in the absence of ATP, K^+, and Na^+, whereas, in the absence of Mg^{2+}, ouabain does not affect Na^+/K^+-ATPase activity. Putting together all these observations, Sen et al. (1969) concluded that the inhibition of the enzyme is associated with a decrease in Na^+-dependent phosphorylation of the enzyme. Both the Na^+-dependent phosphorylation of the enzyme, occurring either in the presence or absence of ouabain, and the ouabain-induced Pi incorporation seem to yield chemically identical phosphorylated proteins. Enzymic or basic hydrolysis of these labeled phosphoproteins gives rise to the labeled peptides identical in their chromatographic properties (Lindenmayer et al., 1968; Post et al., 1969; Siegel et al., 1969).

There have been consistent findings indicating that when the steroid–enzyme complex is formed in the presence of Na^+, Mg^{2+}, and ATP, the dissociation and association rates of the complex are markedly reduced by K^+; when the complex formation occurs in the presence of Mg^{2+} and Pi, both K^+ and Na^+ decrease the rate of association and are essentially without effect on the rate of dissociation (Lindenmayer and Schwartz, 1970; Akera and Brody, 1971; Skou, 1971; Van Winkle et al., 1972).

Radioactively labeled ouabain has been used to investigate the binding sites on Na^+/K^+-ATPase. In addition to a radioactive labeling procedure, a combination of photoaffinity and radiolabeling techniques has revealed that there are several ouabain binding sites on the Na^+/K^+-ATPase molecule. Hansen (1976) reported that ^3H-g-strophanthin (ouabain) binds to Na^+/K^+-ATPase from ox brain. The binding is dependent on inorganic phosphate and Mg^{2+} concentration. Scatchard analysis of the binding data yields a curved line, indicating nonuniform populations of ouabain binding sites. Interestingly, the nonlinear curve can be converted to a linear curve by adding K^+. Hansen concluded that there may be two or more populations of "enzymes" with different affinities for substrates and ligands affecting g-strophanthin binding. The high affinity sites may be less dependent on Pi for binding. The addition of K^+ causes a change in the enzyme's configuration, with a resultant homogeneous binding site for g-strophanthin. Ruoho and Kyte (1974), by using an ethyl-diazomalonyl derivative of cymarin, were able to photolabel Na^+/K^+-ATPase purified from canine renal medulla. A covalent bond is formed between the photoaffinity ligand and the enzyme. The formation of such a bond results in an inactivation of the enzyme.

The ouabain sites are accessible from the outside surface of the plasma membrane, whereas the phosphorylation site is accessible from the inside surface, and both sites are localized on the larger polypeptide subunit of Na^+/K^+-ATPase. These facts have led to the suggestion that the large subunit has its sequence exposed to both sides of the membrane. However, Ruoho and Kyte (1974) did not rule out the possibility that the ligand may bind to a small peptide subunit. The molecular weight of the large chain of Na^+/K^+-ATPase is 130,000 in a folded sphere configuration of approximately 75 Å, a distance long enough to span the plasma membrane. Recently, Collins et al. (1982) examined the photoaffinity labeling of highly purified lamb kidney Na^+/K^+-ATPase using tritiated 2-nitro-5-azidobenzoyl derivatives of ouabain (NAB-ouabain) as ligands. The technique revealed that two proteolipid (γ^1 and γ^2) fractions were labeled by the ligand. Furthermore, it appeared that γ^1 and γ^2 are interchangeable, as estimated by column chromatography. Since the total yield of γ component is consistent with the recovery of the catalytic subunit, they suggested that this component is present in a stoichiometric amount in Na^+/K^+-ATPase. N-terminal analysis by the dansyl method has identified Arg as the N-terminal amino acid of the γ-component. Ouabain binds to a catalytic subunit (molecular weight, 100,000) and to a small proteolipid (molecular weight, 10,000). Racker (1976) suggested that the proteolipid may form an ion channel through the cell membrane. Inhibition of Na^+/K^+-ATPase activity by ouabain correlates to the binding of ouabain to both the subunit and the γ-proteolipid component of the enzyme.

2.3.3. Metal Ions

Na^+/K^+-ATPase activity is activated by Na^+ and K^+ ions. Maximal activation of the enzyme by Na^+ and K^+ requires an optimal ratio between these two monovalent cations. The ratio giving maximal activation of the enzyme is about 5 or 10:1 (Glick, 1972). A large excess of either K^+ or Na^+ will result in an inhibition of enzyme activity because of cross-competition between these two cations for their respective sites (Skou, 1960, 1962; Ahmed et al., 1966). The requirement for Na^+ is absolute, in that the presence of

Na^+ is essential for the activation of the enzyme. K^+ ion can be substituted by various monovalent cations, including Rb^+, Cs^+, Li^+, and Tl^+ (Skou and Hilberg, 1965; Britten and Blank, 1968; Skulskii et al., 1973). The efficiency of these cations in replacing K^+ shows a descending order of effectiveness: $Tl^+ > K^+ > Rb^+ > NH_4{}^+ > Cs^+ > Li^+$ (Dahl and Hokin, 1974). The best demonstration of the cross-competition between K^+ and Na^+ ions was provided by Whittam (1962) using a resealed erythrocyte ghost. In this preparation, the ATP hydrolysis that is activated by intracellular Na^+ and is associated with active cation transport is relatively unaffected by increasing intracellular K^+ concentration over the range of 25–135 m*M*. However, Na^+ inhibits ATP hydrolysis and cation transport in an extracellular medium containing low K^+, suggesting that Na^+ competes with K^+ sites on the enzyme Na^+/K^+-ATPase. Mg^{2+} is essential for Na^+/K^+-ATPase enzymatic activity. The optimal concentrations of Mg^{2+} and ATP for the enzyme activity appear to be interrelated (Skou, 1960), with maximal enzyme activity observed at a Mg^{2+} : ATP ratio of 1 : 1 or 2 : 1. A large excess of either Mg^{2+} or ATP decreases the enzyme activity. Many divalent cations can substitute for Mg^{2+}, but with lower efficiency. For example, Mn^{2+} and Co^{2+} have been used to replace Mg^{2+} in the enzyme preparation, but these two cations exhibit only one tenth of the efficiency of Mg^{2+} (Atkinson et al., 1968; Atkinson and Lowe, 1972; Rendi and Uhr, 1964). Hexum (1974) showed that Cu^{2+}, Zn^{2+}, Fe^{2+}, and Pb^{2+} are potent inhibitors of Na^+/K^+-ATPase prepared from rat brain. The concentrations of divalent cations that exhibited 50% inhibition of the NaI-washed brain Na^+/K^+-ATPase were 1 m*M* (Zn^{2+} and Cu^{2+}), 3 m*M* (Fe^{2+}), and 20 m*M* (Pb^{2+}). Kinetic analysis of the inhibitory mechanism indicated that Pb^{2+} and Fe^{2+} inhibit the enzyme competitively with a K_i value of approximately 2 m*M*, whereas Zn^{2+} and Cu^{2+} inhibit the enzyme by a noncompetitive interaction, showing K_i values of 3.5 m*M* and 1.2 m*M*, respectively. Both Zn^{2+} and Cu^{2+} form complexes with ATP. Since the inhibitory effect of Cu^{2+} is reduced by preincubation in media containing ATP but not enzyme, it is likely that divalent metal ions interact with the enzyme at a site other than the substrate site (Hexum, 1974). Other studies (Ting-Beall et al., 1973) showed that Cu^{2+} inhibits Na^+/K^+-ATPase with both reversible and irreversible actions. At higher concentrations (>6 μM), Cu^{2+} inactivates the enzyme irreversibly, but at lower concentrations (<1.4 μM), it interacts with the enzyme reversibly. The mechanism of Cu^{2+} inhibition of Na^+/K^+-ATPase appears to involve SH-groups because SH-containing compounds could reverse the inhibition (Peters et al., 1966).

Pb^{2+} is a reversible inhibitor of Na^+/K^+-ATPase (Hexum, 1974;

Siegel and Fogt, 1977). The inhibition can be prevented by incubation of the enzyme with EDTA, EGTA, penicillamine, thiols, or ATP (Siegel and Fogt, 1977). The inhibition of Na^+/K^+-ATPase by Pb^{2+} appears to be dependent on the availability of ATP, which forms $PbATP^{2-}$. Treatment of the enzyme with Na^+, K^+, Mg^{2+}, and ATP prior to addition of Pb^{2+} does not alter the extent of subsequent inhibition, indicating that the Pb^{2+} ion interacts with the enzyme directly. The functional groups that may interact with Pb^{2+} have been proposed to be sulfhydryl (Klotz et al., 1952; Madsen, 1963), carboxylate (Gurd and Murray, 1954), and imidazole (Tanford, 1952) groups. Siegel and Fogt (1977) suggested that Pb^{2+} interacts with Na^+/K^+-ATPase to form a modified form of Pb–enzyme complex (Pb-E_1) that cannot undergo the conformational change to the E_2 form that would be necessary for hydrolytic activity.

Purified Na^+/K^+-ATPase from rabbit kidney contains metals, i.e., Al, Ca, Cu, Fe, Mg, Ni, Ag, Na, Si, Be, Cd, Cr, Co, Pb, Mn, Sr, Sn, Zn, and Hg. Among the metals detected in this preparation, Na and Si have the highest molar ratio of metal/enzyme (Specht and Robinson, 1973). Other metal ions inhibit Na^+/K^+-ATPase activity with varying potency. Al^{3+} inhibits the enzyme, having an I_{50} value of 3 m*M*, whereas Ni^{2+}, Ca^{2+}, Cd^{2+}, and Mg^{2+} inhibit the enzyme at an I_{50} value of 0.1 m*M*, 1 m*M*, 0.4 m*M*, and 0.0004 m*M*, respectively (Specht and Robinson, 1973). Although the mechanism for trace metal inhibition of Na^+/K^+-ATPase is not known, the data seem to suggest that these metal ions inhibit the enzyme by influencing stages late in the reaction sequence. Divalent cations that inhibit Na^+/K^+-ATPase activity may be influenced by Na^+ and K^+ concentrations in the medium (Ting-Beall et al., 1973). However, Siegel and Fogt (1977) showed that Pb^{2+} inhibition of the enzyme is not affected by Na^+ or K^+. Preincubation of Na^+/K^+-ATPase with $BeCl_2$ in the absence of ATP results in a marked inhibition of enzyme activity, and the inhibition is dependent on Mg^{2+}, Na^+, and K^+ (Toda, 1968). The I_{50} value for Be^{2+} to inhibit Na^+/K^+-ATPase activity in the presence of Mg^{2+}, Na^+, and K^+ is 1.8 μ *M*. In the presence of Mg^{2+}, increasing K^+ concentrations enhanced the rate of inhibition of the enzyme. In this condition, NH_4^+ or Rb^+ can be substituted for K^+. Na^+ can reverse the inhibition by Be^{2+}, suggesting that Be^{2+} might bind to the Na^+-specific site of Na^+/K^+-ATPase molecule. An alternative explanation has been offered: Be^{2+} may inhibit the change in conformational states of the enzyme (Toda, 1968).

Submillimolar concentrations of Ca^{2+} are known to inhibit Na^+/K^+-ATPase (Dunham and Glynn, 1961; Epstein and Whittam, 1966) and this inhibition can be reversed by increasing the Mg^{2+}

concentration in the medium (Skou, 1957, 1960). Epstein and Whittam (1966) suggested that a Ca^{2+}-ATP complex formed may be the actual inhibitor of Na^+/K^+-ATPase; this Ca^{2+}-ATP will bind to the enzyme-substrate sites that normally bind Mg^{2+}-ATP. Hoffman (1962) showed that the site at which Ca^{2+} exerts its effect is intracellularly oriented. Subsequently, Atkinson and Lowe (1972) confirmed the earlier findings and proposed that Ca^{2+}-ATP may be bound chiefly in a Na^+-independent fashion to yield a K^+-insensitive enzyme–substrate complex, and consequently leading to the inhibition of Na^+/K^+-ATPase. Powis et al. (1981) reported that Ca^{2+} has a stimulatory effect on brain Na^+/K^+-ATPase.

There have been several reports concerning the presence of an inhibitor of Na^+/K^+-ATPase in a commercial source of ATP preparation (Charney et al., 1975; Hudgins and Bond, 1977; Josephson and Cantley, 1977; Cantley et al., 1977; Beauge and Glynn, 1978). The contaminant has been isolated and identified as vanadate (VO_5^{2-}). The compound exerts its inhibitory action in a VO_5^{2-} form. A 50% inhibition of the enzyme activity is achieved at vanadate concentrations of 40 n*M* (Cantley et al., 1977). Wu and Phillis (1979c) observed that vanadate inhibition of Na^+/K^+-ATPase can be potentiated by increasing Mg^{2+} and K^+ concentrations, whereas the inhibition can be reversed by increasing Na^+ concentration, suggesting that vanadate may inhibit the enzyme at its Na^+ sites. Interestingly, Co^{2+} (100 μ*M*) enhances the inhibitory effect of vanadate, whereas Mn^{2+} (100 μ*M*) antagonizes the inhibition. Kinetic analysis of vanadate inhibition at fixed concentrations of Mg^{2+}, K^+, and Na^+ suggested that vanadate inhibits Na^+/K^+-ATPase activity in a noncompetitive manner (Wu and Phillis, 1979c). Vanadate does not inhibit K^+-dependent *p*-nitrophenylphosphatase, suggesting that the noncompetitive inhibition takes place at the Na^+-activated Mg^{2+}-dependent kinase site of the enzyme reaction. Consistent with this report was the finding by Cantley et al. (1978a), who showed that the high-affinity vanadate site is identical with a low-affinity ATP site on the enzyme, whereas the low-affinity vanadate site is responsible for inhibition of Na^+-stimulated adenosine triphosphatase activity and is probably associated with sodium-dependent protein phosphorylation. Bond and Hudgins (1979) also provided evidence to support the mechanism of vanadate inhibition of Na^+/K^+-ATPase proposed by Cantley et al. (1978a) and Wu and Phillis (1979c). Since vanadate is present in tissue at concentrations that might influence Na^+/K^+-ATPase activity in vivo, it has been suggested that vanadate may act as an endogenous regulator of Na^+/K^+-ATPase activity (Cantley et al., 1978a; Bond and Hudgins, 1979).

2.3.4. Neurotropic Agents

Several neurotropic agents have been shown to inhibit Na^+/K^+-ATPase. Chlorpromazine and other phenothiazines were mentioned in a previous section of the text. Diphenylhydantoin may exert its anticonvulsant action by facilitating the movement of Na^+ out of the cell (Woodbury, 1955), but, contrary to this suggested mechanism, Rawson and Pincus (1968) showed that diphenylhydantoin consistently inhibits Na^+/K^+-ATPase isolated from brain homogenates of man and rodent. The effect of diphenylhydantoin on the enzyme activity was later observed to vary under different assay conditions. Festoff and Appell (1968), using rat brain synaptosomes, found that the agent stimulates phosphate release by 92% above the control rate when the experiment is carried out in a medium containing 50 m*M* Na^+ and 0.2 m*M* K^+. However, when the Na^+/K^+ ratio is changed to 5–10:1, diphenylhydantoin produces little or no effect, whereas a significant inhibition of the enzyme activity is observed at a Na^+/K^+ ratio of less than 5:1. The inconsistencies among the various findings and the failure of recent experiments to demonstrate any significant effect of diphenylhydantoin on partially purified enzyme from beef brain, dog kidney, and dog heart under various ratios of Na^+/K^+ suggest that diphenylhydantoin may not have significant effect on Na^+/K^+-ATPase activity (Schwartz et al., 1975).

Ethanol has been shown to inhibit (Jarnefelt, 1961) and stimulate (Israel and Kuriyama, 1971) Na^+/K^+-ATPase. Knox et al. (1972) also showed that cat brain Na^+/K^+-ATPase activity is significantly increased by chronic treatment with ethanol. Consistent with these findings, Banerjee and Sharma (1978) reported that in peripheral organs of cat chronic treatment of ethanol results in an increase in the number of Na^+/K^+-ATPase molecules. This change in enzyme number may represent an adaptive change in the physiology of the body after chronic ethanol. However, Akera et al. (1973) could not find any change in either brain Na^+/K^+-ATPase activity or in [^{3}H]-ouabain binding in homogenates or microsomal fractions during treatment or withdrawl periods. No definitive conclusion can therefore be arrived at regarding the in vitro or in vivo effect of ethanol on Na^+/K^+-ATPase activity. It appears unlikely that the alternation in sodium pump activity seen after chronic ethanol administration is caused directly by a specific effect on Na^+/K^+-ATPase (Schwartz et al., 1975).

Organochlorine insecticides inhibit vertebrate and invertebrate Na^+/K^+-ATPase. Matsumura and Patil (1969) showed that organochlorides inhibit Na^+/K^+-ATPase activity in rat brain synaptosomes. Koch (1969) and Koch et al. (1969) also showed that Na^+/K^+-ATPase activity of rabbit brain, cockroach nerve cord, and honey bee ganglia can be inhibited by a number of chlorinated hydrocarbon insecticides, including chlordane, aldrin, dieldrin, 3,P,P-DDT and 0,P,P-TDE. Cyclodiene insecticides have also been reported to inhibit Na^+/K^+-ATPase (Koch et al., 1969; Cutkomp et al., 1971; Desaiah and Koch, 1975; Yap et al., 1972). The inhibitory effect of the organochlorine insecticides on rat brain Na^+/K^+-ATPase has been shown by Folmar (1978). The results indicated that, at an inhibitor concentration of 20 m*M*, the effectiveness of the insecticides as Na^+/K^+-ATPase inhibitors follow the descending order chlordane > kelthane > kepone > heptachlor > TDE > 3,P,P-DDT > 0,P-DDT > endrin, dieldrin.

Chan and Quastel (1970) studied the effects of various neurotropic agents on Na^+/K^+-ATPase activity of rat brain cortex. They were not able to demonstrate any significant effect of local anesthetics or atropine on the enzyme activity. Schaefer et al. (1975) tested a series of neurotropic drugs on Na^+/K^+-ATPase from rat brain microsomal preparation. Although the drugs tested did not show any significant effect on enzyme activity at a concentration of 100 μ*M*, these workers did observe that a number of neurotropic agents were able to reverse the inhibition of Na^+/K^+-ATPase by 50 μ*M* ascorbic acid. The neurotropic agents tested included phenobarbital, glutethimide, morphine, meperedine, harmine, mescaline, LSD, cocaine, procaine, lidocaine, meprobamate, trimetozine, chlordiazepoxide, diazepam, chlorpromazine, promethazine, reserpine, tetrabenazine, oxypertine, haloperidol, trifluperidol, desipramine, imipramine, trimipramine, amitriptyline, nortripytline, nialamide, pentylenetetrazole, strychnine, trimethadione, diphenylhydantoin, nicotine, tetraethylammonium, guanethidine, bretylium, dibenamine, phentolamine, yohimbine, phenoxybenzamine, propranolol, pronethalol, prenylamine, atropine, methantheline, physostigmine, tubocurarine, decamethorium, theophylline, apomorphine, and tremorine. Of these agents, chlorpromazine, apomorphine, oxypertine, reserpine, promethazine, tetrabenazine, yohimbine, desipramine, imipramine, and LSD were shown to be effective in reversing the inhibition of Na^+/K^+-ATPase by ascorbic acid. The mechanism by which these drugs reverse the inhibition may be related to their ability to inhibit lipid peroxidation. Inhibitors of lipid peroxidation are known to eliminate the inhibitory action of ascorbic acid on Na^+/K^+-ATPase activity (Schaefer et al., 1975).

2.3.5. Stimulation of Na^+/K^+-ATPase by Catechols and Catecholamines

Stimulation of Na^+/K^+-ATPase by noradrenaline was first described in brown adipose tissue (Herd et al., 1970). The effect can be blocked by propranolol. Subsequent to this finding, there have been several reports concerning catecholamine stimulation of Na^+/K^+-ATPase activity in various brain preparations.

The central issue has been the mechanism(s) by which catecholamines stimulate Na^+/K^+-ATPase. Schaefer et al. (1972, 1974) were the first to demonstrate that catecholamines could stimulate particle-bound Na^+/K^+-ATPase from rat brain synaptosomal membranes. Monophenolic and nonphenolic phenylethylamines, such as tyramine, α-methyltyramine, metaraminol, amphetamine, and phenylephrine are not able to stimulate the enzyme. Yoshimura (1973) reported that enhanced activity of Na^+/K^+-ATPase can be obtained by treating the tissue preparation with noradrenaline, dopamine, and serotonin (5-hydroxytryptamine; 5-HT). A number of investigators have since shown that catecholamines can increase the activity of the ouabain-sensitive Na^+- and K^+-activated ATPase in brain homogenates and microsomal preparations (*see* review by Phillis and Wu, 1981).

The mechanism(s) underlying this increased activity of Na^+/K^+-ATPase in brain preparations is not clear, although three different hypotheses have been put forward. Schaefer et al. (1974) demonstrated that catecholamines enhance the activity of Na^+/K^+-ATPase by reversing the action of a soluble inhibitory factor in the brain homogenate. The soluble factor was later identified as ascorbic acid (Schaefer et al., 1974), and it was suggested that a heavy metal ion (possibly iron) bound to membrane structures is responsible for the inhibitory effect of both the soluble factor and ascorbic acid. Catecholamines are thought to act by chelation of the ferric ion, since Schaefer observed that chelating agents have a similar action to catecholamines. The involvement of a soluble tissue factor in the catecholamine's stimulation of Na^+/K^+-ATPase was reported by Rodriguez de Lores Arnaiz and Mistrorigo de Pacheco (1978), who showed that noradrenaline can stimulate rat cerebral cortical membrane Na^+/K^+-ATPase in the presence of a fraction containing the factor, but that, in its absence, noradrenaline ($10^{-4}M$) has a weak inhibitory effect. The results were explained by suggesting that noradrenaline has an inhibitory action on the enzyme, and that the reported stimulation of the enzyme by noradrenaline is probably not mediated by the activation of an adrenergic receptor. Similar results were reported by Godfraind et al. (1974).

Other investigators have been in favor of the hypothesis that catecholamines stimulate Na^+/K^+-ATPase by removal of an inhibitory metal ion. Godfraind et al. (1974) suggested that a disinhibitory action of catecholamines could account for the stimulation of Na^+/K^+-ATPase. Further, Godfraind et al. (1974) suggested the chelation of Ca^{2+} may be the mechanism of catecholamine stimulation of Na^+/K^+-ATPase. Hexum (1977) also proposed that activation of the enzyme could be a result of removal of inhibitory divalent metal ions through the formation of a catecholamine–metal ion complex, since it is known that Na^+/K^+-ATPase activity is inhibited in vitro by such divalent metal ions as Ca^{2+}, Cu^{2+}, Zn^{2+}, and Fe^{2+}. More recently, Cantley et al. (1978b) have shown that noradrenaline stimulation of Na^+/K^+-ATPase in the rat brain may result from the removal of the inhibitory effect of endogenous vanadium in V^{5+} form. Noradrenaline will reduce V^{5+} to V^{+3}, which is ineffective in inhibiting the enzyme. Svoboda and Mosinger (1981a,b) combine most of the above suggestions in saying that it may be conceivable that all the mechanisms, i.e., the direct inhibition by endogenous Fe^{2+} and Ca^{2+} and the lipoperoxidative degradation induced by Fe^{2+} plus ascorbic acid, cooperate in the inhibitory effect on the Na^+/K^+-ATPase. Catecholamines would reverse the inhibition by virtue of their chelating property or antilipoperoxidase activity.

In coming to similar conclusions, both Rodriguez de Lores Arnaiz and Mistrorigo de Pacheco (1978) and Svoboda and Mosinger (1981a,b) employed rather high concentrations of catecholamines ($>10^{-5}M$) to stimulate the enzyme, and both groups of investigators failed to observe any antagonism with α- or β-adrenergic blocking agents. A similar lack of effect by α- and β-adrenergic blocking agents was reported by Van der Krogt and Belfroid (1980) upon noradrenaline stimulation of rat brain striatal Na^+/K^+-ATPase. *Cis*-flupenthixol antagonizes the effects of dopamine stimulation of Na^+/K^+-ATPase, but this action is apparently unrelated to the antagonism of dopamine receptors by the drug, since a pharmacologically inactive isomer, trans-flupenthixol, has equal or higher potency in blocking the activation of Na^+/K^+-ATPase by dopamine. High concentrations of catecholamine (10^{-5}–$10^{-4}M$) were again employed, and these authors concluded that stimulation of brain Na^+/K^+-ATPase by catecholamines is a nonspecific phenomenon unrelated to receptor action and may in fact be related to the removal of the inhibitory effect of divalent metal cations on the enzyme.

The third hypothesis is that the stimulation of brain Na^+/K^+-ATPase activity by catecholamines is mediated through a receptor-related mechanism. If, however, catecholamine stimulation of Na^+/K^+-ATPase is merely the result of the removal of inhibitory divalent metal ions, the stimulation should be abolished in the presence of high concentrations of chelating agents. This does not appear to be the case. Wu and Phillis (1979c) have demonstrated that although EDTA ($10^{-4}M$) has a maximal stimulatory effect on Na^+/K^+-ATPase, the addition of $10^{-5}M$ noradrenaline to an enzyme preparation containing 10^{-4} or $10^{-3}M$ EDTA can further enhance enzyme activity. Even though the amount of activation by noradrenaline is much reduced under these conditions, the presence of a residual component in the response makes it apparent that divalent metal ion inhibition can only partially account for catecholamine stimulation of brain Na^+/K^+-ATPase. A number of investigators have reported antagonism of the stimulatory effects of catecholamines on brain Na^+/K^+-ATPase with both α- and β-adrenergic antagonists. Iwangoff et al. (1974) and Gilbert et al. (1975) were able to demonstrate antagonism of noradrenaline stimulation of Na^+/K^+-ATPase with phentolamine. Hexum (1977) blocked the effects of isoprenaline with the β-antagonist propranolol. The stimulatory effects of dopamine have been antagonized by chlorpromazine and fluphenazine (Desaiah and Ho, 1977; Akagawa and Tsukuda, 1979). Serotonin stimulation of Na^+/K^+-ATPase from rat cerebral cortical homogenates can be effectively blocked by metergoline, a specific serotonin antagonist (Wu and Phillis, 1979b). The evidence from studies with catecholamine antagonists supports the hypothesis that a component of the response of Na^+/K^+-ATPase to catecholamines is mediated by a receptor-linked reaction.

Wu and Phillis (1978, 1979a, 1980) have shown that Na^+/K^+-ATPase in rat cerebral cortical homogenates or synaptosomal preparations is stimulated by noradrenaline, the threshold for this effect being less than $10^{-7}M$. Adrenaline is less potent than noradrenaline. The β-agonist, isoprenaline, activates the enzyme at $10^{-6}M$. α-Agonists are also active in stimulating Na^+/K^+-ATPase. The threshold concentration for methoxamine is $2.5 \times 10^{-7}M$, but phenylephrine has no significant stimulant activity. These results clearly demonstrate that Na^+/K^+-ATPase activation evoked by noradrenaline can be blocked by adrenoceptor antagonists, and raise the possibility of involvement by both α- and β-adrenoceptors (Daly, 1977).

In order to study the structure–activity relationships of catecholamine stimulation of Na^+/K^+-ATPase, agents with structural features analogous to noradrenaline were tested for their ability to stimulate rat brain Na^+/K^+-ATPase in brain homogenates. The results show that monophenolic substances are inactive in stimulating the enzyme. Catechol and pyrogallol are both active in stimulating the enzyme activity. Among catechol analogs, only 1,2-dihydroxybenzene and its derivatives are active, whereas the 1,3-dihydroxybenzene series and 1,4-dihydroxybenzene and its derivatives are inactive. The conclusion seems to be that the catechol moeity is the important feature for the compound to act as a stimulator of Na^+/K^+-ATPase in the rat brain. L-isomers of the adrenergic agonists are more active than D-isomers in eliciting pharmacological responses. Our results demonstrate that Na^+/K^+-ATPase can be activated by both D- and L-isomers of noradrenaline and isoprenaline. L-Noradrenaline is approximately 10 times more potent than D-noradrenaline in stimulating Na^+/K^+-ATPase, whereas the difference in the potency of the D- and L-isomers of isoprenaline is less apparent, since there is only a threefold difference in the abilities of L-isoprenaline and D-isoprenaline to stimulate brain Na^+/K^+-ATPase.

Studies with receptor antagonists showed that whereas the stimulation of Na^+/K^+-ATPase induced by L-noradrenaline can be blocked by phentolamine or propranolol, that evoked by D-noradrenaline cannot be antagonized by these agents, suggesting that although both L- and D-isomers of noradrenaline have the ability to stimulate Na^+/K^+-ATPase activity, the underlying mechanism that elicits the stimulation could be different. Similarly, stimulation of the enzyme by catechol is resistent to antagonism by both α- and β-adrenoceptor antagonists. These findings indicate the possibility that at least two separate mechanisms are operating in catecholamine-evoked stimulation of Na^+/K^+-ATPase. A nonstereoselective activation appears to be involved in the actions of agents such as catechol and D-noradrenaline, since the effects of these agents are not blocked by α- or β-antagonists. The mechanism involved in this action is likely to be caused by nonreceptor-mediated effects such as the chelation of inhibitory metal ions and antilipoperoxidation. High concentrations of noradrenaline, isoprenaline, and their analogs may also elicit this nonreceptor-mediated response, since the effects of these agents, at high concentrations, are not blocked by adrenoceptor antagonists. The ability of α- or β-blockers to antagonize the stimulation of Na^+/K^+-ATPase by low concentrations of noradrenaline, methoxamine, and isoprenaline suggests that there are additional receptor-mediated components in the activation of Na^+/K^+-ATPase.

2.3.6. Detergents

Detergents have been used to solubilize Na^+/K^+-ATPase from the membrane structures during the isolation procedure. Thus the effect of detergents on enzyme activity has been reported, with a variety of different results. High concentrations of detergents like deoxycholate or lauryl sulfate (Glynn, 1963; Jarnefelt, 1961) inhibit Na^+/K^+-ATPase. Banerjee et al. (1970) also reported the inactivation of the brain enzyme after treatment with detergents such as Luberol and Triton X-100. Lower concentrations of detergents have been shown to activate Na^+/K^+-ATPase in microsomal or membranal preparations (Jorgensen and Skou, 1971; Schirachi et al., 1970; Banejee et al., 1970; Wu and Phillis, 1980). Although it seems easy to explain the inhibitory effects of detergents on Na^+/K^+-ATPase, the enhanced activity of the enzyme caused by these agents is not so easily understood. Jorgensen and Skou (1971) suggested that detergents may increase activities of Na^+/K^+-ATPase by exposing latent enzyme sites. This suggestion is consistent with the observation that detergent (deoxycholate) activates the enzyme in the membrane preparation that has low activity. Other approaches employed have been to determine the ouabain binding sites in the enzyme preparation. If the effect of detergents on the enzyme is to expose the latent enzyme sites, one would expect that ouabain binding sites should increase accordingly. Choi and Akera (1978) showed that the specific activity of Na^+/K^+-ATPase from canine brain increases markedly after treatment with deoxycholate, and a corresponding increase in ouabain binding sites was also reported. However, Choi and Akera (1978) cautioned that the effect of deoxycholate on enzyme activity and ouabain binding sites may be different, since, in their study, they observed a decrease in the enzyme turnover number. Such a decrease in the apparent turnover number of the Na^+/K^+-ATPase can be explained by the treatment increasing the concentration of ouabain binding sites to a greater degree than Na^+/K^+-ATPase activity. They indicated that the change in the characteristics of membrane lipids resulting from detergent treatment could also account for the increase in ouabain binding sites and the decrease in the apparent turnover number of the enzyme.

2.3.7. Gangliosides

Gangliosides are sialic acid-containing glycosphingolipids. The membrane-associated enzymes change their properties when the membrane preparations are incubated with gangliosides and when the gangliosides are bound to the membranes. Adenylate cyclase of rat cerebral membranes is activated by gangliosides (Partington and Daly, 1979). Na^+/K^+-ATPase is a membrane-bound enzyme, and gangliosides may also be able to modify the enzyme activity. Breer and Jeserich (1979) reported that gangliosides inhibit ATPase, whereas Caputto et al. (1977) showed that gangliosides strongly stimulate Na^+/K^+-ATPase activity. Leon et al. (1981) reported a more detailed study on the effects of gangliosides on Na^+/K^+-ATPase from rat brain preparation. For gangliosides to be effective on the enzyme preparation, it is necessary to incubate the enzyme with gangliosides for a period of at least 60 min so the gangliosides will have an opportunity to bind to the membranes. Indeed, Seiler et al. (1976), Fishman et al. (1976), and Toffano et al. (1980) have demonstrated that gangliosides interact with cells and subcellular preparations and become associated with membranes. The binding (association) of gangliosides to membranes in vitro is time-dependent (Leon et al., 1981). Similarly, when Leon et al. (1981) examined the effect of gangliosides on Na^+/K^+-ATPase activity, they found that the activation of the enzyme by the gangliosides (concentrations less than 40 n*M*) is also time-dependent. 40 n*M* of 11^3 Neu AC-Gg OSe_4 Cer (GM_1) activates the Na^+/K^+-ATPase activity by over 40% after 2 h of incubation in vitro. In the study (Leon et al., 1981), IV^3 Neu AC, 11^3 Neu AC-Gg OSe_4 Cer (GD_{1a}), 11^3 (Neu $AC)_2$-GgOSe_4 Cer (GD_{1b}), and IV^3 Neu AC, 11^3 $(NeuAc)_2$-GgOSe_4 (GT_{1b}) were also tested for their effects on the rat brain Na^+/K^+-ATPase. The results indicate that GM_1 produces an approximately 43% activation of the enzyme, whereas GD_{1b} produces a 24.8% activation. *N*-Acetylneuraminic acid, the anionic detergent sodium dodecylsulphate, cerebromide, and sulphatide produce slight activation ranging from 4 to 7%. It would appear that GM_1 treatment leads to a stable, irreversible insertion of GM_1 molecules into the membrane bilayer, and the insertion of the ganglioside results in significant changes in the activity of a membrane-integral protein, i.e., Na^+/K^+-ATPase. These workers also indicated that the insertion of gangliosides into the membranes produces an insignificant change in the membrane fluidity. Since Na^+/K^+-ATPase contains glycolipid components, it will be interesting to see whether gangliosides will be associated with the purified form of Na^+/K^+-ATPase. Such a study may yield results indicating the mechanism(s) by which gangliosides modify Na^+/K^+-ATPase activity.

3. Methods of Assay and Detection

3.1. Spectrophotometric Method

This method is based on the measurement of the release of inorganic phosphate during enzymatic reaction. The inorganic phosphate liberated is then converted to a molybdate–phosphate complex that exhibits intense color and can be measured quantitatively by a spectrophotometric method (Fiske and Subbarow, 1925).

Bonting et al. (1961) used the method to determine the quantitative distribution of Na^+/K^+-ATPase activity in several tissues of the cat. The following is their assay procedure: The freeze-dried homogenates are reconstituted with distilled water. Varying quantities of the homogenates are mixed with the standard assay media to give a net weight concentration of the tissue ranging from 0.5 to 35 mg/mL. Six aliquots of 10 μ L of each mixture are placed in microtest tubes, three of which are incubated for 1 h at 37°C, while the others remain in ice; trichloroacetic acid (50 μ L, 10%) is added to them immediately. After mixing and centrifugation, a 30-μL aliquot is transferred to another microtest tube containing 30 μL of color reagent. The color reagent is prepared by dissolving 400 mg $FeSO_4$ in 10 mL of a 1% ammonium molybdate solution in 1.15*N* H_2SO_4, and is used within 2 h after its preparation. The resulting color is read within 2 h at 700 nm in Beckman model DU spectrophotometer. Reagent blanks and inorganic phosphate standards are included in each experiment and are taken into account when converting the optical density to amount of inorganic phosphate formed per unit of protein at a specified incubation period. The reading from the nonincubated samples is subtracted from the results. The assay media are presented in Table 1. Medium A was designed to allow optimum activity of Na^+/K^+-ATPase. Media B, C, and D were designed to inhibit Na^+/K^+-ATPase completely while Mg^{2+}-ATPase activity remained intact. Media E and F were designed to measure Ca^{2+}-ATPase and Mg^{2+}-ATPase activity. As indicated in Table 1, the pH value of the incubation was maintained at 7.5 and the ratio between ATP and Mg^{2+} was kept at 2. Na^+/K^+-ATPase activity can be calculated by subtracting the average activities of reaction occurring in media B, C, and D.

There are a number of modifications that have been made to this procedure. The pH value of the reaction has been changed to 7.2 (Skou, 1957) or 7.4 (Dunham and Glynn, 1961; Tobin et al., 1973). It has been suggested that the Mg^{2+} : ATP ratio in the incubation medium be maintained at approximately 2:1 because high concentrations of Mg^{2+} are known to inhibit Na^+/K^+-ATPase activity (Fagan and Racker, 1977; Wu and Phillis, 1980).

3.2. Coupled Assay Method of Barnett

Na^+/K^+-ATPase hydrolyzes ATP to form inorganic phosphate and ADP. In the spectrophotometric analysis of Na^+/K^+-ATPase

Table 1
Composition of Media Used in the Determination of Na^+/K^+-ATPase Activity[a]

	Ionic Composition, m*M*									
Medium[a,b]	ATP	Mg^{2+}	K^+	Na^+	CN^-	EDTA	Tris	Ouabain	Ca^{2+}	Enzyme activity determined
A	2	1	5	58	10	0.1	92	—	—	Mg^{2+}-ATPase Na^+,K^+-ATPase
B	2	1	5	58	10	0.1	91	0.1	—	Mg^{2+}-ATPase
C	2	1	—	14	10	0.1	146	—	—	Mg^{2+}-ATPase (ouabain-sensitive?)
D	2	1	—	14	10	0.1	144	0.1	—	Mg^{2+}-ATPase (ouabain-sensitive)
E	2	1	5	58	10	0.1	91	—	1	Mg^{2+}-ATPase Na^+,K^+-ATPase Ca^{2+}-ATPase
F	2	1	5	58	10	0.1	90	0.1	1	Mg^{2+}-ATPase Ca^{2+}-ATPase

[a]This table is a modified version of that of Bonting et al., 1961.

[b]The medium prepared according to the list yielded an osmolarity of approximately 312–320 mOsm. The medium was maintained at pH 7.5.

activity, the ATP concentration in the reaction medium is decreasing continually, whereas the concentration of ADP is gradually increasing. Since there is the possibility that ADP may influence the kinetic properties of Na^+/K^+-ATPase, and since the decreasing concentration of ATP in the medium may also alter the kinetic property of the enzyme reaction, Barnett (1970) proposed that an ATP-regenerating system should be included in the assay reaction. The production of ADP is a result of the Na^+/K^+-ATPase reaction, and can be linked to the oxidation of NADH and monitored continuously at 340 nm. The procedure provides not only the continuous recording of ATP hydrolysis, but also offers a nearly constant ATP concentration in the medium even at low ATP concentrations, and ADP accumulation is prevented (Cantley and Josephson, 1976). In a typical assay, the following concentrations of compounds are included in the medium: 20 units of pyruvate kinase, 2 units lactate dehydrogenase, 1.5 m*M* of phosphoenolpyruvate, 0.1 m*M* of NADH, 50 m*M* Tris-HCl (pH 7.4), 100 m*M* NaCl, 10 m*M* KCl, 5 m*M* $MgCl_2$, and 5 m*M* ATP. The assay is usually carried out with about 1 μg of protein/mL of medium at 37°C. The specific activities are two to four times higher by the coupled method of Barnett than by measuring inorganic release (Fagan and Rocker, 1977), as described in the previous section. Slight variation in the components and the concentrations of the medium solution have also been reported (Cantley and Josephson, 1976).

3.3. pH Stat Method

This is a method based on the measurement of ATP hydrolysis. In an unbuffered solution at a pH value greater than 7.6, it is estimated that 1 Eq of H^+ is released upon the hydrolysis of 1 mol of ATP (Green and Mommaerts, 1953). One can use a base to titrate the amount of H^+ ion formed during the enzymatic reaction. The assay solution contains 2 m*M* ATP, 2.5 m*M* $MgCl_2$, 0.25 m*M* ethylenelus (oxyethylenenitroiolo) tetra-acetic acid, 0.5 mg/mL bovine serum albumin, and 1 m*M* dithiothreitol. Two milliliters of this solution are euqilibrated at 25°C in the thermostat-regulated reaction vessel of a radiometer automatic titrator. The solution is adjusted to pH 7.8–7.9 and subsequently controlled by the addition of 0.1*N* amino methylpropanol from a 50-μL Hamilton syringe driven by the micrometer syringe drive of the titrator. The reaction is initiated by the addition of 0.1–0.25 unit of Na^+/K^+-ATPase. The appropriate concentrations of NaCl and KCl are added to the medium according to the design of the experiment (Albers et al.,

1968). In more recent experiments, Cantley and Josephson (1976) used 2 m*M* NaOH to titrate H^+ formation employing a radiometer type TTT1a titrator. They bubbled the dry nitrogen through the solution to reduce CO_2 absorption. The report indicated that the change in ATP concentration was less than 10% and the change in Na^+ concentration was less than 5% during the course of the assay.

Albers et al. (1968) observed that the hydrolysis of ATP by Na^+/K^+-ATPase as measured by pH stat technique is not linear with time. They proposed that this is because of the partial denaturation of the enzyme in a dilute solution and because of the production of inhibitory products during the reaction. It is interesting to note that Cantley and Josephson (1976) observed that the enzyme reaction measured by the pH stat technique is linear with time when KCl is omitted from the incubation medium. The reaction becomes nonlinear upon the addition of KCl. They suggested that the inactivation of Na^+/K^+-ATPase is related to the presence of Mg^{2+}, K^+, and ATP. Indeed, when the enzyme was preincubated for 20 min in the presence of all possible combinations of K^+, Mg^{2+}, and ATP, it was found that the simultaneous presence of all three ligands was required for the inactivation of the enzyme.

3.4. Radioisotope Method

The basic principle for the radioisotope method is the same as that described in a previous section. It is based on the fact that Na^+/K^+-ATPase hydrolyzes ATP to form ADP and inorganic phosphate. In this instance, the ATP used in the reaction is in the form of γ-^{32}P-ATP. The enzyme reaction will result in the release of radiolabeled inorganic phosphate ^{32}Pi into the medium, and the ADP formed during the reaction will not be radioactively labeled. By separating ^{32}Pi from the γ-^{32}P-ATP in the medium, one can determine the amount of ^{32}Pi formed and consequently the amount of ATP that has been hydrolyzed.

A typical example of an application of the technique has been reported by Siegel and Albers (1967). In their assay procedure, the medium contains 2 m*M* γ-^{32}P-ATP, 60 m*M* NaCl, 25 m*M* KCl, 2 m*M* $MgCl_2$, 100 m*M* Tris-HCl (pH 7.4), 0.1% (w/v) bovine serum albumin, 0.1 m*M* Tris-EDTA, and 0.2 mg of enzyme protein preparation, altogether in a final volume of 40 μ L. The assay medium is incubated for 20 min at 26°C, and the reaction is terminated by adding 10 μ L of 25% (w/v) trichloroacetic acid. After centrifugation at 1000*g* for 10 min at 2°C, a 40-μL portion of the supernatant is removed and added to 10 μL of 5% (w/v) ammonium molybdate in 4*N* H_2SO_4. Phosphomolybdate is extracted into 50 μL of isobutanol. The isobutanol phase is separated and used for scintillation counting. The amount of ATP hydrolyzed is calculated from the specific activity of the reaction mixture. When the concentration of ATP approaches that of ammonium molybdate, interference with the extraction of inorganic phosphate is encountered. This can be overcome by increasing the portion of ammonium molybdate. This method represents a very sensitive and convenient way of determining Na^+/K^+-ATPase activity.

3.5. Ligand-Labeling Techniques

Ouabain, a cardiac glucoside, is known to bind to Na^+/K^+-ATPase with high affinity (*see* section 2.3.2). The binding of ouabain to Na^+/K^+-ATPase leads to a decrease in Na^+/K^+-ATPase activity. Despite the fact that [^{3}H]-ouabain may bind to the enzyme at more than one site (Hansen, 1976), the procedure is useful in determining the total Na^+/K^+-ATPase enzyme present in a preparation. The ligand labeling method can only determine the number of enzyme molecules present, but it cannot determine the activity of the enzyme in a preparation. In a typical experiment, Hansen (1976) described the binding procedure as follows. [^{3}H]-g-strophanthin binding is achieved by incubation at 37°C of enzyme (0.1–0.2 mg protein/mL, g-strophanthin-sensitive activity 170–240 μ*M* Pi/mg protein/h) with 3 m*M* Mg^{2+}, 0.2–3 m*M* Pi (buffered with propandial), and 40–48 m*M* Tris-HCl (pH 7.25, 37°C) to keep the ionic strength nearly constant. The binding reaches equilibrium in 10 min or less (Banerjee and Sharma, 1978). The g-strophanthin binding level is determined at equilibrium of binding. Bound and nonbound g-strophanthin are separated by filtration through a membrane filter (pore size, 0.6 mm) with suction. Aliquots of the filtrates are taken to determine the nonbound ^{3}H-radioactivity. The total ^{3}H-radioactivity of the incubation is determined under similar conditions. The difference between the radioactivity of the incubation medium and that of filtrate is taken as bound ^{3}H-g-strophanthin according to results indicated by Banerjee and Sharma (1978). The nonspecific accumulation of radioactivity is determined by assaying parallel incubations in which ATP and Mg^{2+} are omitted from the reaction mixture and is generally less than 12% of the total binding.

Covalent labeling of the enzyme can be achieved by a photoaffinity labeling procedure using derivatives of ouabain. The photoaffinity reaction results in the formation of a complex that is stable to trichloroacetic acid, organic solvents, and heat denaturation. Forbush et al. (1978) reported the photoaffinity labeling of

Na^+/K^+-ATPase using 2-nitro-5-azidobenzoyl (NAB) derivatives of ouabain. The photoaffinity labeling of Na^+/K^+-ATPase involves two steps of reaction. First, NAB-[^{3}H]-ouabain is bound to the enzyme, and second, the NAB-[^{3}H]-ouabain-labeled Na^+/K^+-ATPase is photolyzed using long wave-length UV light irridation. NAB-ouabain binding is carried out in the presence of Na^+, Mg^{2+}, and ATP or in the presence of Mg^{2+} and Pi. The medium usually contains 120 m*M* NaCl, 3 m*M* NaATP, 3 m*M* $MgCl_2$, 0.2% BSA, and 25 m*M* Tris, pH 7.5 (37°C). NAB-[^{3}H]-ouabain concentrations ranging from 10^{-8} to $10^{-5}M$ are included in the incubation. Enzyme concentrations (5–150 μ g/mL) are incubated for 5–60 min at 37°C. The binding of NAB-ouabain reaches equilibrium after 60 min of incubation. The equilibrium of ligand binding corresponds to maximal inhibition of Na^+/K^+-ATPase activity. After washing and pelleting the NAB-[^{3}H]-ouabain-bound, Na^+/K^+-ATPase is resuspended in a binding medium lacking NAB-[^{3}H]-ouabain and photolyzed at 0°C for 15 min with a hand-held lamp UVSL (UV Products). The resulting covalently bound NAB-[^{3}H]-ouabain–enzyme complex can be used for further purification, isolation, or enzyme sequencing (Forbush et al., 1978; Collins et al., 1982). Other photoaffinity probes such as 4-(ethyldiazomalonyl)cymarin (DAMN cymarin) have also been used to label Na^+/K^+-ATPase (Ruoho and Kyte, 1974).

3.6. Immunochemical Methods

The immunochemical method was not developed to measure the enzyme activity. It was designed to localize the Na^+/K^+-ATPase and its subunits in the central nervous system. The basic approach here is to use immunocytochemical methods to localize antigenic sites on the enzyme protein molecule (Kyte, 1976a,b). Wood et al. (1977) have successfully utilized antisera to the enzyme purified from electric eel to localize the Na^+/K^+-ATPase in the brain of the black ghost knifefish *(Sternarchus albifrons).* In this study, the fish are anaesthetized and perfused through the heart with a fixative containing freshly made 4.0% paraformaldehyde and 0.1% glutaraldehyde in 0.12*M* Millonig's phosphate buffer, pH 7.2 (Millonig, 1961). After perfusion, the brain is removed and stored overnight at 4°C in 4% phosphate-buffered paraformaldehyde without glutaraldehyde. The fixed brain is cut parasagittally into 75-μ m slices with a sectioner. The slices are collected in 0.12*M* Millonig's buffer. The antiserum is prepared by injecting the isolated Na^+/K^+-ATPase (*Electrophorus* electric organ) into rabbits to stimulate antibody formation (Jean and Albers, 1976). The antiserum has little activity toward mammalian brain Na^+/K^+-ATPase (Jean and Albers, 1976). Goat anti-rabbit lgG is prepared as described by Whitaker (1975). The 75-μm tissue slices are incubated by immersion for 30–40 min at room temperature in rabbit antisera to the Na^+/K^+-ATPase diluted to either 1:30 or 1:40 in phosphate-buffered saline (PBS). The slices are washed with PBS (five changes) for 3 h at room temperature prior to incubation with a 1:90 dilution of peroxidase-labeled goat anti-rabbit lgG. After more washing for 3 h, the slices are incubated again for 15–20 min at 0–4°C with a solution containing 30 mg diaminobenzidine and 10 mL of 30% hydrogen peroxide in 50 mL PBS. The slices are washed with PBS for 30 min and are dissected for postfixation with 2% osmium tetroxide in 0.12*M* Millong's buffer, pH 7.2. The slices are dehydrated through an ethanol series and propylene oxide and flat embedded in EPon-Araldite. Ultrathin sections are examined by an electron microscopic procedure. Similar techniques have been employed to localize Na^+/K^+-ATPase in the oocytes of *Heterandria formosa Agassiz* (Riehl, 1980) or in the dog area postrema (Klara et al., 1978).

4. Conclusion

Na^+/K^+-ATPase has been suggested to be one of the macromolecular constituents of nervous tissue that plays a critical role in the function of the nervous system. The activity of the enzyme is essential for the active transport of sodium and potassium ions. This process is a prime requisite for the conduction of nerve impulses (Wood et al., 1977). The enzyme may play an important role in regulating the transport of solutes across cellular membranes and the release of neurotransmitters (Vizi, 1980). Recently, inhibition of brain Na^+/K^+-ATPase activity has been implicated in producing coma in patients with Reye's syndrome and hepatic encephalopathy (Trauner, 1980). Since the first discovery of Na^+/K^+-ATPase activity in the crab nerve by Skou (1957), this enzyme has been the object of discussion in thousands of reports. Although we have learned a significant amount about the enzyme structure, the reaction mechanism and the function of this enzyme in the past 26 yr, there is still much that needs to be learned about the transitional steps of the enzyme reaction, and the conformational arrangement of the enzyme molecule. The determination of different components, such as peptide chains, of Na^+/K^+-ATPase remains one of the most challenging areas of research. Finally, the interaction of the Na^+/K^+-ATPase molecule with its surrounding membrane structures will be one of the most interesting aspects of this enzyme.

Acknowledgments

The author would like to thank Mr. John Keller for reading the manuscript. Ms. Cathy Van Der Giessen's assistance in preparing this manuscript is much appreciated. The views expressed in this publication are those of the author and do not necessarily reflect those of the Addiction Research Foundation.

References

Ahmed K. and Judah J. D. (1964) Preparation of lipoproteins containing cation-dependent ATPase. *Biochim. Biophys. Acta* **93,** 603–613.

Ahmed K. and Judah J. D. (1965) Identification of active phosphoprotein in a cation-activated adenosine triphosphatase. *Biochim. Biophys. Acta* **104,** 112–120.

Ahmed K., Judah J. D., and Scholefield P. G. (1966) Interaction of sodium and potassium with a cation-dependent adenosine triphosphatase system from rat brain. *Biochim. Biophys. Acta* **120,** 351–360.

Akagawa K. and Tsukuda Y. (1979) Presence and characteristics of catecholamine-sensitive Na-K ATPase in rat striatum. *J. Neurochem.* **32,** 269–271.

Akera T. and Brody T. M. (1968) Inhibition of brain sodium- and potassium-stimulated adenosine triphosphatase activity by chlorpromazine free radical. *Mol. Pharmacol.* **4,** 600–612.

Akera T. and Brody T. M. (1969) The interaction between chlorpromazine free radical and microsomal sodium- and potassium-activated adenosine triphosphatase from rat brain. *Mol. Pharmacol.* **5,** 605–614.

Akera T. and Brody T. M. (1970) Inhibitory sites on sodium- and potassium-activated adenosine triphosphatase for chlorpromazine free radical and ouabain. *Mol. Pharmacol.* **6,** 557–566.

Akera T. and Brody T. M. (1971) Membrane adenosine triphosphatase. The effect of potassium on the formation and dissociation of the ouabain–enzyme complex. *J. Pharmacol. Exp. Ther.* **176,** 545–557.

Akera T., Rech R. H., Marquis W. J., Tobin T., and Brody T. M. (1973) The lack of relationship between brain Na^+, K^+-activated adenosine triphosphatase and the development of tolerance to ethanol in rats. *J. Pharmacol. Exp. Ther.* **185,** 594–601.

Albers R. W. (1967) Biochemical aspects of active transport. *Ann. Rev. Biochem.* **36,** 727–756.

Albers R. W., Koval G. J., and Siegel G. J. (1968) Studies on the interaction of ouabain and other cardioactive steroids with sodium-potassium-activated adenosine triphosphatase. *Mol. Pharmacol.* **4,** 324–326.

Albers R. W., Rodriguez de Lores G., and De Robertis E. (1965) Sodium-potassium-activated ATPase and potassium-activated p-nitrophenylphosphatase: A comparison of their subcellular localizations in rat brain. *Proc. Natl. Acad. Sci. USA* **53,** 557–564.

Alexander D. R. and Rodnight R. (1970) Separation at neutral pH of a ^{32}P-labelled membrane protein associated with the sodium-plus-magnesium ion-activated adenosine triphosphatase from ox brain. *Biochemistry J.* **119,** 44–45P.

Ames G. F. L. (1974) Resolution of bacterial protein by polyacrylamide gel electrophoresis on slabs membrane, soluble and periplasmic fractions. *J. Biol. Chem.* **249,** 634–644.

Asano Y., Tashima Y., and Matsui H. (1970) (Na plus minus K plus)-ATPase from the frog bladder and its relationship to sodium transport. *Biochim. Biophys. Acta* **219,** 169–178.

Askari A. and Huang W. (1980) Na^+, K^+-ATPase: Half-of-the subunits cross-linking reactivity suggests an oligomeric structure containing a minimum of four catalytic subunits. *Biochem. Biophys. Res. Commun.* **93,** 448–453.

Askari A. and Huang W. (1981) Na^+, K^+-ATPase in (Ca^{2+} + ouabain)-dependent phosphorylation by Pi. *FEBS Lett.* **126,** 215–218.

Atkinson A. and Lowe A. G. (1972) The pH dependence of the effects of Na^+, K^+ and ouabain on the ATPase activity of NaI-treated brain microsomes with Mg^{2+}, Mn^{2+}, Ca^{2+} and Zn^{2+} as divalent metal activators. *Biochim. Biophys. Acta* **266,** 103–115.

Atkinson A., Hunt S., and Lowe A. G. (1968) Manganese activation of a (Na + K)-dependent ATPase in pig brain microsomes. *Biochim. Biophys. Acta* **167,** 469.

Avruch J. and Fairbanks G. (1972) Demonstration of a phosphopeptide intermediate, Na^+- and K^+-stimulated adenosine triphosphatase reaction of the erythrocyte membrane. *Proc. Natl. Acad. Sci. USA* **69,** 1216–1220.

Banerjee S. P. and Sen A. K. (1969) On the mechanism of inhibition of (Na + K)-ATPase by *N*-ethylmaleimide and ethacrynic acid. *Fed. Proc.* **28,** 589.

Banerjee S. P. and Sen A. K. (1970) Inhibition of sodium- and potassium-dependent adenosine triphosphatase by ethacrynic acid: Two modes of action. *Mol. Pharmacol.* **6,** 680–690.

Banerjee S. P. and Sharma V. K. (1978) ^{3}H-ouabain binding to peripheral organs of cats: Effects of ethanol. *Br. J. Pharmacol.* **62,** 475–479.

Banerjee S. P., Wong S. M. E., and Sen A. K. (1970) Differentiation between two conformations of (Na + K)-ATPase by *N*-ethylmaleimide. *Fed. Proc.* **29,** 723.

Barnett R. E. (1970) Effect of monovalent cations on the ouabain inhibition of the sodium and potassium ion activated adenosine triphosphatase. *Biochemistry* **9,** 4644–4648.

Beauge L. A. and Glynn I. M. (1978) Commercial ATP containing traces of vanadate alters the response of ($Na^+ + K^+$) ATPase to external potassium. *Nature* **272,** 551–552.

Bond G. H. and Hudgins P. M. (1979) Kinetics of inhibition of Na^+, K^+-ATPase by Mg^{2+}, K^+ and vanadate. *Biochemistry* **18,** 325–331.

Bonting S. L. (1970) Sodium-Potassium Activated Adenosine Triphosphatase and Cation Transport, in *Membrane Ion Transport,* Vol. I. (Bittar E. E., ed.), Wiley Interscience, New York.

Bonting S. L., Simon K. A., and Hawkins N. M. (1961) Studies on sodium-potassium-activated adenosine triphosphatase. I. Quantitative distribution in several tissues of the cat. *Arch. Biochem. Biophys.* **95,** 416–423.

Breer H. and Jeserich G. (1979) Ganglioside interaction with synaptosomal membranes, in *Glyco-Conjugates, Proceedings of the 5th International Symposium* (Schauer R., Booer P., Buddecke E., Kramer M. F., Vliegenthart J. C. G., and Wiegandt H. eds.), pp. 575–575, Georg Thieme, Stuttgart.

Britten J. S. and Blank M. (1968) Thallium activation of the ($Na^+ + K^+$)-activated ATPase of rabbit kidney. *Biochim. Biophys. Acta* **159,** 160–166.

Brodsky W. A. and Shamoo A. E. (1973) Binding of ATP to and release from microsomal Na^+, K^+-ATPase. *Biochim. Biophys. Acta* **291,** 208–228.

Brodsky W. and Sohn R. J. (1974) Acid-stable and heat-stable phosphoenzyme complex of ($Na^+ + K^+$)-ATPase in the eel electric organ, and the related concept of active Na transport. *Ann. NY Acad. Sci.* **242,** 106–119.

Cantley L. C., Jr. and Josephson L. (1976) A slow interconversion between active and inactive states of (Na-K)-ATPase. *Biochemistry* **15,** 5280–5287.

Cantley L. C., Cantley L. G., and Josephson L. (1978a) A characterization of vanadate interactions with the (Na-K)-ATPase mechanistic and regulatory implications. *J. Biol. Chem.* **253,** 7361–7368.

Cantley L. C., Ferguson J. H., and Kustin K. (1978b) Norepinephrine complexes and reduces vanadium (V) to reverse vanadate inhibition of the (Na-K)-ATPase. *J. Am. Chem. Soc.* **100,** 5210–5212.

Cantley L. C., Jr., Josephson L., Warner R., Yanacisawa M., Lechene C., and Guidotti G. (1977) Vanadate is a potent (Na-K)-ATPase inhibitor found in ATP derived from muscle. *J. Biol. Chem.* **252,** 7421–7423.

Caputto R., Maccioni A. H. R., and Caputto B. I. (1977) Activation of deoxycholate-solubilized adenosine triphosphatase by ganglioside and asialoganglioside preparations. *Biochem. Biophys. Res. Commun.* **74,** 1046–1052.

Chan S. L. and Quastel J. H. (1970) Effects of neurotropic drugs on sodium influx into rat brain cortex in vitro. *Biochem. Pharmacol.* **19,** 1071–1085.

Charney A. N., Silva P., and Epstein F. H. (1975) An in vitro inhibitor of Na-K-ATPase present in an adenosine triphosphate preparation. *J. Appl. Physiol.* **39,** 156–158.

Charnock J. S., Rosenthal A. S., and Post R. L. (1963) Studies of the mechanism of cation transport. II. A phosphorylated intermediate in the cation-stimulated enzymic hydrolysis of adenosine triphosphate. *Aust. J. Exp. Biol. Med. Sci.* **41,** 675–686.

Cheng E. Y. and Cutkomp L. K. (1975) The ATPase system in American cockroach muscle and nerve cord. *Insect. Biochem.* **5,** 421–427.

Choi Y. R. and Akera T. (1978) Membrane (Na^+, K^+)-ATPase of canine brain, heart and kidney. Tissue-dependent differences in kinetic properties and the influence of purification procedures. *Biochim. Biophys. Acta* **508,** 313–327.

Collins R. C. and Albers R. W. (1972) The phosphoryl acceptor protein of Na-K-ATPase from various tissues. *J. Neurochem.* **19,** 1209–1213.

Collins J. H., Forbush B., III, Lane L. K., Ling E., Schwartz A., and Zot A. (1982) Purification and characterization of an (Na^+ + K^+)-ATPase proteolipid labelled with a photoaffinity derivative of ouabain. *Biochim. Biophys. Acta* **686,** 7–12.

Cortas N. and Walser M. (1971) Sodium-potassium dependent ATPase in toad bladder: Novel response to ouabain. *Fed. Proc.* **30,** 331.

Cummins J. and Hyden H. (1962) Adenosine triphosphate levels and adenosine triphosphatases in neurons, glia and neuronal membranes of the vestibular nucleus. *Biochim. Biophys. Acta* **60,** 271–283.

Cutkomp L. K., Yap H. H., Vea E. U., and Koch R. B. (1971) Inhibition of oligomycin-sensitive (mitochondrial) Mg^{2+}-ATPase by DDT and selected analogs in fish and insect tissue. *Life Sci.* **10,** 1201–1209.

Dahl J. L. and Hokin L. E. (1974) The sodium-potassium adenosinetriphosphatase. *Ann. Rev. Biochem.* **43,** 327–356.

Daly J. (1977) *Cyclic Nucleotides in the Nervous System.* Plenum, New York.

Dean R. B. (1941) Theories of electrolyte equilibrium in muscle. *Biol. Symp.* **3,** 331–348.

Desaiah D. and Ho I. K. (1977) Kinetics of catecholamine sensitive Na^+-K^+-ATPase activity in mouse brain synaptosomes. *Biochem. Pharmacol.* **26,** 2029–2035.

Desaiah D. and Koch R. B. (1975) Inhibition of fish brain ATPases by aldrin-transdiol, aldrin, dieldrin and photodieldrin. *Biochem. Biophys. Res. Commun.* **64,** 13–19.

Duggan D. E. and Noll R. M. (1965) Effects of ethacrynic acid and cardiac glycosides upon a membrane adenosine triphosphatase of renal cortex. *Arch. Biochem. Biophys.* **109,** 388–96.

Dunham E. T. and Glynn I. M. (1961) Adenosine triphosphatase activity and the active movements of alkali metal ions. *J. Physiol.* (Lond.) **156,** 274–293.

Eayrs J. T. (1955) The cerebral cortex of normal and hypothyroid rats. *Acta Anat.* (Basel) **25,** 160–163.

Epstein F. H. and Whittam R. (1966) The mode of inhibition by calcium of cell-membrane adenosine triphosphatase activity. *Biochem. J.* **99,** 232–238.

Fagan J. B. and Racker E. (1977) Reversible inhibition of (Na^+-K^+)-ATPase by Mg^{2+}, adenosine triphosphate and K. *Biochemistry* **16,** 152–158.

Fahn S. and Cote L. J. (1968) Regional distribution of sodium-potassium activated adenosine triphosphatase in the brain of the Rhesus monkey. *J. Neurochem.* **15,** 433–436.

Fahn S., Hurley M. R., Koval G. J., and Albers R. W. (1966a) Sodium-potassium-activated adenosine triphosphatase of *Electrophorus* electric organ. *J. Biol. Chem.* **241,** 1890–1895.

Fahn S., Koval G. J., and Albers R. W. (1966b) Sodium-potassium-activated adenosine triphosphatase of *Electrophorus* electric organ. *J. Biol. Chem.* **241,** 1882–1889.

Festoff B. W. and Appel S. H. (1968) Effect of diphenylhydantoin on synaptosome Na^+, K^+-ATPase. *J. Clin. Invest.* **47,** 2752–2758.

Fiske C. H. and Subbarow Y. (1925) The colorimetric determination of phosphorus. *J. Biol. Chem.* **66,** 375–400.

Fishman P. H., Moss J., and Vaughan M. (1976) Uptake and metabolism of gangliosides in transformed mouse fibroblasts. Relationship of ganglioside structure to choleragen response. *J. Biol. Chem.* **251,** 4490–4494.

Folmar L. C. (1978) In vitro inhibition of rat brain ATPase, PNPPase and ATP-^{32}Pi exchange by chlorinated-diphenyl ethanes and cyclodiene insecticides. *Bull. Environ. Contam. Tox.* **19,** 481–487.

Forbush B., III, Kaplan J. H., and Hoffman J. F. (1978) Characterization of a new photoaffinity derivative of ouabain: Labelling of the large polypeptide and of a proteolipid component of the Na,K-ATPase. *Biochemistry* **17,** 3667–3676.

Formby B. (1975) Age-dependent changes in (Na,K)-ATPase in brains of mice susceptible to seizures. *Experientia* **31,** 315–316.

Froehlich J. P., Albers R. W., Koval G. J., Goebel R., and Berman M. (1976) Evidence for a new intermediate state in the mechanism of (Na^+ + K^+)-adenosine triphosphatase. *J. Biol. Chem.* **251,** 2186–2188.

Fukushima Y. and Tonomura Y. (1973) Two kinds of high energy phosphorylated intermediate, with and without bound ADP in the reaction of Na^+-K^+-dependent ATPase. *J. Biochem.* (Tokyo) **74,** 135–142.

Geel S. E., Valcana T., and Timiras P. S. (1967) Effect of neonatal hypothyroidism and of thyroxine on L-(^{14}C)-lucine incorporation in protein in vivo and the relationship to ionic levels in the developing brain of the rat. *Brain Res.* **4,** 143–150.

Gilbert J. C., Wyllie M. G., and Davison D. V. (1975) Nerve terminal ATPase as possible trigger for neurotransmitter release. *Nature* **255,** 237–238.

Giotta G. J. (1975) Native (Na^+ + K^+)-dependent adenosine triphosphatase has two trypsin-sensitive sites. *J. Biol. Chem.* **250,** 5159–5164.

Glick N. B. (1972) Inhibition of Transport Reactions. A. Inhibitors of ATPase: NaK-ATPase and Related Enzymic Activities, in *Metabolic Inhibitors* (Quastel J. H., Kates M., and Hochester R. M., eds.), Academic, New York, London.

Glynn I. M. (1957) The action of cardiac glycosides on sodium and potassium movements in human red cells. *J. Physiol.* (Lond.) **136,** 148–173.

Glynn I. M. (1963) Transport of adenosine triphosphatase in electric organ. The relation between ion transport and oxidative phosphorylation. *J. Physiol.* (Lond.) **169,** 452–465.

Godfraind T., Koch M.-C., and Verbeke N. (1974) The action of EGTA on the catecholamines stimulation of rat brain Na-K-ATPase. *Biochem. Pharmacol.* **23,** 3505–3511.

Grasso A. (1967) A sodium- and potassium-stimulated adenosine triphosphatase in the cockroach nerve cord. *Life Sci.* **6,** 1911–1918.

Green I. and Mommaerts W. F. H. M. (1953) Adenosine triphosphatase systems of muscle. I. An electrotitrimetic method of measurement. *J. Biol. Chem.* **202,** 541–549.

Gurd F. R. N. and Murray G. R., Jr. (1954) Preparation and properties of serum and plasma proteins. XXXIX. The interaction of human serum albumin with plumbous ions. *J. Am. Chem. Soc.* **76,** 187–190.

Hackenberg H. and Krieglstein J. (1972) Comparative study on the inhibition of Na^+, K^+-ATPase by chlorpromazine, promazine, imipramine and their monodesmethyl metabolites. *Naunyn-Schmiedebergs Arch. Pharmacol.* **247,** 63–73.

Hansen O. (1976) Non-uniform populations of g-strophanthin binding sites of (Na^+ + K^+)-activated : ATPase apparent conversion to uniformity by K^+. *Biochim. Biophys. Acta* **433,** 383–392.

Hart W. M., Jr. and Titus E. O. (1973a) Isolation of a protein component of sodium-potassium transport: Adenosine triphosphatase containing ligand-protected sulfhydryl groups. *J. Biol. Chem.* **248,** 1365–1371.

Hart W. M., Jr. and Titus E. O. (1973b) Sulfhydryl groups of Na^+, K^+-ATPase: Protection by physiological ligands and exposure by phosphorylation. *J. Biol. Chem.* **248,** 4674–4681.

Hasie R. J. A. (1965) The localization of adenosine triphosphatases in morphologically characterized subcellular fractions of guinea-pig brain. *Biochem. J.* **96,** 404–412.

Heinz E. and Hoffman J. F. (1965) Phosphate incorporation and Na, K-ATPase activity in human red blood cell ghosts. *J. Cell Comp. Physiol.* **65,** 31–43.

Herd P. A., Horwitz B. A., and Smith R. E. (1970) Norepinephrine-sensitive Na^+/K^+-ATPase activity in brown adipose tissue. *Experientia* **26,** 825–826.

Hertz L., Schousboe A., Formby B., and Lennox-Buchthal M. (1974) Some age-dependent biochemical changes in mice susceptible to seizures. *Epilepsia* **15,** 619–631.

Hexum T. D. (1974) Studies on the reaction catalyzed by transport (Na,K) adenosine triphosphatase. I. Effects of divalent metals. *Biochem. Pharmacol.* **23,** 3441–3447.

Hexum T. D. (1977) The effect of catecholamines on transport (Na,K) adenosine triphosphatase. *Biochem. Pharmacol.* **26,** 1221–1227.

Hoffman J. F. (1962) The active transport of sodium by ghosts of human red blood cells. *J. Gen. Physiol.* **45,** 837–859.

Hokin L. E. (1973) Studies on the characterization of the sodium-potassium transport adenosine triphosphatase. X. Purification of the enzyme from the rectal gland of *Squalus acanthias*. *J. Biol. Chem.* **248,** 2593–2605.

Hokin L. W. and Hokin M. R. (1963) Biological transport. *Ann. Rev. Biochem.* **32,** 553–578.

Hokin L. E., Dahl J. L., Deupree J. D., Dixon J. F., Hackney J. F., and Perdue J. F. (1973) Studies on the characterization of the sodium-potassium transport adenosine triphosphatase. X. Purification of the enzyme from the rectal gland of *Squalus acanthias*. *J. Biol. Chem.* **248,** 2593–2605.

Huang W. and Askari A. (1978) Na,K^+-ATPase: On the nature of the "cross-linked" subunit obtained in the presence of *O*-phenanthroline and cupric ion. *Biochem. Biophys. Res. Commun.* **82,** 1314–1319.

Huang W. and Askari A. (1979) Na^+,K^+-ATPase: Effects of detergents on the cross-linking of subunit in the presence of Cu^{2+} and *O*-phenanthroline. *Biochim. Biophys. Acta* **578,** 547–552.

Hudgins P. M. and Bond G. H. (1977) (Mg^{2+} + K^+)-dependent inhibition of Na,K-ATPase due to a contaminant in equine muscle ATP. *Biochem. Biophys. Res. Commun.* **77,** 1024–1029.

Israel M. A. and Kuriyama K. (1971) Effect of in vivo ethanol administration on adenosine triphosphatase activity of subcellular fractions of mouse brain and liver. *Life Sci.* **10,** 591–599.

Iwangoff P., Enz A., and Chappeus A. (1974) Effect of adrenergic blockers on the activation of brain ATPase by noradrenaline. *Experientia* **30,** 688.

Jarnefelt J. (1961) Inhibition of the brain microsomal ATPase by depolarizing agents. *Biochim. Biophys. Acta* **48,** 111–116.

Jarnefelt J. (1964) Conversion of the Na^+ and K^+ independent part of the brain microsomal ATPase to a form requiring added Na^+ and K^+. *Biochem. Biophys. Res. Commun.* **17,** 330–334.

Jean D. H. and Albers R. W. (1976) Immunochemical studies on the large polypeptide of *Electrophorous electroplax* ($Na^+ + K^+$)-ATPase. *Biochim. Biophys. Acta* **452,** 219–226.

Johnstone R. M. (1963) Sulfhydryl Agents, Arsenicals, in *Metabolic Inhibitors,* Vol. 2. (Hochster R. M. and Quastel J. H., eds.), Academic, New York.

Jones V. D., Lockett G., and Landon E. J. (1965) A cellular action of mercurial diuretics. *J. Pharmacol. Exp. Ther.* **147,** 23–31.

Jorgensen P. L. (1974) Purification of Na^+,K^+-ATPase: Active site determination of criteria of purity. *Ann. NY Acad. Sci.* **242,** 36–52.

Jorgensen P. L. and Skou J. C. (1971) Purification and characterization of ($Na^+ + K^+$)-ATPase. I. The influence of detergents on the activity of ($Na^+ + K^+$)-ATPase in preparations from the outer medulla of rabbit kidneys. *Biochim. Biophys. Acta* **233,** 366–380.

Josephson L. and Cantley L. C. (1977) Isolation of a potent (Na-K)-ATPase inhibitor from striated muscle. *Biochemistry* **16,** 4572–4578.

Kahlenberg A., Dulak N. C., Dixon J. F., Galsworthy P. R., and Hokin L. E. (1969) Studies on the characterization of the sodium-potassium transport adenosine triphosphatase. V. Partial purification of the Lubrol-solubilized beef brain enzyme. *Arch. Biochem. Biophys.* **131,** 253–262.

Kahlenberg A., Galsworthy P. R., and Hokin L. E. (1968) Studies on the characterization of the sodium-potassium transport adenosine triphosphatase. II. Characterization of the acyl phosphate intermediate as an L-glutamyl-gamma-phosphate residue. *Arch. Biochem. Biophys.* **126,** 331–324.

Kanazawa T., Saito M., and Tonomura Y. (1967) Properties of a phosphorylated protein as a reaction intermediate of Na^+-K^+-sensitive ATPase. *J. Biochem.* (Tokyo) **61,** 555–566.

Kanazawa T., Saito M., and Tonomura Y. (1970) Formation and decomposition of a phosphorylated intermediate in the reaction of Na^+-K^+-dependent ATPase. *J. Biochem.* (Tokyo) **67,** 693–711.

Karlish S. J. (1980) Characterization of conformational changes in (Na,K)-ATPase labelled with fluorescein at the active site. *J. Bioenerg. Biomemb.* **12,** 111–136.

Klara P. M., Brizzee K. R., Chen I.-Li, and Yates R. D. (1978) Ultrastructural localization of ATPase activity in the dog area postrema. *Brain Res.* **146,** 165–171.

Klotz I. M., Urquhart J. M., and Fiess H. A. (1952) Interactions of metal ions with the sulfhydryl group of serum albumin. *J. Am. Chem. Soc.* **74,** 5537–5538.

Knox W. H., Perrier R. G., and Sen A. K. (1972) Effect of chronic administration of ethanol on Na^+,K^+-ATPase activity in areas of the cat brain. *J. Neurochem.* **19,** 2881–2884.

Koch R. B. (1969) Fractionation of olfactory tissue homogenates. Isolation of a concentrated plasma membrane fraction. *J. Neurochem.* **16,** 145–157.

Koch R. B., Cutkomp L. K., and Do F. M. (1969) Chlorinated hydrocarbon insecticide inhibition of cockroach and honey bee ATPases. *Life Sci.* **8,** 289–297.

Kurokawa M., Sakamoto T., and Kato M. (1965) Distribution of sodium-plus-potassium-stimulated adenosine-triphosphatase activity in isolated nerve ending particles. *Biochem. J.* **97,** 833–844.

Kyte J. (1971) Phosphorylation of a purified (Na^+ + K^+) adenosine triphosphatase. *Biochem. Biophys. Res. Commun.* **43,** 1259–1265.

Kyte J. (1972) Properties of the two polypeptides of sodium and potassium-dependent adenosine triphosphatase. *J. Biol. Chem.* **247,** 7642–7649.

Kyte J. (1975) Structural studies of sodium and potassium ion-activated adenosine triphosphatase. The relationship between molecular structure and the mechanism of active transport. *J. Biol. Chem.* **250,** 7443–7449.

Kyte J. (1976a) Immunoferritin determination of the distribution of the (Na^+,K^+)-ATPase over the plasma membranes of renal convoluted tubules. I. Distal segment. *J. Cell. Biol.* **68,** 287–303.

Kyte J. (1976b) Immunoferritin determination of the distribution of the (Na^+,K^+)-ATPase over the plasma membranes of renal convoluted tubules. I. Distal segment. *J. Cell. Biol.* **68,** 304–381.

Laemmli U. K. (1970) Cleavage of structural proteins during the assembly of the head of bacteriophage T_4. *Nature* **227,** 680–685.

Leon A., Facci L., Toffano G., Sonnino S., and Tettamanti G. (1981) Activation of (Na^+,K^+)-ATPase by nanomolar concentrations of GM_1 ganglioside. *J. Neurochem.* **37,** 350–357.

Liang S. and Winter C. G. (1977) Digitonin-induced changes in subunit arrangement in relation to some in vitro activities of the Na^+,K^+-ATPase. *J. Biol. Chem.* **252,** 8278–8284.

Lindenmayer G. E., Laughter A. H., and Schwartz A. (1968) Incorporation of inorganic phosphate-32 into a Na^+,K^+-ATPase preparation: A stimulation by ouabain. *Arch. Biochem. Biophys.* **127,** 187–192.

Lindenmayer G. E. and Schwartz A. (1970) Conformational changes induced in Na^+,K^+-ATPase by ouabain through a K^+-sensitive reduction: Kinetic and spectroscopic studies. *Arch. Biochem. Biophys.* **140,** 371–378.

Madsen N. B. (1963) A Comprehensive Treatial, in *Metabolic Inhibitors,* Vol. 2 (Hochster R. M. and Quastel J. H., eds.), Academic, New York.

Matsumura F. and Patil K. C. (1969) Adenosine triphosphatase sensitive to DDT in synapses of rat brain. *Science* **166,** 121–122.

Medzihardsky F., Kline M. H., and Hokin L. E. (1967) Studies on the characterization of the sodium-potassium transport adenosine triphosphatase. I. Solubilization, stabilization and estimation of apparent molecular weight. *Arch. Biochem. Biophys.* **121,** 311–316.

Millonig G. (1961) Advantages of a phosphate buffer for OsO_4 solution in fixation. *J. Appl. Physiol.* **32,** 1637.

Moczydlowski E. G. (1981a) Characterization of 2',3'-*O*-(2,4,6-trinitro-cyclohexadienylidene) adenosine 5'-triphosphate as a fluorescent probe of the ATP site of sodium and potassium transport adenosine triphosphatase. Determination of nucleotide binding stoichiometry and ion induced changes in affinity for ATP. *J. Biol. Chem.* **256,** 2346–2356.

Moczydlowski E. G. (1981b) Inhibition of sodium and potassium adenosine triphosphatase by 2',3'-*O*-(2,4,6-trinitro-cyclohexadienylidene)adenine nucleotides. Implication for the structure and mechanism of the Na:K pump. *J. Biol. Chem.* **256,** 2357–2366.

Nagano K., Kanazawa T., Mizuno N., Tashima Y., Nakao T., and Nakao M. (1965) Some acyl phosphate-like properties of P^{32}-labelled sodium-potassium-activated adenosine triphosphatase. *Biochem. Biophys. Res. Commun.* **19,** 759–764.

Ottolenghi P. and Ellory J. C. (1983) Radiation inactivation of (Na,K)-ATPase, an enzyme showing multiple radiation-sensitive domains. *J. Biol. Chem.* **258,** 14895–14907.

Partington C. R. and Daly J. W. (1979) Effect of gangliosides on adenylate cyclase activity in rat cerebral cortical membranes. *Mol. Pharmacol.* **15,** 484–491.

Peters R. A., Shorthouse M., and Walshe J. M. (1966) Studies on the toxicity of copper. II. The behaviour of microsomal membrane ATPase of the pigeon's brain tissue to copper and some other metallic substance. *Proc. Roy. Soc. Ser. B.* **166,** 285–294.

Petrali E. H. and Sulakhe P. V. (1980) Isolation of plasma membrane fraction highly enriched in ouabain-sensitive Na^+-K^+-ATPase from rat brain white matter. *Int. J. Biochem.* **12,** 407–420.

Phillis J. W. and Wu P. H. (1981) Catecholamines and the sodium pump in excitable cells. *Prog. Neurobiol.* **17,** 141–184.

Portuis H. J. and Repke K. (1964) Versuch einer analyse der beziehunge zwischen chemischer struktur und digitalis-ahnlicher wirksamkeit an der zezeptorebene. *Arzneim. Forsch.* **14,** 1073–1077.

Post R. L., Kume S., Tobin T., Orcutt B., and Sen A. K. (1969) Flexibility of an active center in sodium-plus-potassium adenosine triphosphatase. *J. Gen. Physiol.* **54,** 306–326.

Post R. L., Sen A. K., and Rosenthal A. S. (1965) A phosphorylated intermediate in adenosine triphosphate-dependent sodium and potassium transport across kidney membranes. *J. Biol. Chem.* **240,** 1437–1445.

Powis D. A., Wattus G. D., and Glynn I. (1981) The stimulatory effect of calcium on sodium-potassium ATPase of nervous tissue. *FEBS Lett.* **126,** 285–288.

Quastel J. H. (1970) Transport process at the brain cell membrane. *Neurosci. Res.* **3,** 1–41.

Racker E. (1976) Structure and function of ATP-driven ion pumps. *Trends Biochem. Sci.* **1,** 244–247.

Rawson M. D. and Pincus J. H. (1968) The effect of diphenylhydantoin on sodium, potassium, magnesium-activated adenosine triphosphatase in microsomal fraction of rat and guinea pig brain and on whole homogenates of human brain. *Biochem. Pharmacol.* **17,** 573–579.

Reichert W. H. (1975) Cerebral magnesium and sodium-potassium ATPase following audiogenic seizure in mice. *Exp. Neurol.* **49,** 596–600.

Rendi R. and Uhr M. L. (1964) Sodium, potassium-requiring adenosine triphosphatase activity. I. Purification and properties. *Biochim. Biophys. Acta* **89,** 520–531.

Riehl R. (1980) Ultracytochemical localization of Na^+, K^+-activated ATPase in the oocytes of *Heterandria formosa Agassiz,* 1853 (Pisces, Poecilliadae). *Reprod. Nutr. Develop.* **20,** 191–196.

Robinson J. D., Lowinger J., and Bettinger B. (1968) Chlorpromazine: differential effects on membrane-bound enzymes from rat brain. *Biochem. Pharmacol.* **17,** 1113–1116.

Rodriguez de Lores Arnaiz G. and Mistrorigo de Pacheco M. (1978) Regulation of (Na^+,K^+) adenosine triphosphatase of nerve ending membranes. Action of norepinephrine and a soluble factor. *Neurochem. Res.* **3,** 733–744.

Rosenblatt D. E., Lauter C. J., and Trams E. G. (1976) Deficiency of a Ca^{2+}-ATPase in brains of seizure-prone mice. *J. Neurochem.* **27,** 1299–1304.

Ruoho A. and Kyte J. (1974) Photoaffinity labelling of the ouabain-binding site on (Na^+ + K^+) adenosine triphosphatase. *Proc. Natl. Acad. Sci. USA* **71,** 2352–2356.

Schaefer A., Komlos M., and Seregi A. (1975) Lipid peroxidation as the cause of the ascorbic acid-induced decrease of adenosine triphosphatase activities of rat brain microsomes and its inhibition by biogenic amines and psychotropic drugs. *Biochem. Pharmacol.* **24,** 1781–1786.

Schaefer A., Seregi A., and Komlos M. (1974) Ascorbic acid-like effect of the soluble fraction of rat brain on adenosine triphosphatase and its relation to catecholamines and chelating agents. *Biochem. Pharmacol.* **23,** 2257–2271.

Schaefer A., Unyi G., and Pfeifer A. K. (1972) The effects of a soluble factor and of catecholamines on the activity of adenosine triphosphatase in subcellular fractions of rat brain. *Biochem. Pharmacol.* **21,** 2289–2294.

Schatzmann H. J. (1953) Herzglykoside als hemmstoffe fur der aktiven kalium-und natriumtransport durch dei Erythrocytenmembran. *Helv. Physiol. Pharmac. Acta* **11,** 346–354.

Schirachi D. Y., Allard A. A., and Trevor A. J. (1970) Partial purification and ouabain sensitivity of Lubrol-extracted sodium-potassium transport adenosine triphosphatases from brain and cardiac tissue. *Biochem. Pharmacol.* **19,** 2893–2906.

Schurmans-Stekhovens F. and Bonting S. L. (1981) Transport adenosine triphosphatase: Properties and functions. *Physiol. Rev.* **61,** 1–76.

Schwartz A., Bachelard H. S., and McIlwain H. (1962) The sodium, potassium-stimulated adenosine triphosphatase activity and other properties of cerebral microsomal fractions and subfractions. *Biochem. J.* **84,** 626–637.

Schwartz A., Lindenmayer G. F., and Allen J. C. (1975) The sodium-potassium adenosine triphosphatase: Pharmacological, physiological and biochemical aspects. *Pharmacol. Rev.* **27,** 3–134.

Schwartz A., Matsui H., and Laughter A. H. (1968) Tritrated digoxin binding to Na^+,K^+-activated adenosine triphosphatase: Possible allosteric site. *Science* **159,** 323–325.

Seiler F. R., Sedlacek N. N., Luben G., and Wiegandt H. (1976) Alteration of the lymphocyte surface by vibrio cholerae neuraminidase, gangliosides and lysolecithins. *Behring. Inst. Mitt.* **59,** 22–29.

Sen A. K., Tobin T., and Post R. L. (1969) A cycle for ouabain inhibition of sodium- and potassium-dependent adenosine triphosphate. *J. Biol. Chem.* **244,** 6596–6604.

Shamoo A. E. and Brodsky W. A. (1972) Functions of the E-ATP and E-P complex in the membrane ATPase reaction. *Biochim. Biophys. Acta* **255,** 220–230.

Siegel G. J. and Albers R. W. (1967) Sodium-potassium-activated adenosine triphosphatase of *Electrophorus* electric organ. IV. Modification of response to sodium and potassium by arsenite plus 2,3-dimercaptopropanol. *J. Biol. Chem.* **242,** 4972–4979.

Siegel G. J. and Fogt S. M. (1977) Inhibition by lead ion of *ELectrophorus electroplax* (Na^+ + K^+)-adenosine triophosphatase and K^+-p-nitrophenylphosphatase. *J. Biol. Chem.* **252,** 5201–5205.

Seigel G. J. and Albers R. W. (1967) Sodium-potassium-activated adenosine triphosphatase. VI. Characterization of the phosphoprotein formed from orthophosphate in the presence of ouabain. *J. Biol. Chem.* **244,** 3264–3269.

Skou J. C. (1957) The influence of some cations on an adenosine triphosphatase from peripheral nerves. *Biochim. Biophys. Acta* **23,** 394–401.

Skou J. C. (1960) Further investigations on a Mg^{2+} + Na^+ activated adenosine triphosphatase possibly related to the active linked transport of Na^+ and K^+ across the nerve membrane. *Biochim. Biophys. Acta* **42,** 6–23.

Skou J. C. (1962) Preparation from mammalian brain and kidney of the enzyme system involved in active transport of Na^+ and K^+. *Biochim. Biophys. Acta* **58,** 314–325.

Skou J. C. (1971) The Role of the Phosphorylated Intermediates in the Reaction of the Na^+,K^+-Activated Enzyme System, in *Membrane*

Bound Enzymes (Porcellati G. and diJeso F., eds.), Plenum, New York.

Skou J. C. (1981) Eosin, a fluorescent probe of ATP binding to the (Na^+ + K^+) ATPase. *Biochim. Biophys. Acta* **647,** 232–240.

Skou J. C. and Hilberg C. (1965) The effect of sulphydryl-blocking reagents and of urea on the (Na^+ + K^+)-activated enzyme system. *Biochim. Biophys. Acta* **110,** 359.

Skulskii I. A., Manninen V., and Jarnefelt J. (1973) Interaction of thallous ions with the cation transport mechanism in erythrocytes. *Biochim. Biophys. Acta* **298,** 702–709.

Somogyi J., Budai M., Nyiro L., Kaluza G. A., Nagel W., and Willig F. (1969) Activity and structural changes in Mg^{2+}-dependent and Na^+,K^+-activated adenosine triphosphatase prepared from rat brain following detergent treatment. *Acta Biochem. Biophys. Acad. Sci. (Hung.)* **4,** 219–236.

Specht S. C. and Robinson J. D. (1973) Stimulation of the (Na^+ + K^+)-dependent adenosine triphosphatase by amino acids and phosphatidylserine: Chelation of trace metal inhibitors. *Arch. Biochem. Biophys.* **154,** 314–323.

Squires R. F. (1965) On the interactions of Na^+, K^+, Mg^{2+} and ATP with the (Na^+ + K^+)-activated ATPase from rat brain. *Biochem. Biophys. Res. Commun.* **19,** 27–32.

Stahl W. L., Sattin A., and McIlwain H. (1966) Separation of adenosine diphosphate-adenosine triphosphate-exchange activity from the cerebral microsomal sodium plus potassium ion-stimulated adenosine triphosphatase. *Biochem. J.* **99,** 404–412.

Stefanovic V., Ebel A., Hermetet J. C., and Mandel P. (1974) Na^+-K^+-ATPase activity in brain regions of C_{57} and DBA mice. *J. Neurochem.* **22,** 1139–1141.

Svoboda P. and Mosinger B. (1981a) Catecholamines and the brain microsomal Na,K-adenosine triphosphatase. I. Protection against lipoperoxidative damage. *Biochem. Pharmacol.* **30,** 427–432.

Svoboda P. and Mosinger B. (1981b) Catecholamines and the brain microsomal Na,K-adenosine triphosphatase. II. The mechanism of action. *Biochem. Pharmacol.* **30,** 433–439.

Sweadner K. J. (1977) Cross-linking and modification of Na,K-ATPase by ethyl acetimidate. *Biochem. Biophys. Res. Commun.* **78,** 962–969.

Sweadner K. J. (1978) Purification from brain of an intrinsic membrane protein fraction enriched in (Na^+ + K^+)-ATPase. *Biochim. Biophys. Acta* **508,** 486–499.

Sweadner K. J. (1979) Two molecular forms of (Na^+ + K^+)-stimulated ATPase in brain. Separation and difference in affinity for strophanthidin. *J. Biol. Chem.* **254,** 6060–6067.

Tanaka R. (1974) Role of lipids in activation of Na^+,K^+-dependent ATPase and K^+-dependent phosphatase of the brain. *Rev. Neurosci.* **1,** 181–230.

Tanford C. (1952) The effect of pH on the combination of serum albumin with metals. *J. Am. Chem. Soc.* **74,** 211–215.

Ting-Beall H. P., Clark D. A., Snelter C. H., and Wells W. W. (1973) Studies on the interaction of chick brain microsomal (Na^+ + K^+)-ATPase with copper. *Biochim. Biophys. Acta* **291,** 229–236.

Tobin T., Akera T., Baskin S. I., and Brody T. M. (1973) Calcium ion and sodium- and potassium-dependent adenosine triphosphatase: Its mechanism of inhibition and identification of the E_1-P intermediate. *Mol. Pharmacol.* **9,** 336–349.

Tobin T. and Sen A. K. (1970) Stability and ligand sensitivity of ^{3}H-ouabain binding to Na^+,K^+-ATPase. *Biochim. Biophys. Acta* **198,** 120–131.

Toda G. (1968) The effects of cations on the inhibition of sodium- and potassium-activated adenosine triphosphatase by beryllium. *J. Biochem.* (Tokyo) **64,** 457–464.

Toffano G., Benvegnu D., Bonetti A. C., Leon A., Orlando P., Ghidoni R., and Tettamanti G. (1980) Interactions of GM_1 ganglioside with crude rat brain neuronal membranes. *J. Neurochem.* **35,** 861–866.

Trauner D. A. (1980) Regional cerebral Na^+,K^+-ATPase activity following octanoate administration. *Pediatr. Res.* **14,** 844–845.

Uesugi S., Kahlenberg A., Medzihrodsky F., and Hokin L. E. (1969) Studies on the characterization of the sodium-potassium transport adenosine triphosphatase. IV. Properties of a Lubrol-solubilized beef brain microsomal enzyme. *Arch. Biochem. Biophys.* **130,** 156–163.

Uesugi S., Dulak N. C., Dixon J. F., Hexum T. I., Dahl J. L., Perdue J. F., and Hokin L. E. (1971) Studies on the characterization of the sodium-potassium transport adenosine triphosphatase. I. Large-scale partial purification and properties of a Lubrol-solubilized bovine brain enzyme. *J. Biol. Chem.* **269,** 531–543.

Valcana T. and Timiras P. S. (1969) Effect of hypothyroidism on ionic metabolism and Na-K-activated ATP phosphohydrolase activity in the developing rat brain. *J. Neurochem.* **16,** 935–943.

Van der Krogt J. A. and Belfroid R. D. (1980) Characterization and localization of catecholamine-susceptible Na-K ATPase activity of rat striatum: Studies using catecholamine receptor (ant)agonists and lesion techniques. *Biochem. Pharmacol.* **29,** 857–868.

Van Winkle W. B., Allen J. C., and Schwartz A. (1972) The nature of the transport ATPase-digitalis complex. III. Rapid binding studies and effects of ligands on the formation and stability of magnesium plus phosphate-induced glycoside–enzyme complex. *Arch. Biochem. Biophys.* **151,** 85–92.

Vizi E. S. (1979) Pre-synaptic modulation of neurochemical transmission. *Prog. Neurobiol.* **12,** 181–290.

Vizi E. S. (1980) Modulation of cortical release of acetylcholine by noradrenaline released from nerves arising from the rat locus coeruleus. *Neuroscience* **5,** 2139–2144.

Whitaker J. N. (1975) The antigenicity of myelin encephalitogenic protein: Production of antibodies to encephalitogenic protein with deoxyribonucleic acid–encephalitogenic protein complexes. *J. Immuno.* **114**, 823–828.

Whittaker V. P. (1959) The isolation and characterization of acetylcholine-containing particles from brain. *Biochem. J.* **72**, 694–706.

Whittam R. (1962) The asymmetrical stimulation of a membrane adenosine triphosphatase in relation to active cation transport. *Biochem. J.* **84**, 110–118.

Whittam R. and Chipperfield A. R. (1973) Ouabain binding to the sodium pump in plasma membranes isolated from ox brain. *Biochim. Biophys. Acta* **307**, 563–577.

Whittam R., Wheeler K. P., and Blake A. (1964) Oligomycin and active transport reactions in cell membranes. *Nature* **203**, 720–724.

Wilson W. E., Sivitz W. I., and Hanna L. T. (1970) Inhibition of calf brain membranal sodium- and potassium-dependent adenosine triphosphatase by cardioactive sterols. A binding site model. *Mol. Pharmacol.* **6**, 449–459.

Winter C. G. (1972) Differential effects of digitonin on some enzyme activities of the sodium pump. *Biochim. Biophys. Acta* **266**, 135–143.

Wood J. G., Jean D. H., Whitaker J. N., McLaughlin B. J., and Albers W. (1977) Immunocytochemical localization of the sodium, potassium activated ATPase in knifefish brain. *J. Neurocyto.* **6**, 571–581.

Woodbury D. M. (1955) Effect of diphenylhydantoin on electrolytes and radio sodium turnover in brain and other tissues of normal hyponatremic and postictal rats. *J. Pharmacol. Exp. Ther.* **115**, 74–95.

Wu P. H. and Phillis J. W. (1978) Effects of α- and β-adrenergic blocking agents on the biogenic amine stimulated (Na^+,K^+) ATPase of rat cerebral cortical synaptosomal membrane. *Gen. Pharmacol.* **9**, 421–424.

Wu P. H. and Phillis J. W. (1979a) Receptor-mediated noradrenaline stimulation of (Na^+-K^+) ATPase in rat brain cortical homogenates. *Gen. Pharmacol.* **10**, 189–192.

Wu P. H. and Phillis J. W. (1979b) Metergoline antagonism of 5-hydroxytryptamine-induced activation of rat cortical (Na^+-K^+) ATPase. *J. Pharm. Pharmacol.* **31**, 782–784.

Wu P. H. and Phillis J. W. (1979c) Effects of vanadate on brain (Na^+,K^+) ATPase and p-nitrophenylphosphatase. Interactions with mono- and di-valent ions and with noradrenaline. *Intl. J. Biochem.* **10**, 629–635.

Wu P. H. and Phillis J. W. (1980) Characterization of receptor-mediated catecholamine activation of rat brain cortical Na^+-K^+-ATPase. *Intl. J. Biochem.* **12**, 353–359.

Yamaguchi M. and Tonomura Y. (1979) Simultaneous binding of three Na^+ and two K^+ ions to Na^+,K^+-dependent ATPase and changes in its affinities for the ions induced by the formation of a phosphorylated intermediate. *J. Biochem.* (Tokyo) **86,** 509–523.

Yamaguchi M. and Tonomura Y. (1980a) Binding of monovalent cations to Na^+,K^+-dependent ATPase purified from porcine kidney. I. Simultaneous binding of three sodium and two potassium or rubidium ions to the enzyme. *J. Biochem.* (Tokyo) **88,** 1365–1375.

Yamaguchi M. and Tonomura Y. (1980b) Binding of monovalent cations to Na^+,K^+-dependent ATPase purified from porcine kidney. II. Acceleration of transition from a K^+-bound form to a Na^+-bound form by binding of ATP to a regulatory site of the enzyme. *J. Biochem.* (Tokyo) **88,** 1377–1385.

Yamaguchi M. and Tonomura Y. (1980c) Binding of monovalent cations to Na^+,K^+-dependent ATPase purified from porcine kidney. III. Marked changes in affinities for mono-valent cations induced by formation of an ADP-insensitive but not an ADP-sensitive phosphoenzyme. *J. Biochem.* (Tokyo) **88,** 1387–1397.

Yap H. H., Desaiah D., Koch R. B., and Cutkomp L. K. (1972) Sensitivity of fish ATPase to polychlorinated biphenyls. *Nature* **233,** 61–62.

Yoda A. and Hokin L. E. (1970) On the reversibility of binding of cardiotonic steroids to a partially purified (Na + K)-activated adenosine triphosphatase from beef brain. *Biochem. Biophys. Res. Commun.* **40,** 880–886.

Yoshimura K. (1973) Activation of Na-K activated ATPase in rat brain by catecholamine. *J. Biochem.* (Tokyo) **74,** 389–391.

Chapter 14
Brain Adenylate Cyclase

Prakash V. Sulakhe, Ramesh C. Gupta, and Gowraganahalli Jagadeesh

1. Introduction

Adenylate cyclase (AC) from the mammalian central nervous system (CNS) has been the subject of intensive examination (*see* Drummond, 1983, for a recent review). The presence of AC that is sensitive to numerous neurotransmitters has been documented, and further, both stimulation and inhibition of the brain enzyme activity by neurotransmitters have been observed. The minimum components comprising neurotransmitter-sensitive AC in the brain appear identical to those discerned from the detailed studies carried out using lymphoma cells (Smiegel et al., 1984), platelets (Limbird, 1983), liver (Smiegel et al., 1984; Codina et al., 1984), heart (Drummond, 1984; Sulakhe et al., 1985), and red blood cells (Stiles et al., 1984).

So far, the available evidence documents the presence of three components—a receptor protein specific for each neurotransmitter, a catalytic unit of AC, and guanine nucleotide-binding proteins called N_s and N_i that mediate stimulation and inhibition of the enzyme, respectively (Gilman, 1984). Although there has been considerable success in the purification of some receptors (for example, β-adrenergic receptors) (*see* Stiles et al., 1984), there is not as yet a homogeneous preparation of an AC catalytic unit from either brain or peripheral tissues. There are, however, reports of purification of the guanosine triphosphate (GTP)-binding coupling proteins, both N_s and N_i, from tissues such as liver (Northup et al., 1983), heart (Malbon et al., 1985), and brain (Neer et al., 1984). It is also now evident that GTP-binding coupling proteins possess GTPase activity (Gilman, 1984; Ross et al., 1983), and this enzymatic characteristic has made their coupling roles even more crucial in the overall mechanisms underlying stimulatory and inhibitory actions of neurotransmitters on AC. The availability of two bacterial toxins (cholera toxin and pertussis toxin, called islet-activating protein, IAP) has allowed identification of N_s and N_i protein subunits (α) in the cellular fractions, since they undergo ADP ribosylation in the presence of ^{32}P-NAD and cholera toxin or pertussis toxin (IAP), respectively (Gilman, 1984). Photoaffinity labeling of the receptors using appropriate ligands has also allowed identification of the receptors in the membrane fractions derived from various tissues (Stiles et al., 1984).

Aside from the presence of neurotransmitter-sensitive AC in the CNS, divalent cation-sensitive, notably Ca^{2+}-sensitive, AC activity is present in this tissue. This particular form of AC is detected in (neuro)secretory tissues. The evidence strongly suggests that calmodulin, a multifunctional Ca^{2+}-binding protein, mediates the stimulatory action of Ca^{2+} on brain AC activity (*see* review by Drummond, 1983 and references in articles by Sulakhe, 1985 and MacNeil et al., 1985). There exists a strong likelihood that the catalytic unit of brain AC that is sensitive to Ca^{2+}-calmodulin may indeed be different from that sensitive to regulation by neurotransmitters via coupling proteins N_s and N_i. Since a Ca^{2+} ion is involved in many critical functions of the brain (excitability, presynaptic neurotransmitter release), there is growing interest in understanding how Ca^{2+} controls neuronal cyclic adenosine monophosphate (AMP) metabolism, since this cation not only regulates the synthesis of cyclic AMP, but, via a mediatory role of calmodulin, regulates cyclic AMP degradation as well (Drummond, 1983).

The main purpose of the following article is to describe the methodology employed in our laboratory for assaying AC activity in cell-free preparations. We have additionally referenced a relevant methodology that has been used successfully for isolation of hormone receptors (specifically, β-adrenergic receptors) and the coupling proteins, N_s and N_i.

2. Methodology

2.1. Preparation of an AC-Enriched Plasma Membrane Fraction From Rat Cerebral Cortex

Rats are killed by decapitation, and brain tissue is quickly removed and the cerebral cortices separated. The tissue (about 3 g/cortex) is homogenized in 10 vol (v/w) of cold (4°C) 10 m*M* Tris-Cl (or 10 m*M* Imidazole) buffer, pH 7.8, containing 0.2 m*M* dithiothreitol, and 10% w/v sucrose with a Potter-Elvehjem homogenizer (glass tube, Teflon pestle; 0.1–0.15 mm clearance) using a motor-driven pestle (300 rev/min, 15 strokes). The homogenate is centrifuged at 8000*g* (RC-5 Superspeed refrigerated centrifuge—SS-34 rotor, radius 4.25 in.) for 20 min, and the resulting supernatant is again centrifuged at 9000*g* for 20 min. This pellet is washed once with homogenizing buffer, then suspended in sucrose-free homogenizing buffer. The suspension is incubated at 0°C for 30 min, then homogenized by 6 strokes in a hand-operated glass-Teflon homogenizer. The resulting suspension is made up to 34% (w/w) sucrose. A 28.5% (w/w) sucrose solution is layered above the 34% sucrose. A final 10% w/w sucrose solution is layered above this to obtain a total volume of 35 mL. These density gradients are centrifuged (B60 Ultracentrifuge, IEC) at 60,000*g* for 2 h in a swing-out rotor (SB 283).

The gradient fractions are recovered—one from each interface and a lower fraction (the pellet). The fraction of the middle interface, the desired fraction, is found to be enriched in AC, as well as ouabain-sensitive (Na^+/K^+)ATPase. All operations are carried out at 4°C.

Many methods for isolation of neuronal plasma membranes have been described in the literature. The procedure described above is adopted from Jones and Matus (1974).

2.2. Assay Procedures

Adenylate cyclase catalyzes the conversion of ATP to cAMP. In order to determine the enzyme activity, several procedures, as listed below, have been used to quantify the product (cAMP) formation. These procedures, for convenience, can be divided into paper chromatography and column chromatography. In our laboratory, the column chromatographic method of Salomon (1979) has been used routinely and is therefore described in detail.

2.2.1. Paper Chromatography; Method 1

One of the earliest methods to quantitate the cAMP formed as a result of AC activity has been described by Rabinowitz et al. (1965) and Ho et al. (1967). After termination of the reaction, samples are spotted on Whatman 3 MM chromatography paper along with the unlabeled carrier cAMP. The chromatogram is developed overnight (14–20 h) in isopropyl alcohol : concentrated ammonia : water (7 : 1 : 2) (additional details are described by Drummond and Duncan, 1970). The sheets are dried and the spots visualized under UV light. The spots are then cut out and placed in scintillation vials to be counted in 10 mL of toluene scintillation cocktail.

2.2.2. Paper Chromatography; Method 2

Meruelo et al. (1975) have combined the above procedure (method 1) with electrophoresis. In this method, samples are spotted on Whatman 3 MM chromatography paper sprayed with 1% sodium borate and blotted dry. The chromatogram is subjected to electrophoresis in 1% sodium borate for 40–45 min at 50 ± 5 V/cm. The paper is dried and the cAMP spots are identified by UV light. The spots are cut from each sheet and sewn to a second Whatman 3 MM sheet with an electric sewing machine. The chromatogram is developed as previously described. It is possible to use ^3H-, ^{14}C-, or ^{32}P-labeled ATP as substrates for the enzyme assay in both methods 1 and 2.

2.2.3. Single-Step Column Chromatography; Method 1

Krishna et al. (1968) developed the separation of cAMP using barium-sulfate precipitation; this was further modified by Rodbell and Krishna (1974). In this procedure samples obtained after termination of the reaction are applied to Dowex columns (AG 50W-X4, 200–400 mesh, H^+ form) previously washed with water. The cAMP is eluted from the columns by sequential washing with water (3 × 1 mL). The first two eluates contain primarily ATP and adenosine diphosphate (ADP), and are discarded. The third eluate contains cAMP and is collected in test tubes. To this, 12 mL of 0.15*M* $ZnSO_4$ is added, followed by 0.2 mL of 0.15*M* $Ba(OH)_2$. The supernatant obtained after centrifuging is decanted into nascent $BaSO_4$ columns (i.e., columns containing a plug of $BaSO_4$ precipitate). The eluate is collected in scintillation vials and the radioactivity is counted in a 12-mL scintillation cocktail. This method gives lower blank values (0.0015% of total ^{32}P added), and allows for a greater number of samples to be processed than does paper chromatography. It is nonetheless time consuming, since it requires daily preparation of the $BaSO_4$ columns [also *see* Bär and Hechter (1969) for an additional critique of this method].

2.2.4. Single-Step Column Chromatography; Method 2.

Ramachandran (1971) and White and Zenser (1971) have described the measurement of cAMP using alumina column chromatography. This procedure is based on the fact that ATP, ADP, AMP, and inorganic phosphate (P_i) are present in a multivalent anionic form at neutral pH; they are preferentially adsorbed, whereas cAMP, existing in the univalent negative ionic form under the same condition, is eluted out. After termination of the reaction, the samples are decanted into neutral alumina columns previously equilibrated with Tris-Cl buffer. The flow-through fraction that contains cAMP is collected in scintillation vials and counted for radioactivity. This procedure yields high blank values (about 0.03% of the total ^{32}P added).

2.2.5. Two-Step Column Chromatography

Salomon et al. (1974) and Wincek and Sweat (1975) have independently described the separation of cAMP using coupling Dowex cation exchange with alumina column chromatography. The method eliminates the inconsistent and high blank values routinely obtained with the above procedures (also *see* Salomon, 1979). The following procedure is routinely used in our laboratory.

2.2.5.1. Reagents

1. Adenosine 5'-triphosphate (ATP) disodium salt (99–100%, Sigma A-5394, MW 551.2)
2. [α-^{32}P]-ATP (18–30 Ci/mmol, New England Nuclear)
3. Adenosine 3',5'-cyclic monophosphate (cAMP) (Sigma A-9501, MW 329.2)
4. [2,8-^{3}H]-cAMP, ammonium salt (31.5 Ci/mmol, New England Nuclear)
5. Alumina, chromatographic (neutral, activity grade I, type WN-3, Sigma A-9003)
6. Bovine serum albumin (Sigma A-6003)
7. Bray's solution (New England Nuclear)
8. Creatine phosphate (disodium salt, MW 255.1, Sigma P-6502; di Tris salt, MW 453.4, Sigma P-4635)
9. Creatine phosphokinase (from rabbit muscle, type I, Sigma C-3755, 120–190 U/mg)
10. Dowex AG 50W-X4 (200–400 mesh, H^+ form, Bio-Rad Laboratories)
11. Ethyleneglycol-bis-(β-aminoethyl ether)*N,N'*-tetraacetic acid (EGTA) (Sigma E-4378, MW 380.4)
12. Glycylglycine (free base) (Sigma G-1002, MW 132.1)
13. Guanosine 5'-triphosphate (GTP) sodium salt (type IIS, Sigma, MW 523.3)
14. Hydrochloric acid (BDH)
15. 3-Isobutyl-1-methylxanthine (IBMX) (Aldrich or Sigma, MW 222.25)
16. Imidazole (Sigma I-0125, MW 68.1)
17. Magnesium chloride ($MgCl_2 \cdot 6H_2O$, J. T. Baker, MW 203.3)

18. Sodium chloride (BDH, MW 58.44)
19. Sodium hydroxide (BDH, MW 40)
20. Sodium lauryl sulfate (SDS) (Fisher)
21. Sodium phosphate, dibasic (Na_2HPO_4, (Fisher, MW 141.96)
22. Sodium phosphate, monobasic ($NaH_2PO_4 \cdot H_2O$, McArthur, MW 137.99)
23. Sucrose (Anachemia, MW 342.3)
24. Tris (hydroxymethyl)aminomethane (Sigma T-1503)

2.2.5.2. Preparation of Solutions

1. ATP (20 m*M*, pH 7.5): Dissolve 55.1 mg ATP in 4 mL double-distilled water (DDW). Adjust the pH to 7.5 by adding 1*M* Tris base. Determine the concentration of the solution by measuring absorbence at 259 nm, assuming the molar absorbency index A_m = 15,400. Adjust the final volume to 5.0 mL with DDW to get a final concentration of 20 m*M*.
2. cAMP (15 m*M*, pH 7.5): Dissolve 123.5 mg cAMP in about 20 mL DDW. Adjust the pH to 7.5 with 1*M* Tris. Add [^{3}H]-cAMP (stock supplied 1 mCi/mL) to give about 10^6 cpm/mL, and make up the volume to 25 mL with DDW.
3. BSA (10 mg/mL): Dissolve 50 mg BSA in 5 mL DDW.
4. EGTA (30 m*M*, pH 7.5): Dissolve 1.1 g in about 80 mL DDW. Adjust the pH to 7.5 with 1*M* Tris and make up the volume to 100 mL.
5. GTP (7.5 m*M*, pH 7.0): Dissolve 39.2 mg in about 8 mL DDW. Adjust the pH to 7.0 with 1*M* Tris and adjust the final concentration (with DDW) by assuming a molar absorbency index A_m = 13,700 at 252 nm.
6. Glycylglycine buffer (750 m*M*, pH 7.5): Dissolve 9.9 g glycylglycine base in about 70 mL DDW. Adjust the pH to 7.5 by adding 1*M* Tris. Adjust the volume to 100 mL with DDW.
7. IBMX (15 m*M*): Dissolve 333.4 mg in 100 mL DDW. Preheat in boiling waterbath until the solution is clear.
8. Imidazole HCl (1*M* stock, pH 7.5): Dissolve 272.4 g in about 3 L DDW. Adjust the pH to 7.5 with 5*N* HCl. Add 800 mg (0.02%) sodium azide (to prevent fungal growth) and adjust the volume with DDW to 4 L.
9. Magnesium chloride (75 m*M*): Dissolve 1.5 g

$MgCl_2 \cdot 6H_2O$ in DDW and adjust the volume to 100 mL.

10. Phosphocreatine (300 m*M*): Dissolve 765.3 mg as disodium salt or 1.36 g as di Tris salt in 8 mL DDW. Adjust the volume to 10 mL and the pH to 7.5.
11. Phosphocreatine kinase: Dissolve lyophilized powder in 1 mg/mL BSA solution or introduce it as a solid.
12. Stop solution:
    a. 4% SDS: Dissolve 4 g SDS in glycylglycine buffer (200 m*M*, pH 7.5). Adjust the volume to 100 mL with glycylglycine buffer.
    b. On the day of the assay, the stop solution is prepared by mixing 2 mL SDS with 0.4 mL 10 m*M* ATP and 1.6 mL DDW at room temperature.
13. Tris base (1*M* stock): Dissolve 12.11 g Tris base in DDW and adjust the volume to 100 mL. Adjust the pH to 7.5 by adding 1*N* HCl.

2.2.5.3. Stability of Solutions All buffer and salt solutions are stored at 4°C and are stable for several months. The ATP, GTP, and creatine phosphate are stored at –20°C. Drugs are usually prepared once a month unless otherwise specified, and are stored at –20°C. The stop solution and scintillation fluid are used at laboratory temperature. The assay cocktail is prepared fresh each day, and kept at 4°C until used.

2.2.5.4. Other Materials

1. Polypropylene columns, 0.8 × 4 cm (Bio-Rad Econo-column, catalog #731–1550)
2. Glass wool
3. Perspex rack to hold 100 columns. These racks can be conveniently placed over large polypropylene trays for washing or scintillation vials placed in cardboard boxes. One rack can easily be mounted on another during elution.
4. Liquid scintillation counter (Nuclear Chicago ISOcap 300A or Beckman LS 9800)

2.2.5.5. Preparation of Columns

1. Preparation of 50% Dowex columns: Suspend 100 g Dowex AG 50W-X4 (200–400 mesh) in about 800 mL DDW. Stir for 10 min and allow to settle. When the resin has settled, carefully decant the supernatant. Repeat this washing procedure several times until the supernatant is color-free. Adjust the volume of the resin suspension to 200 mL with DDW to attain

a slurry of 50% Dowex. Add 1.4 mL of the slurry into polypropylene columns containing a tight glass-wool plug at the bottom, to obtain a 1-mL resin bed. Pass DDW through these columns to allow the beads to pack. Store the columns at room temperature in a tray containing DDW to prevent drying. On the day of use, wash the columns once with 1*N* NaOH, twice with DDW, once with 1*N* HCl, and 4–6 times with DDW, in that order.

2. Preparation of alumina columns: Weigh out 0.6 g neutral (dry) alumina into polypropylene columns containing glass wool at the bottom as a support. Wash columns with about 15 mL of 100 m*M* imidazole HCl (pH 7.5). The columns are now ready for use or may be stored at room temperature in a tray containing DDW. The columns should be washed once with 100 m*M* imidazole HCl prior to use.

2.2.5.6. ASSAY PROCEDURE See Table 1 for preparing the assay cocktail.

Weigh the required quantity of creatine phosphokinase in a test tube; add BSA, then the remaining reagents. Fractions (10–30 μL) are preincubated for 5 min at 30°C. The reaction (final vol, 0.15 mL) is started by the addition of [α-^{32}P]-ATP (0.5 m*M*, specific activity, 20–40 cpm/pmol). The tubes are incubated in a shaking water bath at 30°C for 10 min. The reaction is terminated by adding 150 μL of stopping solution.

2.2.5.7. COLUMN PROCEDURES

1. To each reaction tube add 1 mL H_2O and transfer the contents to the Dowex columns.
2. Wash the columns with 1.2 mL H_2O.
3. Place the Dowex columns over the alumina columns previously equilibrated with 10 mL 100 m*M* imidazole HCl, pH 7.5. Add 4 mL to the Dowex columns, allowing them to drain into the alumina columns.
4. Place alumina columns over the scintillation vials and add 2.5 mL 100 m*M* imidazole HCl buffer. Measure the radioactivity in 15 mL of Bray's solution.

2.2.5.8. CALCULATIONS The simultaneous counting of ^{32}P and ^{3}H usually gives a spillover of counts from one channel to another. However, with the particular counter we used, the spillover was approximately 2.5% in the ^{3}H channel (channel A) that originated from ^{32}P channel (channel B).

The corrected ^{3}H counts in ^{3}H channel = $A - (B \times CR)$, where A

Table 1
Assay Cocktail

Reagents	Stock conc., m*M*	Vol for 10 assays, μL	Final concentration in assay, m*M*
Creatine phosphokinase	Solid	80 U	53 U/mL
BSA	10 mg/mL	15	0.1 mg/mL
Creatine phosphate di Tris or disodium salt	150	100	10
$MgCl_2$	150	50	5
Assay buffer (Tris HCl)	750	100	50
$[^3H]$-cAMP (with cold cAMP)	15	100	1 m*M* and ~10,000 cpm
Dithioerythritol	60	50	2

equals count in the ^3H channel for the sample, and *B* equals counts in the ^{32}P channel for the sample, and the channel ratio *(CR)* equals B″ ÷ B′ (see below).

The specific radioactivity *(SRA)* of $[\alpha\text{-}^{32}P]$-ATP (cpm/pmol) is:

$$SRA = \frac{B' \times \text{dilution (here, 201)}}{\text{ATP concentration (pmol/assay tube)}}$$

$$= \frac{\text{cpm} \times 201}{2500\ \text{pmol}/\mu\text{L} \times 30\ \mu\text{L in counting}}$$

cAMP total formed (pmol of cAMP/min/mg protein =

$$\frac{A' \times B}{[A' - (B \times CR) \times SRA \times \text{mg protein/tube} \times \text{incubation tube}]}$$

A' = cpm for $[^3H]$-cAMP standard in the ^3H channel
B' = cpm for $[\alpha\text{-}^{32}P]$-ATP standard in the ^{32}P channel
B'' = cpm for $[\alpha\text{-}^{32}P]$-ATP standard in the ^3H channel

The blank value for the boiled fraction is calculated in a similar way and finally $cAMP_{net}$ formed = $cAMP_{total}$ – blank value (pmol cAMP/min/mg protein).

To check the column performance, monitor the percent recovery of added [^{3}H]-cAMP. This is calculated as: $A/A' \times 100$. If the value drops to 50%, replace the columns; first the alumina and then the Dowex.

2.3. Isolation of the Components of AC

2.3.1. Solubilization of Membrane-Associated Proteins

A synaptosomal membrane fraction is first isolated from brain (e.g., *see* Enomoto and Asakawa, 1983). The membrane is washed with low ionic salt buffer in order to remove peripheral proteins; integral proteins are extracted by detergents. Of the several detergents examined (Lubrol PX, cholate, digitonin, CHAPS, and so on), Lubrol PX has been used successfully by various investigators to solubilize the guanine nucleotide-binding proteins (N_s and N_i) and catalyst (C).

2.3.2. Isolation of Receptors

The β-adrenergic receptors from hamster lung and guinea pig lung membranes are purified by affinity chromatography on Sepharose-alprenolol gel (*see* Stiles et al., 1984). The guinea pig lung receptors are purified to apparent homogeneity by further high-pressure liquid chromatography (HPLC) and appear as a single band of approximately 64,000 on SDS gels. Because of the presence of a variety of receptors and their wide heterogeneity in brain, little effort has been made to purify receptors from brain to homogeneity and then study them in reconstitution studies.

2.3.3. Isolation of the Catalytic Subunit

Because of the extreme lability of the catalytic subunit of adenylate cyclase, it has so far not been possible to purify it to homogeneity despite several efforts made by different investigators (Bradham and Hegazy, 1983). What has been achieved is the partially purified catalytic subunit that is obtained by subjecting the detergent-solubilized proteins from synaptosome-rich membranes to gel filtration (Salter et al., 1981). This procedure has separated the catalytic subunit (C) from guanine nucleotide proteins. The fraction rich in C is not stimulated by Gpp(NH)p, GTP, or NaF. The fraction rich in guanine nucleotide does not have any cyclase activity. The GTP response is restored when these two fractions are mixed together. Resolution of C from N can be further enhanced by exposing the membrane to increased ionic strength (Bradham and Hegazy, 1983) prior to solubilization by detergent. Drummond (1984) has described a convenient method of partial purification of AC from heart and skeletal muscle involving solubilization with CHAPS and $(NH_4)_2SO_4$ fractionation.

2.3.4. Isolation of Guanine Nucleotide-Binding Proteins N_i and N_s

Detergent-solubilized proteins from synaptosome-rich membrane are chromatographed sequentially on DEAE Sephacel, Ultrogel ACA 34, and heptylamine Sepharose. The proteins N_i and N_s are eluted together up to the Ultrogel ACA 34 column and then resolved by heptylamine Sepharose column chromatography into N_i and N_s. N_i and N_s are identified by ribosylating them with IAP and cholera toxin, respectively (Hildebrandt et al., 1984; Enomoto and Asakawa, 1984). N_s is also identified by the procedure described by Sternweis et al. (1981). Neer et al. (1984) have described the isolation of N_i from brain.

2.3.5. Functional Reconstitution of the Components of AC

The goal of reconstituting a hormonally responsive AC system from isolated components has seemed a distant one for many years. Such a complete reconstituted system could provide exciting mechanistic insights into the molecular basis of hormone activation of AC. Although C has yet to be purified, preparations of C resolved from both the receptor and N_s have been described (Ross, 1981; Levitzki, 1982). Two of the components of this system have been purified to homogeneity from a variety of sources. The pure receptor proteins from frog erythrocytes (Shorr et al., 1982), mammalian lungs (Benovic et al., 1984; Homey et al., 1983), and rat adipose tissue (Cubero and Malbon, 1984) have been shown to consist of single polypeptides (M_r = 55,000–65,000). They have been reconstituted into lipid vesicles and shown to be fully functional with regard to both ligand binding and stimulation of AC activity following fusion into cells lacking the receptor (Cerione et al., 1983). N_s has also been purified from several sources (Hildebrandt et al., 1984; Northup et al., 1980; Sternweis et al., 1981; Hanski et al., 1981; Codina et al., 1984). Human erythrocyte N_s, and very likely N_s from other sources as well, is an $\alpha\beta\gamma$ heterotrimer with subunit molecular weights of 42,000 (α), 35,000 (β), and 5000 (γ) (Hildebrandt et al., 1984).

In terms of the resolution and reconstitution of component parts, probably the most thoroughly studied system is the stimulatory AC complex. Reconstitution of the inhibitory AC complex from its components has been studied less extensively. Hormone and guanine nucleotide regulation of AC involves two separate guanine nucleotide-binding regulatory proteins, N_s and N_i, mediating AC stimulation and inhibition, respectively. N_i, like N_s, has recently been purified to homogeneity and is a heterodimeric protein composed of α (M_r, 41,000) and β (M_r, 35,000) (Katada et al., 1984), and possibly γ (M_r, 5000–15,000) subunits.

Acknowledgments

This work was supported by a grant from the Medical Research Council of Canada. We are grateful to Susan Sulakhe, Janice MacKay, and D. Gregg Rokosh for critical reading of this manuscript.

References

Bär H-P. and Hechter O. (1969) Adenyl cyclase assay in fat cell ghosts. *Anal. Biochem.* **29,** 476–489.

Benovic J. L., Shorr R. G. L., Caron M. G., and Lefkowitz R. J. (1984) The mammalian β_2 adrenergic receptor: Purification and characterization. *Biochemistry* **23,** 4510–4518.

Bradham L. S. and Hegazy M. G. (1983) Separation of the catalytic and stimulatory regulatory subunits of brain adenylate cyclase. *J. Cyclic Nucleotide Protein Phosphor. Res.* **9,** 331–340.

Cerione R. A., Shulovice B., Benovic J. L., Lefkowitz R. L., and Caron M. G. (1983) Pure β adrenergic receptor: The single polypeptide confers catecholamine responsiveness to adenylate cyclase. *Nature* **306,** 562–566.

Codina J., Hildebrandt J. D., Sekura R. D., Birnbaumer M., Bryan J., Manclark C. R., Iyengar R., and Birnbaumer L. (1984) N_s and N_i: The stimulating and inhibiting regulatory components of adenylcyclase. *J. Biol. Chem.* **259,** 5871–5886.

Cubero A. and Malbon C. C. (1984) The fat cell β adrenergic receptor. *J. Biol. Chem.* **259,** 1344–1350.

Drummond G. I. (1983) Cyclic nucleotides in the nervous system. *Adv. Cyclic Nucleotide Res.* **15,** 373–494.

Drummond G. I. (1984) Properties of the resolved catalytic unit of skeletal muscle adenylate cyclase. *Arch. Biochem. Biophys.* **235,** 427–437.

Drummond G. I. and Duncan L. (1970) Adenylate cyclase in cardiac tissue. *J. Biol. Chem.* **245,** 976–984.

Enomoto K. and Asakawa T. (1983) Evidence for the presence of a GTP-dependent regulatory component of adenylate cyclase in myelin from rat brain. *J. Neurochem.* **40,** 434–439.

Enomoto K. and Asakawa T. (1984) GTPase activity associated with the inhibitory GTP-binding regulatory component of adenylate cyclase purified from rat brain. *FEBS Lett.* **171,** 233–239.

Gilman A. G. (1984) G proteins and dual control of adenylate cyclase. *Cell* **36,** 577–579.

Hanski E., Sternweis P. C., Northup J. K., Dromerick A. W., and Gilman A. G. (1981) The regulatory component of adenylate cyclase. *J. Biol. Chem.* **256,** 12911–12919.

Hildebrandt J. D., Codina J., Risinger R., and Birnbaumer L. (1984) Identification of a γ subunit associated with the adenylyl cyclase regulatory proteins N_s and N_i. *J. Biol. Chem.* **259,** 2039–2042.

Ho R. J., Jeanrenaud B. J., Posternak Th., and Renold E. (1967) Insulin-like action of ouabain. II. Primary antilipolytic effect through inhibition of adenyl-cyclase. *Biochim. Biophys. Acta* **144,** 74–82.

Homey C. J., Rockson J. G., Countaway J., and Egan D. A. (1983) Purification and characterization of the mammalian β_2 adrenergic receptor. *Biochemistry* **22,** 660–668.

Jones D. H. and Matus A. I. (1974) Isolation of synaptic plasma membrane from brain by combined flotation-sedimentation density gradient centrifugation. *Biochim. Biophys. Acta* **356,** 276–287.

Katada T., Bokosh G. M., Smigel M. D., Ui M., and Gilman A. G. (1984) The inhibitory nucleotide-binding regulatory component of adenylate cyclase. *J. Biol. Chem.* **259,** 3586–3595.

Krishna G., Weiss B., and Brodie B. B. (1968) A simple sensitive method for the assay of adenyl cyclase. *J. Pharmacol. Exp. Ther.* **163,** 379–385.

Levitzki A. (1982) Activation and inhibition of adenylate cyclase by hormones: Mechanistic aspects. *Trends Pharmacol. Sci.* **3,** 203–208.

Limbird L. E. (1983) α_2-Adrenergic systems: Models for exploring hormonal inhibition of adenylate cyclase. *Trends Pharmacol. Sci.* **4,** 135–138.

MacNeil S., Lakey T., and Tomlinson S. (1985) Calmodulin regulation of adenylate cyclase activity. *Cell Calcium* **6,** 213–226.

Malbon C. C., Mangano T. J., and Watkins D. C. (1985) Heart contains two substrates (M_r = 40,000 and 41,000) for pertussis toxin-catalyzed ADP-ribosylation that co-purify with N_s. *Biochem. Biophys. Res. Commun.* **128,** 809–815.

Meruelo D., Bromberg F. G., and Edidin M. (1975) An assay for adenyl cyclase based on a combination of sodium borate electrophoresis and paper chromatography. *Anal. Biochem.* **67,** 30–43.

Neer E. J., Lok J. M., and Wolf L. G. (1984) Purification and properties of the inhibitory guanine nucleotide regulatory unit of brain adenylate cyclase. *J. Biol. Chem.* **259,** 14222–14229.

Northup J. K., Smigel M. D., Sternweis P. C., and Gilman A. G. (1983) The subunits of the stimulatory regulatory component of adenylate cyclase. Resolution of the activated 45,000 dalton (α) subunit. *J. Biol. Chem.* **258,** 11369–11376.

Northup J. K., Sternweis P. C., Smigel M. D., Schleifer L. S., Ross E. M., and Gilman A. G. (1980) Purification of the regulatory component of adenylate cyclase. *Proc. Natl. Acad. Sci. USA* **77,** 6516–6520.

Rabinowitz M., Desalles L., Meisler J., and Loraud L. (1965) Distribution of adenyl-cyclase activity in rabbit skeletal muscle fractions. *Biochim. Biophys. Acta* **97,** 29–36.

Ramachandran J. (1971) A new simple method for separation of adenosine 3',5'-cyclic monophosphate from other nucleotides and its use in the assay of adenyl cyclase. *Anal. Biochem.* **43,** 227–239.

Rodbell M. and Krishna G. (1974) Preparation of isolated fat cells and fat cell "ghosts": Methods for assaying adenylate cyclase activity and levels of cyclic AMP. *Meth. Enzymol.* **31,** 103–114.

Ross E. M. (1981) Physical separation of the catalytic and regulatory proteins of hepatic adenylate cyclase. *J. Biol. Chem.* **256,** 1949–1953.

Ross E. M., Pedersen S. E., and Florio V. A. (1983) Hormone-Sensitive Adenylate Cyclase: Identity, Function and Regulation of the Protein Components, in *Current Topics in Membranes and Transport* vol. 18 (Kleinzeller A. and Martin B. R., eds.), Academic, New York.

Salomon Y. (1979) Adenylate cyclase assay. *Adv. Cyclic Nucleotide Res.* **10,** 35–55.

Salomon Y., Londos C., and Rodbell M. (1974) A highly sensitive adenylate cyclase assay. *Anal. Biochem.* **58,** 541–548.

Salter R. S., Krinks M. H., Klee C. B., and Neer E. J. (1981) Calmodulin activates the isolated catalytic unit of brain adenylate cyclase. *J. Biol. Chem.* **256,** 9830–9833.

Shorr R. G. L., Strohsacker M. W., Lavin T. N., Lefkowitz R. J., and Caron M. G. (1982) The β_1-adrenergic receptor of turkey erythrocyte: Molecular heterogeneity revealed by purification and photoaffinity labeling. *J. Biol. Chem.* **257,** 12341–12350.

Smiegel M., Katada T., Northup J. K., Bokoch G. M., Ui M., and Gilman A. G. (1984) Mechanisms of guanine nucleotide mediated regulation of adenylate cyclase activity. *Adv. Cyclic Nucleotide Res.* **17,** 1–18.

Sternweis P. C., Northup J. K., Smigel M. D., and Gilman A. G. (1981) The regulatory component of adenylate cyclase. *J. Biol. Chem.* **256,** 11517–11526.

Stiles G. L., Caron M. G., and Lefkowitz R. J. (1984) β-Adrenergic receptors: Biochemical mechanisms of physiological regulation. *Physiol. Rev.* **64,** 661–743.

Sulakhe P. V. (1985) EGTA-sensitive and -insensitive forms of particulate adenylate cyclase in rat cerebral cortex: Regulation by divalent cations and GTP. *Can. J. Physiol. Pharmacol.* **63,** 1007–1016.

Sulakhe P. V., Jagadeesh G., and Braun A. P. (1985) Cardiac Autonomic Receptors and Adenylate Cyclase in Health and Disease, in *Regulation of Heart Function—Basic Concepts and Clinical Applications* (Rupp H., ed.), Thieme-Stratton, New York (in press).

White A. A. and Zenser T. V. (1971) Separation of cyclic 3',5'-nucleotide monophosphates from other nucleotides on aluminum oxide columns: Application to the assay of adenyl cyclase and guanyl cyclase. *Anal. Biochem.* **41,** 372–396.

Wincek T. J. and Sweat F. W. (1975) The assay of adenyl cyclase by sequential Dowex-alumina chromatography. *Anal. Biochem.* **64,** 631–635.

Chapter 15
Characterization of a Multifunctional Ca^{2+}-Calmodulin-Dependent Protein Kinase in the Brain

Eishichi Miyamoto

1. Introduction

Calcium (Ca^{2+}) is involved in several types of neurofunctions, including biosynthesis of neurotransmitters (Patrick and Barchas, 1974; Morgenroth et al., 1975; Yamauchi et al., 1981), stimulus-secretion coupling of neurotransmitters and hormones (Katz and Miledi, 1967; Douglas, 1968; Rubin, 1970), microtubule assembly–disassembly (Weisenberg, 1972), and many metabolic reactions. Although the precise molecular mechanism mediating the actions of Ca^{2+} in the brain remains to be elucidated, accumulating evidence suggests that many of the Ca^{2+} effects are mediated through calmodulin, a ubiquitous calcium-binding protein (Cheung, 1980; Klee et al., 1980; Means and Dedman, 1980). The protein was first discovered as a cyclic nucleotide phosphodiesterase activator (Cheung, 1970; Kakiuchi et al., 1970). Since then, more than 10 enzymes have been reported to be regulated by calmodulin in the presence of Ca^{2+}. Furthermore, the action of calmodulin is regulated by calmodulin-binding proteins (Kakiuchi, 1983).

Protein phosphorylation is involved in the regulation of many cellular processes, such as enzymatic reactions and protein–protein interactions (Greengard, 1978; Krebs and Beavo, 1979; Cohen, 1982). More than 70 neuron-specific phosphoproteins occur in mammalian brain (Greengard, 1983). In the brain, as in other tissues, there occur many protein kinases, such as types I and II cyclic AMP-dependent protein kinases, cyclic GMP-dependent protein kinase, casein kinases I and II, phospholipid-dependent protein kinase, calmodulin-dependent protein kinases, and tyrosine kinase.

In an attempt to elucidate Ca^{2+} actions through calmodulin in the central nervous system (CNS), we studied Ca^{2+}-calmodulin-dependent protein kinase in the brain (Miyamoto et al., 1981a,b; 1982). We purified and characterized two kinds of Ca^{2+}-calmodulin-dependent protein kinase with different molecular weights in the cytosol fraction of the rat brain (Fukunaga et al., 1982a,b) and rat anterior pituitary gland (Hatada et al., 1983). Ca^{2+}-calmodulin-dependent protein kinases have been isolated from the cytosol fraction of the brain (Yamauchi and Fujisawa, 1980, 1983b; Kennedy and Greengard, 1981; Bennett et al., 1983; Goldenring et al., 1983; Kennedy et al., 1983; Kuret and Schulman, 1984). In addition, Ca^{2+}-calmodulin-dependent protein phosphorylation has been identified in the particulate fractions of the brain, including membrane fractions (Schulman and Greengard, 1978a,b; Sobue et al., 1982; Sorensen and Mahler, 1983; Waalas et al., 1983), postsynaptic densities (Grab et al., 1981; Kelly et al., 1984; Goldenring et al., 1984), synaptic vesicles (DeLorenzo et al., 1979; Burke and DeLorenzo; 1982a,b; Moskowitz et al., 1983), synaptosomal fraction (Sieghart et al., 1979; O'Callaghan et al., 1980), and microtubules (Burk and DeLorenzo, 1981; Yamauchi and Fujisawa, 1982, 1983a; Fukunaga et al., 1982b; Yamamoto et al., 1983; Schworer and Soderling, 1983). These results indicate the occurrence of endogenous Ca^{2+}-calmodulin-dependent protein kinase in membrane fractions of the brain and the molecular basis for the regulatory effects of Ca^{2+} on neuronal functions.

Recently, the occurrence of Ca^{2+}-calmodulin-dependent protein kinases distinct from myosin light-chain kinase in muscle tissues was reported in many tissues, including the liver (Ahmad et al., 1982; Payne et al., 1983), skeletal muscle (Woodgett et al., 1982), the pancreas (Gorelick et al., 1983), and *Torpedo* electric organ (Palfrey et al., 1983). This chapter describes purification and characterization of a brain Ca^{2+}-calmodulin-dependent protein kinase, and discusses the functional significance of the enzyme in the brain.

2. Assay and Purification of a Ca^{2+}-Calmodulin-Dependent Protein Kinase From the Cytosol and Particulate Fractions of Rat Brain

2.1. Assay

The standard assay system for Ca^{2+}-calmodulin-dependent protein kinase activity contained, in a final volume of 0.2 mL, 5 μmol of Tris-HCl buffer, pH 7.5, 2 μmol of Mg^{2+}, 0.2 μmol of dithiothreitol, either 0.2 μmol of EGTA or 0.02 μmol of Ca^{2+} as indicated, either 3.6 nmol or 0.1 μmol of [γ-^{32}P]-ATP containing 7–20 × 10^5 cpm, 41 μg of chicken gizzard myosin light chain as usual substrate or other substrate as indicated, and the indicated amount of protein kinase with or without 1.0 μg of calmodulin. Incubation was initiated by the addition of [γ-^{32}P]-ATP and carried out at 30°C for 10 min in a shaking waterbath. The reaction was terminated by addition of 4 mL of 5% trichloroacetic acid, 0.25% sodium tungstate, and 1.2*N* H_2SO_4, and 0.2 mL of 0.63% bovine serum albumin was added as a carrier protein. After standing at 0°C for 5 min, the mixture was centrifuged, and the supernatant solution was removed by aspiration. The precipitate was dissolved in 0.1 mL of 1*N* NaOH, and 2 mL of 5% trichloroacetic acid, 0.25% sodium tungstate solution, and 1.2*N* H_2SO_4 were added. The procedure of centrifugation, removing the supernatant, dissolving the protein in alkali, and reprecipitating the protein was repeated. The protein was finally collected by centrifugation and dissolved in 0.1 mL of 1*N* NaOH. The radioactivity incorporated into protein was measured in a liquid scintillation counter.

Calmodulin was purified from bovine brain by the method of Yazawa et al. (1980). The final preparation of calmodulin showed a single protein band with a molecular weight of 17,000 on SDS-PAGE. Myosin was prepared from chicken gizzard by the method of Nonomura and Ebashi (1975). Calmodulin-deficient myosin light chains used as substrate were obtained from chicken gizzard myosin according to the method of Perrie and Perry (1970) and Matsuda et al. (1977) at the final step of the DEAE-cellulose column.

2.2. Purification

The cytosol enzyme was purified by the method described previously (Fukunaga et al., 1982b). The purification procedures included the preparation of the crude extract, acid precipitation, ammonium sulfate fractionation, affinity chromatography on calmodulin-Sepharose, the first DEAE-cellulose column chromatography, gel filtration on Sephacryl S-300, and second DEAE-cellulose column chromatography, in that order.

In our modified method for enzyme purification, the steps of acid treatment and ammonium sulfate precipitation were omitted. The crude extract was directly applied to DEAE-cellulose column. Then, the enzyme was further purified by column chromatography of calmodulin-Sepharose affinity, Affi-Gel Blue, Phenyl Sepharose, and gel filtration on Sephacryl S-300.

The specific activity and physicochemical properties of the enzyme purified by both methods were identical.

Ca^{2+}-calmodulin-dependent protein kinase was purified from the particulate fraction of rat brain, as described previously (Fukunaga et al., 1984). The extensively washed particulate fraction was homogenized in buffer containing 0.5% Triton and 0.1*M* NaCl. The homogenate was centrifuged and the supernatant was decanted. From the supernatant, Ca^{2+}-calmodulin-dependent protein kinase was purified. The procedure for the purification of the enzyme consisted of column chromatography of DEAE-cellulose, calmodulin-Sepharose 4B affinity, second DEAE-cellulose, and Sephacryl S-300 gel filtration, in that order, essentially as described for the purification of the soluble enzyme (Fukunaga et al., 1982b). Triton X-100 (0.1%) was included only in the buffer for the first DEAE-cellulose column chromatography and was omitted in the rest of the procedure.

3. Physicochemical Properties of Ca^{2+}-Calmodulin-Dependent Protein Kinase

3.1. Polyacrylamide Gel Electrophoresis

When the purified enzyme (640,000 enzyme) from the cytosol and particulate fractions was subjected to analytical gel electrophoresis in a 5.0% polyacrylamide gel under conditions of nondenaturation, a single protein band was observed by protein staining. Protein kinase activity was detected in the position corresponding to the protein staining band.

The purified enzyme from the particulate fraction was subjected to sodium dodecyl sulfate-polyacrylamide gel electrophoresis (SDS-PAGE) (Fig. 1). A single protein band was observed by staining. The densitometrical analysis of the stained protein band using 10 μg of the purified particulate enzyme revealed that the purity of the enzyme was more than 91%. The molecular weight was estimated to be about 49,000.

3.2. Molecular Weight, Stokes Radius, and S Value

The molecular weight of the enzyme was estimated to be about 640,000 by gel filtration analysis using Sephacryl S-300. On the basis of the similar experiments (Fukunaga et al., 1982b), it was

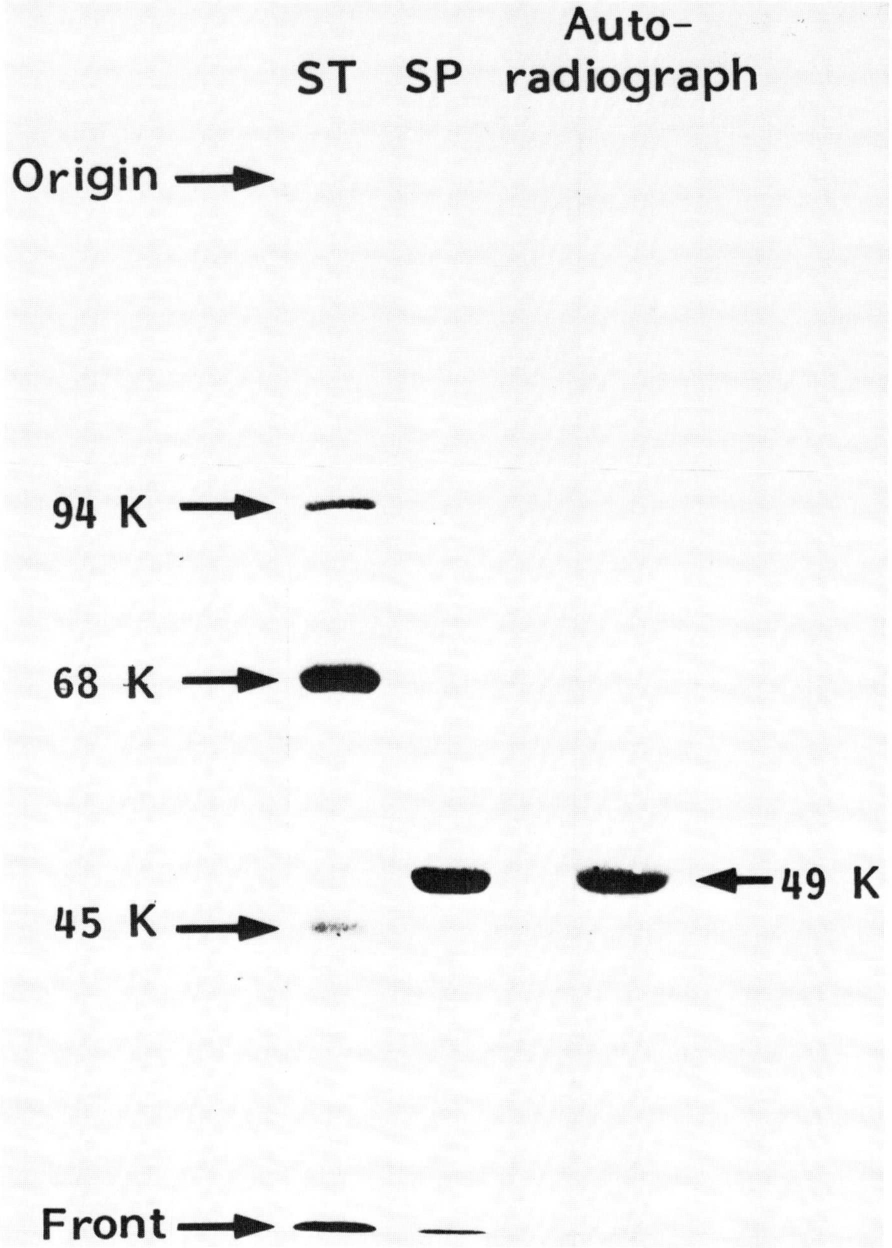

Fig. 1. SDS-polyacrylamide slab gel electrophoresis and autoradiography of the gel. The purified particulate enzyme (26 μg of protein) was incubated with 25 m*M* Tris-HCl buffer, pH 7.5, 10 m*M* $MgCl_2$, 0.1 m*M* $CaCl_2$, 0.5 μg of calmodulin, and 0.04 m*M* [γ-^{32}P]-ATP in a total volume of 0.1 mL. Incubation was conducted at 30°C for 1 min. SDS-PAGE was carried out by the method of Laemmli (1970). The gel was stained, destained, dried, and subjected to autoradiography (Fukunaga et al., 1984). Abbreviations: SP, staining pattern; ST, protein standards.

concluded that the enzyme is not in an aggregate form. The Stokes radius was estimated to be about 81 Å by Sephacryl S-300 gel filtration analysis.

Upon sucrose density gradient centrifugation, the purified protein kinase holoenzyme migrated as a single peak of activity with a sedimentation coefficient of 15.3 S.

3.3. Autophosphorylation of the Purified Enzyme

The purified enzyme from the cytosol and particulate fractions were each incubated with [γ-^{32}P]-ATP without added substrate or other kinases under standard assay conditions. The autoradiograph revealed that incorporation of phosphate into protein coincided well with that of the enzyme protein (Fig. 1). The reaction was dependent on Ca^{2+} and calmodulin. The catalytic subunit (5.0 μg/mL) obtained from rat brain type II cyclic AMP-dependent protein kinase or protein kinase inhibitor (0.5 mg/mL) had no effect on phosphorylation of the enzyme, indicating that the enzyme is autophosphorylated.

3.4. Effect of Increasing Amounts of Calmodulin on Enzyme Activity

A titration curve was plotted for the enzyme activity as a function of increasing amounts of calmodulin to a given amount of the enzyme from the particulate fraction by the method of Adelstein and Klee (1981) (Fig. 2). The enzyme concentration was $5 \times 10^{-8}M$, assuming that the molecular weight is 640,000. The results indicate that 14 mol of calmodulin are needed to fully activate 1 mol of the purified enzyme.

The molecular weight of the enzyme was estimated to be 640,000 by gel filtration analysis, 15.3 S by sucrose density gradient centrifugation, and 49,000 by SDS-PAGE. These results suggest that the holoenzyme consists of 11–14 subunits. The titration curve of the enzyme activity using increasing amounts of calmodulin indicated that 14 mol of calmodulin were needed to fully activate the enzyme (Fig. 2). Adelstein and Klee (1981) reported that the stoichiometry of the interaction is 1 mol of calmodulin/mol of turkey gizzard myosin light-chain kinase, and that the enzyme has a single peptide chain. Our data on the stoichiometry of the enzyme to calmodulin may correlate well with the subunit structure.

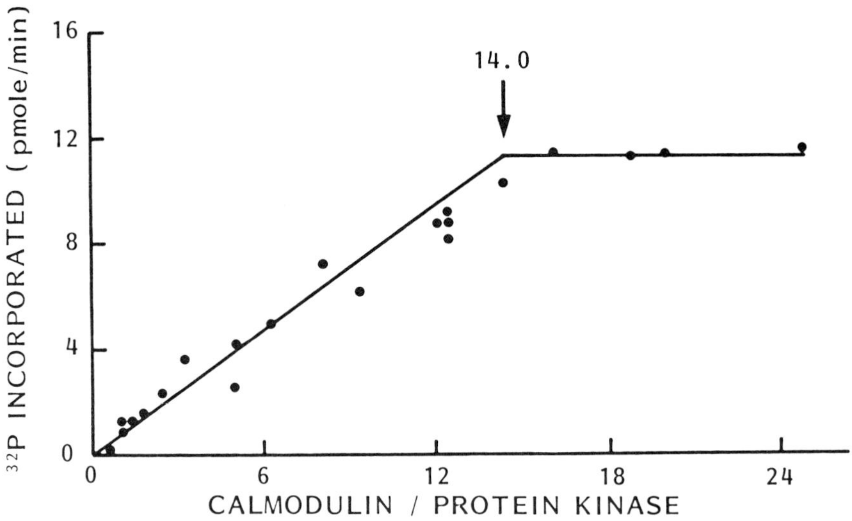

Fig. 2. Titration of Ca^{2+}-calmodulin-dependent protein kinase activity using increasing amounts of calmodulin. The concentration of the enzyme was $5 \times 10^{-8}M$. Assays were carried out at 0°C for 15 s in the presence of increasing amounts of calmodulin. The arrow indicates a molar ratio of calmodulin/Ca^{2+}-calmodulin-dependent protein kinase at about 14 (Fukunaga et al., 1984).

4. Kinetic Properties of the Purified Enzyme

4.1. Dependence on Ca^{2+} and Calmodulin

The activity of the enzyme was absolutely dependent on Ca^{2+} and calmodulin, since removal of either compound inactivated the enzyme. Phospholipid (50 μg of phosphatidylserine/mL) could not substitute for calmodulin. Protein kinase inhibitor (500 μg/mL) had no effect on the activity. We then investigated whether the purified enzyme would have any of the activity of cyclic AMP-dependent protein kinase, casein kinase, phospholipid-dependent protein kinase, and phosphorylase b kinase with whole histone, phosvitin, lysine-rich histone, and phosphorylase b, respectively, as substrates under the respective optimal assay conditions. None of these activities was detected.

4.2. Effect of Mg^{2+} and Mn^{2+} and Ionic Strength

The maximal enzyme activity was obtained at 5–20 m*M* Mg^{2+} and at 3–10 m*M* Mn^{2+} in the presence of 0.1 m*M* Ca^{2+}. At con-

Table 1
Kinetic Properties of Brain Ca^{2+}- and Calmodulin-Dependent Protein Kinase[a]

K_m for ATP	109 μM
K_m for myosin light chain	30 μM
K_a for calmodulin	12 nM
K_a for Ca^{2+}	1.9 μM

[a]From Fukunaga et al., 1982b.

centrations of each cation over these ranges, the enzyme activity gradually decreased. The V_{max} was higher with Mg^{2+} than with Mn^{2+}, but the concentration required for half-maximal activity was lower with Mn^{2+} than with Mg^{2+}. The removal of either Ca^{2+} or Mg^{2+} and Mn^{2+} resulted in complete abolishment of enzyme activity. When the effect of increasing the ionic strength on the activity was examined, the maximal activity was obtained at less than 0.1 of ionic strength (μ), and over this value, activity decreased gradually.

4.3. Effects of Various Compounds

Table 1 shows kinetic parameters of the purified cytosol enzyme for various compounds. Similar values were obtained with the purified particulate enzyme.

The K_a values for calmodulin and Ca^{2+} of the partially purified myosin light-chain kinases from chicken gizzard and rabbit myometrium were 0.6 nM and 0.4 μM, and 0.8 nM and 0.4 μM, respectively, under our standard assay conditions (Higashi et al., 1983). These results were comparable to those seen in the case of turkey gizzard myosin light-chain kinase (Hathaway et al., 1981), indicating that brain Ca^{2+}-calmodulin-dependent protein kinase had a lower sensitivity to Ca^{2+} and calmodulin.

4.4. Substrate Specificity

In addition to myosin light chain, a variety of proteins were examined as possible substrates for the purified enzyme (Table 2). Myosin light chain proved to be the most appropriate substrate for the enzyme. Moreover, myelin basic protein, casein, and arginine-rich histone served as good substrates, in decreasing order, for the enzyme. The reaction was dependent on the addition of Ca^{2+} and calmodulin.

Table 2
Substrate Specificity of Brain Ca^{2+}-Calmodulin-Dependent Protein Kinase

Substrate	$+Ca^{2+a,b}$	
	–CaM	+CaM
Myosin light chain	0.5	100.0
Microtubule protein	12.0	50.7
Myelin basic protein	0.9	45.3
Casein	0.5	33.7
Arginine-rich histone	0.8	14.7
Protamine	2.3	4.2
Phosphorylase b	0	0
Phosvitin	0	0
Lysine-rich histone	0	0
Slightly lysine-rich histone	0	0

[a]Values are given as percentages.

[b]The cytosol enzyme (1.6 μg) was incubated with 0.5 m*M* ATP with 108 μg of myosin light chains and of microtubule protein or 100 μg of other substrates as indicated in the presence or absence of 1.0 μg of calmodulin (CaM) with 0.1 m*M* Ca^{2+} under standard conditions. Values have been corrected for those determined without substrate (Fukunaga et al., 1982b).

The physicochemical and kinetic properties of the enzyme from the cytosol and particulate fractions described above were very similar. Our recent study (Isobe and Miyamoto, manuscript in preparation) showed that the elution time on HPLC, amino acid analysis, and the migration position of the enzyme subunit on two-dimensional polyacrylamide slab gel electrophoresis were identical for both enzymes. Furthermore, the enzyme showed a different distribution from lactic dehydrogenase, a marker enzyme for the soluble fraction, indicating that the cytosol enzyme is not contaminated in the particulate fraction and therefore the enzyme occurs in both cytosol and particulate fractions.

5. Endogenous Substrates

5.1. Endogenous Phosphorylation of Proteins in Subcellular Fractions

The endogenous phosphorylation of substrates in subcellular fractions was investigated without the addition of exogenous

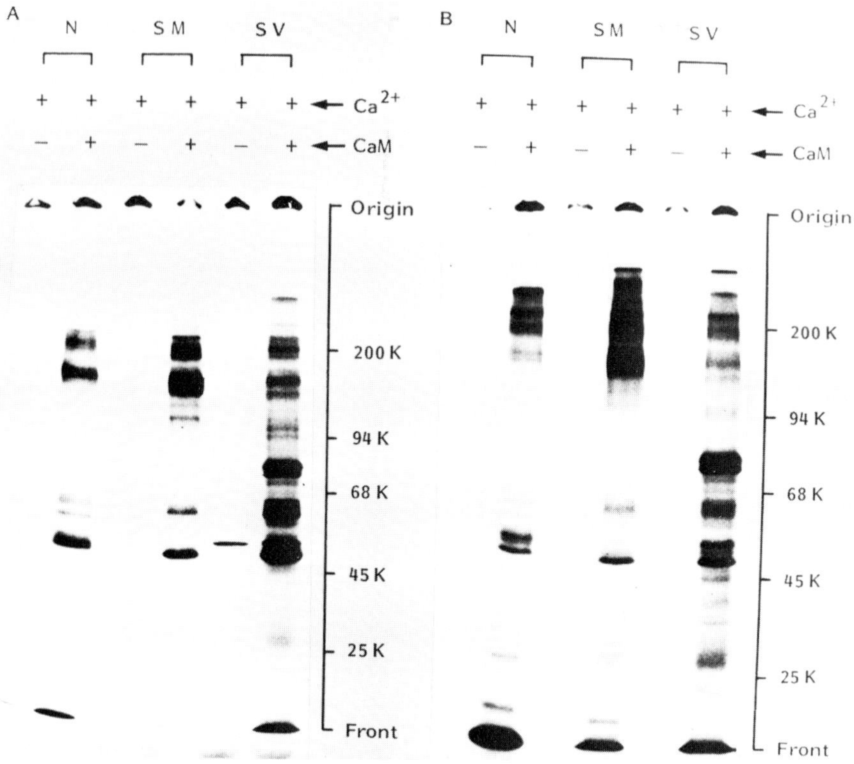

Fig. 3. A: Autoradiograph showing endogenous Ca^{2+}-calmodulin-dependent phosphorylation of proteins in nuclei (N), synaptic membranes (SM), and synaptic vesicles (SV). Each of the fractions (150 μg of protein) was incubated in the presence of 0.1 m*M* $CaCl_2$ and 0.04 m*M* [γ-^{32}P]-ATP with or without 2 μg of calmodulin (CaM) in a total volume of 0.1 mL at 30°C for 30 s, under standard conditions. B: Autoradiograph showing Ca^{2+}-calmodulin-dependent phosphorylation of proteins in nuclei (N), synaptic membranes (SM), and synaptic vesicles (SV) by the exogenous purified protein kinase. After heat treatment, each of the fractions (150 μg of protein) was incubated with 0.35 μg of the purified Ca^{2+}-calmodulin-dependent protein kinase from the particulate fraction of rat brain in the same conditions and treated as described above (Fukunaga et al., 1984).

Ca^{2+}-calmodulin-dependent protein kinase and substrates. Figure 3 shows the autoradiograph performed for the endogenous phosphorylation of the nucleus, synaptic membrane, and synaptic vesicle fractions. In all three fractions, various substrate proteins were phosphorylated in a Ca^{2+}-calmodulin-dependent manner.

At least five proteins with a molecular weight of 49,000–200,000 in the nuclear fraction, eight proteins with a molecular weight of 49,000–220,000 in the synaptic membrane fraction, and 15 proteins with a molecular weight of 49,000–220,000 in the synaptic vesicle fraction were phosphorylated and visible on the autoradiograph.

Phosphorylation of proteins in the further purified nuclear fraction was investigated. The pattern of protein phosphorylation was identical with that of Fig. 3A.

Synapsins I_a and I_b (duplet) were phosphorylated in a Ca^{2+}-calmodulin-dependent manner in the synaptic vesicle fraction, as reported (Huttner et al., 1981).

The phosphorylation of proteins with molecular weights of 220,000, 200,000, 190,000, 140,000, and 130,000 in the synaptic membrane fraction, and with molecular weights of 80,000, 78,000, 64,000, and 62,000 in the synaptic vesicle fraction was prominent.

The myelin, mitochondrial, and microsomal fractions were also investigated. In these fractions, the extent of phosphorylation was much less and the number of phosphorylated proteins were fewer. Six proteins with a molecular weight of 49,000–220,000 in the myelin fraction, one protein with a molecular weight of 49,000 in the mitochondrial fraction, and seven proteins with a molecular weight of 49,000–220,000 in the microsomal fraction were phosphorylated in a Ca^{2+}-calmodulin-dependent manner.

The purified Ca^{2+}-calmodulin-dependent protein kinase from the particulate fraction of rat brain was added to the nucleus, synaptic membrane, and synaptic vesicle fractions. Prior to incubation, each fraction was heated at 100°C for 1 min in order to destroy the endogenous enzyme activity. The patterns of phosphorylated proteins on the autoradiograph (Fig. 3B) were in good agreement with those of the endogenous phosphorylation of proteins in the fractions (Fig. 3A). These results confirmed that the purified Ca^{2+}-calmodulin-dependent protein kinase was capable of phosphorylating various endogenous substrate proteins. This also suggests that the purified enzyme is responsible for the endogenous phosphorylation of proteins in the fractions and is therefore the same as the endogenous enzyme.

5.2. Phosphorylation of Microtubule Proteins and the Effects on Microtubule Assembly

Microtubules were prepared from porcine brains by three temperature-dependent cycles of assembly and disassembly, according to the method of Shelanski et al. (1973). The microtubule preparation contained endogenous protein kinase activity, but this activity was independent of Ca^{2+} and calmodulin. Incubation of microtubules with the cytosol 640,000 enzyme resulted in the Ca^{2+}-calmodulin-dependent phosphorylation of α- and β-tubulins, microtubule-associated protein 2 (MAP2), and other proteins (Fukunaga et al., 1982b).

Microtubule-associated proteins (MAP) were prepared from microtubules by the method of Sobue et al. (1981). MAPs were phosphorylated by the 640,000 enzyme. The maximal incorporation of phosphate into MAPs by incubation for 30 min with 0.5 m*M* ATP was 25 nmol/mg of MAPs. The phosphorylated MAPs were then subjected to SDS-PAGE and autoradiography. Several components, including MAP2, tau factor, and other high molecular weight compounds, were phosphorylated in a Ca^{2+}-calmodulin-dependent manner (Fig. 4). The degree of phosphorylation differed among four components of tau factor, although phosphate was incorporated into all of them. MAP1 and tubulin were not found in this MAP's fraction. To confirm that tau factor itself is phosphorylated, and that tubulin or other proteins are not contaminated in this fraction, the following experiments were performed. Tau factor was further purified from the crude MAPs fraction by gel filtration. The purified tau factor showed the same pattern of phosphorylation on autoradiograph. Under the conditions of SDS-PAGE used, tubulin migrated slightly faster than the smallest component of tau factor and therefore did not comigrate with any component of tau factor. A similar pattern on the electrophoresis was presented by Sobue et al. (1981). The amount of phosphate incorporated was more into tau factor than into tubulin. Therefore, under the conditions used, tubulin phosphorylation was not detected in a comparable degree with that of tau factor.

To determine which is more phosphorylated in MAPs between MAP2 and tau factor, each stained protein band of the electrophoresis was cut and directly counted in a liquid scintillation counter. The ratio of phosphate incorporation into MAP2 to tau factor was 0.5, indicating that tau factor is a better substrate under these conditions, since the protein ratio of MAP2 to tau factor determined by a densitometer was 0.8.

The effect of MAPs phosphorylation on microtubule assembly was examined with or without the addition of the 640,000 enzyme to the preincubation mixture. Figure 5 shows that the MAPs phosphorylation with the enzyme during preincubation exerted an inhibitory effect on microtubule assembly, in contrast to the case of unphosphorylated MAPs without the enzyme (the boiled enzyme). The inhibition was observed over the entire incubation time for microtubule assembly testing. However, the inhibition of the

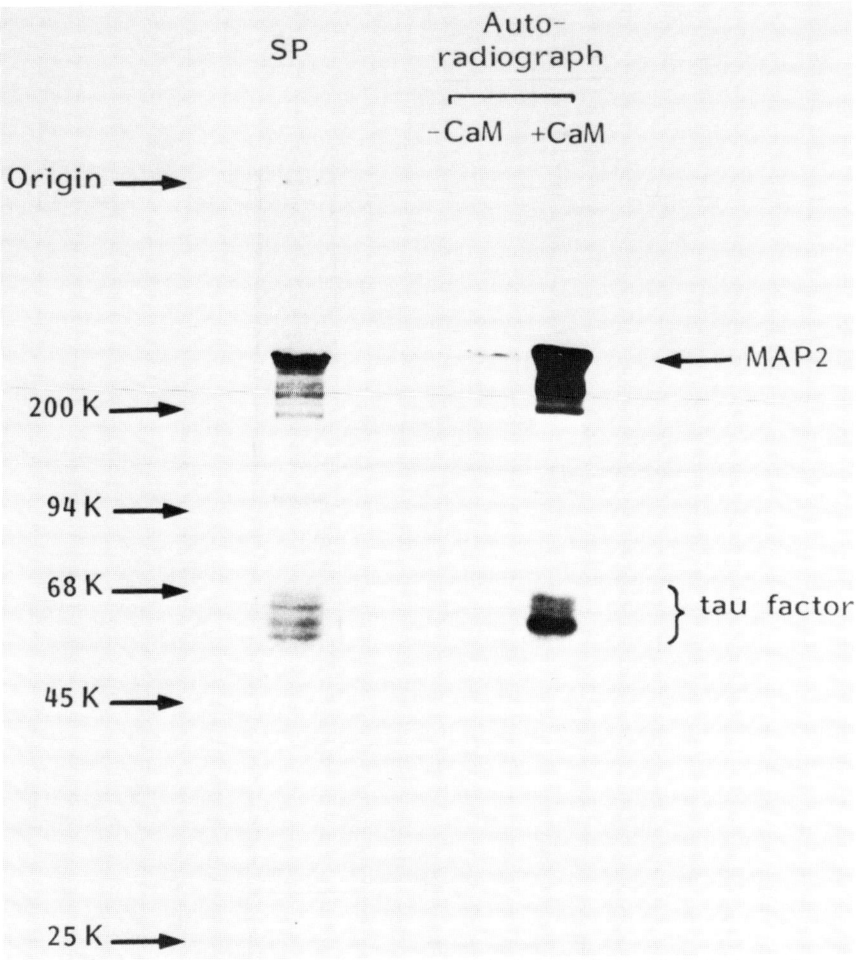

Fig. 4. SDS-polyacrylamide slab gel electrophoresis of phosphorylated MAPs and autoradiography of the gel. MAPs (11 μg) was phosphorylated with 1.0 μg of the 640,000 enzyme in the presence of 0.1 m*M* Ca^{2+} and 0.02 m*M* [γ-^{32}P]-ATP with or without 1.0 μg of calmodulin (CaM) in 25 m*M* Tris-HCl buffer, pH 7.5, under standard conditions. SDS-PAGE was carried out by the method of Laemmli (1970). The gel was stained, destained, dried, and subjected to autoradiography (Yamamoto et al., 1983). Abbreviation: SP, staining pattern.

initial rate (77.3%) of microtubule assembly was more prominent than that of the final extent (33.2%).

To determine a dose-dependence of the inhibition of microtubule assembly by MAPs phosphorylation, different amounts of the 640,000 enzyme in the preincubation mixture were tested. When

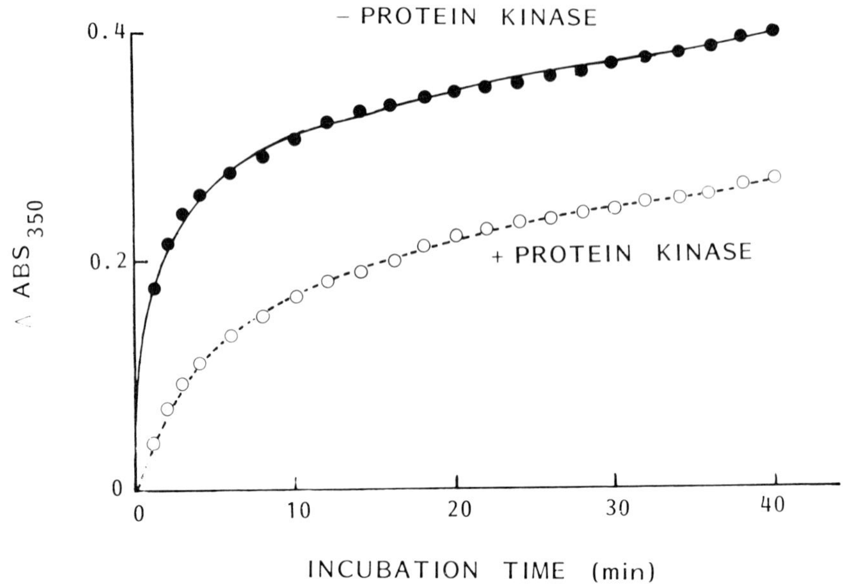

Fig. 5. Effect of the 640,000 enzyme on microtubule assembly. MAPs (0.24 mg) were preincubated at 30°C for 30 min with (○---○) or without (+ boiled enzyme) (●——●) 9.0 μg of the 640,000 enzyme under standard conditions. Under these preincubation conditions, 21.6 nmol of phosphate were incorporated into 1 mg of MAPs with the enzyme. Then, incubation for microtubule assembly was started at 0 time with 3.0 mg/mL of tubulin (Yamamoto et al., 1983).

1.4 and 25.0 nmol were incorporated into 1 mg of MAPs with given amounts of the enzyme, the initial rates were inhibited by 6.1 and 60.0%, respectively. In this experiment, the inhibition of the final extent was 9.5 and 17.0%, respectively.

Since the 640,000 enzyme absolutely required calmodulin for activity, the removal of this protein from the preincubation mixture resulted in no phosphorylation of the MAPs. Thus, the addition of calmodulin to the preincubation mixture had a greater inhibitory effect on microtubule assembly.

Similarly, the removal of Ca^{2+} by adding EGTA from the preincubation mixture led to a lack of phosphorylation of MAPs and therefore had no inhibitory effect on microtubule assembly, whereas the MAPs phosphorylation with Ca^{2+} comparatively induced an inhibition of the microtubule assembly during incubation.

The effect of MAP phosphorylation on critical tubulin concentration was examined using the method of Jameson and Caplow (1981). Figure 6 shows amounts of microtubules formed as a function of tubulin concentration. No significant difference of critical

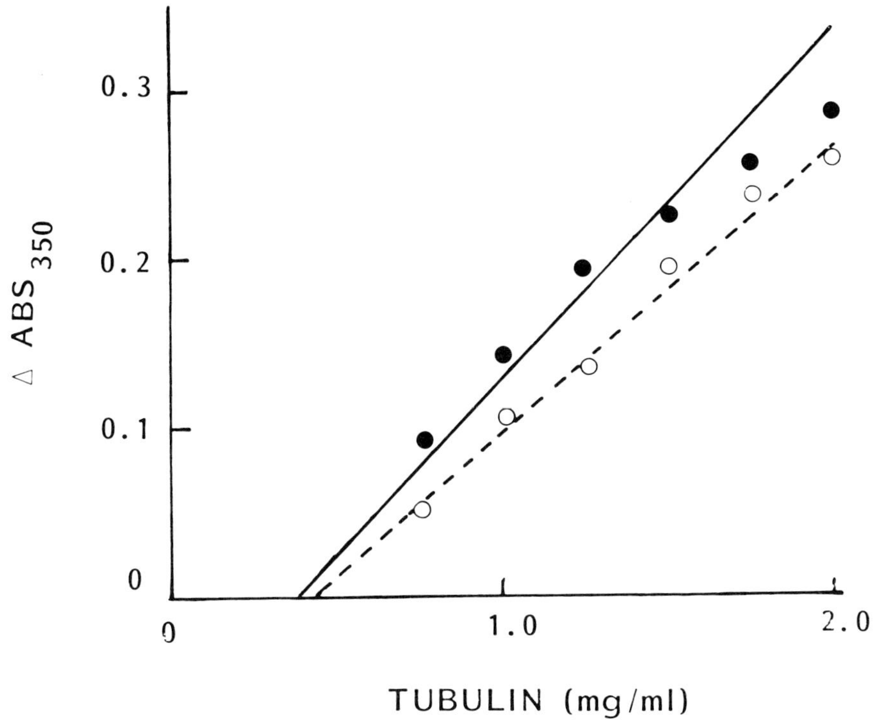

Fig. 6. Effect of MAPs phosphorylation on the critical tubulin concentration for microtubule assembly. MAPs (0.16 mg) were preincubated at 30°C for 30 min with (○- - -○) or without (+ boiled enzyme) (●——●) 10.0 μg of the 640,000 enzyme in the presence of 500 μ*M* ATP under standard conditions. Under these conditions, 21.6 nmol of phosphate were incorporated into 1 mg of MAPs. Incubation for microtubule assembly was then started. The assembled microtubules were diluted into 10 m*M* MES buffer, pH 6.8, containing 0.5 m*M* Mg^{2+}, 50 m*M* KCl, 1 m*M* EGTA, and 2 m*M* GTP at 37°C, and the final absorbance at 350 nm was determined after a new steady state had been attained. The change in absorbance at 350 nm was determined at each protein concentration by subtracting the absorbance of blanks that consisted of an identical reaction mixture diluted without previous polymerization and kept at 0°C (Yamamoto et al., 1983).

tubulin concentration was observed between unphosphorylated and phosphorylated MAPs. On the other hand, the slope of the line for microtubule formation differed slightly, indicating that the final extent of microtubule assembly is affected by MAPs phosphorylation. These results suggest that the inhibitory effect of MAPs phosphorylation on microtubule assembly is not caused by a change in critical tubulin concentration, but rather by the change in the amount of tubulin polymerized.

Marcum et al. (1978) reported that calmodulin inhibits microtubule assembly in a Ca^{2+}-dependent manner. The mechanism of the inhibition is that Ca^{2+}-calmodulin complex binds directly to tubulin (Nishida et al., 1979; Kumagai and Nishida, 1979) or MAPs (Sobue et al., 1981; Lee and Wolff, 1982, 1984), thereby blocking microtubule assembly. The concentration of calmodulin required to exert the effect was of a micromolar to millimolar order. In the present study, calmodulin was needed only to activate the 640,000 enzyme, therefore the concentration required was of a nanomolar order. The reaction system through the activation of the Ca^{2+}-calmodulin-dependent protein kinase would be more specific.

5.3. Phosphorylation of Glycogen Synthase

Glycogen synthase is known to be phosphorylated by multiple protein kinases (Roach, 1981). The phosphorylation of the enzyme is accompanied in most cases by enzyme inactivation. Recently, the calmodulin-dependent glycogen synthase kinases were detected in skeletal muscle (Woodgett et al., 1982, 1983) and in the liver (Payne and Soderling, 1980; Ahmad et al., 1982; Payne et al., 1983; Imazu et al., 1984; Camici et al., 1984).

Dephosphorylated glycogen synthase was prepared from rabbit skeletal muscle (Nimmo et al., 1976). Figure 7 shows the staining pattern of glycogen synthase by SDS-PAGE and the autoradiograph (Iwasa et al., 1983). Glycogen synthase migrated as one major and several minor bands (Takeda et al., 1975). Only limited endogenous phosphorylation of glycogen synthase was observed. The addition of the 640,000 enzyme increased the Ca^{2+}-calmodulin-dependent phosphorylation of glycogen synthase. The stained band of glycogen synthase coincided well with phosphate incorporation (Fig. 7). The 640,000 enzyme did not phosphorylate phosphorylase b.

Figure 8 shows the time course of glycogen synthase phosphorylation by the 640,000 enzyme. A small amount of endogenous phosphorylation occurred in the absence of the 640,000 enzyme. At 30 min incubation, the reaction reached a plateau. The amount of phosphate incorporated into glycogen synthase was about 1.4 mol/mol subunit, indicating that at least two sites/subunit are phosphorylated.

The sites on glycogen synthase phosphorylated by the 640,000 enzyme were examined. The autoradiograph of the sample treated by cyanogen bromide after glycogen synthase phosphorylation was examined. Phosphate was incorporated into both CB-1 (molecular weight, 9000) and CB-2 (molecular weight, 24,000–29,000), and CB-1 was more extensively phosphorylated. These

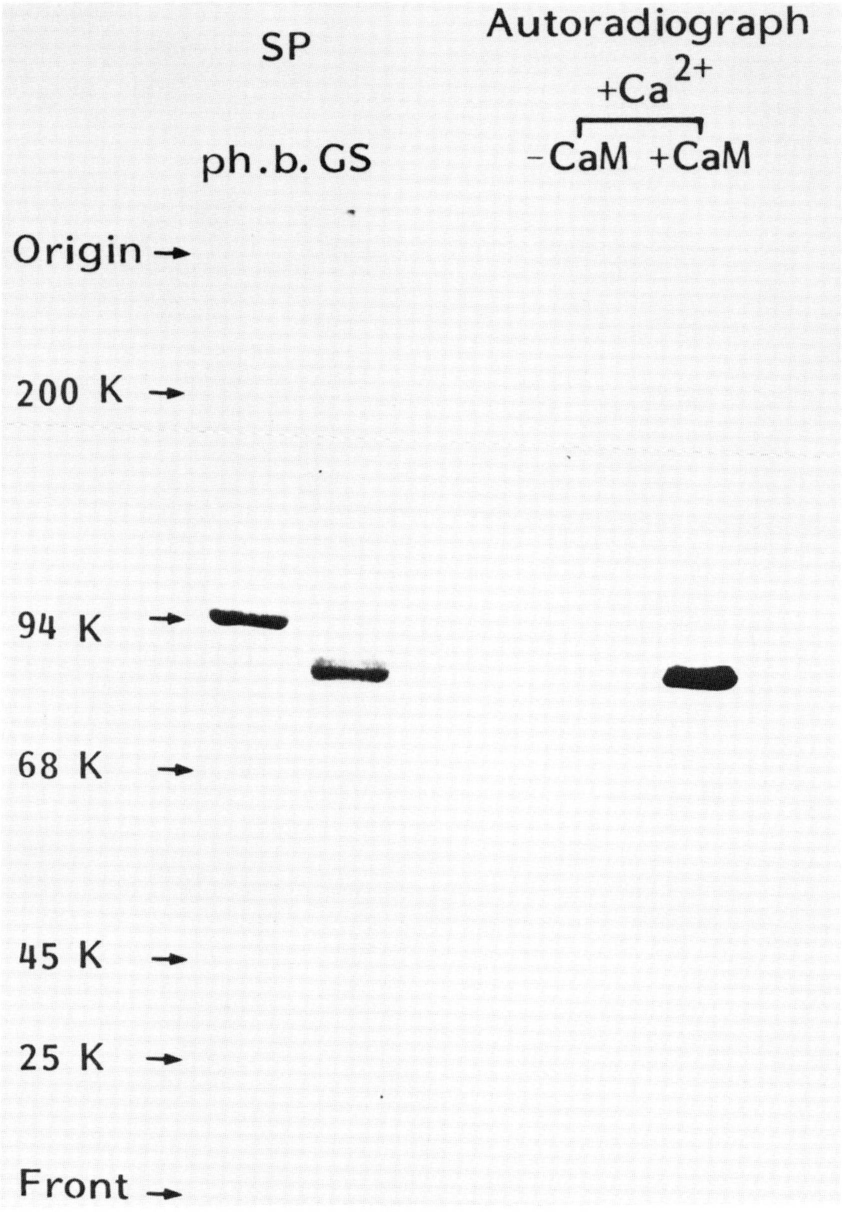

Fig. 7. SDS-PAGE of the phosphorylated glycogen synthase and autoradiography of the gel. Glycogen synthase (27 μg) was incubated with 0.25 μg of the 640,000 enzyme and 0.02 m*M* [γ-^{32}P]-ATP in the presence or absence of calmodulin (CaM), under standard conditions. After termination of the reaction and heat treatment, aliquots (0.04 mL) were subjected to SDS-PAGE, followed by autoradiography (Iwasa et al., 1983). Abbreviations: SP, a staining pattern of phosphorylase b (ph. b) and glycogen synthase (GS).

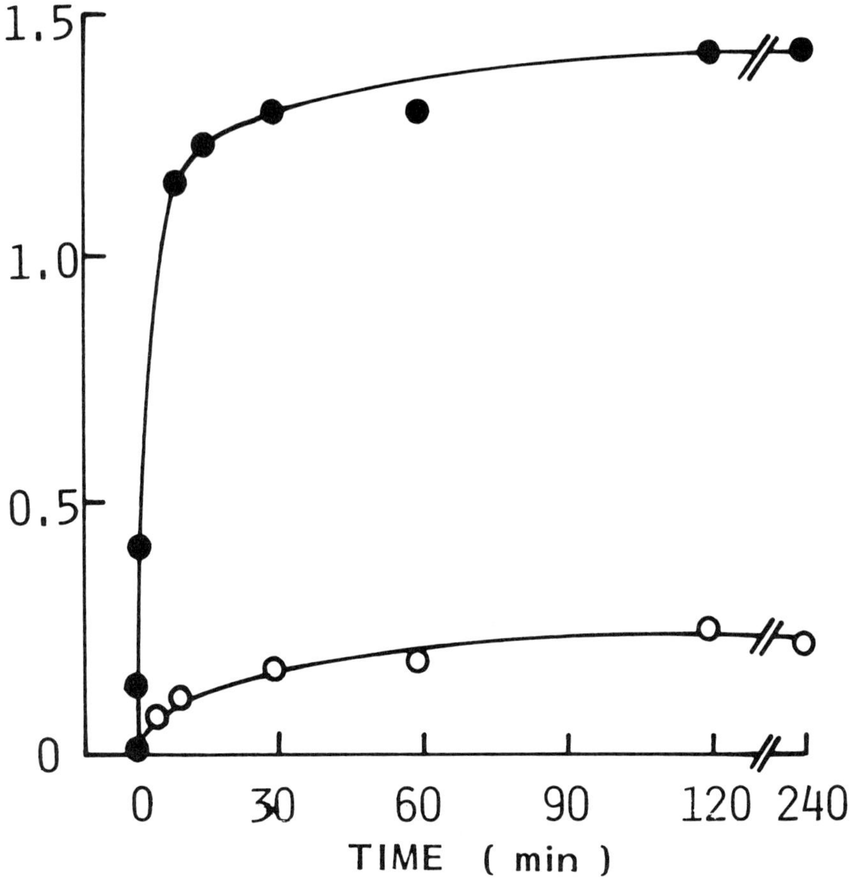

Fig. 8. Time course of glycogen synthase phosphorylation by the 640,000 enzyme. Glycogen synthase (17.0 μg) was phosphorylated with 2 μg of the 640,000 enzyme, 0.5 m*M* [γ-^{32}P]-ATP, 0.1 m*M* Ca^{2+}, and 1 μg of calmodulin, under standard conditions (●——●). The value was calculated using a M_r of the subunit as 85,000, and was corrected for that of the 640,000 enzyme alone. The phosphorylation of glycogen synthase without the addition of the 640,000 enzyme was represented (○---○) (Iwasa et al., 1983).

results are consistent with those described above, in that at least two sites/subunit are phosphorylated. Phosphorylation of site 2 (CB-1) or sites 1a and 3a + 3b + 3c (CB-2) are known to influence the enzyme activity, and the effects are additive. This suggests that the phosphorylation of glycogen synthase by the brain Ca^{2+}-calmodulin-dependent protein kinase affects the enzyme activity.

The 640,000 enzyme phosphorylated both seryl and threonyl residues in tubulin- and microtubule-associated proteins, whereas cyclic AMP-dependent protein kinase phosphorylated only seryl residues (Yamamoto et al., 1985). We therefore examined whether seryl and threonyl residues in glycogen synthase are phosphorylated by the 640,000 enzyme. Phosphate was only incorporated into seryl residues in glycogen synthase.

Synapsin I kinase from the brain was also reported to phosphorylate glycogen synthase from the skeletal muscle (McGuiness et al., 1983).

Our recent results indicate that the phosphorylation of the glycogen synthase by the 640,000 enzyme was accompanied by inactivation of the enzyme (Iwasa et al., 1984). The relationship between phosphorylation and inactivation of glycogen synthase was further investigated. Less (about 0.7 mol) than 1 mol phosphate/mol subunit incorporated led to a plateau of enzyme inactivation. The prolonged incubation had no effect on further inactivation of the enzyme. Thus, the site of phosphorylation of glycogen synthase related to activity is rapidly saturated, and further incorporation of phosphate into other sites seems to have no effect on the enzyme activity.

5.4. Phosphorylation of Neurofilaments and Glial Fibrillary Acidic Protein (GFA Protein)

Many of the intermediate filaments, including neurofilaments, desmin, vimentin, keratin, and neurofilaments can be phosphorylated by protein kinases (for review, *see* Lazarides, 1982). Endogenous phosphorylation of neurofilament triplet was reported to be independent of cyclic AMP (Julien and Mushynski, 1981; Julien et al., 1983). Cyclic AMP-dependent protein kinase associated with microtubules is capable of phosphorylating neurofilaments (Leterrier et al., 1981; Julien et al., 1983).

Neurofilaments were prepared from the rat spinal cord (Czosnek et al., 1981; Ishizaki et al., 1983) and from white matter of the porcine brain stem (Julien and Mushynski, 1981).

The neurofilament triplet in the cytoskeleton isolated from the spinal cord can be phosphorylated endogenously, i.e., in the absence of the 640,000 enzyme or of the catalytic subunit (Fig. 9) (Tanaka et al., 1984). Addition of Ca^{2+} or calmodulin, or both, had no stimulative effect on the endogenous phosphorylation of neurofilaments. The triplet was phosphorylated in a Ca^{2+}-calmodulin-dependent manner by the 640,000 enzyme (Fig. 9). The catalytic subunit of brain type II cyclic AMP-dependent protein kinase was also capable of phosphorylating neurofilaments (Fig. 9). The

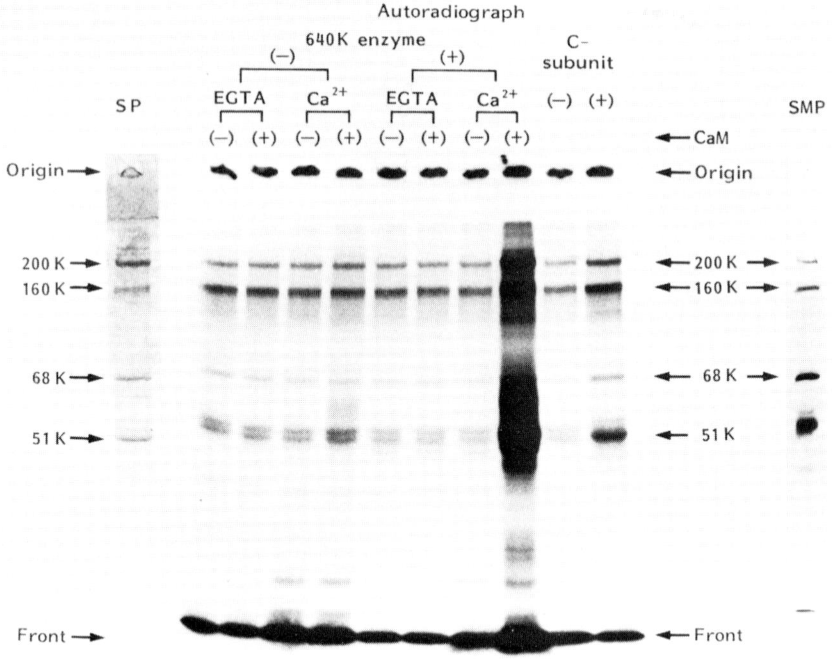

Fig. 9. Endogenous and exogenous phosphorylation of spinal cord neurofilaments and GFA protein. Neurofilaments (50 μg of protein) from rat spinal cord were incubated in a final volume of 0.2 mL, with 25 m*M* Tris-HCl buffer, pH 7.5, 10 m*M* $MgCl_2$, 1 m*M* dithiothreitol, either 0.1 m*M* $CaCl_2$ or 1 m*M* EGTA, 0.04 m*M* [γ-^{32}P]-ATP (5–20 × 10^5 cpm), with or without 1 μg of calmodulin and 2.2 μg of the 640,000 enzyme for the Ca^{2+}-calmodulin-dependent reaction and with 25 m*M* Tris-HCl buffer, pH 7.5, 10 m*M* $MgCl_2$, 0.04 m*M* [γ-^{32}P]-ATP (5–20 × 10^5 cpm), and with or without 4.8 μg of the catalytic subunit for the cyclic AMP-dependent reaction. After incubation at 30°C for 10 min, the reaction was stopped with the sample buffer (Laemmli, 1970). Aliquots (0.04 mL) of samples were subjected to SDS-PAGE (9% acrylamide), followed by autoradiography (Tanaka et al., 1984). SMP represents the fluorograph of slowly migrating axonal polypeptides in the ventral root and sciatic nerve (52 μg of protein). Abbreviations: SP, staining pattern; CaM, calmodulin; C-subunit, catalytic subunit.

160,000 and 200,000 components appeared to be phosphorylated more intensively than the 68,000 component.

To confirm whether neurofilaments themselves are phosphorylated, slowly migrating axonal polypeptides labeled with L-[^{35}S]-methionine *in situ* were simultaneously electrophoresed with neurofilaments phosphorylated in vitro. Three peptides

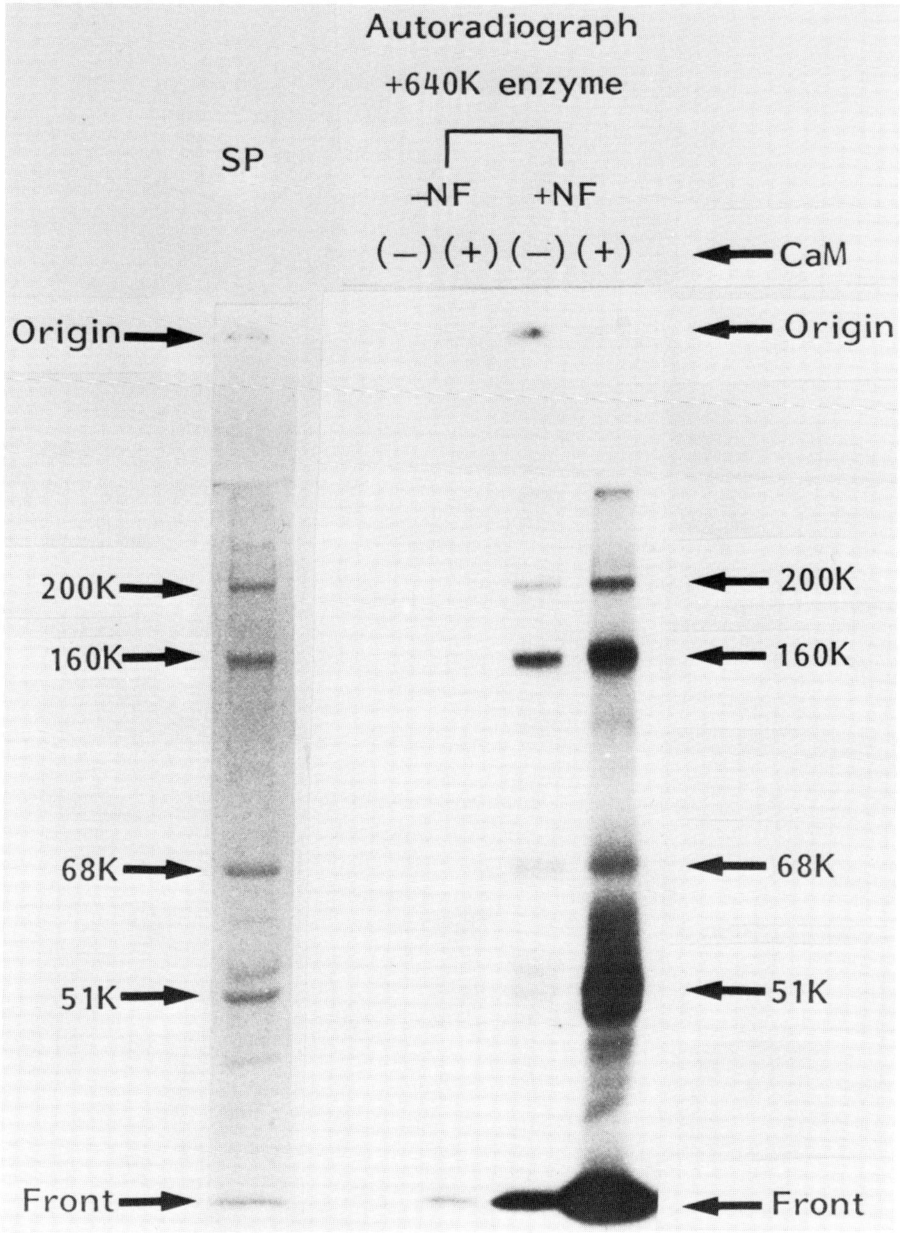

Fig. 10. Phosphorylation of spinal cord neurofilaments and GFA protein by the 640,000 enzyme. Neurofilaments (15 μg of protein) (NF) from rat spinal cord were incubated with 0.74 μg of the 640,000 enzyme, under standard conditions. The treatment of samples and symbols were described in the legend to Fig. 9. The concentration of acrylamide for SDS-PAGE was 7.5%.

Table 3
Phosphorylation of Brain Neurofilaments by the Catalytic Subunit of Brain Type II Cyclic AMP-Dependent Protein Kinase

Catalytic subunit	Neurofilament[a]	Phosphate incorporated, pmol
(+)	(–)	1.1
(–)	(+)	72.2
(+)	(+)[b]	34.4
(+)	(+)	105.9

[a]Neurofilaments (28.7 μg of protein) from the porcine brain were incubated with 2.4 μg of the catalytic subunit. After incubation at 30°C for 10 min, phosphate incorporated into protein was determined by counting radioactivity in trichloroacetic acid-insoluble material by scintillation spectrophotometry, as described previously (Miyamoto et al., 1973; Miyamoto et al., 1978).

[b]Prior to incubation, the neurofilament fraction was heated at 100°C for 2 min to inactivate endogenous protein kinase activity (Tanaka et al., 1984).

(200,000, 160,000, and 68,000) phosphorylated in the presence of the Ca^{2+}-calmodulin-dependent protein kinase or of the catalytic subunit were in a complete agreement with those of the methionine-labeled neurofilament triplet (Fig. 9), indicating that neurofilaments are in fact phosphorylated.

GFA protein (molecular weight, 51,000) was endogenously phosphorylated; the reaction was independent of Ca^{2+} plus calmodulin, or of cyclic AMP. The catalytic subunit is capable of phosphorylating GFA protein (Fig. 9). Whether or not GFA protein is phosphorylated by the 640,000 enzyme is not clear in Fig. 9, because of intensive phosphorylation of other proteins around this molecular weight region. However, phosphorylation of GFA protein and the neurofilament triplet by the 640,000 enzyme is clearly shown in Fig. 10, where a smaller amount of the enzyme was used.

Neurofilaments isolated from the brain stem also contained endogenous protein kinase activity that was independent of Ca^{2+} plus calmodulin, or of cyclic AMP. Incubation of neurofilaments with the 640,000 enzyme resulted in Ca^{2+}-calmodulin-dependent phosphorylation of the neurofilament triplet.

The brain neurofilament triplet was phosphorylated by the catalytic subunit (Table 3). The phosphate incorporation was observed in the absence of the enzyme, indicating that endogenous protein kinase activity occurs.

Our study (Yamamoto et al., 1983) indicated that the phosphorylation of MAPs inhibited microtubule assembly. Microtubules are shown to depolymerize by a Ca^{2+}-calmodulin-dependent protein kinase from the brain in the presence of ATP and Mg^{2+} (Yamauchi and Fujisawa, 1983a). Phosphorylation of neurofilaments may affect the dynamic state of the assembly–disassembly equilibrium in neurons, and their subtle interaction with other cytoskeletal components, particularly microtubules.

6. Interaction of Purified Enzyme With Various Psychotropic Drugs

It was earlier reported that many psychotropic drugs have calmodulin-binding ability, by which the activity of calmodulin-activated enzymes is inhibited (Levin and Weiss, 1979). Therefore, the interaction of the enzyme with psychotropic drugs through calmodulin was examined in the present study (Fukunaga and Miyamoto, unpublished observation). Among drugs tested, phenothiazine derivatives were most potent inhibitors, with IC_{50} values of 10–50 μM (Table 4). Benzodiazepine derivatives were less potent, and some of them had little effect on the enzyme. Trifluperidol, a butyrophenone derivative, inhibited the enzyme slightly, giving a higher IC_{50} value.

7. Conclusions

Calcium-calmodulin-dependent protein kinase with a high molecular weight occurs in the cytosol and particulate fractions of the brain. The enzymes in both fractions may be identical (Isobe and Miyamoto, manuscript in preparation). However, the role of the enzymes in the cytosol and particulate fractions may differ, since the localization of the endogenous substrates for each enzyme is different.

The Ca^{2+}-calmodulin-dependent enzyme with a molecular weight of 640,000 migrates at a position of 49,000 on SDS-PAGE. An extremely fine protein band was observed at a position of 60,000. The ratio of the amount of the 49,000 protein to that of the 60,000 protein was 17 when the densitometric analysis was made with the stained SDS-PAGE gel of the enzyme. On the other hand, it was reported that the final preparation of the enzyme showed three protein bands with molecular weights of 50,000 (α), 58,000

Table 4
Effect of Psychotropic Agents on Ca^{2+}-Calmodulin-Dependent Protein Kinase

Class of drug	Drug	IC_{50}, μM[a]
Phenothiazine	Trifluoperazine	11
	Fluphenazine	15
	Chlorpromazine	48
	Perazine	50
Benzodiazepine	Diazepam	66
	Chlordiazepoxide	130
	Nitrazepam	140
	Flunitrazepam	190
	Clonazepam	>200
	Bromazepam	>200
Butyrophenone	Trifluperidol	>200

[a]The purified cytosol enzyme was used (Fukunaga and Miyamoto, unpublished observation).

(β'), and 60,000 (β) (Bennett et al., 1983) and two protein bands with molecular weights of 52,000 and 63,000 (Goldenring et al., 1983). The main subunit of the enzyme may be the same, but the subunit compositions differ among the investigators. Further studies must be conducted to determine the enzyme structure.

It has been reported that myosin light-chain kinase from skeletal muscle (Pires and Perry, 1977), human platelets (Hathaway and Adelstein, 1979), cardiac muscle (Walsh et al., 1979), and turkey gizzard (Adelstein and Klee, 1981) has a strict substrate specificity and utilizes exclusively myosin light chain as substrate. These results indicate that the 640,000 enzyme differs from myosin light-chain kinase with respect to substrate specificity. The 640,000 enzyme appears to be multifunctional and can phosphorylate several exogenous and endogenous proteins in a Ca^{2+}-calmodulin-dependent manner. In addition to the substrates described above, synapsin I (Kennedy and Greengard, 1981), tryptophan 5'-monooxygenase and tyrosine 3-monooxygenase (Yamauchi and Fujisawa, 1979a; Yamauchi et al., 1981), fodrin (calspectin) (Sobue et al., 1982), and cytosol proteins (Yamauchi and Fujisawa, 1979b; Wrenn et al., 1980) have been reported to serve as substrates for brain Ca^{2+}-calmodulin-dependent protein kinase. Thus, the enzyme may be involved in many Ca^{2+}-regulated reactions in the

CNS through calmodulin-dependent phosphorylation of endogenous proteins. These are (a) biosynthesis of neurotransmitters, (b) stimulus-secretion coupling of neurotransmitters and hormones, (c) axoplasmic transport, (d) enzyme activation and inactivation, and (e) neuronal shape and motility.

Psychotropic drugs such as antipsychotic drugs and some minor tranquilizers can inhibit the activity of the enzyme. These results may suggest that the actions or side effects of the drugs are caused by the inhibition of the enzyme in the brain.

The findings provide some insight into molecular mechanisms underlying the effects of Ca^{2+} and calmodulin on neuronal functions.

Acknowledgments

This work was supported in part by a Grant-in-Aid for Scientific Research from the Ministry of Education, Science and Culture of Japan. I am grateful to T. Muramatsu for secretarial services.

References

Adelstein R. S. and Klee C. B. (1981) Purification and characterization of smooth muscle myosin light chain kinase. *J. Biol. Chem.* **256,** 7501–7509.

Ahmad Z., DePaoli-Roach A. A., and Roach P. J. (1982) Purification and characterization of a rabbit liver calmodulin-dependent protein kinase able to phosphorylate glycogen synthase. *J. Biol. Chem.* **257,** 8348–8355.

Bennett M. K., Erondu N. E., and Kennedy M. B. (1983) Purification and characterization of a calmodulin-dependent protein kinase that is highly concentrated in brain. *J. Biol. Chem.* **258,** 12735–12744.

Burke B. E. and DeLorenzo R. J. (1981) Ca^{2+}- and calmodulin-stimulated endogenous phosphorylation of neurotublin. *Proc. Natl. Acad. Sci. USA* **78,** 991–995.

Burke B. E. and DeLorenzo R. J. (1982a) Ca^{2+}- and calmodulin-dependent phosphorylation of endogenous synaptic vesicle tubulin by a vesicle-bound calmodulin kinase system. *J. Neurochem.* **38,** 1205–1218.

Burke B. E. and DeLorenzo R. J. (1982b) Ca^{2+}- and calmodulin-regulated endogenous tubulin kinase activity in presynaptic nerve terminal preparations. *Brain Res.* **236,** 393–415.

Camici M., Ahmad Z., DePaoli-Roach A. A., and Roach P. J. (1984) Phosphorylation of rabbit liver glycogen synthase by multiple protein kinases. *J. Biol. Chem.* **259,** 2446–2473.

Cheung W. Y. (1970) Cyclic 3',5'-nucleotide phosphodiesterase. Demonstration of an activator. *Biochem. Biophys. Res. Commun.* **38,** 533–538.

Cheung W. Y. (1980) Calmodulin plays a pivotal role in cellular regulation. *Science* **207,** 19–27.

Cohen P. (1982) The role of protein phosphorylation in neural and hormonal control of cellular activity. *Nature* **296,** 613–620.

Czosnek H., Soifer D., Mack K., and Wisniewski H. M. (1981) Similarity of neurofilament proteins from different parts of the rabbit nervous system. *Brain Res.* **216,** 387–398.

DeLorenzo R. J., Freedman S. D., Yohe W. B., and Maurer S. C. (1979) Stimulation of Ca^{2+}-dependent neurotransmitter release and presynaptic nerve terminal protein phosphorylation by calmodulin and a calmodulin-like protein isolated from synaptic vesicles. *Proc. Natl. Acad. Sci. USA* **76,** 1838–1842.

Douglas W. W. (1968) Stimulus-secretion coupling: The concept and clues from chromaffin and other cells. *Brit. J. Pharmacol.* **34,** 451–474.

Fukunaga K., Yamamoto H., Iwasa Y., and Miyamoto E. (1982a) Multiple specificities of brain Ca^{2+}- and calmodulin-dependent protein kinase for substrate. *Life Sci.* **30,** 2019–2024.

Fukunaga K., Yamamoto H., Matsui K., Higashi K., and Miyamoto E. (1982b) Purification and characterization of a Ca^{2+}- and calmodulin-dependent protein kinase from rat brain. *J. Neurochem.* **39,** 1607–1617.

Fukunaga K., Yamamoto H., Tanaka E., and Miyamoto E. (1984) A Ca^{2+}-calmodulin-dependent protein kinase in the particulate fraction of rat brain and endogenous phosphorylation of particulate-bound substrates. *Biomed. Res.* **5,** 165–176.

Goldenring J. R., Gonzalez B., McGuire J. S., Jr., and DeLorenzo R. J. (1983) Purification and characterization of a calmodulin-dependent kinase from rat brain cytosol able to phosphorylate tubulin and microtubule-associated proteins. *J. Biol. Chem.* **258,** 12632–12640.

Goldenring J. R., McGuire J. S., Jr., and DeLorenzo R. J. (1984) Identification of the major postsynaptic density protein as homologous with the major calmodulin-binding subunit of a calmodulin-dependent protein kinase. *J. Neurochem.* **42,** 1077–1084.

Gorelick F. S., Cohn J. A., Freedman S. D., Delahunt N. G., Gershoni J. M., and Jamieson J. D. (1983) Calmodulin-stimulated protein kinase activity from rat pancreas. *J. Cell Biol.* **97,** 1294–1298.

Grab D. J., Carlin R. K., and Siekevitz P. (1981) Function of calmodulin in postsynaptic densities. II. Presence of a calmodulin-activatable protein kinase activity. *J. Cell Biol.* **89,** 440–448.

Greengard P. (1978) *Cyclic Nucleotides, Phosphorylated Proteins, and Neuronal Function.* Raven, New York.

Greengard P. (1983) Neuron-specific phosphoproteins in mammalian brain. *Proc. 5th Int. Conf. Cyclic Nucleotides and Protein Phosphorylation,* p. 92.

Hatada Y., Munemura M., Fukunaga K., Yamamoto H., Maeyama M., and Miyamoto E. (1983) Calmodulin and Ca^{2+}- and calmodulin-dependent protein kinase in rat anterior pituitary gland. *J. Neurochem.* **40,** 1082–1089.

Hathaway D. R. and Adelstein R. S. (1979) Human platelet myosin light chain kinase requires the calcium binding protein, calmodulin, for activity. *Proc. Natl. Acad. Sci. USA* **76,** 1653–1657.

Hathaway D. R., Adelstein R. S., and Klee C. B. (1981) Interaction of calmodulin with myosin light chain kinase and cAMP-dependent protein kinase in bovine brain. *J. Biol. Chem.* **256,** 8183–8189.

Higashi K., Fukunaga K., Matsui K., Maeyama M., and Miyamoto E. (1983) Purification and characterization of myosin light-chain kinase from porcine myometrium and its phosphorylation and modulation by cyclic AMP-dependent protein kinase. *Biochim. Biophys. Acta* **747,** 232–240.

Huttner W. B., DeGennaro L. J., and Greengard P. (1981) Differential phosphorylation of multiple sites in purified protein I by cyclic AMP-dependent and calmodulin-dependent protein kinases. *J. Biol. Chem.* **256,** 1482–1488.

Imazu M., Strickland W. G., Chrisman T. D., and Exton J. H. (1984) Phosphorylation and inactivation of liver glycogen synthase by liver protein kinases. *J. Biol. Chem.* **259,** 1813–1821.

Ishizaki Y., Tashiro T., and Kurokawa M. (1983) A calcium-activated protease which preferentially degrades the 160,000-kDa component of the neurofilament triplet. *Eur. J. Biochem.* **131,** 41–45.

Iwasa T., Fukunaga K., Yamamoto H., Tanaka E., and Miyamoto E. (1983) Ca^{2+}, calmodulin-dependent phosphorylation of glycogen synthase by a brain protein kinase. *FEBS Lett.* **161,** 28–32.

Iwasa T., Fukunaga K., Yamamoto H., Tanaka E., and Miyamoto E. (1984) Ca^{2+}, calmodulin-dependent phosphorylation, and inactivation of glycogen synthase by a brain protein kinase. *Arch. Biochem. Biophys.* **235,** 212–217.

Jameson L. and Caplow M. (1981) Modification of microtubule steady-state dynamics by phosphorylation of the microtubule-associated proteins. *Proc. Natl. Acad. Sci. USA* **78,** 3413–3417.

Julien J.-P. and Mushynski W. E. (1981) A comparison of *in vitro*- and *in vivo*-phosphorylated neurofilament polypeptides. *J. Neurochem.* **37,** 1579–1585.

Julien J.-P., Smoluk G. D., and Mushynski W. E. (1983) Characteristics of the protein kinase activity associated with rat neurofilament preparations. *Biochim. Biophys. Acta* **755,** 25–31.

Kakiuchi S. (1983) Calmodulin-binding proteins in brain. *Neurochem. Int.* **5,** 159–169.

Kakiuchi S., Yamazaki R., and Nakajima H. (1970) Properties of a heat-stable phosphodiesterase activity factor isolated from brain extract. *Proc. Jpn. Acad.* **46,** 587–592.

Katz B. and Miledi R. (1967) A study of synaptic transmission in the absence of nerve impulses. *J. Physiol.* (Lond.) **192,** 407–436.

Kelly P. T., McGuinness T. L., and Greengard P. (1984) Evidence that the major postsynaptic density protein is a component of a Ca^{2+}/calmodulin-dependent protein kinase. *Proc. Natl. Acad. Sci. USA* **81,** 945–949.

Kennedy M. B. and Greengard P. (1981) Two calcium/calmodulin-dependent protein kinases, which are highly concentrated in brain, phosphorylate protein I at distinct sites. *Proc. Natl. Acad. Sci. USA* **78,** 1293–1297.

Kennedy M. B., McGuinness T., and Greengard P. (1983) A calcium/calmodulin-dependent protein kinase from mammalian brain that phosphorylates synapsin. I. Partial purification and characterization. *J. Neurosci.* **3,** 818–831.

Klee C. B., Crouch T. H., and Richman P. G. (1980) Calmodulin. *Ann. Rev. Biochem.* **49,** 489–515.

Krebs E. G. and Beavo J. A. (1979) Phosphorylation-dephosphorylation of enzymes. *Ann. Rev. Biochem.* **48,** 923–939.

Kumagai H. and Nishida E. (1979) The interactions between calcium-dependent regulator protein of cyclic nucleotide phosphodiesterase and microtubule proteins. II. Association of calcium-dependent regulator protein with tubulin dimer. *J. Biochem.* **85,** 1267–1274.

Kuret J. and Schulman H. (1984) Purification and characterization of a Ca^{2+}/calmodulin-dependent protein kinase from rat brain. *Biochemistry* **23,** 5495–5504.

Laemmli U. K. (1970) Cleavage of structural proteins during the assembly of the head of bacteriophage T4. *Nature* **227,** 680–685.

Lazarides E. (1982) Intermediate filaments: A chemically heterogenous, developmentally regulated class of proteins. *Ann. Rev. Biochem.* **51,** 219–250.

Lee Y. C. and Wolff J. (1982) Two opposing effects of calmodulin on microtubule assembly depend on the presence of microtubule-associated proteins. *J. Biol. Chem.* **257,** 6306–6310.

Lee Y. C. and Wolff J. (1984) Calmodulin binds to both microtubule-associated protein 2 and τ proteins. *J. Biol. Chem.* **259,** 1226–1230.

Leterrier J.-F., Liem R. K. H., and Shelanski M. L. (1981) Preferential phosphorylation of the 150,000 molecular weight component of neurofilaments by a cyclic AMP-dependent, microtubule-associated protein kinase. *J. Cell Biol.* **90,** 755–760.

Levin R. M. and Weiss B. (1979) Selective binding of antipsychotics and other psychoactive agents to the calcium-dependent activator of cyclic nucleotide phosphodiesterase. *J. Pharmacol. Exp. Ther.* **208,** 454–459.

Marcum J. M., Dedman J. R., Brinkley B. R., and Means A. R. (1978)

Control of microtubule assembly–disassembly by calcium-dependent regulator protein. *Proc. Natl. Acad. Sci. USA* **75,** 3771–3775.

Matsuda G., Suzuyama Y., Maita T., and Umegane T. (1977) The L-2 light chain of chicken skeletal muscle myosin. *FEBS Lett.* **84,** 53–56.

McGuiness T. L., Lai Y., Greengard P., Woodgett J. R., and Cohen P. (1983) A multifunctional calmodulin-dependent protein kinase. Similarities between skeletal muscle glycogen synthase kinase and a brain synapsin I kinase. *FEBS Lett.* **163,** 329–334.

Means A. R. and Dedman J. R. (1980) Calmodulin—an intracellular calcium receptor. *Nature* **285,** 73–77.

Miyamoto E., Fukunaga K., and Matsui K. (1982) Demonstration of Two Types of Ca^{2+} Dependent Protein Kinases in the Brain and Other Tissues, in *Calmodulin and Intracellular Ca^{2+} Receptors* (Kakiuchi S., Hidaka H., and Means A. R., eds.), pp. 255–265, Plenum, New York.

Miyamoto E., Fukunaga K., Matsui K., and Iwasa Y. (1981a) Occurrence of two types of Ca^{2+}-dependent protein kinases in the cytosol fraction of the brain. *J. Neurochem.* **37,** 1324–1330.

Miyamoto E., Matsui K., Fukunaga K., Nishime S., and Iwasa Y. (1981b) The occurrence of myosin light chain kinase in non-muscle tissues. *Biomedical Res.* **2,** 341–346.

Miyamoto E., Miyazaki K., Hirose R., and Kashiba A. (1978) Multiple forms of protein kinases in myelin and microsomal fractions of bovine brain. *J. Neurochem.* **31,** 269–275.

Miyamoto E., Petzold G. L., Kuo J. F., and Greengard P. (1973) Dissociation and activation of adenosine 3',5'-monophosphate-dependent and guanosine 3',5'-monophosphate-dependent protein kinase by cyclic nucleotides and by substrates. *J. Biol. Chem.* **248,** 179–189.

Morgenroth V. H., Boadle-Biber M. C., and Roth R. H. (1975) Activation of tyrosine hydroxylase from central noradrenergic neurons by calcium. *Mol. Pharmacol.* **11,** 427–435.

Moskowitz N., Glassman A., Ores C., Schook W., and Puszkin S. (1983) Phosphorylation of brain synaptic and coated vesicle proteins by endogenous Ca^{2+}/calmodulin- and cAMP-dependent protein kinases. *J. Neurochem.* **40,** 711–718.

Nimmo H. G., Proud C. G., and Cohen P. (1976) The purification and properties of rabbit skeletal muscle glycogen synthase. *Eur. J. Biochem.* **68,** 21–30.

Nishida E., Kumagai H., Ohtsuki I., and Sakai H. (1979) The interactions between calcium-dependent regulator protein of cyclic nucleotide phosphodiesterase and microtubule proteins. I. Effect of calcium-dependent regulator protein on the calcium sensitivity of microtubule assembly. *J. Biochem.* **85,** 1257–1266.

Nonomura Y. and Ebashi S. (1975) Isolation and Identification of Smooth Muscle Contractile Proteins, in *Methods in Pharmacology,* Vol. 3:

Smooth Muscle (Daniel E. E. and Paton D. M., eds.), pp. 141–162, Plenum, New York.

O'Callaghan J. P., Dunn L. A., and Lovenberg W. (1980) Calcium-regulated phosphorylation in synaptosomal cytosol: Dependence on calmodulin. *Proc. Natl. Acad. Sci. USA* **77,** 5812–5816.

Palfrey H. C., Rothlein J. E., and Greengard P. (1983) Calmodulin-dependent protein kinase and associated substrates in *Torpedo* electric organ. *J. Biol. Chem.* **258,** 9496–9505.

Patrick R. L. and Barchas J. D. (1974) Stimulation of synaptosomal dopamine synthesis by veratridine. *Nature* **250,** 737–739.

Payne M. E., Schworer C. M., and Soderling T. R. (1983) Purification and characterization of rabbit liver calmodulin-dependent glycogen synthase kinase. *J. Biol. Chem.* **258,** 2376–2382.

Payne M. E. and Soderling T. R. (1980) Calmodulin-dependent glycogen synthase kinase. *J. Biol. Chem.* **255,** 8054–8056.

Perrie W. T. and Perry S. V. (1970) An electrophoretic study of the low-molecular weight components of myosin. *Biochem. J.* **119,** 31–38.

Pires E. M. V. and Perry S. V. (1977) Purification and properties of myosin light chain kinase from fast skeletal muscle. *Biochem. J.* **167,** 137–146.

Roach P. J. (1981) Glycogen synthase and glycogen synthase kinases. *Curr. Top. Cell. Reg.* **20,** 45–105.

Rubin R. P. (1970) The role of calcium in the release of neurotransmitter substances and hormones. *Pharmacol. Rev.* **22,** 389–428.

Schulman H. and Greengard P. (1978a) Stimulation of brain membrane protein phosphorylation by calcium and an endogenous heat-stable protein. *Nature* **271,** 478–479.

Schulman H. and Greengard P. (1978b) Ca^{2+}-dependent protein phosphorylation system in membranes from various tissues, and its activation by "calcium-dependent regulator." *Proc. Natl. Acad. Sci. USA* **75,** 5432–5436.

Schworer C. M. and Soderling T. R. (1983) Substrate specificity of liver calmodulin-dependent glycogen synthase kinase. *Biochem. Biophys. Res. Commun.* **116,** 412–416.

Shelanski M. L., Gaskin F., and Cantor C. R. (1973) Microtubule assembly in the absence of added nucleotides. *Proc. Natl. Acad. Sci. USA* **70,** 765–768.

Sieghart W., Forn J., and Greengard P. (1979) Ca^{2+} and cyclic AMP regulate phosphorylation of same two membrane-associated proteins specific to nerve tissue. *Proc. Natl. Acad. Sci. USA* **76,** 2475–2479.

Sobue K., Fujita M., Muramoto Y., and Kakiuchi S. (1981) The calmodulin-binding protein in microtubules is tau factor. *FEBS Lett.* **132,** 137–140.

Sobue K., Kanda K., and Kakiuchi S. (1982) Solubilization and partial purification of protein kinase systems from brain membranes that phosphorylate calspectin. A spectrin-like calmodulin-binding protein (fodrin). *FEBS Lett.* **150,** 185–190.

Sorensen R. G. and Mahler H. R. (1983) Calcium-stimulated protein phosphorylation in synaptic membranes. *J. Neurochem.* **40,** 1349–1365.

Takeda Y., Brewer H. B., and Larner J. (1975) Structural studies on rabbit muscle glycogen synthase. I. Subunit composition. *J. Biol. Chem.* **250,** 8943–8950.

Tanaka E., Miyamoto E., Tashiro T., Komiya Y., and Kurokawa M. (1984) Ca^{2+}-calmodulin-dependent and cyclic AMP-dependent phosphorylation of neurofilaments and glial fibrillary acidic protein. *Biomedical Res.,* **5,** 239–244.

Walaas S. I., Nairn A. C., and Greengard P. (1983) Regional distribution of calcium- and cyclic adenosine 3':5'-monophosphate-regulated protein phosphorylation systems in mammalian brain. I. Particulate systems. *J. Neurosci.* **3,** 291–301.

Walsh M. P., Vallet B., Autric F., and Demaille J. G. (1979) Purification and characterization of bovine cardiac calmodulin-dependent myosin light chain kinase. *J. Biol. Chem.* **254,** 12136–12144.

Weisenberg R. C. (1972) Microtubule formation in vitro in solutions containing low calcium concentrations. *Science* **177,** 1104–1105.

Woodgett J. R., Davison M. T., and Cohen P. (1983) The calmodulin-dependent glycogen synthase kinase from rabbit skeletal muscle. Purification, subunit structure and substrate specificity. *Eur. J. Biochem.* **136,** 481–487.

Woodgett J. R., Tonks, N. K., and Cohen P. (1982) Identification of a calmodulin-dependent glycogen synthase kinase in rabbit skeletal muscle, distinct from phosphorylase kinase. *FEBS Lett.* **148,** 5–11.

Wrenn R. W., Katoh N., Wise B. C., and Kuo J. F. (1980) Stimulation by phosphatidylserine and calmodulin of calcium-dependent phosphorylation of endogenous proteins from cerebral cortex. *J. Biol. Chem.* **255,** 12042–12046.

Yamamoto H., Fukunaga K., Goto S., Tanaka E., and Miyamoto E. (1985) Ca^{2+}, calmodulin-dependent regulation of microtubule formation via phosphorylation of microtubule-associated protein 2, τ factor, and tubulin, and comparison with the cyclic AMP-dependent phosphorylation. *J. Neurochem.* **44,** 759–768.

Yamamoto H., Fukunaga K., Tanaka E., and Miyamoto E. (1983) Ca^{2+}- and calmodulin-dependent phosphorylation of microtubule-associated protein 2 and τ factor, and inhibition of microtubule assembly. *J. Neurochem.* **41,** 1119–1125.

Yamauchi T. and Fujisawa H. (1979a) Activation of tryptophan 5-monooxygenase by calcium-dependent regulator protein. *Biochem. Biophys. Res. Commun.* **90,** 28–35.

Yamauchi T. and Fujisawa H. (1979b) Most of the Ca^{2+}-dependent endogenous phosphorylation of rat brain cytosol proteins requires Ca^{2+}-dependent regulator protein. *Biochem. Biophys. Res. Commun.* **90,** 1172–1178.

Yamauchi T. and Fujisawa H. (1980) Evidence for three distinct forms of calmodulin-dependent protein kinases from rat brain. *FEBS Lett.* **116,** 141–144.

Yamauchi T. and Fujisawa H. (1982) Phosphorylation of microtubule-associated protein 2 by calmodulin-dependent protein kinase (kinase II) which occurs only in the brain tissues. *Biochem. Biophys. Res. Commun.* **109,** 975–981.

Yamauchi T. and Fujisawa H. (1983a) Disassembly of microtubules by the action of calmodulin-dependent protein kinase (kinase II) which occurs only in the brain tissues. *Biochem. Biophys. Res. Commun.* **110,** 287–291.

Yamauchi T. and Fujisawa H. (1983b) Purification and characterization of the brain calmodulin-dependent protein kinase (kinase II), which is involved in the activation of tryptophan 5-monooxygenase. *Eur. J. Biochem.* **132,** 15–21.

Yamauchi T., Nakata H., and Fujisawa H. (1981) A new activator protein that activates tryptophan 5-monooxygenase and tyrosine 3-monooxygenase in the presence of Ca^{2+}-, calmodulin-dependent protein kinase. Purification and characterization. *J. Biol. Chem.* **256,** 5404–5409.

Yazawa M., Sakuma M., and Yagi K. (1980) Calmodulins from muscles of marine invertebrates, scallop and sea anemone. Comparison with calmodulins from rabbit skeletal muscle and pig brain. *J. Biochem.* **87,** 1313–1320.

Chapter 16
S-Adenosyl-L-Homocysteine Hydrolase

O. Z. Sellinger

1. Historical Comments

The enzymatic activity that catalyzes the reversible hydrolysis of *S*-adenosyl-L-homocysteine (AdoHcy) to L-homocysteine (Hcy) and adenosine (Ado), or, in reverse, the formation of AdoHcy from Hcy and Ado, is known as AdoHcy hydrolase (AHH) (EC 3.3.1.1). Another name for it, obviously, although rarely used, is AdoHcy synthase. The activity was first described as a single enzymatic entity by de la Haba and Cantoni (1959), although it was known since 1955 that AdoHcy undergoes enzymatic breakdown when incubated with crude rat liver extracts (Ericson et al., 1955). In that same year AdoHcy was chemically characterized as the product derived from *S*-adenosyl-L-methionine (AdoMet) via transmethylation (Baddiley and Jamieson, 1955), a reaction first revealed by the pioneering studies of Cantoni and Scarano (1954). The reaction catalyzed by AHH has been extensively examined and its role in cellular biology across species and phyla (Finkelstein, 1971; Duerre and Walker, 1977) dealt with on several occasions. The role of this enzyme in transmethylation reactions taking place in cells of the nervous system has also been examined (Schatz et al., 1979), although no comprehensive treatment of this facet of AHH action exists. Yet, information on brain AHH has been summarized (Sellinger and Schatz, 1979) in a monograph devoted to the role of AdoMet in the CNS (Zappia et al., 1979). The present chapter will cover matters related to AHH under several headings, both to pool knowledge and to update the reader. It cannot be hoped, in the limited space available, to improve on or to surpass the coverage of the total AHH and AdoHcy literature provided by Ueland in 1982.

2. Occurrence

The presence of AHH in a varied universe of biological systems has been documented (Walker and Duerre, 1975; Eloranta, 1977). Since transmethylation, as a process involved in the transfer of the methyl group from AdoMet to a variety of methyl acceptor molecules, occurs universally, it is natural to expect that AHH should be ubiquitous. This expectation is apparently fulfilled, inasmuch as the absence of AHH activity has not been reported. Given that AHH assures the catabolism of AdoHcy, a most effective natural inhibitor of every transmethylation reaction known (Borchardt, 1977), it is reasonable that all cells should possess and utilize this activity readily, since there is no evidence of endogenous inhibitors of the enzyme.

2.1. Neural Systems

There is agreement that rat brain tissue exhibits rather low amounts of AHH activity when compared to more abundantly endowed peripheral organs, such as the pancreas and the liver (Finkelstein and Harris, 1973; Walker and Duerre, 1975; Eloranta, 1977). AHH activity levels have been determined on several occasions and by a variety of assay procedures in both the synthetic and the hydrolytic direction of catalysis, i.e., by measuring the synthesis or the hydrolysis of AdoHcy. Activities have usually been compared by relating the activity capable of synthesizing or hydrolyzing 1 nmol of AdoHcy/mg protein/min. In this chapter the synthetic activity will be expressed in s units and the hydrolytic activity in h units. Thus, 1 s and 1 h unit of AHH activity refer to 1 nmol of AdoHcy synthesized or hydrolyzed by 1 mg of protein/min. The following have been reported: very low s and h levels of activity in the brain of the rat, dog, and chicken and moderate levels in the brain of the rabbit (Walker and Duerre, 1975); subsequently, a developmental study using a more sensitive assay procedure revealed 11 s units in rat brain (Eloranta, 1977) and 5–7 times that in rat liver. A regional breakdown of AHH activity revealed 0.4–0.5 h units among seven rat brain regions and the spinal cord (Schatz et al., 1977b), whereas a later survey of 13 rat brain regions showed a range between a high of 0.46 h units in the hypothalamus and a low of 0.2 h units in the midbrain (Broch and Ueland, 1980). Hydrolytic activity toward [^{14}C]-AdoHcy has also been reported in N-18 neuroblastoma cells incubated in monolayer culture (Crooks et al., 1979), and high synthetic activity was reported in the rat hippocampus (Reddington and Pusch, 1983).

2.2. Nonneural Mammalian Systems

Few authors have compared AHH activities in both h and s units and within the context of a single experimental design. Walker and Duerre (1975) measured AHH in s and h units in seven tissues of four warm-blooded vertebrates and found that in the rat, dog, and rabbit the liver contains the highest level of AHH s units and in the chicken the highest level was found in the kidney. The ratio of AHH s to h units favors by far the former in all tissues hitherto examined, except for dog brain, in which AHH s units are barely detectable. It is not clear, however, what portion of the dog's brain was used in the assay. In a later study (Eloranta, 1977) the pancreas proved to be the richest in AHH s units of all rat tissues. A recent survey of AHH in the heart of five mammalian species revealed the highest h activity in the guinea pig and the lowest in the dog, with the hearts of the cat, rat, and rabbit exhibiting intermediate levels (Schrader et al., 1981). Human and rat liver were compared as well, and 100-fold lower AHH h activity values were noted in the former (Ueland et al., 1979). AHH activity was also reported in human red blood cells (Ferro et al., 1981), bovine adrenal cortex (Doskeland and Ueland, 1982), rat intestinal mucosa (Finkelstein and Harris, 1975), mouse lymphocytes (Hershfield et al., 1979), rat prostate and testis (Liau et al., 1981), L 1210 cells (Glazer and Hartman, 1980), Chinese hamster-human hybrid cells (Hershfield and Francke, 1982), mouse macrophages (Aksamit et al., 1982), human lymphoblasts (Hershfield, 1979), human placenta (Hershfield et al., 1979), rat hepatomas (Finkelstein et al., 1978), and chick embryo fibroblasts, normal and virus-transformed (Pierré et al., 1977).

2.3. Other Occurrences

AHH activity has been measured and the enzyme thoroughly characterized in two plant sources; the spinach beet leaf (Poulton and Butt, 1976) and the yellow lupin seed (Guranowski and Pawelkiewicz, 1977). The synthetic activity of AHH in the spinach beet leaf was found to be about 28 s units, whereas in the lupin seed it appeared to be about 10-fold lower. Yet, both activities could be successfully purified, the spinach beet leaf AHH by about a 100-fold and the lupin seed enzyme by about 1000-fold.

2.4. Developmental, Physiological, and Pharmacological Studies

Eloranta (1977) examined the influence of age on AHH of rat brain and liver, noting no increase in the former over the activity assayed at birth. Hepatic AHH, on the other hand, increased less than twofold over the 29-wk period tested. Walker and Duerre (1975) had obtained similar findings, yet found a marked drop in AHH activity in rat cerebellum from about 2.5 s units early postnatally to a stable activity of about 1 s unit beginning at about 1 mo and remaining constant throughout adult life. Hypophysectomy caused marked reductions of AHH activity in the prostate, testes, and adrenal (Liau et al., 1981; Wong and Ciaranello, 1982). Testosterone administration to hypophysectomized rats (Liau et al., 1981) restored the activity, gradually culminating in complete return to normal values by 10 wk. Dexamethasone therapy had no restorative power (Wong and Ciaranello, 1982). According to Liau et al. (1981), AHH in the prostate and testes exists as part of a ternary methylation complex, consisting of AHH, tRNA methyltransferase enzymes (EC 2.1.1), and AdoMet synthase (EC 2.5.1.6). These workers also reported the presence of two distinct AHH isozymes in the prostate and testicular tissue, a finding that has remained unreplicated. Decreased AHH activity was also noted in six of seven rat brain regions tested following the administration of a single dose of the chemical convulsant agent L-methionine-*d,l*-sulfoximine (Schatz et al., 1977b). In a further attempt to modulate AHH activity of rat liver, Eloranta (1977) injected rats with 100 mg/kg of L-methionine and assayed the enzyme at 0.5, 2, 12, and 72 h. No effect was noted. Hepatic AHH increased, however, following a regimen leading to pyridoxine deficiency; hepatic AdoHcy levels increased 500% above control values under these circumstances, as well (Eloranta et al., 1976). AHH activity also significantly decreased in rat liver following the administration of the antiviral and antitumor agent 9-β-D-arabinofuranosyladenine (Eloranta et al., 1982). Perfusion of the liver with Ado also led to a fall in AHH activity (Duerre and Briske-Anderson, 1981).

2.5. Intracellular Localization of AHH

AHH is recovered as a soluble enzyme activity, i.e., a vast proportion of the total tissue activity appears nonsedimentable after prolonged centrifugation at high *g* values. This has been documented for rat liver (Ueland et al., 1979) and rat brain (Sellinger and Schatz, 1979). Broch and Ueland (1980) reported that of the 25% of the total rat brain AHH activity recovered in the crude particulate fraction, 54% associated with the sucrose density subfraction B, known to contain synaptosomes. These workers erroneously reported that Sellinger and Schatz (1979) failed to find AHH activity in synaptosomes. The centrifugation scheme used by the latter workers (Sellinger and de Balbian Verster, 1962) to assess the intracellular distribution of AHH in rat brain was intended to partition brain tissue into one soluble and four particulate fractions. The choice of such a centrifugation scheme was warrented by the possibility that AHH, being a hydrolytic enzyme, could have a lysosomal localization. Thus, one of the four particulate fractions isolated, fraction L, was highly enriched in acid phosphatase, the lysosomal marker enzyme.

2.6. Gene Location

The location of the gene for AHH has been mapped and found to be on the same chromosome as the enzyme Ado deaminase, which metabolizes Ado, one of the products of AHH action on AdoHcy (Hershfield and Francke, 1982). The chromosome bearing the genes for both enzymes has been identified as number 20 by the procedure of concordant segregation and the utilization of (a) human-Chinese hamster cell hybrids and (b) monoclonal antibodies to the human placental AHH. The precise distance on chromosome 20 of the AHH and Ado deaminase genes has not been determined. Although their copresence on the same chromosome was calculated to have a reasonable chance of not being a random event, no hard evidence exists that this synteny is anything but chance. It is of interest in this context that Kamatami et al. (1982) reported finding a mutant clone of the murine lymphoid cell line with only 11–13% wild type AHH activity.

3. Studies of the Purified Enzyme

3.1. Purification Procedures

Table 1 lists the results of 14 studies designed to produce purified AHH. Except for two plant sources, the enzyme has been obtained largely from mammalian sources, with liver as the preferred organ. Classical enzyme purification procedures, blending specific precipitation and the use of an assortment of traditional column chromatographic support media, have been followed, most recently, by the use of affinity chromatographic methods. These have been loosely defined according to the affinity material employed as involving "AdoHcy affinity gel chromatographic steps" (Briske-Anderson and Duerre, 1982) or "utilizing Sepharose-$NH(CH_2)_6NH_2$" (Guranowski and Pawelkiewicz, 1977), aminohexyl-Sepharose 4B (Schatz et al., 1979), mercaptopurine-

Table 1
The Purification of *S*-Adenosylhomocysteine Hydrolase

Source	Purification (X-fold)	Molecular weight, $\times 10^{-3}$	Subunit molecular weight, $\times 10^{-3}$	p*I*	K_m AdoHcy, μM	K_m Ado, μM	K_m Hcy, μM	Reference
Rat liver	880	—	57.5	—	12.3	0.94	164 (L)	Briske-Anderson and Duerre, 1982
Rat liver	300	180	47	5.7	15.2	1.0	155 (D,L)	Fujioka and Takata, 1981
Rat liver	135	220	47	5.6	0.9	0.6	60 (L)	Kajander and Raina, 1981
Rat liver	240	—	48	—	—	—	—	Chabannes et al., 1979b
Mouse liver	—	180–190	45–46.5	5.65–5.75	0.75	0.2	—	Doskeland and Ueland, 1982
Hamster liver	4666	200	48	—	1.4	1.0	—	Kim et al., 1983
Beef liver	—	180–190	50	5.35–5.50	0.75	0.2	—	Doskeland and Ueland, 1982
Beef adrenal cortex	—	180–190	50	5.35–5.50	0.75	0.2	—	Doskeland and Ueland, 1982
Human red blood cell	1846	—	—	—	—			Ferro et al., 1981
Beef (calf) liver	756	237.5	50–60	5.8–6.0	63	420; 1.8	—	Richards et al., 1978, Guranowski et al., 1981
Beef (calf) liver	1444	192	48	—	10.5	45	—	Palmer and Abeles, 1979
Rat brain	2371	180	48	5.6	37	—	—	Sellinger and Schatz, 1979
Spinach beet leaf	100	—	—	—	41	12.8	1200 (L)	Poulton and Butt, 1976
Yellow lupin seed	1193	110–200	55	4.9	12	2.3	4600 (D,L)	Guranowski and Pawelkiewicz, 1977

riboside-aminohexyl Sepharose (Chabannes et al., 1979a,b), and, most correctly, 8-(3-aminopropylamine)-Ado linked to CH-Sepharose 4B (Kajander and Raina, 1981). The Table lists the efficiency of all methods and the general properties of the product. It may be noted that, just as most roads lead to Rome, most procedures lead to a fairly similar final product, i.e., a tetra- or pentameric molecule with a p*I* between 5.35 and 5.7 and with greatly varying apparent K_m values for all three substrates. The latter finding is not surprising given the diversity of the individual assays used all over the globe to determine the kinetic constants (*see* Tables 2 and 3). It is noteworthy that severe discrepancies may occur within the same laboratory and with the same enzyme source with regard to a given K_m value. Thus, a K_m value for Ado using beef (calf) liver enzyme has been reported as 420 (Richards et al., 1978) and 1.8 μmol/1 (Guranowski et al., 1981), respectively.

3.2. Characterization of the Purified AHH

3.2.1. Physical Properties

Only a few examples are given here. The characteristic treatment of AHH as a molecule whose physical constants, once determined, could help understand its catalytic function in the cell, has yielded a sedimentation coefficient of 8.78 (Palmer and Abeles, 1979) or 8.8 (Ueland and Doskeland, 1977) for the hepatic enzyme and an $E^{1\%}_{1cm}$ value of 14.2 at 280 nm. The enzyme exhibits an absorption peak at 277 nm and weak absorption in the 300–400 nm range (Palmer and Abeles, 1976; Fujioka and Takata, 1981). A Stokes radius of 48 Å has been determined (Ueland and Doskeland, 1977). The energy of activation was computed from an Arrhenius plot as 14.4 kcal/mol, with a temperature coefficient of 2.4 (Guranowski and Pawelkiewicz, 1977). On HPLC, the enzymes from bovine and mouse tissues (Doskeland and Ueland, 1982) gave identical retention times (12.43–12.46 min), yet inexplicable differences in the affinity of the different AHH proteins toward DEAE-cellulose were also noted.

3.2.2. Chemical Properties

The activity of purified AHH decays in the absence of protective sulfhydryl reagents. The plant enzyme loses all activity when incubated for 20 min in the presence of 0.05 m*M* *p*-hydroxymercuribenzoate or 1 m*M* *N*-ethylmaleimide, and activity cannot be restored by sulfhydryl agents (Guranowski and Pawelkiewicz, 1977). The amino acid compositions of the purified beef and rat liver AHH are available (Richards et al., 1978; Fujioka and

Table 2
The Assay of *S*-Adenosyl-L-Homocysteine:Hydrolase Synthetic Activity: A Historical Review

Assay number	Year	Authors	Buffer	pH	Ado, m*M*	L-Hcy, n*M*	Other	Workup
1	1959	De la Haba and Cantoni	Na-succinate	6.3	10	10 (D,L)	—	Assay disappearing SH groups
2	1973	Finkelstein and Harris	KPO_4	7.3	2.5[a]	2	—	Ion-exchange chromatography
3	1975	Walker and Duerre	PO_4	7.0	1.5	0.6	—	Assay disappearing SH groups
4	1976	Poulton and Butt	Tris-HCl	8.5	2.2[a]	11 (D,L)	—	Paper chromatography of ^{14}[8-C]-Ado-Hcy formed
5	1976	Kajander et al.	KPO_4	7.3	0.2[a]	3	—	Paper electrophoresis or ^{14}[8-C]-AdoHcy
6	1977	Guranowski and Pawelkiewicz	Tris-HCl	8.5	2	10 (D,L)	—	As in 4
7	1978	Saebo and Ueland	KPO_4 or Hepes[b]	7.3	0.01[a]	3 (D,L)	—	TLC of products
8	1979	Hershfield	KPO_4	7.0	0.05–0.2	5	5 μ*M* EHNA[c]	TLC of products
9	1979	Hoffman et al.	PO_4	7.0	1.5	0.6	+1 m*M* EDTA	Ion-exchange chromatography

10	1979a	Chabannes et al.	PO_4	?	2.5	5 (D,L)	5 μ*M* thiodiglycol	HPLC on Bondapak C_{18}
11	1981	Fujioka and Takata	KPO_4	7.2	0.01[a]	5 (D,L)		Paper chromatography
12	1981	Liau et al.	*see* 5				1 μ*M* EHNA	Electrophoresis at 1800
13	1981	Schrader et al.	Hepes	7.4	0.002–0.05[a]	2	20 μ*M* EHNA[c] 2 m*M* Mg	Thin layer chromatography
14	1981	Guranowski et al.	Tris-HCl	8.0	1	10[d]	2 m*M* DTT	Chromatography
15	1982	Ueland	Hepes	7.0	0.005	3 (D,L)	5 m*M* Mg 150 m*M* KCl 0.25% BSA ± 5 m*M* mercaptethanol	As in 7
16	1983	Kim et al.	KPO_4	7.6	≤0.02	10 (D,L)	5 m*M* DTT 1 m*M* EDTA 10% glycerol	SP-Sephadex-C25 isolation of inosine

[a] ^{14}[8-C].
[b] *N*-2-hydroxyethylpiperazine *N*'-2-ethanesulfonic acid.
[c] Erythro-9-(2-hydroxy-3-nonyl)adenine, an adenosine deaminase inhibitor.
[d] ^{35}[S]-Hcy.

Table 3
The Assay of *S*-Adenosyl-L-Homocysteine Hydrolase Hydrolytic Activity: A Historical Review

Assay number	Year	Authors	Buffer	pH	AdoHcy[a], mM	Other	Workup
1	1974	Cortese et al.	KPO_4	7.0	3	—	Paper chromatography
2	1975	Walker and Duerre	KPO_4	7.0	3	—	Ion exchange chromatography
3	1976	Poulton and Butt	Tris	8.5	1.3	1.3 m*M* thiodiglycol, AD[b]	Paper chromatography of inosine
4	1977b	Schatz et al.	PO_4	7.4	0.12	AD[b]	HPLC on Bondapak C_{18}
5	1977a	Schatz et al.	PO_4	7.4	0.32		Paper chromatography
6	1977	Trewyn and Kerr	PO_4	7.4	0.25	AD[b] + thetin-Hcy methyltransferase	Counting [^{3}H]-Met after removal of reactants with dextran-coated charcoal
7	1979	Palmer and Abeles	PO_4	7.4	0.06	AD[b], 1 m*M* K-EDTA	Inosine production at 265 n*M*
8	1979	Ueland and Saebo	Hepes[c]	7.0	0.02–0.15	AD[b], 0.2% BSA, 150 m*M* KCl, 5 m*M* Mg[b], 2 m*M* mercaptoethanol	TLC of AdoHcy, inosine, and hypoxanthine
9	1981	Fujioka and Takata	KPO_4	7.2	0.05	AD[b]	Decrease in A_{265}

10	1981	Schrader et al.	Hepes[c]	7.4	0.005–0.02	AD[b], 2 m*M* Mg, purine nucleoside phosphorylase, xanthine oxidase	Urate formation at 295 n*M*
11	1981	Trewyn and Kerr	PO_4	7.2	0.05	5 m*M* dimethylthetin, thetin-Hcy methyltransferase, AD[b], purine nucleoside phosphorylase, xanthine oxidase, BSA	Urate formation at 292 n*M*
12	1983	Guranowski and Jakubowski	Hepes[c]	8.0	0.02	AD[b]	TLC of inosine
13	1983	Kim et al.	PO_4	7.6	0.02	AD[b], 5 m*M* DTT, 1 m*M* EDTA, 10% glycerol	SP-Sephadex C25, isolation of inosine

[a][^{14}-C]-AdoHcy (assay No. 1); ^{3}[H]-AdoHcy or *S*-adenosyl-L-[2(*n*)-^{3}H]-homocysteine (assay No. 6).
[b]AD, adenosine deaminase.
[c]*N*-2-hydroxyethylpiperazine *N*'-2-ethanesulfonic acid.

Takata, 1981). In addition to all the common amino acids, there are 39.2 half-Cys residues/mol and four disulfide bonds in rat liver protein. Beef liver protein contains, in addition to all the common amino acids, 1.2 residues of glucosamine/mol. Cystine residues were not determined.

It was learned in 1976 (Palmer and Abeles) that the enzyme from calf liver contains NAD that is tightly bound, 1 mol/subunit of 48 kdaltons. NAD can be released when the protein is precipitated with $HCl0_4$. Furthermore, in the presence of Ado, absorbance in the 300–370 nm range increases appreciably, consistent with the production of NADH. The presence of NAD tightly bound to AHH has been confirmed in statement form for placental enzyme (Hershfield and Kredich, 1978; Hershfield et al., 1979) and with data for rat liver enzyme (Fujioka and Takata, 1981). Fujioka and Takata estimated the presence of 0.81 mol of NAD/subunit. They also noted that bisulfite reversibly inhibits AHH and that the degree of inhibition is directly proportional to the amount of adduct formed between the bisulfite and enzyme-bound NAD. The presence of tightly bound NAD in AHH from the other sources listed in Table 1 has not been reported. Apparently, relatively high concentrations of the enzyme protein (at least 0.67 mg/mL) are required to demonstrate a significant change in the difference spectrum that ensues following the addition of Ado, which, as mentioned above, reflects the formation of AHH-bound NADH. Such large amounts of pure AHH are routinely unavailable when AHH is purified, e.g., from rat brain.

3.3. Other Properties

AHH is not a substrate for a cAMP-dependent protein kinase, either before or after removal of any putative phosphate groups that it might contain in the native state (Doskeland and Ueland, 1982). AHH is, however, identical to the c-AMP–Ado binding protein of liver, fully described by Ueland (1982). The recognition of this identity was reported as early as 1978 (Saebo and Ueland). An AHH–Ado complex was also reported to form when the enzyme from yellow lupin seeds was incubated with [^{14}C]-Ado (Jakubowski and Guranowski, 1978). Further analysis of this reaction showed that the lupin seed AHH, which is a dimer (*see* Table 1), forms a 1:2 enzyme:Ado complex. Interestingly, while undergoing the two-step binding reactions with Ado, AHH does not lose its ability to catalyze the synthesis of AdoHcy (Jakubowski and Guranowski, 1981).

4. AHH Reaction

4.1. Substrates

The observation that the reaction catalyzed by AHH is reversible was originally reported by de la Haba and Cantoni (1959), who, after having identified AdoHcy as the product of the condensation of Ado and Hcy by "the purified condensing enzyme," used the thioether (AdoHcy) as substrate, noting that the extent of hydrolysis "was very small." Hydrolysis could be accelerated, however, by "trapping" Ado with Ado deaminase and Hcy with thetin-Hcy methylpherase (EC 2.1.16), an enzyme that methylates Hcy to Met in the presence of thetin (methylthioglycollate). A K_{eq} of 1.4×10^{-6} was calculated for the reaction by de la Haba and Cantoni (1959). Since 1959, many compounds, natural and synthetic, have been evaluated as potential AHH substrates. Many found ineffective at one time were later included in the substrate list. Cysteine is a good example. It was reported as ineffective at 50 m*M* in 1959 (de la Haba and Cantoni) and as effective at 30 m*M* in 1983 (Guranowski and Jakubowski), (K_m: 15 and 35 m*M* with the bovine and lupin AHH, respectively). Other effective substrates, including N^6-methylAdo (Hoffman, 1978) and 3-deazaAdo (Richards et al., 1978) are listed in the papers by Chiang et al. (1977), Guranowski et al. (1981), and Guranowski and Jakubowski (1983). Table 2 of the review of Ueland (1982) gives additional details.

4.2. Catalytic Mechanism

Palmer and Abeles (1976) noted that during the course of the AHH reaction, the 4' proton of Ado exchanges with solvent water. On the basis of this and other relevant experimental findings, these workers proposed a mechanism for the reversible hydrolysis of AdoHcy involving oxidation of the position 3' of Ado, followed by elimination of Hcy to give an intermediate that reacts with water to yield 3'-ketoadenosine. The latter is then reduced to Ado. The correctness of the proposed mechanism was verified by Palmer and Abeles (1979) in a series of elegant experiments, including one that required the chemical synthesis of 4',5'-dehydroAdo and its subsequent oxidation by AHH. The catalytic properties of AHH necessary for the protein to perform its physiological role in the cell have been examined (Ueland and Helland, 1980). It was found that under the "physiological" conditions of the measurements, with AHH, Ado, and AdoHcy concentrations set at 10, 2, and 50 μ*M*, respectively, not all of the Ado participates in AdoHcy synthesis or undergoes deamination. The findings led to the formulation of a hypothesis that postulated the existence of two stable complexes, AdoHcy–E and Ado–E, as cellular catalytic intermediates. The former would be susceptible to increase as the concentration of AHH increased and as Hcy entered the system, whereas the latter would be represented by the Ado sequestered by AHH that, by so doing, became inactive. The authors freely recognized the tentative nature of this scheme, but pointed out that catalytic mechanisms not taking into account physiological concentrations of reactants are also of limited biological significance.

4.3. Inhibition

As with the substrates of AHH, the list of AHH inhibitors also includes a large number of compounds (Chiang et al., 1977, 1978; Guranowski et al., 1981; Ueland, 1981; Borchardt et al., 1982). There is particular interest in the inhibition of this enzyme because this results in increases in the intracellular levels of AdoHcy, which, in turn, may be a regulatory element of cellular methylation rates. Alternatively, the inability to utilize Ado in AdoHcy synthesis could cause a shift in its metabolism and in its role as a putative neurotransmitter, particularly in brain tissue. The observation that AHH may be inhibited by AdoMet, adenine, and Hcy has been made repeatedly (Poulton and Butt, 1976; Schatz et al., 1977b; Briske-Anderson and Duerre, 1982). 3-DeazaAdo also inhibits AHH (Chiang et al., 1977; Guranowski et al., 1981; Kim et al., 1983). Interestingly, it was noted that, in the rat, the administration of 3-deazaAdo results in the accumulation of both AdoHcy and 3-deazaAdoHcy, whereas in the hamster only 3-deazaAdoHcy forms (Chiang and Cantoni, 1979). An explanation of this species difference was provided when it was shown that AHH from hamster liver has a 10 times smaller K_m for AdoHcy relative to the rat enzyme. The inhibitory power of periodate-oxidized Ado has been described both in vitro and in vivo (Hoffman et al., 1979; Hoffman, 1980), and it was pointed out that transmethylation rates may be estimated "from the subsequent accumulation of AdoHcy." A note of caution was voiced, however, because it was recognized that some of the transmethylating enzymes, particularly hepatic Gly *N*-methyltransferase, are relatively insensitive to inhibition by AdoHcy (Heady and Kerr, 1973). Hence, only methylations catalyzed by enzymes strongly inhibited by AdoHcy should become less operational following treatment with periodate-oxidized Ado. A similar argument applies, it is believed, to the situation in which high AdoHcy levels are quickly and transiently generated by the joint administration of Ado and Hcy thiolactone or of Hcy alone (Schatz et al., 1981a,b; Gharib et al., 1983), except, that in this circumstance, AdoHcy levels may be increased not only in the liver and other peripheral organs, but equally in the brain. Additional successful AHH inhibitors are the antibiotic 9-β-D-arabinofuranosyladenine and its mono- and triphosphate derivatives (Trewyn and Kerr, 1977; Hershfield, 1979; Helland and Ueland 1981), AMP and ADP (Ueland and Saebo, 1979), and 5'-deoxyAdo (Palmer and Abeles, 1979). Finally, the inhibitory effect of 5'-deoxy-5'-S-isobutyl thioAdo (SIBA) on the synthesis of AHH and on its activity should be mentioned (Robert-Gero et al., 1979).

4.4. *Inactivation*

AHH may be inactivated by a variety of means, and many workers have examined this phenomenon. Simple incubation of the protein at 0° and 37°C, and in the presence of Ado, AdoHcy, and of several Ado analogs results in inactivation (Kajander and Raina, 1981). Ueland (1982), Hershfield and associates (1982), and others have examined AHH inactivation in the presence of adenine, adenine nucleotides, 2'-deoxyAdo, 9-β-D-arabinofuranosyladenine, cordycepin, and 5-deoxy-5'-methylthioAdo (Hershfield, 1979; Hershfield et al., 1979; Ueland and Saebo, 1979; Helland and Ueland, 1981; Ferro et al., 1981; Abeles et al., 1980). In general, the following were noted: (1) The enzyme requires the presence of reducing agents for maximum stability; (2) Adenine nucleotides inactivate the enzyme irreversibly; (3) Adenine inactivates the enzyme reversibly; (4) Inactivation by 9-β-D-arabinofuranosyladenine and 2'-deoxyAdo is also irreversible, and these compounds form tight, although not covalent, bonds with the enzyme, involving the enzyme-bound NAD, which becomes reduced; and (5) Inactivation by cordycepin and 5'-deoxymethylthioAdo, compounds that lack, respectively, the 3' and the 5' hydroxyls, does not involve the enzyme-bound NAD and its reduction to NADH.

5. Role of AHH in Metabolism

In this section some general biochemical points relevant to the role of AHH in the metabolic disposition of several classes of compounds will be briefly dealt with. Obviously the place to begin is AdoHcy itself, since this thioether is the primary "raison d'être" for the existence of the enzyme, or . . . is it the other way around?

5.1. AdoHcy, Adenosine, and Homocysteine

In "higher animal tissues" (Cortese et al., 1974) AdoHcy is the metabolite of choice in the transsulfuration pathway, i.e., the process that converts Met to Cys with the loss of one carbon atom. It is important to note that this pathway is partly fueled by the crucial step that causes one of the products of AHH catalysis, Hcy, to condense with Ser to form cystathionine, instead of remethylating back to Met and on to AdoMet. It may be said, therefore, that the role of AHH in the direction of AdoHcy hydrolysis is energized by reactions that metabolize Hcy, an amino acid whose levels in tissue are in fact so low that an adequate quantitative determination of them had to await the development of a "sensitive radiochemical enzyme assay" (Kredich et al., 1981) and could not be handled by the ever more sophisticated technology of automated amino acid analysis. Another item of interest is the relative tissue diversity of the purinic metabolites formed from AdoHcy. Although, for instance, brain tissue forms no uric acid (Schatz et al., 1977a) and mouse kidney forms a 10th of what rat kidney is able to form, livers of both species put out uniformly large amounts of urate. All in all, there is agreement that the principal metabolic pathway for AdoHcy in mammals is its hydrolysis and that other pathways are quantitatively minor (Duerre et al., 1969; Walker and Duerre, 1975).

5.2. AdoMet

AHH may intervene in the metabolism of AdoMet, i.e., in its ability to donate methyl groups to appropriate acceptors, every time its rate of AdoHcy hydrolysis diminishes or that of AdoHcy synthesis augments. Indeed, the maintenance of the AdoMet/AdoHcy ratio in cells is probably as dependent on the rate of performance of AHH as on that of the AdoHcy-sensitive methyltransferases. However, it is also generally accepted that, in healthy cells and tissues, a dynamic equilibrium between AdoMet and AdoHcy prevails and that moderate fluctuations of the AdoMet/AdoHcy ratio do not create havoc, but are rather irrelevant. Methylations slow down only when the concentration of AdoHcy increases dramatically, and they accelerate only when the levels of AdoMet significantly decrease (Schatz et al., 1978, 1983). Indeed, we have interpreted the decreases in cerebral AdoMet levels noted after the administration of the convulsant agent L-methionine-*d,l*-sulfoximine to mice and rats (Schatz and Sellinger, 1975) as indicators of an accelerated utilization of the methyl group of AdoMet in different transmethylation reactions, involving as methyl acceptors His (Schatz et al., 1978), tRNA (Dainat et al., 1978;

Der and Sellinger, 1980; Sellinger, 1982), phospholipids (Schatz et al., 1981b), and the carboxyl group of protein-bound Asp (Schatz et al., 1981b, 1983). The levels of activity of cerebral AHH in h units were also determined after methionine sulfoximine, and found to be significantly decreased in four out of seven rat brain regions (Schatz et al., 1977b). It is noteworthy that the activity of AHH was found significantly decreased in the very brain regions in which the levels of AdoMet also fell significantly. Conversely, in the hypothalamus no significant AHH decrease and no significant drop in AdoMet levels were noted.

5.3. cAMP

Zimmerman et al. (1980) incubated mouse lymphocytes with micromolar concentrations of Ado or 3'-deazaAdo and noted an accumulation of "supramillimolar" concentrations of AdoHcy and 3'-deazaAdoHcy that resulted in an increase of cAMP release upon stimulation with prostaglandin E_1, Ado, 2'-chloroAdo, isoproterenol, and cholera toxin. The authors suggested that the increased cAMP response was attributable to the direct inhibition by the high levels of intracellular AdoHcy or 3'-deazaAdoHcy of the two enzymes forming and metabolizing cAMP, namely adenylate cyclase and cAMP phosphodiesterase. Yet, most recently, Zimmerman et al. (1984) reported that AHH was most likely not directly involved in another process previously attributed to its action, thus putting a damper on the overall importance of AdoHcy hydrolysis in the biology of the lymphocyte.

5.4. Biogenic Amines

Although the ability of AdoHcy to inhibit transmethylations has been known practically since its discovery, the relevance of this property as a possible regulatory element in the metabolism and physiological action of biogenic amines in brain tissue stems from the work of Deguchi and Barchas (1971), who identified the "transmethylation stimulatory factor" of brain tissue as brain AHH. Lovenberg (1982) has more recently commented on the significance of the control of AdoHcy removal by AHH for optimal "small molecule methylation."

6. Assay Protocols

Over the years, many technological twists have been reported, all designed to improve the various aspects of the AHH assay, both in the synthetic (preferred in early times) and hydrolytic (preferred in modern times) modes. Although we have decided to tabulate as many varieties of the assay as have been adequately described, rather than recommend one standard assay, we still have no guarantee that an assay was not missed or that, in the future, new protocols will not emerge. At any rate, Tables 2 and 3 illustrate the multiplicity of approaches employed by workers (so far) on three continents who have found it necessary not just to assay the enzyme activity but, in addition, to continue modernizing the existing assay technologies all the way to the present.

References

Abeles R. H., Tashjian A. H., Jr., and Fish S. (1980) The mechanism of inactivation of *S*-adenosylhomocysteinase by 2'-deoxyadenosine. *Biochem. Biophys. Res. Commun.* **95,** 612–617.

Aksamit R. R., Falk W., and Cantoni G. L. (1982) Inhibition of chemotaxis by *S*-deazadenosylhomocysteine in a mouse macrophage cell line. *J. Biol. Chem.* **257,** 621–625.

Baddiley J. and Jamieson G. A. (1955) Synthesis of *S*-(5'-deoxyadenosine-5') homocysteine, a product from enzymic methylations involving "Active Methionine". *J. Chem. Soc.* (Lond.), 1085–1089.

Borchardt R. T. (1977) Synthesis and Biological Activity of Analogs of Adenosylhomocysteine as Inhibitors of Methyltransferases, in *The Biochemistry of Adenosylmethionine* (Salvatore F., Borek E., Zappia V., Williams-Ashman H. G., and Schlenck F., eds.), Columbia University, New York.

Borchardt R. T., Patel U. G., and Bartel R. L. (1982) Adenosine Dialdehyde: A Potent Inhibitor of *S*-Adenosylhomocysteine Hydrolase, in *Biochemistry of S-Adenosylmethionine and Related Compounds.* (Usdin E., Borchardt R. T., and Creveling C. R., eds.), Macmillan, London.

Briske-Anderson M. and Duerre J. A. (1982) *S*-adenosylhomocysteine hydrolase from rat liver. *Can. J. Biochem.* **60,** 118–123.

Broch O. J. and Ueland P. M. (1980) Regional and subcellular distribution of *S*-adenosylhomocysteine hydrolase in the adult rat brain. *J. Neurochem.* **35,** 484–488.

Cantoni G. L. and Scarano E. (1954) The formation of *S*-adenosylhomocysteine in enzymatic transmethylation reactions. *J. Am. Chem. Soc.* **76,** 4744–4750.

Chabannes B. E., Bidard J. N., Sarda D., and Cronenberger L. A. (1979a) Nouvelle méthode de séparation des nucléosides dérivés de l'adénosine par chromatographie liquide à haute performance. Application au dosage de la *S*-adénosyl-L-homocystéine hydrolase du foie de rat. *J. Chromatogr.* **170,** 430–436.

Chabannes B., Cronenberger L., and Pacheco H. (1979b) Purification de la *S*-adénosyl-L-homocysteine hydrolase du foie de rat par chromatographie d'affinité. *Experientia* **35,** 1014–1015.

Chiang P. K., Richards H. H., and Cantoni G. L. (1977) *S*-adenosyl-L-homocysteine hydrolase: Analogues of *S*-adenosyl-L-homocysteine as potential inhibitors. *Mol. Pharmacol.* **13,** 939–947.

Chiang P. K. and Cantoni G. L. (1979) Perturbation of biological transmethylations by 3-deazaadenosine in vivo. *Biochem. Pharmacol.* **28,** 1897–1902.

Chiang P. K., Cantoni G. L., Bader J. P., Shannon W. M., Thomas J. J., and Montgomery J. A. (1978) Adenosylhomocysteine hydrolase inhibitors: Synthesis of 5'-deoxy-5'-(isobutylthio)-3-deazaadenosine and its effect on Rous sarcoma virus and Cross murine leukemia virus. *Biochem. Biophys. Res. Commun.* **82,** 417–423.

Cortese R., Perfetto E., Arcari P., Prota G., and Salvatore F. (1974) Formation of uric acid from adenosylhomocysteine in rat liver. *Int. J. Biochem.* **5,** 535–545.

Crooks P. A., Dreyer R. N., and Coward J. K. (1979) Metabolism of *S*-adenosylhomocysteine and *S*-tubercidinylhomocysteine in neuroblastoma cells. *Biochemistry* **18,** 2601–2609.

Dainat J., Salas C. E., and Sellinger O. Z. (1978) Alteration of the specificity of brain tRNA methyltransferases and of the pattern of brain tRNA methylation in vivo by methionine sulfoximine. *Biochem. Pharmacol.* **27,** 2655–2658.

Deguchi T. and Barchas J. (1971) Inhibition of transmethylations of biogenic amines by *S*-adenosylhomocysteine. *J. Biol. Chem.* **246,** 3175–3181.

De la Haba G. and Cantoni G. L. (1959) The enzymatic synthesis of *S*-adenosyl-L-homocysteine from adenosine and homocysteine. *J. Biol. Chem.* **234,** 603–608.

Der O. and Sellinger O. Z. (1980) Transfer Ribonucleic Acids in the Developing Brain: The Effect of the Convulsant Methionine Sulfoximine on tRNA Function, in *Circulatory and Developmental Aspects of Brain Metabolism* (Spatz M., Mršulja B. B., Rakić Lj., and Lust W. D., eds.), Plenum, New York.

Doskeland S. O. and Ueland P. M. (1982) Comparison of some physicochemical and kinetic properties of *S*-adenosylhomocysteine hydrolase from bovine liver, bovine adrenal cortex, and mouse liver. *Biochim. Biophys. Acta* **708,** 185–193.

Duerre J. A. and Briske-Anderson M. (1981) Effect of adenosine metabolites on methyltransferase reactions in isolated rat livers. *Biochim. Biophys. Acta* **678,** 275–282.

Duerre J. A. and Walker R. D. (1977) Metabolism of Adenosylhomocysteine, in *The Biochemistry of Adenosylmethionine* (Salvatore F., Borek E., Zappia V., Williams-Ashman H. G., and Schlenck F., eds.), Columbia University, New York.

Duerre J. A., Miller C. H., and Reams G. G. (1969) Metabolism of S-adenosyl-L-homocysteine in vivo by the rat. *J. Biol. Chem.* **244,** 107–111.

Eloranta T. O. (1977) Tissue distribution of S-adenosylmethionine and S-adenosylhomocysteine in the rat. Effect of age, sex, and methionine administration on the metabolism of S-adenosylmethionine, S-adenosylhomocysteine and polyamines. *Biochem. J.* **166,** 521–529.

Eloranta T. O., Kajander E. O., and Raina A. M. (1976) A new method for the assay of tissue S-adenosylhomocysteine and S-adenosylmethionine. *Biochem. J.* **160,** 287–294.

Eloranta T. O., Kajander E. O., and Raina A. M. (1982) Effect of 9-β-D-arabinofuranosyladenine and erythro-9(2-hydroxy-3-nonyl)adenine on the metabolism of S-adenosylhomocysteine, S-adenosylmethionine, and adenosine in rat liver. *Med. Biol.* **60,** 272–277.

Ericson L. E., Williams J. N., Jr., and Elvehjem C. A. (1955) Enzymatic cleavage of S-adenosylhomocysteine and the transfer of labile methyl groups. *Acta Chem. Scand.* **9,** 859–860.

Ferro A. J., Vandenbark A. A., and McDonald M. R. (1981) Inactivation of S-adenosylhomocysteine hydrolase by 5'deoxy-5'-methylthioadenosine. *Biochem. Biophys. Res. Commun.* **100,** 523–531.

Finkelstein J. D. (1971) Inherited Disorders of Sulphur Metabolism (Carson N. A. J. and Raine D. N., eds.), Churchill Livingstone, London.

Finkelstein J. D. and Harris B. (1973) Methionine metabolism in mammals: Synthesis of S-adenosylhomocysteine in rat tissue. *Arch. Biochem. Biophys.* **159,** 160–165.

Finkelstein J. D. and Harris B. (1975) Methionine metabolism in mammals: S-adenosylhomocysteine hydrolase in rat intestinal mucosa. *Arch. Biochem. Biophys.* **171,** 282–286.

Finkelstein J. D., Harris B. J., Grossman M. R., and Morris H. P. (1978) S-adenosylhomocysteine metabolism in rat hepatomas. *Proc. Soc. Exp. Biol. Med.* **159,** 313–316.

Fujioka M. and Takata Y. (1981) S-adenosylhomocysteine hydrolase from rat liver. Purification and some properties. *J. Biol. Chem.* **256,** 1631–1635.

Gharib A., Chabannes B., Sarda N., and Pacheco H. (1983) In vivo elevation of mouse brain S-adenosyl-L-homocysteine after treatment with L-homocysteine. *J. Neurochem.* **40,** 1110–1112.

Glazer R. L. and Hartman K. D. (1980) Evidence that the inhibitory effect of adenosine, but not cordycepin, on the methylation of nuclear RNA is mediated by S-adenosylhomocysteine hydrolase. *Mol. Pharmacol.* **18,** 483–490.

Guranowski A. and Jakubowski H. (1983) Substrate specificity of S-adenosylhomocysteinase. Cysteine as a substrate of the plant and mammalian enzymes. *Biochim. Biophys. Acta* **742,** 250–256.

Guranowski A. and Pawelkiewicz J. (1977) Adenosylhomocysteinase from yellow lupin seeds. Purification and properties. *Eur. J. Biochem.* **80,** 517–523.

Guranowski A., Montgomery J. A., Cantoni G. L., and Chiang P. K. (1981) Adenosine analogs as substrates and inhibitors of *S*-adenosylhomocysteine hydrolase. *Biochemistry* **20,** 110–115.

Heady J. E. and Kerr S. J. (1973) Purification and characterization of glycine *N*-methyltransferase. *J. Biol. Chem.* **248,** 69–72.

Helland S. and Ueland P. M. (1981) Interaction of 9-β-D-arabinofuranosyladenine, 9-β-D-arabinofuranosyladenine 5'monophosphate and 9-β-D-arabinofuranosyladenine 5'-triphosphate with *S*-adenosylhomocysteinase. *Cancer Res.* **41,** 673–678.

Hershfield M. S. (1979) Apparent suicide inactivation of human lymphoblast *S*-adenosylhomocysteine hydrolase by 2'-deoxyadenosine and adenine arabinoside. *J. Biol. Chem.* **254,** 22–25.

Hershfield M. S. and Francke U. (1982) The human genes for *S*-adenosylhomocysteine hydrolase and adenosine deaminase are syntenic on chromosome 20. *Science* **216,** 739–742.

Hershfield M. S. and Kredich N. M. (1978) *S*-Adenosylhomocysteine hydrolase is an adenosine binding protein: A target for adenosine toxicity. *Science* **202,** 757–760.

Hershfield M. S., Kredich N. M., Small W. C., and Fredericksen M. L. (1979) Suicide-Like Inactivation of Human *S*-Adenosylhomocysteine Hydrolase by Analogs of Adenosine, in *Transmethylation* (Usdin E., Borchardt R. T., and Creveling C. R., eds.), Elsevier/North Holland, New York.

Hoffman J. L. (1978) Biosynthesis of *S*-N^6-methyladenosylhomocysteine, an inhibitor of RNA methyltransferases. *J. Biol. Chem.* **253,** 2905–2907.

Hoffman J. L. (1980) The rate of transmethylation in mouse liver, as measured by trapping *S*-adenosylhomocysteine. *Arch. Biochem. Biophys.* **205,** 132–135.

Hoffman D. R., Cornatzer W. E., and Duerre J. A. (1979) Relationship between tissue levels of *S*-adenosylmethionine, *S*-adenosylhomocysteine, and transmethylation reactions. *Can. J. Biochem.* **57,** 56–64.

Jakubowski H. and Guranowski A. (1978) Adenosylhomocysteinase–adenosine complex. *Biochem. Biophys. Res. Commun.* **84,** 1060–1068.

Jakubowski H. and Guranowski A. (1981) *S*-adenosylhomocysteinase from yellow lupin seeds: Stoichiometry and reactions of the enzyme–adenosine complex. *Biochemistry* **20,** 6877–6881.

Kajander O. and Raina A. M. (1981) Affinity-chromatographic purification of *S*-adenosyl-L-homocysteine hydrolase. Some properties of the enzyme from rat liver. *Biochem. J.* **193,** 503–512.

Kajander O., Eloranta T., and Raina A. (1976) A sensitive isotopic assay method for *S*-adenosylhomocysteine hydrolase. Some properties of the enzyme from rat liver. *Biochim. Biophys. Acta* **438,** 522–531.

Kamatani N., Willis E. H., and Carson D. A. (1982) Selection of Mutant Murine Lymphoid Cells Partially Deficient in S-Adenosylhomocysteine Hydrolase. in *Biochemistry of S-Adenosylmethionine and Related Compounds.* (Usdin E., Borchardt R. T., and Creveling C. R., eds.), Macmillan, London.

Kim I-K., Zhang C-Y, Chiang P. K., and Cantoni G. L. (1983) S-adenosylhomocysteine hydrolase from hamster liver: Purification and kinetic properties. *Arch. Biochem. Biophys.* **236,** 65–72.

Kredich N. M., Kendall H. E., and Spence F. J., Jr. (1981) A sensitive radiochemical enzyme assay for S-adenosylhomocysteine and L-homocysteine. *Anal. Biochem.* **116,** 503–510.

Liau M. C., Chang C. F., Saunders G. F., and Tsai Y. H. (1981) S-adenosylhomocysteine hydrolases as the primary target enzymes in androgen regulation of methylation complexes. *Arch. Biochem. Biophys.* **208,** 261–272.

Lovenberg W. (1982) Methylation of Small Molecules: An Overview, in *Biochemistry of S-Adenosylmethionine and Related Compounds.* (Usdin E., Borchardt R. T., and Creveling C. R., eds.), Macmillan, London.

Palmer J. L. and Abeles R. H. (1976) Mechanism for enzymatic thioether formation. Mechanism of action of S-adenosylhomocysteinase. *J. Biol. Chem.* **251,** 5817–5819.

Palmer J. L. and Abeles R. H. (1979) The mechanism of action of S-adenosylhomocysteinase. *J. Biol. Chem.* **254,** 1217–1226.

Pierré A., Richou M., Lawrence F., Robert-Géro M., and Vigier P. (1977) Decreased rate of S-adenosyl-L-homocysteine metabolism: An early event related to transformation in cells infected with Rous sarcoma virus. *Biochem. Biophys. Res. Commun.* **76,** 813–819.

Poulton J. E. and Butt V. S. (1976) Purification and properties of S-adenosyl-L-homocysteine hydrolase from leaves of spinach beet. *Arch. Biochem. Biophys.* **172,** 135–142.

Reddington M. and Pusch R. (1983) Adenosine metabolism in a rat hippocampal slice preparation: Incorporation into S-adenosylhomocysteine. *J. Neurochem.* **40,** 285–290.

Richards H. H., Chiang P. K., and Cantoni G. L. (1978) Adenosylhomocysteine hydrolase. Crystallization of the purified enzyme and its properties. *J. Biol. Chem.* **253,** 4476–4480.

Robert-Géro M., Blanchard P., Lawrence F., Pierré A., Vuilhorgne M., and Lederer E. (1979) Antiviral and oncostatic properties of synthetic and natural analogues of S-adenosylhomocysteine (SAH). Transmethylation (Usdin E., Borchardt R. T., and Creveling C. R., eds.), Elsevier/North Holland, New York.

Saebo J. and Ueland P. M. (1978) An adenosine 3' : 5'-monophosphate adenosine-binding protein from mouse liver. Association with S-adenosylhomocysteinase activity. *FEBS Lett.* **96,** 125–128.

Schatz R. A. and Sellinger O. Z. (1975) Effect of methionine and methionine sulfoximine on rat brain *S*-adenosylmethionine levels. *J. Neurochem.* **24,** 63–66.

Schatz R. A., Vunnam C. R., and Sellinger O. Z. (1977a) *S*-Adenosyl-L-homocysteine in brain. *Neurochem. Res.* **2,** 27–38.

Schatz R. A., Vunnam C. R., and Sellinger O. Z. (1977b) Species and tissue differences in the catabolism of *S*-adenosyl-L-homocysteine: A quantitative, chromatographic study. *Life Sci.* **20,** 375–384.

Schatz R. A., Frye K., and Sellinger O. Z. (1978) Increased in vivo methylation of [^{3}H]-histamine in the methionine sulfoximine epileptogenic brain. *J. Pharmacol. Exp. Ther.* **207,** 794–800.

Schatz R. A., Vunnam C. R., and Sellinger O. Z. (1979) *S*-Adenosyl-L-homocysteine Hydrolase From Rat Brain: Purification and Properties, in *Transmethylation* (Usdin E., Borchardt R. T., and Creveling C. R., eds.), Elsevier/North Holland, New York.

Schatz R. A., Wilens T. E., and Sellinger O. Z. (1981a) Decreased in vivo protein and phospholipid methylation after in vivo elevation of brain *S*-adenosyl-L-homocysteine. *Biochem. Biophys. Res. Commun.* **98,** 1097–1107.

Schatz R. A., Wilens T. E., and Sellinger O. Z. (1981b) Decreased transmethylation of biogenic amines after in vivo elevation of brain *S*-adenosyl-L-homocysteine. *J. Neurochem.* **36,** 1739–1748.

Schatz R. A., Wilens T. E., Tatter S. B., Gregor P., and Sellinger O. Z. (1983) Possible role of increased brain methylation in methionine sulfoximine epileptogenesis: Effects of administration of adenosine and homocysteine thiolactone. *J. Neurosci. Res.* **10,** 437–447.

Schrader J., Schutz W., and Bardenheuer H. (1981) Role of *S*-adenosylhomocysteine hydrolase in adenosine metabolism in mammalian heart. *Biochem. J.* **196,** 65–70.

Sellinger O. Z. (1982) Methionine sulfoximine in vivo modifies the $tRNA^{Lys}$ pool of developing rat brain. *J. Neurochem.* **38,** 1676–1685.

Sellinger O. Z. and Schatz R. A. (1979) Cerebral Utilization of Adenosylmethionine and Adenosylhomocysteine: Effects of Methionine Sulfoximine, in *Biochemical and Pharmacological Roles of Adenosylmethionine in the Central Nervous System* (Zappia V., Usdin E., and Salvatore F., eds.), Pergamon, New York.

Sellinger O. Z. and de Balbian Verster F. (1962) Glutamine synthetase of rat cerebral cortex: Intracellular distribution and structural latency. *J. Biol. Chem.* **237,** 2836–2844.

Trewyn R. W. and Kerr S. J. (1977) The enzymatic synthesis of *S*-adenosyl-L-[2(*n*)-^{3}H]-homocysteine. *Anal. Biochem.* **82,** 310–316.

Trewyn R. W. and Kerr S. J. (1981) An improved rapid assay for *S*-adenosyl-L-homocysteine hydrolase. *J. Biochem. Biophys. Methods* **4,** 299–307.

Ueland P. M. (1981) *S*-Adenosylhomocysteinase from mouse liver. Inactivation of the enzyme in the presence of metabolites. *Int. J. Biochem.* **14,** 207–213.

Ueland P. M. (1982) Pharmacological and biochemical aspects of *S*-adenosylhomocysteine and *S*-adenosylhomocysteine hydrolase. *Pharmacol. Rev.* **34,** 223–253.

Ueland P. M. and Doskeland S. O. (1977) An adenosine 3':5'-monophosphate adenosine binding protein from mouse liver. Purification and partial characterization. *J. Biol. Chem.* **252,** 677–686.

Ueland P. M. and Helland S. (1980) *S*-adenosylhomocysteinase from mouse liver. Catalytic properties at cellular enzyme level. *J. Biol. Chem.* **255,** 7722–7727.

Ueland P. M. and Saebo J. (1979) *S*-adenosylhomocysteinase from mouse liver. Effect of adenine and adenine nucleotides on the enzyme catalysis. *Biochemistry* **18,** 4130–4135.

Ueland P. M., Berge R. K., Saebo J., and Farstad M. (1979) *S*-adenosylhomocysteine hydrolase in human and rat liver is localized to the cytosol fraction of the tissue homogenate. *FEBS Lett.* **101,** 184–186.

Walker R. D. and Duerre J. A. (1975) *S*-adenosylhomocysteine metabolism in various species. *Can. J. Biochem.* **53,** 312–319.

Wong D. L. and Ciaranello R. D. (1982) Regulatory Aspects of *S*-Adenosylmethionine in the Periphery and the CNS, in *Biochemistry of S-Adenosylmethionine and Related Compounds* (Usdin E., Borchardt R. T., and Creveling C. R., eds.), Macmillan, London.

Zappia V., Usdin E., and Salvatore F. (eds.) (1979) Biochemical and Pharmacological Roles of Adenosylmethionine in the Central Nervous System. Pergamon, New York.

Zimmerman T. P., Schmitges C. J., Wolberg G., Deeprose R. D., Duncan G. S., Cuatrecasas P., and Elion G. B. (1980) Modulation of cyclic AMP metabolism by *S*-adenosylhomocysteine and *S*-3-deazaadenosylhomocysteine in mouse lymphocytes. *Proc. Natl. Acad. Sci. USA* **77,** 5639–5643.

Zimmerman T. P., Iannone M., and Wolberg G. (1984) 3-Deazaadenosine. *S*-Adenosylhomocysteine hydrolase-independent mechanism of action in mouse lymphocytes. *J. Biol. Chem.* **259,** 1122–1126.

Chapter 17
Phenol Sulfotransferase

Jerome A. Roth

1. Introduction

Sulfate conjugation is one of the major catabolic processes involved in the inactivation and elimination of a variety of phenolic drugs and other structurally related exogenous agents. Sulfation of pharmacologically active xenobiotics is assumed to promote their biological inactivation and, subsequently, facilitate excretion of these agents in the kidney by increasing their water solubility. What has generally been overlooked and disregarded as of major physiological consequence is the fact that a variety of endogenous bioactive substances can also be sulfated in vivo, although in many cases the significance of these reactions has not been appropriately evaluated. Included in the group of endogenous compounds that are known to undergo extensive sulfate conjugation are the catecholamine neurotransmitters dopamine (DA), norepinephrine (noradrenaline, NA), and epinephrine (adrenaline, A) (Roth and Rivett, 1982), as well as several neuropeptides including enkephalin (Unsworth and Hughes, 1982) and cholecystokinin (Dockray et al., 1978). As will be pointed out in greater detail below, sulfation of these endogenous compounds requires a common sulfate donor, 3'-phosphoadenosine-5'-phosphosulfate (PAPS), but is catalyzed by distinct forms of the enzyme phenol sulfotransferase (PST). Because of the apparent broad substrate specificity demonstrated, this reaction may represent a common metabolic pathway by which a variety of endogenous neuroactive agents and other physiologically active substances can be biochemically modified.

Examination of biological activity of the sulfated neuroactive agents that have been identified in vivo reveals that the functional consequence of this reaction is dependent on the particular substance being esterified. In the case of the neurotransmitters or neuromodulators, sulfation has been found to lead to either activation or inactivation of the endogenous agent in question. For example, in the case of the peptide cholecystokinin the sulfate conjugate has been shown to be the physiologically active form of the neurotransmitter (Dockray et al., 1978; Hermansen, 1984). In contrast, the sulfate conjugate of leucine enkephalin has been found to be essentially biologically inert (Unsworth and Hughes, 1982; LaBella et al., 1982) and thus represents an inactive form of the peptide that may be reactivated by hydrolysis of the sulfate moiety by arylsulfatase.

Clearly the evidence presented over the past several years suggests that sulfoconjugation constitutes an important physiological process that has the potential to regulate the function of a variety of endogenous bioactive substances in vivo. This chapter will briefly summarize the recent findings in regard to the properties, function, and assay of the various forms of PST. Special emphasis will be placed on the properties of human PST since the properties of this enzyme appear to be biochemically different from those in commonly used laboratory animals.

2. Properties of Phenol Sulfotransferase

Probably one of the most complex issues regarding the function and properties of PST relates to the multiplicity of this enzyme and the selective substrate specificities demonstrated by the different forms. Unfortunately, much of the early literature has inadequately addressed this issue and inadvertently perpetuated this confusion. The nomenclature initially used to describe the different forms of PST was based on the substrate or substrates used to measure enzymatic activity. Accordingly, the forms of PST that esterify phenolic compounds were designated as the phenol or aryl sulfotransferases, whereas those that conjugated steroids were appropriately classified as the steroid sulfotransferases. As is now known, the issue is somewhat more complex since multiple forms within each class of sulfotransferase have been identified and in some cases these enzymes have been shown to have overlapping substrate specificity.

The first demonstration that multiple forms of PST exist was published by Banerjee and Roy in 1966. They identified two forms of guinea pig liver sulfotransferase, one of which was capable of esterifying both phenol and the steroids aldosterone and estrone. In 1970, Mattock and Jones (1970) revealed the existence of two distinct forms of PST in rat liver, one of which conjugated phenol and the other one of which esterified tyramine and tyrosine methyl ester. In a recent series of papers, Jakoby and coworkers (Sekura and Jakoby, 1979, 1981; Jakoby, 1982) have extended these studies and demonstrated that rat liver actually contains at least four distinct forms of PST that are separable by ion exchange chromatography with DEAE cellulose. Two of the forms (Types I–II) have similar substrate specificity, amino acid composition, and preferentially conjugate neutral aromatic alcohols such as phenol and naphthol. The fourth form of the transferase, Type IV, appears to be somewhat unique in that it is also capable of sulfating the catecholamines DA and A as well as tyramine, tyrosine methyl ester, and several small peptides that contain a tyrosyl residue on the N-terminal end. Type III sulfotransferase is similar to that of Type IV, although it has not been as well characterized as the other three forms of the enzymes. Borchardt and Schasteen (1982) have also reported purifying a form of rat liver PST that resembles the Type IV PST isolated by Jakoby (1982), although several differences were observed, including the K_m values for PAPS binding, molecular weight, and K_i for PAP binding. Jakoby and coworkers (Marcus et al., 1980; Lyon and Jacoby, 1980) have also purified a form of sulfotransferase that is capable of sulfating primary and secondary alcohols, including ascorbic acid, vitamin A, propranolol, and several steroids, such as vitamin D, estradiol, testosterone, and cortisol. However, phenolic compounds and phenolic steroids were not substrates for this latter enzyme.

Multiple forms of PST have also been identified in human tissues. Based on differences in substrate and inhibitor specificities, Rein et al. (Rein et al., 1981, 1982a,b, 1984) were the first to demonstrate the presence of at least two functionally distinct forms of PST in a number of human tissues. The M form of PST, as classified by Sandler and coworkers (Rein et al., 1981, 1982a), selectively esterifies the catecholic and phenolic amines, whereas the P form of the sulfotransferase selectively conjugates phenol and a variety of other nonamine containing acceptor substrates, such as the NA metabolite 3-methoxy-4-hydroxyphenylethylene glycol (MOPEG, MHPG). However, there is overlapping specificity within the different forms of human PST in that the M form of the transferase can also esterify phenol and other neutral phenolic substrates. In general the M form possesses a lower affinity for these neutral phenolic substrates than that of the P form of the enzyme. Weinshilboum and coworkers (Anderson and Weinshilboum, 1979; Reiter et al., 1983; Anderson et al., 1981; Reiter and Weinshilboum, 1982; Campbell et al., 1984; Young et al., 1984) have reported that the M form of human PST is considerably more heat-labile than the P form and they have used this difference to

Table 1
Purification of Human Brain M-PST

Step	Total volume, mL	Protein, mg/mL	Specific activity, pmol/min/mg	Yield, %	Purification
12,000*g* supernatant	700	2.92	1.44	100	1.0
100,000*g* supernatant	650	2.13	2.11	99	1.5
DEAE cellulose	294	0.25	25.17	63	17.4
AffiGel Blue	297	0.0034	353	12	244
Sephacryl-S-200	55	3.1×10^{-4}	7735	4	5371

analyze the activities of the two enzyme species in human platelets. Recent data by this group suggest that these enzymes may be under separate genetic control (VanLoon and Weinshilboum, 1984).

Whittemore et al. (1985) have recently demonstrated the presence of three forms of PST in human brain by chromatography on a DEAE cellulose column. Two of the forms demonstrated substrate specificities similar to those of P-PST described previously by Sandler and coworkers (Rein et al., 1981, 1982a), and thus have been designated P_I and P_{II} PST. The third form eluting from the column resembled the M species and is capable of sulfating a variety of catecholamines and structurally related compounds. The M and P_I forms of PST were purified approximately 5000- and 400-fold, respectively, using Affi-gel Blue affinity and Sephacryl S-200 gel chromatography (*see* Tables 1 and 2). The pH optimum for the M form of PST was approximately 7.0 in triethanolamine buffer, whereas the pH optimum for the P_I species was 8.5. In addition, the two P forms of the human brain sulfotransferase were more sensitive to inhibition by dichloro-*p*-nitrophenol, *p*-nitrophenol sulfate, and ATP than was the M form.

Whether the two forms of PST isolated from human brain actually represent two distinct molecular entities or different aggregate states of the same enzyme species is not known at this time. The only difference between the two enzymes reported is

Table 2
Purification of Human Brain P_I-PST

Step	Total volume, mL	Protein, mg/mL	Specific activity, pmol/min/mg	Yield, %	Purification
12,000*g* supernatant	830	1.97	6.17	100	1.0
100,000*g* supernatant	755	0.58	20.6	91	3.4
DEAE cellulose	495	0.53	24.4	64	4.0
AffiGel Blue	240	0.0135	320	10	51.9
Sephacryl S-200	50	0.0041	2,412	5	391

their sensitivity to NaCl; P_I being inhibited approximately 90% at 200 m*M* NaCl and P_{II} inhibited less than 5% at this salt concentration. The effects of NaCl on P_I- and P_{II}-PST suggest that salt inhibition may, in fact, be a useful tool in differentiating the two forms of PST in human brain.

Weinshilboum (personal communication) has recently confirmed the presence in human platelets of two P forms of PST similar to those observed in brain. Interestingly, recent studies by Yu et al. (1985) have revealed the presence of two M forms of PST in human brain, again based on elution profiles with DEAE cellulose. Interestingly, these two enzymes may be comparable to the Type III and IV forms of rat liver PST. However, as indicated above for the two P species, the nature of the two M forms isolated by Yu et al. (1985) and whether these enzyme species actually represent distinct proteins or different aggregate states of the same enzyme, have not been determined.

A very important point to note is the fact that significant differences exist in the properties of human PST and that identified in a number of animal species. As will be indicated in detail below, these differences are clearly of the magnitude that can potentially alter the function of these enzymes in vivo. As a consequence of these differences, the rat or other common laboratory animals may be unreliable models with which to accurately assess the relevance and function of this enzyme in humans.

Before discussing the differences in the biochemical properties between human and animal PST, it is first necessary to briefly describe some of the similarities that exist. First, the molecular weight of these enzymes isolated from a number of animal species including humans are all around 60,000–70,000, although aggregate forms of the enzyme possessing molecular weights greater than 200,000 have also been reported (Jakoby, 1982; Romain et al., 1982; Borchardt and Schasteen, 1982; Carroll and McEvoy, 1970; Whittemore et al., 1985). Second, PAPS appears to be the common sulfate donor for all forms of the sulfotransferases identified regardless of the species in which the enzyme is identified (Mulder, 1981).

With respect to the differences in the biochemical properties of human and animal PST, there are two significant differences that likely have bearing on the functionality of these enzymes in vivo. The first of these is the observation that the K_m for DA binding to human M-PST is approximately 1 μM, whereas this kinetic constant is almost 100 times greater in a number of commonly used animal species (Table 3) (Roth et al., 1982). Interestingly, the only animal species examined to possess a comparable K_m value for dopamine was another primate, the African green monkey. This possibly suggests that primates have a unique form of PST that is highly specific for the catecholamine neurotransmitters. Consistent with differences in the kinetic constant for catecholamine binding are the differences in the mechanism of the two reactions. Human M-PST proceeds via an ordered reaction mechanism

Table 3
Kinetic Constants for Dopamine Binding to PST[a]

Animal	K_m, μM
Beef	183
Mouse	154
Dog	150
Rat	130
Guinea pig	115
A. Green monkey	2.2
Human	1.7

[a]Assayed with 100,000*g* supernatant solutions.

(Whittemore et al., 1985) and the rat enzyme (amine selective form; Type IV) proceeds via a reversible rapid equilibrium random reaction mechanism (see below) (Duffel and Jakoby, 1981).

The second important functional difference between human and animal PST is the ability of the M and P forms to sulfate the tyrosyl moiety of small peptides. Sekura and Jakoby (1981) reported that the Type IV (M form) arylamine-selective PST isolated from rat liver only conjugated small peptides that contain an N-terminal tyrosyl group, whereas the other three forms of rat liver PST failed to catalyze sulfation of peptides. In contrast, the M form of PST isolated from human brain was incapable of sulfoconjugating a variety of small peptides; however, as will be described in greater detail later in the chapter, the two P forms of human brain sulfotransferase readily conjugated the tyrosyl moiety of a variety of small peptides (leu-enkephalin, angiotensin II, oxytocin, and so on), regardless of the position of the tyrosyl residue within the amino acid chain (Whittemore et al., 1984).

Considerable heterogeneity also exists in the salt sensitivity of multiple forms of PST between rat and human sulfotransferases. Sekura and Jakoby (1979) have demonstrated that aryl sulfotransferases I and II from rat liver exhibit significant activation over a broad range of NaCl concentrations, with inhibition becoming apparent only above 800 m*M* NaCl. Aryl sulfotransferase IV from rat liver is activated at 100 m*M* NaCl, and is inhibited 50% at 200 m*M* (Sekura and Jakoby, 1981). As briefly discussed above, human M- and P-PST activities are also sensitive to salt, although in contrast to the rat enzymes, little activation of human PST activity is observed in the presence of salt.

When the relationship between the pH optima for human M and P_I PST is compared with that of other phenol-sulfating enzymes from other animal species, substantial differences are also noted. Sekura and Jakoby (1979), Borchardt and Schasteen (1982), and Romain et al. (1982) all report optima in the range of pH 5.5–6.5 for rat and dog liver phenol sulfotransferases, whereas Butler et al. (1983) and Rein et al. (1984) report that partially purified P-PST from human platelets and brain have a pH optimum of approximately 7.5. Studies by Whittemore et al. (1985) have determined the pH optima for M and P_I PST to be approximately 7.0 and 8.0, respectively.

Consistent with the differences noted above between rat and human PST are the data obtained concerning the reaction mechanism. The reaction mechanism analyzed by steady-state kinetic methods for both human brain M and P_I PST is illustrated in the equation below.

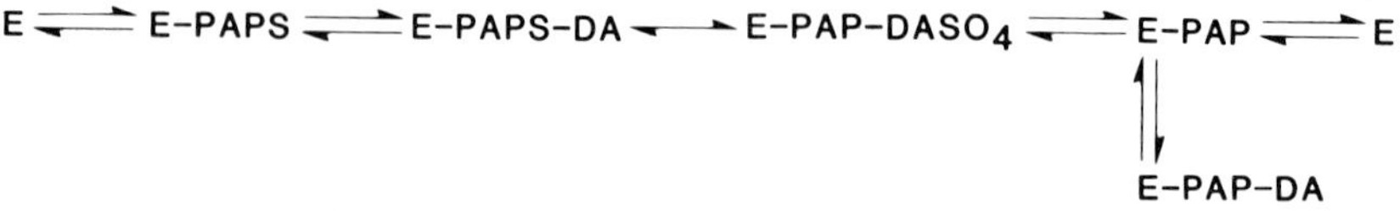

The reaction proceeds via an ordered mechanism with the sulfate donor, PAPS, being the leading substrate. The K_m value for PAPS is calculated to be 0.1 μM. The product of the reaction, 3'-phosphoadenosine-5'-phosphate (PAP), was shown to possess a high affinity for the enzyme and acts as an end product inhibitor by effectively competing with PAPS for binding. Similarly, substrate concentrations approximately 10-fold above the K_m value produce substrate inhibition of PST. Although the product, DA sulfate, was incapable of inhibiting the M form of human brain PST, the product p-NO_2-catechol sulfate was found to be a potent inhibitor of P_I and P_{II} activity, and suggests this latter reaction may be reversible.

Duffel and Jakoby (1981) demonstrated that the rat liver enzyme (Type IV; aryl amine sulfotransferase), in contrast to human PST, proceeds via a reversible rapid equilibrium random mechanism with formation of two dead-end complexes, E-PAPS-ASO_4 and E-PAP-A. This reaction is illustrated in Eq. (2) (A, acceptor substrate):

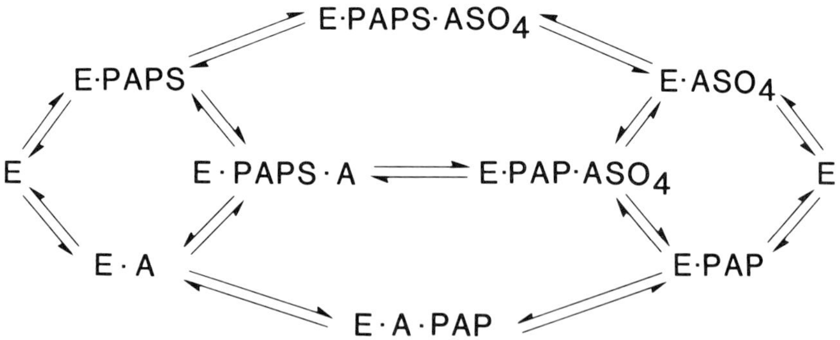

The K_m value for PAPS binding to rat liver Type IV PST was approximately 15 μM, or some 100 times greater than that determined for the comparable human enzyme.

3. Localization of Phenol Sulfotransferase

It is somewhat difficult to accurately describe the tissue distribution of the different forms of PST since the majority of the studies utilize only a single substrate at a fixed concentration and fail to account for the multiplicity of this enzyme. For example, in rat, PST activity (Type I–IV), as measured with 50 μ*M* phenol, was detected in all tissues examined, with liver possessing the highest specific activity and heart the least (Foldes and Meek, 1974). At this concentration of phenol, it is difficult to ascribe which form of PST is responsible for conjugation since activity can occur with all forms of the rat enzyme. PST activity in rat brain represents slightly greater than 10% of the activity found in liver. Under these conditions, highest activity in brain was observed in the hypothalamus, hippocampus, midbrain, and striatum and the lowest activity was associated with cerebellum and pineal gland.

In humans, both M- and P-PST activity have been detected in a number of organs. Highest activity of the M form is associated with intestine and platelets, whereas greatest activity of the P enzyme species is observed in the jejunum and adrenal glands (<1% of jejunum) (Rein et al., 1982a). Both M- and P-PST activities are discernible in human brain, but the activities are considerably lower than in other organs. Highest activity in brain is found in the cerebellum, medulla, cortex, and pituitary. Similarly, the greatest P-PST activity is observed in the medulla, cortex, and pituitary (Rein et al., 1984; Young et al., 1984). Interestingly, P-PST activity in the anterior pituitary is almost six times greater than that in any other area of brain. As mentioned previously, relatively high M- and P-PST activity can be observed in human platelets. This activity has been shown to correlate with the respective PST activities in other tissues, although only P-PST activity appears to correlate with brain P activity (Young et al., 1985). These data suggest that platelet PST may serve as a useful marker for PST activity in vivo. As will be discussed later in this chapter, this is especially relevant in light of recent studies demonstrating a decrease in platelet PST activity in a number of neurological and behavioral disorders.

With regard to the localization of PST in brain, sulfation of the catecholamines and phenol in the CNS occurs predominantly in neurons. This is based on the observation that approximately 60% of the M- and P-PST activity was lost after lesioning rat striatum with the selective neurotoxic agent kainic acid. This loss of PST activity was proportional to the decrease in activity of the neuronal marker enzyme glutamate decarboxylase (Rivett et al., 1984). Since kainic acid is known to selectively destroy neuronal cell bodies at the site of injection, these data imply that both M- and P-PST are localized in neurons intrinsic to the rat striatum. In addition, unpublished observations in our laboratory further demonstrate a 10–20% loss of PST activity in rat striatum after lesioning the substantia nigra with kainic acid. Collectively these data suggest that the major portion of M- and P-PST in the rat striatum is localized in both the dopaminergic nerve endings and neuronal cells intrinsic to the striatum.

4. Sulfate Conjugation of Catecholamines

Sulfate conjugation of catecholamines has been known for over four decades, but this reaction has only recently gained recognition as having any significance in humans. Numerous studies have demonstrated significant urinary excretion of sulfoconjugated DA and NA and their metabolites in human volunteers (*see* review by Roth and Rivett, 1982). The contribution of this reaction to the overall inactivation of the catecholamine DA can be discerned by the study demonstrating that almost 6% of the total exogenously administered dose of L-3,4-dihydroxyphenylalanine (DOPA) to human subjects was excreted as the sulfate conjugate of DA. Relative to other DA catabolites, it is significant that this percentage was similar to that of the excretion of the deaminated product, 3,4-dihydroxyphenylacetic acid (DOPAC), in the urine (Goodall and Alton, 1972). The abundance of DA sulfate excreted in the urine is not surprising and is consistent with the fact that greater than 99% of all DA in the human circulation is present as the sulfate ester (Imai et al., 1970; Tyce et al., 1974; Buu and Kuchel, 1977; Johnson et al., 1980). Similarly, 70–80% of NA and A are present in human plasma as the sulfate conjugates.

With respect to sulfation of catecholamines in CNS, sulfate esters of both NA and DA have been detected in human CSF (Dencker et al., 1967; Tyce et al., 1980). In the particular case of DA, only the conjugated metabolite, and not free DA, was identified in the CSF. Upon administration of L-DOPA to Parkinsonian patients, DA sulfate was found to be significantly elevated. There is substantial variation in the regional distribution of both the absolute amount and the percentage of DA sulfate found in primate brain (Elchisak, 1983; Elchisak et al., 1982, 1983). The brain areas of rhesus monkey that possessed the greatest percentage of DA sulfate were the frontal cortex (74%), cingulate gyrus (31%), cerebellum (16%), and occipital cortex (11%). Surprisingly, the areas of brain that possess the highest levels of dopamine had, in contrast, the lowest levels of conjugated metabolite. For example, in rhesus monkey, the caudate nucleus had approximately 1% of the DA conjugated and the substantia nigra and amygdala had less than 0.5% conjugated. Thus, an apparent inverse relationship exists between the levels of DA and DA sulfate in the CNS of primates.

It is tempting to speculate, based on the kainic acid lesioning studies in rat brain, that sulfation of DA also occurs exclusively in neuronal cells of human and that the presence of DA sulfate in the CSF may reflect neuronal turnover of dopamine. Consistent with this speculation are the observations that DA sulfate is the principal metabolite of DA in cultured mouse neuroblastoma cells (Michelot et al., 1977) and that it is a major catabolite formed in superfused slices of rat striatum (Tyce and Rorie, 1982). However, recent studies by Yu and Walz (1985) have revealed the presence of PST in primary glial cultured cells of newborn rats.

Although DA sulfate is present in human CSF, it is difficult to establish the importance or relative contribution of sulfoconjugation to the overall inactivation and elimination of the catecholamine neurotransmitters in the CNS. Sulfation has been estimated to account for up to 15% of the total metabolism of DA in crude preparations of human frontal cortex when compared to the total PST, monoamine oxidase (MAO) A and B, and soluble and membrane-bound catechol-*O*-methyltransferase (COMT) activity in this tissue (Rivett et al., 1982). These studies, however, do not take into account the selective localization of these enzymes in the CNS and accordingly may not reflect the actual relative contribution of these activities to catecholamine degradation. It should be remembered, as described above, that the M form of PST that is responsible for sulfation of catecholamines, is selectively compartmentalized in neural cells. Therefore, its contribution to catabolism of catecholamines may actually be greater since neuronal PST may have access to the catecholamines selectively taken up by these cells. The contribution of sulfation to the overall inactivation of catecholamines in the periphery is also unknown. However, since sulfate esters of the catecholamines appear to be major urinary metabolites, the expectation is that it contributes significantly to degradation of these neurotransmitters.

5. Sulfate Conjugation of Peptides

In contrast to the relatively extensive studies performed concerning the sulfation of the catecholamine neurotransmitters, relatively little information is available regarding the conjugation of neuropeptides or other proteins. The study of sulfation of peptides and proteins can be traced to the work of Bettelheim (1954), who identified the presence of tyrosine sulfate in human fibrinogen. In the following year Tallan et al. (1955) further demonstrated that approximately 50% of the total urinary excretion of tyrosine was accounted for as the sulfate ester. Although sulfation of the hydroxyl group of tyrosine suggests that the enzymatic reaction is catalyzed by PST, initial attempts to conjugate free tyrosine and proteins with crude preparations of PST were unsuccessful (Nose and Lipmann, 1958; Dodgson et al., 1959; Segal and Mologne, 1959). In contrast, sulfation of the dipeptide tyrosyl-glycine, as well as of tyrosine methyl ester, has been shown to occur readily in vivo (Segal and Mologne, 1959; Dodgson et al., 1959; Jones and Dodgson, 1965; Jones et al., 1966).

Mattock and Jones (1970) were the first to partially purify a form of rat liver PST that is capable of conjugating both tyrosine methyl ester and tyramine (TA). These data are consistent with the more recent studies of Jakoby and coworkers (Sekura and Jakoby, 1981), which demonstrate that only the Type IV (amine or M-sulfating form) of rat liver PST is capable of esterifying a variety of small peptides in which the tyrosyl moiety is present only on the N-terminal position. Peptides that contain an endo-tyrosyl group such as angiotensin are apparently not sulfated by any form of rat liver PST. However, this issue is confused by the observation of Meek and coworkers who have demonstrated that the P-PST selective inhibitor 2,6-dichloro-*p*-nitrophenol, when injected directly into the CNS, effectively inhibits the conjugation of cholecystokinin, a peptide that contains an endo-tyrosyl group (Osvaldo and Meek, 1985). Thus, the question remains as to which form of PST, the M- or P- form, is responsible for the conjugation of cholecystokinin. It is possible that rat brain contains a unique form of PST that is different from the sulfotransferases (Types I–IV) characterized previously in the liver.

In contrast to the findings in the rat, recent studies in our laboratory, illustrated in Table 4, have demonstrated that both the P_I and P_{II} forms of human brain PST are capable of sulfating a variety of small peptides regardless of the position of the tyrosyl group within the amino acid chain (Whittemore et al., 1984). Both the N-terminal tyrosyl-containing peptides leucine enkephalin and growth hormone releasing factor were conjugated, as well as oxytocin, a peptide that contain an endo-tyrosyl residue. Interestingly, the M species of human brain PST, the form that selectively sulfates catecholamines, failed to esterify these peptides. Thus, human M and rat Type IV PST appear to be different in regard to the specificity of the tyrosyl moiety that can be conjugated within the peptide chain. These data further demon-

Table 4
Sulfation of Peptides[a]

Peptide	mM	Sulfated P_I	Product P_{II}	Formed M^b
Leucine enkephalin	0.1	3.73	2.38	ND
	1.0	9.48	4.20	ND
Growth hormone releasing factor	0.1	4.45	2.85	ND
	1.0	9.58	—	ND
Oxytocin	0.1	3.00	2.68	ND
	1.0	2.58	1.22	—

[a]All velocities are expressed as pmol/60 min.

[b]To determine *M* activity, incubations were performed in the presence of the selective P_I and P_{II} inhibitor 2,6-dichloro-*p*-nitrophenol (10 μ*M*). ND, not detected.

strate differences in the properties of human and rat PST and reveal that PSTs from different species are remarkably selective in terms of their ability to sulfate catecholamines and biologically active proteins.

As briefly described above, the sulfate esters of a number of biologically active proteins, including leucine enkephalin (Unsworth and Hughes, 1982), proenkephalin (Unsworth and Hughes, 1982), cholecystokinin (Mutt and Jorpes, 1967, 1971), gastrin (Gregory et al., 1964), the fourth component of complement (C4) (Karp, 1983), phyllokinin (Anastasi, 1966), cerulein (Anastasi, 1967), and fibrinogen (Doolittle and Blombeck, 1964) have all been shown to exist in vivo. In specific cases of the last three proteins, the sulfated species actually functions as the biochemically active species. For example, the biological activity of the sulfate ester of cholecystokinin is appropriately 100 times more potent than the desulfated derivative. In contrast, the sulfate esters of leucine and proenkephalin, although biologically inert, may also have profound biological consequences since it has been suggested that the presence of the sulfate moiety may prevent proteolytic digestion of the peptide and thus provide a more stable form of the neurotransmitter.

Recent studies by Huttner and coworkers (Huttner, 1982; Lee and Huttner, 1983) have revealed that the tyrosyl moieties of a number of large molecular weight proteins are sulfated in vivo and in pheochromocytoma cells grown in culture by a membrane-bound form of PST. Which proteins or polypeptides are capable of undergoing sulfoconjugation and how the amino acid composition or primary or secondary structure of a polypeptide or protein influence the rate and extent of sulfation is not clear. The form of PST responsible for sulfoconjugation of large proteins may be different from that which conjugates the smaller peptides as described above. It is possible that sulfation represents a much more ubiquitous process than previously recognized; a process capable of modulating the activity of enzymes as well as hormones and neurotransmitters. In this regard, sulfation may depict a reaction similar to phosphorylation involved in the regulation of a variety of peptides and proteins. Biochemically, the sulfate molecule is similar to that of phosphate and, therefore, has the potential to regulate enzyme activity or protein function in a fashion similar to phosphate. The two processes are also similar in that there is potential for regulation of the sulfate system, since there are hydrolytic enzymes—the arylsulfatases—that may remove the sulfate group from the peptide chain to regulate biological activity. Accordingly, sulfation may represent an important posttranslational modification of proteins.

6. Platelet Phenol Sulfotransferase

Studies by Hart et al. (1979) were the first to demonstrate that human platelets contain PST. In these studies, it was demonstrated that the enzyme or enzymes in platelets responsible for conjugating the catecholamines and metabolites were similar to PST in human brain. Studies in the laboratories of Sandler (Rein et al., 1981, 1982a, 1982b), Weinshilboum (Reiter, 1983), and others (Butler et al., 1983) have further demonstrated there are at least two forms of PST in human platelets, M- and P-PST. Weinshilboum (personal communication) has recently demonstrated the presence of two P species of the sulfotransferase and one M form consistent with that observed in human brain. The activities of the M and P enzymes vary in different individuals and there is no correlation with age, sex, or platelet MAO activity (Bonham Carter et al., 1981; Reveley et al., 1983; vanLoon and Weinshilboum, 1984). Most importantly, Anderson et al. (1981) found that the activity measured for platelet sulfation of TA and MOPEG significantly correlated with the activity in several human tissues obtained at biopsy. These data imply that platelet PST activity can likely be used as a marker for the activity of this enzyme throughout the body.

However, a recent study by Young et al. (1985) reveals that only the activity associated with the P form or thermostable form of platelet PST and not the M form significantly correlated with human brain PST activity. Nevertheless, detection of PST activity in platelets may provide us with a minimally invasive procedure with which to evaluate the levels of PST in individuals inflicted with a variety of diseases possibly relating to abnormal levels of catecholamines.

7. Assay of Phenol Sulfotransferase

Numerous techniques are available by which PST activity can be assessed, but most possess inherent limitations that restrict their applicability in the routine determination of sulfoconjugation by the different forms of PST. In general, sulfotransferase assays may be divided into two groups: those capable of measuring the sulfation of a wide variety of substrates, and those with very specific substrate requirements. These two general methodologies may be further subdivided into those requiring the use of [^{35}S]-PAPS, and those using unlabeled PAPS. Substrate-specific assays have been developed for *m*-aminophenol (Bernstein and McGilvery, 1952), *p*-nitrophenol (Robins and Lipmann, 1957), 5-hydroxytryptamine (Hidaka et al., 1969; Hidaka and Austin, 1972), 4-methylumbelliferone (van Kempen and Jansen, 1972), and harmol (Wong, 1974), and none of these methods require radiolabeled PAPS.

The more widely applicable PST assays are characterized by a variety of methods for product isolation and measurement. Banerjee and Roy (1968) have described an assay procedure that exploits the ability of various sulfate monoesters to form ion pairs with the phenothiazine dye methylene blue. In this method, the methylene blue salts are extracted into organic solvent and the amount of sulfated products is determined spectrophotometrically. This procedure allows for the measurement of numerous nonpolar steroid and aryl sulfates, as well as aryl sulfamates (Banerjee and Roy, 1966, 1968) and has been employed successfully for the measurement of sulfated napthols as well (Sekura and Jakoby, 1979). This assay is limited, however, in its ability to measure more polar compounds such as the sulfate esters of tyrosine and biogenic amines, since the methylene blue salts of these compounds are much less soluble in organic solvents (Roy, 1981). A more versatile assay system has been reported recently by Sekura et al. (1979). In this procedure, many polar and nonpolar sulfate esters can be quantified using thin layer chromatography (TLC) to separate radiolabeled reactants and products. The use of this methodology is limited by the time-consuming TLC method and the necessity of using different solvent systems for different substrates.

Descriptions of several current methods, including spectrophotometric analysis, ion-exchange chromatography, electrophoretic separation, and a selective salt precipitation assay, are presented below.

7.1. Spectrophotometric Assay

The two major spectrophotometric assays for assessing sulfoconjugation are those of Gregory and Lipman (1957) and Van Kempen and Jansen (1972).

7.1.1. Method of Gregory and Lipman (1957)

This assay utilizes the fact that certain forms of PST are reversible and can catalyze the sulfation of 3'-phosphoadenosine-5'-phosphate (PAP) starting with the appropriate sulfated substrate donor. The assay mixture contains 6.5 μmol of *p*-nitrophenolsulfate as the sulfate donor, 0.5 μmol of phenol, and 1 nmol PAP in 0.9 mL of 100 μmol Tris-HCl buffer, pH 7.8. Enzyme (0.1 mL) prepared in neutralized 0.1*M* glutathione solution was added to the cuvet to start the reaction. Sulfation of phenol is assessed by the increase in the absorption at 400 mμ resulting from the formation of *p*-nitrophenol from its sulfate ester. Sulfation of other phenol derivatives can be assessed relative to the activity observed with phenol.

Although this assay is rapid and inexpensive, there are several potential problems that greatly limit its utility. First, the assay requires that PST be reversible and, as described above, this kinetic requirement does not appear to be the case for all forms of PST. Second, dead-end complexes may form with either substrate and/or product, thus limiting the range of substrate concentrations that can be used to measure sulfotransferase activity. Also, as with any coupled assay, the rate of sulfation of PAP must be greater than the rate of sulfation of the acceptor substrate in order to assess the true rate of PST activity. Regardless of these limitations, this assay can be effectively employed under specific conditions to assess PST activity.

7.1.2. Method of Van Kempen and Jansen (1972)

This fluorometric assay employs the use of 4-methylumbelliferone as the acceptor substrate for PST. The assay mixture for maximal activity routinely contains 1.5 m*M* $MgCl_2$, PAPS, 0.2 m*M* acceptor substrate (see below regarding substrate inhibition), 19 m*M* potassium phosphate, and appropriate enzyme preparation in 3.0 mL of Tris-HCl buffer, pH 7.4. Reactions are incubated for 30 min and terminated by heating in a boiling waterbath for 2 min. After centrifugation, 2 mL of the supernatant solution are placed on a Dowex-50×4 column (1 × 3 cm) and the sulfated product eluted with 3 mL of water. This eluate is acidified with 1 mL of 2*N* HCl and heated at 80°C for 30 min in a waterbath. The fluorescence (E_{max} – 324 mμ; E_{max} – 480 mμ) of the 4-methylumbelliferone is measured in the resulting solution and activity is expressed relative to a standard curve generated with the substrate. The authors indicate, that at concentrations of 4-methylumbelliferone greater than 0.1 m*M*, substrate inhibition may be observed. Blank values are determined in the absence of PAPS.

The utility of this assay is that it is both rapid and inexpensive. The limitation of this assay is that 4-methylumbelliferone is not a selective substrate for either the M or P forms of PST; thus, when measuring sulfation in a crude preparation, total PST is assessed.

7.2. Salt Precipitation Assay

Of all the enzyme assays presently available for assessing PST activity, the most widely used is the salt precipitation assay of Foldes and Meek (1973). This assay utilizes the differences in the selective solubilities of the barium salts of PAPS and sulfated acceptor molecules to separate the product from starting material. The reaction mixture contains [^{35}S]-PAPS, various phenolic acceptor substrates, and enzyme in 1 mL of phosphate buffer of various pH values. This assay can be used with a variety of other buffers, including Tris-HCl, TEA, or glycine-NaOH, to optimize the assay for activity. The reaction is terminated by addition, in sequence, of 0.2 mL of 0.1*M* barium acetate, barium hydroxide, and zinc sulfate. The precipitate formed is removed by low-speed centrifugation, and the precipitation procedure is repeated. An aliquot of the final supernatant solution is removed and assayed for total radioactivity. This precipitation procedure was reported to remove protein and nucleotides and at least 95% of the radioactive PAPS added to the reaction mixture in the absence of enzyme. Blanks are assessed by incubations performed in the absence of the phenolic substrate. Modifications of this assay using similar volumes have been reported.

Although this assay has had great popularity in recent years because it is both very simple and rapid, its application is limited to the degree of solubility of the complex formed between barium and sulfated products after addition of the salts to the reaction mixture. Recovery of the neutral and basic phenol sulfate esters in the supernatant solution ranged between 82 and 100%, whereas recovery of the acid products such as homovanillic acid, dihydroxyphenylacetic acid, and vanillylmandelic acid ranged from only 25 to 42%. Clearly, the poor recovery with acid products makes application of this method extremely limited. Also, high salt concentrations such as those encountered during elution of PST from ion exchange or affinity columns during purification, can affect the recovery of sulfated product. The other major problem with this assay is that blank values are very high and are proportional to the enzyme added to the assay mixture. This is probably a result of sulfation of endogenous substrates that are not precipitable by the barium salt.

7.3. Ion-Exchange Chromatography Assay

Recent studies by Baranczyk-Kuzma et al. (1981) and Borchardt et al. (1983) have utilized anion-exchange Ecteola cellulose to assay for the sulfated products formed by PST. However, both of these procedures are extremely time-consuming in that large volumes of eluant are required to isolate the sulfated metabolites formed in the reaction. A modified Ecteola cellulose assay has recently been described that eliminates much of the time-consuming elution process and is capable of measuring sulfation of catechol and phenolic amines, neutral phenolic substrates, and various small peptides (Whittemore and Roth, 1985).

7.3.1. Method of Borchardt and Coworkers

Ecteola cellulose chromatography has been used previously in the purification of PAPS. Borchardt et al. (1983) has adapted this ion exchange procedure and developed a sensitive assay for sulfation of neutral and acid metabolites of catecholamines. The Ecteola (approximately 100g) is prepared by washing, in succession, with 2 L of 0.01*M* HCl, 4 L of water, 2 L of 0.1*M* NaOH, 4 L of water, 2 L of 0.5*M* formic acid, 4 L of water, and 2 L of 0.02*M* ammonium bicarbonate under gentle vacuum. Columns 1 × 6 cm are prepared for elution of the assay mixture.

The assay mixture consists of 0.29 μ*M* [^{35}S]-PAPS, enzyme, and acceptor substrate in a total of volume of 1 mL of 10 m*M* sodium phosphate buffer, pH 6.4. The reaction is terminated by cooling to 4°C and by addition of 1 mL of 0.02*M* ammonium bicarbonate to adjust the pH to approximately 8.2. The total reaction mixture is placed over an Ecteola cellulose column and the radioactively labeled sulfated product and PAPS are eluted using stepwise gradients consisting of 20 mL each of 0.02, 0.05, 0.1, 0.2, and 0.35*M* ammonium bicarbonate. Elution of the sulfated products occurs between 0.05 and 0.2*M* ammonium bicarbonate, whereas PAPS elutes at 0.35*M* salt. The assay requires that the exact elution profile be determined for each sulfated product.

There are several advantages of the procedure described by Borchardt et al. (1983) over those described above for PST. First, the recovery of the sulfated products is quantitative regardless of whether a neutral or acid acceptor substrate is used. Second, this method yields extremely low blank values.

However, there are several major drawbacks to this procedure that make it of limited application. The time for elution of product is relatively long. For example, elution of homovanillic alcohol sulfate requires a two-step gradient of 20 and 50 m*M* ammonium bicarbonate, whereas homovanillic acid sulfate requires an additional gradient of 100 m*M* ammonium bicarbonate. Under these conditions, the respective elution volumes reported were approximately 30 and 55 mL. In this system, elution of the sulfate ester of homovanillic alcohol is further complicated by coelution with inorganic sulfate.

7.3.2. Method of Whittemore and Roth (1985)

The assay developed by Whittemore and Roth (1985) is a modification of the procedure described by Borchardt et al. (1983). The Ecteola cellulose is prepared by washing in 20 vol/g dry weight in 0.1*M* HCl for at least 30 min with frequent but gentle agitation. The cellulose is subsequently washed with water followed by treatment with 20 vol/g dry weight of 0.1*M* NaOH for an additional 30 min. After repeating this latter step once, the cellulose is converted to the formate counter ion by equilibrating in 0.5*M* formic acid followed by treatment with 5 m*M* ammonium bicarbonate.

Reaction mixtures contain the appropriate acceptor substrate [^{35}S]-PAPS, 1 m*M* pargyline (an MAO inhibitor), and enzyme in 0.5 mL of 50 m*M* triethanolamine buffer, pH 7.4. Pargyline is added to the reaction mixture to prevent deamination of the biogenic amine by mitochondrial monoamine oxidase and is used only when crude enzyme preparations are used to assess catecholamine-sulfating PST activity. This assay can be used with a variety of buffers, including phosphate and Tris-HCl. The reaction is terminated by placing the test tubes in a boiling waterbath for 2 min and immediately placing the samples on ice. Aliquots are removed and chromatographed over Ecteola cellulose columns prepared in 5¼ in. Pasteur pipets (0.5 × 2 cm). The radioactive products are eluted from the column in successive washes with water and 5 m*M*, 20 m*M*, and 200 m*M* ammonium bicarbonate solutions. DA sulfate is completely eluted from the Ecteola cellulose in the first 3 mL of the water wash, whereas inorganic SO_4 and unreacted PAPS are eluted from the column with 6 mL each of 20 and 200 m*M* NH_4HCO_3, respectively. For the blank reaction performed in the absence of the substrate DA, the amount of radiolabel present in the initial water wash is only slightly above background (60–120 dpm/1 mL fraction), accounting for only 0.1–0.3% of the total ^{35}S initially present.

In contrast to the sulfate conjugate of DA, phenyl sulfate is not removed from the column with water, but elutes in the final 6 mL of an 11 mL wash with 5 m*M* ammonium bicarbonate. When phenol is omitted from the reaction mixture, only 100–420 dpm/fraction is detected in the 5 m*M* ammonium wash and accounts for approximately 0.2–0.7% of the total PAPS present in the starting assay mixture. Recovery of radiolabel for both the DA and phenol by this method is essentially quantitative, ranging from 96 to 99%.

Similar to the studies above, this assay is also applicable for the sulfoconjugation of small peptides. The sulfated peptides of leu-enkephalin and angiotensin II were shown to elute in the 5 m*M* ammonium bicarbonate eluate. Similar results are obtained with oxytocin and growth hormone releasing factor, which are sulfated by the P form of human brain PST.

The major advantage of the assay procedure developed by Whittemore and Roth (1985) is that it is rapid and allows for the direct determination of both product formation as well as substrate utilization and inorganic sulfate production. Determination of sulfate production and PAPS utilization have proved to be useful in explaining why the linear time period for sulfate ester formation is relatively short even though total ester formed represented less than 15% of the total PAPS in the assay mixture. Whittemore and Roth (1985) reported that DA sulfate formation was linear for approximately 30 min under standard assay conditions, plateauing rapidly thereafter. Using the Ecteola cellulose assay, it was demonstrated that inorganic sulfate production in this experiment was linear for approximately 40 min and accounted for almost 45% of the PAPS utilized. At the end of 60 min they found that approximately 70% of the PAPS was degraded. Thus, by being able to measure sulfate production and PAPS utilization, these investigators were able to explain why formation of DA sulfate was linear for only 30 min and demonstrated that, in crude preparations of human brain PST, sulfoconjugation of acceptor substrates accounts for only a small percentage of the total PAPS consumed over the course of the incubation.

As described above, the Ecteola assays developed by both Borchardt et al. (1983) and Whittemore and Roth (1985) have the potential benefit of providing quantitative assessment of sulfation of a variety of phenolic substrates and are not restricted by variation of product recovery or excessively high blank values. However, the extremely long times required to assay multiple samples, such as those needed to determine kinetic variables, make the procedure by Borchardt and coworkers (1982, 1983) somewhat limited. The modification of the Ecteola assay described by Whittemore and Roth (1985) clearly offers the advantage of being more rapid, in that the sulfate esters of various biogenic amine acceptor molecules elute from the Ecteola-cellulose within the first 3 mL of an initial water wash, whereas phenyl sulfate requires an 11-mL wash of 5 m*M* NH_4HCO_3 for complete elution from the Ecteola cellulose column.

In addition, the use of Ecteola cellulose to determine PST activity also offers significant improvements over the performance of the widely used method of Foldes and Meek (1973). Recovery of radiolabel is quantitative in the new procedure, which circumvents the difficulties associated with determination of the solubility and recovery of the barium salts of sulfated reaction products. Furthermore, blank values in the new assay are extremely low, often approaching background levels. These extremely low blank values have the added advantage of extending the sensitivity of the assay without the necessity of increasing the specific activity of PAPS. In contrast to the barium precipitation procedure of Foldes and Meek (1973), which cannot differentiate between two simultaneously sulfated substrates, the new assay is highly selective and capable of isolating products with different net charges. Using appropriate substrates at proper concentrations, this property allows for the simultaneous measurement of two forms of PST in one sample, as well as facilitating alternative substrate inhibition analysis of the enzyme.

The assay developed by Whittemore and Roth (1985) is subject to some limitations in that high salt concentration and pH may alter the elution pattern of the neutral sulfated phenolic compounds formed in the reaction. For example, upon elution of PST on DEAE cellulose or on Affi-Gel Blue columns, the high salt concentration must be removed to quantitatively assess P_I, P_{II}, and M activity.

An interesting observation in the course of the studies utilizing Ecteola cellulose was the finding that a major portion of PAPS was actually converted to inorganic sulfate and presumably PAP or breakdown products of PAP that are capable of inhibiting PST (Whittemore and Roth, 1985). Incubations performed with 100,000*g* supernatant fraction of human brain demonstrated that over the course of a 60-min incubation almost 70% of the PAPS was converted to inorganic sulfate. Inorganic SO_4 could be produced by two alternative two-step pathways, the first involving the action of 5'-nucleotidase (Spencer, 1960) to form a mixed sulfate-phosphate anhydride that can then be hydrolyzed in inorganic SO_4. The second pathway involves adenosine-3'-nucleotidase (Brunngraber, 1958) to form adenosine-5'-phosphosulfate (APS), which can subsequently be degraded by phosphatase or APS-sulfohydrolase (Bailey-Wood et al., 1969, 1970) to generate SO_4. The former pathway seems unlikely since 5'-nucleotidase is a plasma membrane-bound enzyme (Bosman and Pike, 1971; Kreutzberg et al., 1978) and the studies described utilized the cytosolic fraction of human brain. In addition, the PST assays were performed at pH 7.4 and this nucleotidase has been found to have a substantially more acidic pH optimum (Denner et al., 1973).

Whittemore and Roth (1985) have demonstrated that inclusion of the phosphatase inhibitor phosphonoacetic acid in the reaction mixtures utilizing 100,000*g* supernatant solution of human brain as a source of PST increases the amount of dopamine sulfation and decreases PAPS catabolism. The linear period of DA sulfation can be greatly extended, resulting in a substantial increase in the amount of sulfated product formed. Concomitant with the increase in dopamine sulfate was decreased utilization of PAPS and decreased formation of sulfate. This effect of phosphonoacetic acid on PAPS degradation appears to be saturable, with maximal effect occurring at approximately 2 m*M*. These data suggest that phosphonoacetic acid should be added to all assays utilizing PAPS as the sulfate donor, especially when crude tissue preparations that may contain nucleotidases or phosphatases are employed.

8. Sulfate Conjugation: Neurological and Behavioral Consequences

As described above in this chapter, there is now sufficient evidence in the literature demonstrating that sulfation of catecholamines, various peptides, and large proteins occurs extensively in vivo and that sulfation in some cases is obligatory for promoting or expressing optimal biological activity.

Several studies have suggested that altered or impaired PST activity may, in fact, have profound physiological consequences in relation to neurological and behavioral disorders. Several reports have appeared in the literature indicating that individuals suffering from dietary migraine headaches when challenged with oral doses of tyramine (TA) excrete significantly lower levels of the sulfate conjugate of this amine in their urine compared to that in a normal population (Youdim et al., 1971; Glover et al., 1983). It was also noted that the endogenous urinary levels of sulfate ester of TA are almost 70% lower than those in normal individuals. These data imply there may be a defect in the conjugation mechanism in patients with a history of TA-sensitive migraine headaches. In support of this are recent studies (Glover et al., 1982; Littlewood, et al., 1982) demonstrating that platelet PST is lower in a subgroup of individuals whose migraine headaches are believed to be precipitated by diet.

In addition to these studies, Sandler and coworkers (Bonham Carter et al., 1978, 1980a,b; Sandler, et al., 1983) have recently demonstrated that TA conjugation in vivo (after oral TA administration) is decreased greater than twofold in depressed individuals. These observations were recently confirmed by Harrison et al. (1983). The low conjugation levels of TA were found to be unrelated to whether the patients were experiencing a depressive episode at the time these measurements were made. In contrast to the migraine patients described above, platelet PST activity in these depressed patients, however, was not found to be significantly different from that of a control population.

Other recent studies have also implied that sulfation of catecholamines may be important in regulating blood pressure. A recent paper by Racz et al. (1984) has demonstrated that DA-3-sulfate is equipotent as dopamine in its ability to inhibit angiotensin II-induced aldosterone secretion in cultured bovine adrenal cells. In contrast, the 4-sulfate ester of DA has no effect on aldosterone secretion. These authors conclude that DA-3-sulfate may be effective in complementing the dopaminergic inhibition of aldosterone secretion at the level of the adrenal glands.

In yet another recent article, Kuchel et al. (1985) reported that total DA (free plus conjugated) increases in orthostatic hypertensive patients in response to postural stimulation. The change in free DA varies considerably and is not a consistent reliable marker for this response. In contrast, the sulfate conjugate of DA increases in essentially all patients with orthostatic hypertension and thus may be a more reliable indicator of the disease. Davidson et al. (1984) have observed an inverse relationship between the levels of free and conjugated catecholamine released into plasma upon exercise and have suggested that the degree of sulfoconjugation may be a factor in determining the concentration of free amines in plasma.

In summary, the data presented in this chapter reveal that

sulfoconjugation of bioactive agents has the potential to have profound physiological consequences in humans. Sulfation appears to be a ubiquitous process, occurring with a variety of endogenous neuroactive agents and, depending on the chemical involved, has the potential to activate or inactivate these compounds. The fact that multiple forms of the sulfotransferase exist with distinct substrate selectivity and distinct cellular sites implies a specific regulatory role for these enzymes in vivo.

References

Anastasi A., Bertaccini G., and Erspamer V. (1966) Pharmacological data on phyllokinin (bradykinyl-isoleucyl-tyrosine *O*-sulphate) and bradykinyl-isoleucyl-tyrosine. *Br. J. Pharmacol.* **27,** 479–485.

Anastasi A., Erspamer V., and Endean R. (1967) Isolation and structure of caerulein, an active decapeptide from the skin of *Hyla caerulea. Experentia* **23,** 699–700.

Anderson R. J. and Weinshilboum R. M. (1979) Phenolsulphotransferase: Enzyme activity and endogenous inhibitors in the human erythrocyte. *J. Lab. Clin. Med.* **94,** 158–171.

Anderson R. J., Weinshilboum R. M., Phillips A. F., and Broughton D. D. (1981) Human platelet phenol sulfotransferase: Assay procedure, substrate, and tissue correlations. *Clin. Chim. Acta.* **110,** 157–167.

Bailey-Wood R., Dodgson K. S., and Rose F. A. (1969) A rat liver sulphohydrolase enzyme acting on adenylyl sulphate. *Biochem. J.* **112,** 257–258.

Bailey-Wood R., Dodgson K. S., and Rose F. A. (1970) Purification and properties of two adenosine 5'-phosphosulfate sulphohydrolases from rat liver and their possible role in the degradation of 3'-phosphoadenosine 5'-phosphosulphate. *Biochim. Biophys. Acta.* **220,** 284–289.

Banerjee R. K. and Roy A. B. (1966) The sulfotransferases of guinea pig liver. *Mol. Pharmacol.* **2,** 56–66.

Banerjee R. K. and Roy A. B. (1968) Kinetic studies for the phenol sulfotransferase reaction. *Biochim. Biophys. Acta.* **151,** 573–586.

Baranczyk-Kuzma A., Borchardt R. T., Schasteen S. C., and Pinnick C. L. (1981) Phenol Sulfotransferase: Purification and Characterization of the Rat Brain Enzyme, in *Phenosulfotransferase in Mental Health Research* (Sandler M. and Usdin E., eds.), Macmillan, London.

Bernstein S. and McGilvery R. W. (1952) The enzymatic conjugation of *m*-aminophenol. *J. Biol. Chem.* **198,** 195–203.

Bettelheim F. R. (1954) Tyrosine-*O*-sulfate in a peptide from fibrinogen. *J. Am. Chem. Soc.* **76,** 2838–2839.

Bonham Carter S. M., Goodwin B. L., and Sandler M. (1980a) Decreased conjugated tyramine output in depression: The effect of oral L-cysteine. *Brit. J. Clin. Pharmacol.* **10,** 305–308.

Bonham Carter S. M., Reveley M. A., Sandler M., Dewhurst J., Little B. C., Hayworth J., and Priest R. G. (1980b) Decreased urinary output of conjugated tyramine is associated with lifetime vulnerability to depressive illness. *Psychiat. Res.* **3,** 13–21.

Bonham Carter S., Sandler M., Goodwin B. L., Sepping P., and Bridges P. K. (1978) Decreased urinary output of tyramine and its metabolites in depression. *Br. J. Psychiat.* **132,** 125–132.

Bonham Carter S. M., Glover V., Sandler M., Gillman P. K., and Bridges P. K. (1981) Human platelet phenolsulphotransferase: Separate control of the two forms and activity range in depressive illness. *Clin. Chim. Acta* **117,** 333–334.

Borchardt R. T. and Schasteen C. S. (1982) Phenol sulfotransferase. I. Purification of a rat liver enzyme by affinity chromatography. *Biochim. Biophys. Acta* **708,** 272–279.

Borchardt R. T., Baranczyk-Kuzma A., and Pinnick C. L. (1983) An Ecteola-cellulose chromatography assay for 3'-phosphoadenosine-5'-phosphosulfate: phenol sulfotransferase. *Anal. Biochem.* **130,** 334–338.

Bosman H. B. and Pike G. L. (1971) Membrane marker enzymes: Isolation, purification, and properties of 5'-nucleotidase from rat cerebellum. *Biochim. Biophys. Acta* **227,** 402–412.

Brunngraber E. G. (1958) Nucleotides involved in the enzymatic conjugation of phenols with sulfate. *J. Biol. Chem.* **233,** 472–477.

Butler P. R., Anderson R. J., and Venton D. L. (1983) Human platelet phenol sulfotransferase: partial purification and detection of two forms of the enzyme. *J. Neurochem.* **41,** 630–639.

Buu N. T. and Kuchel O. (1977) A new method for the hydrolysis of conjugated catecholamines. *J. Lab. Clin. Med.* **90,** 680–685.

Campbell N. R. C., Dunnettee J. H., Mwaluko G., van Loon J., and Weinshilboum R. M. (1984) Platelet phenol sulfotransferase and erythrocyte catechol-*O*-methyltransferase activities: Correlation with methyldopa metabolism. *Clin. Pharmacol. Ther.* **35,** 55–63.

Carroll J. and McEvoy F. A. (1970) The activation of liver phenolsulfotransferase. *Biochem. J.* **119,** 27P.

Davidson L., Vandongen R., Beilin L. J., and Arkwright P. D. (1984) Free and sulfate-conjugated catecholamines during exercise in man. *J. Clin. Endocrinol. Metabl.* **58,** 415–418.

Dencker S. J., Haggendahl J., and Ilves-Haggendahl M. (1967) Presence of free and conjugated noradrenalin in human cerebrospinal fluid. *Acta Physiol. Scand.* **69,** 140–146.

Denner W. H. B., Stokes A. M., Rose F. A., and Dodgson K. S. (1973) Separation and properties of the soluble 3'-phosphoadenosine 5'-

phosphosulphate-degrading enzymes of bovine liver. *Biochim. Biophys. Acta* **315,** 394–401.

Dockray G. J., Gregory R. H., Hutchison J. B., Harris J. I., and Runswick M. J. (1978) Isolation, structure and biological activity of two cholecystokinin octapeptides from sheep brain. *Nature* **274,** 711–713.

Dodgson K. S., Rose F. A., and Tudball N. (1959) Studies on sulphatases 23. The enzymic desulphation of tyrosine *O*-sulphate. *Biochem. J.* **71,** 10–15.

Doolittle R. F. and Blomback B. (1964) Amino-acid sequence investigations of fibrinopeptides from various mammals: Evolutionary implications. *Nature* **202,** 147–152.

Duffel M. W. and Jakoby W. B. (1981) On the mechanism of aryl sulfotransferase. *J. Biol. Chem.* **256,** 11123–11127.

Elchisak M. A. (1983) Distribution of free and conjugated dopamine in human caudate nucleus, hypothalamus and kidney. *J. Neurochem.* **41,** 893–896.

Elchisak M. A., Cosgrove S. E., Ebert M. H., and Burns R. S. (1983) Distribution of free and conjugated dopamine in monkey brain, peripheral tissues and cerebrospinal fluid determined by high performance liquid chromatography. *Brain Res.* **279,** 171–176.

Elchisak M. A., Powers K. H., and Ebert M. H. (1982) Demonstration of conjugated dopamine in monkey CSF by gas chromatography–mass spectrometry. *J. Neurochem.* **39,** 726–728.

Foldes A. and Meek J. L. (1973) Rat brain phenolsulfotransferase—partial purification and some properties. *Biochim. Biophys. Acta* **327,** 365–374.

Foldes A. and Meek J. L. (1974) Occurrence and localization of brain phenolsulphotransferase. *J. Neurochem.* **23,** 303–307.

Glover V., Bonham Carter S., Littlewood J., Rein G., and Sandler M. (1982) Phenolsulphotransferase M and P in Human Platelets: Nature and Clinical Implications of the Two Forms, in *Sulfate Metabolism and Sulfate Conjugation* (Mulder G. J., Caldwell J., van Kempen G. M. J., and Vonk R. J., eds.), Taylor & Francis, London.

Glover V., Littlewood J., Sandler M., Peatfield R., Petty R., and Rose F. C. (1983) Biochemical predisposition to dietary migraine: The role of phenolsulphotransferase. *Headache* **23,** 53–58.

Goodall McC. and Alton H. (1972) Metabolism of 3,4-dihydroxyphenylalanine (L-Dopa) in human subjects. *Biochem. Pharmacol.* **21,** 2401–2408.

Gregory H. and Lipman F. (1957) The transfer of sulfate among phenolic compounds with 3',5'-diphosphoadenosine as coenzyme. *J. Biol. Chem.* **229,** 1081–1090.

Gregory H., Hardy P. M., Jones D. S., Kenner G. W., and Sheppard R. C. (1964) The antral hormone gastrin. *Nature* **204,** 931–933.

Harrison W. M., Cooper T. B., Quitkin F. M., Liebowitz M. R., McGrath

P. J., Stewart J. W., and Klein D. F. (1983) Tyramine sulfate excretion test in depressive illness. *Psychopharmacol. Bull.* **19,** 503–504.

Hart R. F., Renskers K. J., Nelson E. B., and Roth J. A. (1979) Localization and characterization of phenol sulfotransferase in human platelets. *Life Sci.* **24,** 125–130.

Hermansen K. (1984) Effects of cholecystokinin (CCK)-4, nonsulfated CCK-8, and sulfated CCK-8 on pancreatic somatostatin, insulin, and glucagon secretion in the dog: Studies *in vitro*. *Endocrinology* **114,** 1770–1775.

Hidaka H. and Austin J. (1972) Occurrence and distribution of serotonin-*O*-sulfotransferase in human brain; a new radioisotopic assay. *Biochim. Biophys. Acta.* **268,** 132–137.

Hidaka H., Nagatsu T., and Yagi K. (1969) Occurrence of serotonin sulfotransferase in the brain. *J. Neurochem.* **16,** 783–785.

Huttner W. B. (1982) Sulphation of tyrosine residues—a widespread modification of proteins. *Nature* **299,** 273–276.

Imai K., Sugiura M., and Tamura Z. (1970) The presence of conjugated dopamine in normal human plasma. *Chem. Pharm. Bull.* **18,** 2134.

Jakoby W. B. (1982) Aryl and Hydroxysteroid Sulfotransferases, in *Sulfate Metabolism and Sulfate Conjugation* (Mulder G. J., Caldwell J., van Kempen G. M. J., and Vonk R. J., eds.), Taylor and Francis, London.

Johnson G. A., Baker C. A., and Smith R. T. (1980) Radioenzymatic assay of sulfate conjugates of catecholamines and dopa in plasma. *Life Sci.* **26,** 1591–1598.

Jones J. G. and Dodgson K. S. (1965) Biosynthesis of L-tyrosine *O*-sulphate from methyl and ethyl esters of L-tyrosine. *Biochem. J.* **94,** 331–336.

Jones J. G., Scotland S. M., and Dodgson K. S. (1966) The biological sulphation of L-tyrosyl peptides. *Biochem. J.* **98,** 138–141.

Karp D. R. (1983) Post-translational modification of the fourth component of complement. Sulfation of the *alpha*-chain. *J. Biol. Chem.* **258,** 12745–12748.

Kreutzberg G. W., Barron K. D., and Schubert P. (1978) Cytochemical localization of 5'-nucleotidase in glial plasma membranes. *Brain Res.* **158,** 247–257.

Kuchel O., Buu N. T., Hamet P., Larochelle P., Gutkowski J., Schiffrin E. L., Bourque M., and Genest J. (1985) Orthostatic hypotension: A posture-induced hyperdopaminergic state. *Am. J. Med. Sci.* **289,** 3–11.

LaBella F. S., Pinsky C., Christy M. E., and Colton D. (1982) Hyperactivity versus explosive motor behavior induced by opioids in the rat. Mechanisms elucidated with enkephalin-tyrosine-*O*-sulfate and morphine congeners. *Life Sci.* **30,** 1565–1571.

Lee R. W. H. and Huttner W. B. (1983) Tyrosine-*O*-sulfated proteins of

PC12 pheochromocytoma cells and their sulfation by a tyrosyl protein sulfotransferase. *J. Biol. Chem.* **258,** 11326–11334.

Littlewood J., Glover V., and Sandler M. (1982) Platelet phenolsulphotransferase deficiency in dietary migraine. *Lancet* **1,** 983–985.

Lyon E. S. and Jakoby W. B. (1980) The identity of alcohol sulfotransferases with hydroxysteroid sulfotransferases. *Arch Biochem. Biophys.* **202,** 474–481.

Marcus C. J., Sekura R. D., and Jakoby W. B. (1980) A hydroxysteroid sulfotransferase from rat liver. *Anal. Biochem.* **107,** 296–304.

Mattock P. and Jones J. G. (1970) Partial purification and properties of an enzyme from rat liver that catalyzes the sulphation of L-tyrosyl derivatives. *Biochem. J.* **116,** 797–803.

Michelot R. J., Lesko N., Stout R. W., and Coward J. K. (1977) Effect of *S*-adenosylhomocysteine and *S*-tubercidinylhomocysteine on catecholamine methylation in neuroblastoma cells. *Mol. Pharmacol.* **13,** 368–373.

Mulder G. J., ed. (1981) *Sulfate Activation,* in *Sulfation of Drugs and Related Compounds* (Mulder G. J., ed.), CRC Press, Boca Raton, Florida.

Mutt V. and Jorpes J. E. (1967) Isolation of aspartyl-phenylalanine amide from cholecycstokinin-pancreozymin. *Biochem. Biophys. Res. Commun.* **26,** 392–297.

Mutt V. and Jorpes J. E. (1971) Hormonal polypeptides of the upper intestine. *Biochem. J.* **125,** 57P–58P.

Nose Y. and Lipmann F. (1958) Separation of steroid sulfokinase. *J. Biol. Chem.* **233,** 1348–1351.

Osvaldo G. and Meek J. L. (1985) Sulfation of peptides and simple phenols by rat brain phenolsulfotransferase. Inhibition by dichloronitrophenol. *Biochem. Pharmacol.* **34,** 45–49.

Racz K., Buu N. T., Kuchel O., and DeLean A. (1984) Dopamine 3-sulfate inhibits aldosterone secretion in cultured bovine adrenal cells. *Am. Phys. Soc.* **247,** 431–435.

Rein G., Glover V., and Sandler M. (1981) Phenolsulphotransferase in Human Tissue: Evidence for Multiple Forms, in *Phenolsulfotransferase in Mental Health Research* (Sandler M. and Usdin E., eds.), Macmillan, London.

Rein G., Glover V., and Sandler M. (1982a) Multiple forms of phenolsulphotransferase in human tissues: Selective inhibition by dichloronitrophenol. *Biochem. Pharmacol.* **31,** 1893–1897.

Rein G., Glover V., and Sandler M. (1982b) Model Systems for Human Phenolsulphotransferase, in *Sulfate Metabolism and Conjugation* (Mulder G. J., Caldwell J., van Kempen G. M. J., and Vonk R. J., eds.), Taylor & Francis, London.

Rein G., Glover V., and Sandler M. (1984) Characterization of human brain phenolsulphotransferase. *J. Neurochem.* **42,** 80–85.

Reiter C. and Weinshilboum R. (1982) Platelet phenol sulfotransferase activity: Correlation with sulfate conjugation of acetaminophen. *Clin. Pharmacol. Ther.* **32,** 612–621.

Reiter C., Mwaluko G., Dunnettee J., Van Loon J., and Weinshilboum R. (1983) Thermolabile and thermostable human platelet phenol sulfotransferase: Substrate specificity and physical separation. *Naunyn Schmiedeberg's Arch. Pharmacol.* **324,** 140–147.

Reveley A. M., Bonham Carter S. M., Reveley M. A., and Sandler M. (1983) A genetic study of platelet phenolsulphotransferase activity in normal and schizophrenic twins. *J. Psychiat. Res.* **17,** 303–307.

Rivett A. J., Eddy B. J., and Roth J. A. (1982) Contribution of sulfate conjugation deamination and *O*-methylation to metabolism of dopamine and norepinephrine in human brain. *J. Neurochem.* **39,** 1009–1016.

Rivett A. J., Francis A., Whittemore R., and Roth J. A. (1984) Sulfate conjugation of dopamine in rat brain: Regional distribution of activity and evidence for neuronal localization. *J. Neurochem.* **42,** 1444–1449.

Robins P. W. and Lipmann, F. (1957) Isolation and identification of active sulfate. *J. Biol. Chem.* **229,** 837–851.

Romain Y., Demassienux S., and Carriere S. (1982) Partial purification and characterization of two isozymes involved in the sulfurylation of catecholamines. *Biochem. Biophys. Res. Commun.* **106,** 999–1005.

Roth J. A. and Rivett J. A. (1982) Does sulfate conjugation contribute to the metabolic inactivation of catecholamines in humans? *Biochem. Pharmacol.* **31,** 3017–3021.

Roth J. A., Rivett J. A., and Renskers K. J. (1982) Properties of Human Brain Phenolsulfotransferase and Its Role in the Inactivation of Catecholamine Neurotransmitters, in *Sulfate Metabolism and Conjugation* (Mulder G. J., Caldwell J., van Kampen G. M. J., and Vonk R. J., eds.), Taylor & Francis, London.

Roy A. B. (1981) The Chemistry of Sulfated Esters and Related Compounds, in *Sulfation of Drugs and Related Compounds* (Mulder G. J., ed.), CRC Press, Boca Raton, Florida.

Sandler M., Bonham Carter S. M., and Walker M. Sc. (1983) Tyramine conjugation deficit as a trait-marker in depression. *Psychopharmacol. Bull.* **19,** 501–502.

Segal H. L. and Mologne L. A. (1959) Enzymatic sulfurylation of tyrosine derivatives. *J. Biol. Chem.* **234,** 909–911.

Sekura R. D. and Jakoby W. B. (1981) Aryl sulfotransferase IV from rat liver. *Arch. Biochem. Biophys.* **211,** 352–359.

Sekura R. D. and Jakoby W. F. (1979) Phenol sulfotransferases. *J. Biol. Chem.* **254,** 5658–5663.

Sekura R. D., Marcus C. J., Lyon E. S., and Jakoby W. B. (1979) Assay of sulfotransferases. *Anal. Biochem.* **95,** 82–86.

Spencer B. (1960) Endogenous sulphate acceptors in rat liver. *Biochem. J.* **77,** 294–304.

Tallan H. H., Bella S. T., Stein W. H., and Moore S. (1955) Tyrosine-*O*-sulfate as a constituent of normal human urine. *J. Biol. Chem.* **217,** 703–708.

Tyce G. M. and Rorie D. K. (1982) Conjugated dopamine in superfusates of slices of rat striatum. *J. Neurochem.* **39,** 1333–1339.

Tyce G. M., Sharpless N. S., and Meunter M. D. (1974) Free and conjugated dopamine in plasma during levodopa therapy. *Clin. Pharmacol. Ther.* **16,** 782–788.

Tyce G. M., Sharpless N. S., Kerr F. W. L., and Meunter M. D. (1980) Dopamine conjugated in cerebrospinal fluid. *J. Neurochem.* **34,** 210–212.

Unsworth C. D. and Hughes J. (1982) *O*-sulphated Leu-enkephalin in brain. *Nature* **295,** 519–522.

van Kempen G. M. J. and Jansen G. S. I. M. (1972) Quantitative determination of phenolsulfotransferase using 4-methylumbelliferone. *Anal. Biochem.* **46,** 438–442.

vanLoon J. and Weinshilboum R. (1984) Human platelet phenol sulfotransferase: Familial variation in thermal stability of the TS form. *Biochem. Gen.* **22,** 997–1014.

Whittemore R. M. and Roth J. A. (1985) A modified ecteola cellulose assay for M and P phenol sulfotransferase. *Biochem. Pharmacol.* **43,** 1647–1652.

Whittemore R. M., Jeffery D. R., and Roth J. A. (1984) Sulfation of neuroactive peptides by human brain phenol sulfotransferase. *Pharmacol.* **26,** 139.

Whittemore R. M., Pearce L. B., and Roth J. A. (1985) Purification and kinetic characterization of a dopamine-sulfating form of phenol sulfotransferase from human brain. *Biochemistry* **24,** 2477–2482.

Wong K. P. (1974) A new method for measuring sulfotransferase. *Anal. Biochem.* **62,** 149–156.

Youdim M. B. H., Bonham Carter S., and Sandler M. (1971) Conjugation defect in tyramine-sensitive migraine. *Nature* **230,** 127–128.

Young W. F., Jr., Laws E. R., Jr., Sharbrough F. W., and Weinshilboum R. M. (1985) Human phenol sulfotransferase: correlation of brain and platelet activities. *J. Neurochem.* **44,** 1131–1137.

Young W. F., Jr., Okazaki H., Laws E. R., Jr., and Weinshilboum R. M. (1984) Human brain phenol sulfotransferase: Biochemical Properties and regional localization. *J. Neurochem.* **43,** 706–715.

Yu P. H., Rozdilsky B., and Boulton A. A. (1985) The sulfate conjugation of monoamines in human brain: Purification and some properties of an arylamine sulfotransferase from cerebral cortex. *J. Neurochem.* **45,** 836–843.

Yu P. H. and Walz W. (1985) Occurrence of phenolsulfotransferase in primary glial culture cells of rat. *Neurochem. Res.* **16,** 983–992.

Index